MW00620651

# OPTIMIZATION OF POWER SYSTEM OPERATION

**IEEE Press**
445 Hoes Lane
Piscataway, NJ 08854

**IEEE Press Editorial Board**
Tariq Samad, *Editor in Chief*

George W. Arnold     Mary Lanzerotti     Linda Shafer
Dmitry Goldgof       Pui-In Mak          MengChu Zhou
Ekram Hossain        Ray Perez           George Zobrist

Kenneth Moore, *Director of IEEE Book and Information Services (BIS)*


**Technical Reviewers**

Malcom Irving, University of Birmingham

Kit Po Wong, The University of Western Australia

# OPTIMIZATION OF POWER SYSTEM OPERATION

## Second Edition

**JIZHONG ZHU**

**IEEE PRESS**

# WILEY

Copyright © 2015 by The Institute of Electrical and Electronics Engineers, Inc.

Published by John Wiley & Sons, Inc., Hoboken, New Jersey. All rights reserved
Published simultaneously in Canada

No part of this publication may be reproduced, stored in a retrieval system, or transmitted in any form or
by any means, electronic, mechanical, photocopying, recording, scanning, or otherwise, except as
permitted under Section 107 or 108 of the 1976 United States Copyright Act, without either the prior
written permission of the Publisher, or authorization through payment of the appropriate per-copy fee to
the Copyright Clearance Center, Inc., 222 Rosewood Drive, Danvers, MA 01923, (978) 750-8400, fax
(978) 750-4470, or on the web at www.copyright.com. Requests to the Publisher for permission should
be addressed to the Permissions Department, John Wiley & Sons, Inc., 111 River Street, Hoboken, NJ
07030, (201) 748-6011, fax (201) 748-6008, or online at http://www.wiley.com/go/permission.

Limit of Liability/Disclaimer of Warranty: While the publisher and author have used their best efforts in
preparing this book, they make no representations or warranties with respect to the accuracy or
completeness of the contents of this book and specifically disclaim any implied warranties of
merchantability or fitness for a particular purpose. No warranty may be created or extended by sales
representatives or written sales materials. The advice and strategies contained herein may not be suitable
for your situation. You should consult with a professional where appropriate. Neither the publisher nor
author shall be liable for any loss of profit or any other commercial damages, including but not limited to
special, incidental, consequential, or other damages.

For general information on our other products and services or for technical support, please contact our
Customer Care Department within the United States at (800) 762-2974, outside the United States at
(317) 572-3993 or fax (317) 572-4002.

Wiley also publishes its books in a variety of electronic formats. Some content that appears in print may
not be available in electronic formats. For more information about Wiley products, visit our web site at
www.wiley.com.

*Library of Congress Cataloging-in-Publication Data is available:*

Zhu, Jizhong, 1961-
  Optimization of power system operation / Jizhong Zhu. – Second edition.
      pages cm – (IEEE Press series on power engineering)
  Summary: "Addresses advanced methods and optimization technologies and their applications in power
systems"– Provided by publisher.
  ISBN 978-1-118-85415-0 (hardback)
1. Electric power systems–Mathematical models. 2. Mathematical optimization. I. Title.
  TK1005.Z46 2015
  621.3101′5196–dc23
                                                                                        2014023096

Printed in the United States of America

10 9 8 7 6 5 4 3 2 1

TK
1005
Z46
2015

*To My Wife and Son*

# CONTENTS

**CHAPTER 10** *APPLICATION OF RENEWABLE ENERGY* 407

**CHAPTER 13** *UNCERTAINTY ANALYSIS IN POWER SYSTEMS* 529

**CHAPTER 14** *OPERATION OF SMART GRID* 579

# PREFACE

It has been five years since the first edition was published. Some developments have taken place in the power industry. The renewable energy and smart grid include many fresh and vital technologies that are needed to make enormous progress in power grid development. With the development of information technology and computer-based remote control and automation, the systems and technologies for the smart grid are made possible by two-way communication technology and computer processing. This modernized electricity network, which sends electricity from power suppliers to consumers using digital technology to save energy, reduce cost, and increase reliability and transparency, is being promoted by many governments as a way of addressing energy independence, global warming, environment protection, and emergency resilience issues.

In this new edition, *Optimization of Power System Operation*, continues to provide engineers and academics with a complete picture of the optimization techniques used in modern power system operation. It offers a practical, hands-on guide to theoretical developments and to the application of advanced optimization methods to realistic electric power engineering problems. Although the topic areas and depth of coverage remain about the same, the book has been updated to reflect the changes that have taken place in the electric power industry since the First Edition was published five years ago. The research and application of renewable energy and smart grid have being widely addressed in recent years, which have brought a host of new opportunities and challenges to modern power system operation. Thus, in this edition two new Chapters have been added—Chapter 10 on "Application of renewable energy" and Chapter 14 on "Operation of smart grid." The original Chapter 10 on "Reactive power optimization" in the first edition is removed because of limitation of the space. But some contents related to reactive power optimization can still be found in Chapter 8 on "Optimal power flow" and Chapter 13 on "Uncertainty analysis in power systems". In the new Chapter 10, in addition to the introduction of renewable energy resources and the corresponding mathematical models, the optimization operation of renewable energy in power systems, such as maximum power point tracking, voltage calculation for the grid-connected PV system, and voltage analysis in power system with wind energy, is focused. In the new Chapter 14, applications of optimization techniques to smart grid are addressed and the following topics are included: smart grid economic dispatch, two-stage-approach for optimal operation of a smart grid, optimal operation of virtual power plant, smart distribution operation, microgrid operation with wind and PV resources, optimal power flow for smart microgrid, renewable energy and distributed generation technologies, and a new phase angle measurement algorithm.

The author appreciated the suggestions and feedback offered by professors and engineers who have used the first edition. Some professors commented that this book comprehensively applies all kinds of optimization methods to solve power system operation problems, but it needs to provide some problems or exercises at the end of each chapter so that it can be used as a textbook. Some students remarked that they like the examples in the book, and they even have tried to use different methods or written some programs to resolve them. Some readers did an excellent job to find some errors and typos. I have gone through the book and made necessary corrections. Over ten exercises and problems at the end of each chapter have been included in the second edition.

I wish to express my gratitude to IEEE book series editor, Wiley Acquisitions Editor, Project Editor, and the reviewers of the book for their valuable comments and suggestions.

*Jizhong Zhu*

# PREFACE TO THE FIRST EDITION

I have been undertaking the research and practical applications of power system optimization since the early 1980s. In the early stage of my career, I worked in universities such as Chongqing University (China), Brunel University (UK), National University of Singapore, and Howard University (USA). Since 2000 I have been working for ALSTOM Grid Inc. (USA). When I was a full-time professor at Chongqing University, I wrote a tutorial on power system optimal operation, which I used to teach my senior undergraduate students and postgraduate students in power engineering until 1996. The topics of the tutorial included advanced mathematical and operations research methods and their practical applications in power engineering problems. Some of these were refined to become part of this book.

This book comprehensively applies all kinds of optimization methods to solve power system operation problems. Some contents are analyzed and discussed for the first time in detail in one book, although they have appeared in international journals and conferences. These can be found in Chapter 9 "Steady-State Security Regions", Chapter 11 "Optimal Load Shedding", Chapter 12 "Optimal Reconfiguration of Electric Distribution Network", and Chapter 13 "Uncertainty Analysis in Power Systems."

This book covers not only traditional methods and implementation in power system operation such as Lagrange multipliers, equal incremental principle, linear programming, network flow programming, quadratic programming, nonlinear programming, and dynamic programming to solve the economic dispatch, unit commitment, reactive power optimization, load shedding, steady-state security region, and optimal power flow problems, but also new technologies and their implementation in power system operation in the last decade. The new technologies include improved interior point method, analytic hierarchical process, neural network, fuzzy set theory, genetic algorithm, evolutionary programming, and particle swarm optimization. Some new topics (wheeling model, multiarea wheeling, the total transfer capability computation in multiareas, reactive power pricing calculation, congestion management) addressed in recent years in power system operation are also dealt with and put in appropriate chapters.

In addition to the rich analysis and implementation of all kinds of approaches, this book contains considerable hands-on experience for solving power system operation problems. I personally wrote my own code and tested the presented algorithms and power system applications. Many materials presented in the book are derived from my research accomplishments and publications when I worked at Chongqing

University, Brunel University, National University of Singapore, and Howard University, as well as currently with ALSTOM Grid Inc. I appreciate these organizations for providing me such good working environments. Some IEEE papers have been used as primary sources and are cited wherever appropriate. The related publications for each topic are also listed as references, so that those interested may easily obtain overall information.

I wish to express my gratitude to IEEE book series editor Professor Mohammed El-Hawary of Dalhousie University, Canada, Acquisitions Editor Steve Welch, Project Editor Jeanne Audino, and the reviewers of the book for their keen interest in the development of this book, especially Professor Kit Po Wong of the Hong Kong Polytechnic University, Professor Loi Lei Lai of City University, United Kingdom, Professor Ruben Romero of Universidad Estadual Paulista, Brazil, and Dr. Ali Chowdhury of California Independent System Operator, who offered valuable comments and suggestions for the book during the preparation stage.

Finally, I wish to thank Professor Guoyu Xu, who was my PhD advisor twenty years ago at Chongqing University, for his high standards and strict requirements for me ever since I was his graduate student. Thanks to everyone, including my family, who has shown support during the time−consuming process of writing this book.

*Jizhong Zhu*

# ACKNOWLEDGMENTS

I would like to express my appreciation to the IEEE Press Power Engineering book series editor, Professor Mohamed El-Hawary of Dalhousie University, Canada, Wiley-IEEE Press Acquisitions Editor Mary Hatcher, and the technical reviewers of the book for their keen interest and valuable comments in the development of this new edition. I would also like to thank all editors and technical reviewers of the first edition of the book for their constructive suggestions and encouragement during the preparation stage of the book.

I would also like to extend my thanks to Professor Guoyu Xu of Chongqing University, who was my PhD advisor 25 years ago, for his patient guidance, enthusiastic encouragement, and useful critiques of my research work related to the book.

Finally, I wish to thank my family for their support and encouragement throughout the process of writing this book.

# AUTHOR BIOGRAPHY

**Jizhong Zhu** is currently working at ALSTOM Grid Inc. as a senior principal power systems engineer, as well as a Fellow of the ALSTOM Expert Committee. He received his Ph.D. degree from Chongqing University, P.R. China, in February 1990. Dr. Zhu was a full professor in Chongqing University. He won the "Science and Technology Progress Award of State Education Committee of China" in 1992 and 1995, respectively, "Sichuan Provincial Science and Technology Advancement Award" in 1992, 1993, and 1994, respectively, as well as the "Science and Technology Invention Prize of Sichuan Province Science and Technology Association" in 1992. In recognition of Dr. Zhu's work, the Chongqing City Government conferred on him the award of Excellent Young Teacher by in 1992. He was selected as an Outstanding Science and Technology Researcher and won the annual Science and Technology Medal of Sichuan Province in 1993. He was also selected as one of four outstanding young scientists working in China by The Royal Society of UK and China Science and Technology Association and awarded the Royal Society Fellowship in 1994 and the national research prize "Fok Ying-Tong Young Teacher Research Medal" in 1996. He has worked in a number of different institutions all over the world, including Chongqing University in China, Brunel University in the United Kingdom, the National University of Singapore, and the Howard University in the United States, and has been with ALSTOM Grid Inc. (since 2000). He is also an advisory professor at Chongqing University. His research interest is in the analysis, operation, planning, and control of power systems as well as application of renewable energy. He has published six books as an author and co-author, and has published over 200 papers in international journals and conferences.

# CHAPTER 1

# INTRODUCTION

The electric power industry is being relentlessly pressured by governments, politicians, large industries, and investors to privatize, restructure, and deregulate. Before deregulation, most elements of the power industry, such as power generation, bulk power sales, capital expenditures, and investment decision, were heavily regulated. Some of these regulations were at the state level, and some at the national level. Thus new deregulation in the power industry meant new challenges and huge changes. However, despite changes in different structures, market rules, and uncertainties, the underlying requirements for power system operations to be secure, economical, and reliable remain the same.

This book attempts to cover all areas in power system operations. It also introduces some new topics and new applications of the latest new technologies that have appeared in recent years. This includes the analysis and discussion of new techniques for solving old problems as well as the new ones arising as a result of deregulation.

According to the different characteristics and types of the problems as well as their complexity, power system operation is divided into the following aspects that are addressed in this new edition of the book:

- Power flow analysis (Chapter 2)
- Sensitivity calculation (Chapter 3)
- Classical economic dispatch (Chapter 4)
- Security-constrained economic dispatch (Chapter 5)
- Multiarea systems economic dispatch (Chapter 6)
- Unit commitment (Chapter 7)
- Optimal power flow (Chapter 8)
- Steady-state security regions (Chapter 9)
- Application of renewable energy (Chapter 10)
- Optimal load shedding (Chapter 11)
- Optimal reconfiguration of electric distribution networks (Chapter 12)
- Uncertainty analysis in power systems (Chapter 13)
- Operation of smart grids (Chapter 14)

*Optimization of Power System Operation*, Second Edition. Jizhong Zhu.
© 2015 The Institute of Electrical and Electronics Engineers, Inc. Published 2015 by John Wiley & Sons, Inc.

From the viewpoint of optimization, various techniques including traditional and modern optimization methods, which have been developed to solve these power system operation problems, are classified into three groups [1–13]:

**(1)** Conventional optimization methods including
- Unconstrained optimization approaches
- Nonlinear programming (NLP)
- Linear programming (LP)
- Quadratic programming (QP)
- Generalized reduced gradient method
- Newton method
- Network flow programming (NFP)
- Mixed integer programming (MIP)
- Interior point (IP) methods.

**(2)** Intelligence search methods such as
- Neural network (NN)
- Evolutionary algorithms (EAs)
- Tabu search (TS)
- Particle swarm optimization (PSO).

**(3)** Nonquantitative approaches to address uncertainties in objectives and constraints including
- Probabilistic optimization
- Fuzzy set applications
- Analytic hierarchical processes (AHPs).

Power systems basics are introduced first in the following sections, followed by brief descriptions of various optimization techniques that are used to solve power system operation problems.

## 1.1 POWER SYSTEM BASICS

### 1.1.1 Physical Components

A power system can be broadly divided into the generation system that supplies the power, the transmission network that carries the power from the generating centers to the load centers, and the distribution system that feeds the power to nearby homes and industries. Figure 1.1 is a simple power system that shows some basic components.

***Generating Unit*** All power systems have one or more generating units, which are sources of power. Direct current (DC) power can be supplied by batteries, fuel cells, or photovoltaic cells. Alternating current (AC) power is typically supplied by a rotor that spins in a magnetic field in a device known as a turbo generator in a

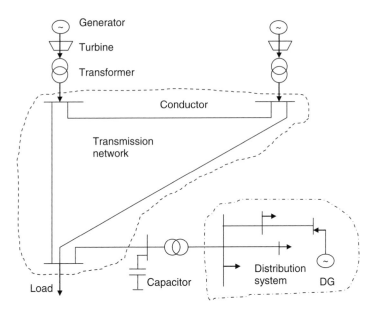

Figure 1.1    A simple power system.

power station. There have been a wide range of techniques used to spin a turbine's rotor, from superheated steam heated using fossil fuel (including coal, gas, and oil) to water itself (hydroelectric power), and wind (wind power). Even nuclear power typically depends on water heated to steam using a nuclear reaction.

The speed at which the rotor spins in combination with the number of generator poles determines the frequency of the AC produced by the generator. All generators on a single system rotate synchronously (i.e., at an identical speed) and will target a set frequency—in China and European countries, this is 50 Hz, and in the United States, 60 Hz. If the load on the system increases, the generators will require more torque to spin at that speed and, in a typical power station, more steam must be supplied to the turbines driving them. Thus the steam used and the fuel expended are directly dependent on the quantity of electrical energy supplied.

***Transformer***    A transformer is a pair of mutually inductive coils used to convey AC power from one coil to the other. It is a static device that transfers electrical energy from one circuit to another through inductively coupled conductors—the transformer's coils. A varying current in the first or *primary* winding creates a varying magnetic flux in the transformer's core and thus a varying magnetic field through the *secondary* winding. This varying magnetic field induces a varying electromotive force (EMF) or "voltage" in the secondary winding. This effect is called mutual induction.

Transformers provide an efficient means of changing voltage and current levels, and make the bulk power transmission system practical. The transformer primary is the winding that accepts power, and the transformer secondary is the winding that delivers power. In an ideal transformer, the induced voltage in the secondary wind-ing ($V_s$) is proportional to the primary voltage ($V_p$), and is given by the ratio of the number of turns in the secondary ($N_s$) to the number of turns in the primary ($N_p$) as follows:

$$\frac{V_s}{V_p} = \frac{N_s}{N_p} \tag{1.1}$$

***Transmission Line or Conductor*** Transmission lines are used to transfer power/energy from sources to loads such as an overhead power line, which is an electric power transmission line suspended by towers or utility poles. Since most of the insulation is provided by air, overhead power lines are generally the lowest-cost method of transmission of large quantities of electrical energy. Towers for support of the lines are made of wood (as-grown or laminated), steel (either lattice structures or tubular poles), concrete, aluminum, and, occasionally, reinforced plastics. The bare wire conductors on the line are generally made of aluminum (either plain or reinforced with steel or sometimes composite materials), although some copper wires are used in medium-voltage distribution and low-voltage connections to customer premises.

An object of uniform cross section has a resistance proportional to its resis-tivity and length and inversely proportional to its cross-sectional area. All materials show some resistance, except for superconductors, which have a resistance of zero. The resistance of an object is defined as the ratio of voltage across it to current through it:

$$R = \frac{V}{I} \tag{1.2}$$

For a wide variety of materials and conditions, the electrical resistance $R$ is con-stant for a given temperature; it does not depend on the amount of current through or the potential difference (voltage) across the object. Such materials are called ohmic materials. For objects made of ohmic materials, the definition of the resistance, with $R$ being a constant for that resistor, is known as Ohm's law.

***Load*** Loads are also called energy consumptions, which use the electrical energy to perform a function. These loads range from household appliances to industrial machinery. Loads are supplied by the energy sources such as generating units through the transmission system (or the grid). The change in the power system load over time—that is, the change in the power consumed or the current in the network as a function of time—is called the load curve. Loads determined by the rated power of the users are random quantities that may assume various values with a certain probability.

The real power $P$ of an individual load, a load group, or the entire system is defined as

$$P = S \cos \phi \tag{1.3}$$

where $S = VI$ is the apparent power ($V$ is the voltage, and $I$ is the current), $\cos \phi$ is the power factor, and $\phi = \arc \tan(Q/P)$, where $Q$ is the reactive power of the load.

**Capacitor**  A capacitor (formerly known as condenser) is a device for storing electrical charge. The forms of practical capacitors vary widely, but all of them contain at least two conductors separated by a non-conductor. Capacitors used as parts of electrical systems, for example, consist of metal foils separated by a layer of insulating film.

The current associated with capacitors leads the voltage because of the time it takes for the dielectric material to charge up to full voltage from the charging current. Therefore, it is said that the current in a capacitor *leads* the voltage. The units (measurement) of capacitance are called farads.

**Fundamental Properties of Circuits**  Electric power is a measurable quantity that is the time rate of increase or decrease in energy. Power is also the mathematical product of two quantities: current and voltage. These two quantities can vary with respect to time (alternating current, AC power) or can be kept at constant levels (direct current, DC power).

An instantaneous power supplied, or consumed by a component of a circuit can be expressed as follows.

$$P = \frac{dE}{dt} = \frac{dE}{dQ}\frac{dQ}{dt} = VI \tag{1.4}$$

It means that the power supplied at any instant by a source, or consumed by a load, is given by the current through the component times the voltage across the component. When current is given in amperes, and voltage in volts, the units of power are watts (W).

There are two fundamental properties of circuits, one is about the current, which is Kirchhoff's first law, and another is about voltage, which is Kirchhoff's second law. The former is also called as Kirchhoff's current law (abbreviated KCL). The latter is also called as Kirchhoff's voltage law (abbreviated KVL). KCL states that, at every instant of time, the sum of the currents flowing into any node of a circuit must equal the sum of the currents leaving the node, where a node is any spot where two or more conductors/wires are joined. KCL can be written as below.

$$\sum_{b \to n} I_b = 0 \tag{1.5}$$

where $n$ is a node of a circuit and $b$ is a collection of conductor branches. The symbol "$b \to n$" means the branch $b$ connects to the node $n$. The direction of the current is defined as positive if the current flows into the node; it is negative if the current leaves the node.

The second of Kirchhoff's fundamental laws, that is KVL, states that the sum of the voltages around any loop of a circuit at any instant is zero.

KVL can be written as below.

$$\sum_{k \in l} V_k = 0 \tag{1.6}$$

where $l$ is a closed circuit (or loop), and $k$ is one of branches in the loop $l$. The symbol "$k \in l$" means the branch $k$ belongs to loop $l$.

## 1.1.2   Renewable Energy Resources

Traditionally, power plants in the power system produce electricity by use of conventional energy sources, which consist primarily of coal, natural gas, and oil. Once a deposit of these fuels is depleted, it cannot be replenished. Thus, renewable energy is now receiving considerable attention. Renewable energy is energy that comes from natural resources such as sunlight, wind, rain, tides, and geothermal heat, which are renewable. Renewable energy sources differ from conventional sources in that, generally, they cannot be scheduled, and they are often connected to the electricity distribution system rather than the transmission system.

Most renewable energy sources originate either directly or indirectly from the sun. They are continually replenished, literally, as long as the sun continues to shine. The following five renewable sources are used most often:

- Solar
- Wind
- Water (hydropower)
- Biomass—including wood and wood waste, municipal solid waste, landfill gas, and biogas, ethanol, and biodiesel
- Geothermal.

## 1.1.3   Smart Grid

A smart grid, also called smart electrical/power grid, intelligent grid, future grid, inter-grid, or intra-grid, is an enhancement of the twentieth century power grid. Traditional power grids are generally used to carry power from a few central generators to a large number of users or customers. In contrast, the smart grid is a modernized electrical grid that uses information and two-way, cyber-secure communications technology to gather and act on information, such as information about the behaviors of suppliers and consumers, in an automated fashion to improve the efficiency, reliability, economics, and sustainability of the production and distribution of electricity. As a globally emerging industry, smart grids include many fresh and vital technologies that are needed to make enormous progress in power grid development. With the development of information technology and computer-based remote control and automation, the systems and technologies for the smart grid are made possible by two-way communication technology and computer processing that has been used for decades in other industries. They are beginning to be used on electricity networks, from the power plants and wind farms all the way to the consumers of electricity in homes and businesses. They offer many benefits to utilities and consumers—mostly seen in big improvements in energy efficiency on the electricity grid and in the energy users' homes and offices. This modernized electricity network, which sends electricity from power suppliers to consumers using digital technology to save energy, reduce cost, and increase reliability and transparency is being promoted by many governments as a way of addressing energy independence, global warming, and emergency resilience issues.

## 1.2  CONVENTIONAL METHODS

### 1.2.1  Unconstrained Optimization Approaches

Unconstrained optimization approaches are the basis of the constrained optimization algorithms. In particular, most of the constrained optimization problems in power system operation can be converted into unconstrained optimization problems. The major unconstrained optimization approaches that are used in power system operation are the gradient method, line search, Lagrange multiplier method, Newton-Raphson optimization, trust-region optimization, quasi-Newton method, double dogleg optimization, conjugate gradient optimization, and so on. Some of these approaches are used in Chapters 2–4, 7, 9, and 14.

### 1.2.2  Linear Programming

Linear programming (LP)-based techniques are used to linearize nonlinear power system optimization problems so that objective functions and constraints of power system optimization problems have linear forms. The simplex method is known to be quite effective for solving LP problems. The LP approach has several advantages. Firstly, it is reliable, especially in regard to the convergence properties. Secondly, it can quickly identify infeasibility. Thirdly, it accommodates a large variety of power system operating limits, including the very important contingency constraints. The disadvantages of LP-based techniques are inaccurate evaluation of system losses and insufficient ability to find an exact solution compared with an accurate nonlinear power system model. However, a large number of practical applications have shown that LP-based solutions generally meet the requirements of engineering precision. Thus LP is widely used to solve power system operation problems such as security-constrained economic dispatch, optimal power flow, steady-state security regions, and so on.

### 1.2.3  Nonlinear Programming

Power system operation problems are nonlinear. Thus nonlinear programming (NLP)-based techniques can easily handle power system operation problems such as the optimal power flow (OPF) problem with nonlinear objective and constraint functions. To solve a NLP problem, the first step in this method is to choose a search direction in the iterative procedure, which is determined by the first partial derivatives of the equations (the reduced gradient). Therefore, these methods are referred to as first-order methods, an example being the generalized reduced gradient (GRG) method. NLP-based methods have higher accuracy than LP-based approaches, and also have global convergence, which means convergence can be guaranteed independent of the starting point, but a slow convergent rate may occur because of zigzagging in the search direction. NLP methods are used in this book in Chapters 5–10, as well as in Chapter 14.

### 1.2.4   Quadratic Programming

Quadratic programming (QP) is a special form of NLP. The objective function of the QP optimization model is quadratic, and the constraints are in linear form. QP has higher accuracy than LP-based approaches. The most-used objective function in power system optimization is the generator cost function, which generally is a quadratic. Thus there is no simplification for such an objective function for power system optimization problem solved by QP. QP is used in Chapters 5 and 8.

### 1.2.5   Newton's Method

Newton's method requires the computation of the second-order partial derivatives of the power-flow equations and other constraints (the Hessian) and is therefore called a second-order method. The necessary conditions of optimality commonly are the Kuhn-Tucker conditions. Newton's method, which is used in Chapters 2, 4, and 8, is favored for its quadratic convergence properties.

### 1.2.6   Interior Point Methods

The interior point (IP) method was originally used to solve LP problems. It is faster and is perhaps better than the conventional simplex algorithm in LP. IP methods were first applied in 1990s to solve OPF problems, and the method has been extended and improved recently to solve OPF problems in QP and NLP forms. The analysis and implementation of IP methods are discussed in Chapter 8.

### 1.2.7   Mixed-Integer Programming

The power system problem can also be formulated as a mixed-integer programming (MIP) optimization problem with integer variables such as transformer tap ratio, phase shifter angle, and unit on or off status. MIP is extremely demanding of computer resources and the number of discrete variables is an important indicator of how difficult an MIP will be to solve. MIP methods that are used to solve OPF problems are the recursive MIP technique using an approximation method and the branch-and-bound (B&B) method, which is a typical method for integer programming. A decomposition technique is generally adopted to decompose the MIP problem into a continuous problem and an integer problem. Decomposition methods such as Benders decomposition method (BDM) can greatly improve the efficiency in solving a large-scale network by reducing the dimensions of the individual subproblems. The results show a significant reduction in the number of iterations, required computation time, and memory space. In addition, decomposition allows the application of a separate method for the solution of each subproblem, which makes the approach very attractive. MIP can be used to solve the unit commitment, OPF, as well as optimal reconfiguration of the electric distribution network.

## 1.2.8   Network Flow Programming

Network flow programming (NFP) is a special form of LP. NFP was first applied to solve optimization problems in power systems in the 1980s. The early applications of NFP were mainly on a linear model. Recently, nonlinear convex NFP has been used in power system optimization problems. NFP-based algorithms have the features of fast speed and simple calculation. These methods are efficient for solving simplified OPF problems such as security-constrained economic dispatch, multiarea systems economic dispatch, and optimal reconfiguration of an electric distribution network.

## 1.3   INTELLIGENT SEARCH METHODS

### 1.3.1   Optimization Neural Network

The optimization neural network (ONN) was first used to solve LP problems in 1986. Recently, ONN was extended to solve NLP problems. ONN is completely different from traditional optimization methods. It changes the solution of an optimization problem into an equilibrium point (or equilibrium state) of a nonlinear dynamic system, and changes the optimal criterion into energy functions for dynamic systems. Because of its parallel computational structure and the evolution of dynamics, the ONN approach appears superior to traditional optimization methods. The ONN approach is applied to solve the classical economic dispatch and multiarea systems economic dispatch in this book.

### 1.3.2   Evolutionary Algorithms

Natural evolution is a population-based optimization process. The evolutionary algorithms (EAs) are different from the conventional optimization methods, and they do not need to differentiate cost function and constraints. Theoretically, similarly to simulated annealing, EAs converge to the global optimum solution. EAs, including evolutionary programming (EP), evolutionary strategy (ES), and GA, are artificial intelligence methods for optimization based on the mechanics of natural selection, such as mutation, recombination, reproduction, crossover, selection, and so on. Since EAs require all information to be included in the fitness function, it is very difficult to consider all OPF constraints. Thus EAs are generally used to solve a simplified OPF problem such as the classic economic dispatch, security-constrained economic power dispatch, or reactive optimization problem, as well as optimal reconfiguration of an electric distribution network.

### 1.3.3   Tabu Search

The Tabu search (TS) algorithm is mainly used for solving combinatorial optimization problems. It is an iterative search algorithm, characterized by the use of a flexible memory. It is able to eliminate local minima and to search areas beyond a local minimum. The TS method is also mainly used to solve simplified OPF problems such as the unit commitment and reactive optimization problems.

### 1.3.4   Particle Swarm Optimization

Particle swarm optimization (PSO) is a swarm intelligence algorithm, inspired by the social dynamics and an emergent behavior that arises in socially organized colonies. The PSO algorithm exploits a population of individuals to probe promising regions of the search space. In this context, the population is called a swarm and the individuals are called particles or agents. In recent years, various PSO algorithms have been successfully applied in many power-engineering problems including OPF. These are analyzed in Chapter 8.

## 1.4   APPLICATION OF THE FUZZY SET THEORY

The data and parameters used in power system operation are usually derived from many sources, with a wide variance in their accuracy. For example, although the average load is typically applied in power system operation problems, the actual load should follow some uncertain variations. In addition, generator fuel cost, volt-ampere reactive (VAR) compensators, and peak power savings may be subject to uncertainty to some degree. Therefore, uncertainties as a result of insufficient information may generate an uncertain region of decisions. Consequently, the validity of the results from average values cannot represent the uncertainty level. To account for the uncertainties in information and goals related to multiple and usually conflicting objectives in power system optimization, the use of probability theory, fuzzy set theory, and analytic hierarchical process (AHP) may play a significant role in decision making.

The probabilistic methods and their application in power systems operation with uncertainty are discussed in Chapter 13. Fuzzy sets may be assigned not only to objective functions but also to constraints, especially the nonprobabilistic uncertainty associated with the reactive power demand in constraints. Generally speaking, the satisfaction parameters (fuzzy sets) for objectives and constraints represent the degree of closeness to the optimum and the degree of enforcement of constraints, respectively. With the maximization of these satisfaction parameters, the goal of optimization is achieved and simultaneously the uncertainties are considered. The application of fuzzy sets to OPF problems is also presented in Chapter 13. The AHP is a simple and convenient method to analyze a complicated problem (or complex problem). It is especially suitable for problems that are very difficult to analyze wholly quantitatively, such as OPF with competitive objectives or uncertain factors. The details of the AHP algorithm are given in Chapter 7. AHP is employed to solve unit commitment, multiarea economic dispatch, OPF, VAR optimization, optimal load shedding, and uncertainty analysis in power systems.

## REFERENCES

1. Kirchamayer LK. *Economic Operation of Power Systems*. New York: Wiley; 1958.
2. El-Hawary ME, Christensen GS. *Optimal Economic Operation of Electric Power Systems*. New York: Academic; 1979.
3. Gross C. *Power System Analysis*. New York: Wiley; 1986.

4. Wood AJ, Wollenberg B. *Power Generation Operation and Control*. 2nd ed. New York: Wiley; 1996.
5. Heydt GT. *Computer Analysis Methods for Power Systems*. Stars in a circle publications, AR; 1996.
6. Lee TH, Thorne DH, Hill EF. A transportation method for economic dispatching—application and comparison. IEEE Trans. on Power Syst. 1980;99:2372–2385.
7. Zhu JZ, Momoh JA. Optimal VAR pricing and VAR placement using analytic hierarchy process. Electr. Pow. Syst. Res. 1998;48(1):11–17.
8. Zhang WJ, Li FX, Tolbert LM. Review of reactive power planning: objectives, constraints, and algorithms. IEEE Trans. Power Syst. 2007;22(4):2177–2186.
9. Zhu J.Z, Hwang D, and Sadjadpour A "Real Time Congestion Monitoring and Management of Power Systems," IEEE/PES T&D 2005 Asia Pacific, Dalian, August 14–18, 2005.
10. Nocedal J, Wright SJ. *Numerical Optimization*. Springer; 1999.
11. Luenberger DG. *Introduction to Linear and Nonlinear Programming*. USA: Addison-wesley Publishing Company, Inc.; 1973.
12. Kennedy J and Eberhart R, "Particle swarm optimization," in Proceedings of IEEE International Conference on Neural Networks, Perth, Australia, vol. 4, 1995, pp. 1942–1948.
13. Hopfield JI. Neural networks and physical systems with emergent collective computational abilities. Proc. Natl. Acad. Sci. U.S.A. 1982;79:2554–2558.

# CHAPTER 2

# POWER FLOW ANALYSIS

This chapter deals with the power flow problem. The power flow algorithms include the Newton–Raphson method in both polar and rectangular forms, the Gauss–Seidel method, the DC power flow method, and all kinds of decoupled power flow methods such as fast decoupled power flow, simplified BX and XB methods, as well as decoupled power flow without major approximation.

## 2.1 MATHEMATICAL MODEL OF POWER FLOW

Power flow is well known as "load flow." This is the name given to a network solution that shows currents, voltages, and real and reactive power flows at every bus in the system. Since the parameters of the elements such as lines and transformers are constant, the power system network is a linear network. However, in the power flow problem, the relationship between voltage and current at each bus is nonlinear, and the same holds for the relationship between the real and reactive power consumption at a bus or the generated real power and scheduled voltage magnitude at a generator bus. Thus power flow calculation involves the solution of nonlinear equations. It gives us the electrical response of the transmission system to a particular set of loads and generator power outputs. Power flows are an important part of power system operation and planning.

Generally, for a network with $n$ independent buses, we can write the following $n$ equations.

$$\left.\begin{aligned}
Y_{11}\dot{V}_1 + Y_{12}\dot{V}_2 + \cdots + Y_{1n}\dot{V}_n &= \dot{I}_1 \\
Y_{21}\dot{V}_1 + Y_{22}\dot{V}_2 + \cdots + Y_{2n}\dot{V}_n &= \dot{I}_2 \\
\cdots \\
Y_{n1}\dot{V}_1 + Y_{n2}\dot{V}_2 + \cdots + Y_{nn}\dot{V}_n &= \dot{I}_n
\end{aligned}\right\} \tag{2.1}$$

The matrix form is

$$\begin{bmatrix} Y_{11} & Y_{12} & \cdots & Y_{1n} \\ Y_{21} & Y_{22} & \cdots & Y_{2n} \\ \vdots & \vdots & & \vdots \\ Y_{n1} & Y_{n2} & \cdots & Y_{nn} \end{bmatrix} \begin{bmatrix} \dot{V}_1 \\ \dot{V}_2 \\ \vdots \\ \dot{V}_n \end{bmatrix} = \begin{bmatrix} \dot{I}_1 \\ \dot{I}_2 \\ \vdots \\ \dot{I}_n \end{bmatrix} \tag{2.2}$$

*Optimization of Power System Operation*, Second Edition. Jizhong Zhu.
© 2015 The Institute of Electrical and Electronics Engineers, Inc. Published 2015 by John Wiley & Sons, Inc.

or

$$[Y][V] = I \tag{2.3}$$

where $I$ is the bus current injection vector, $V$ is the bus voltage vector, $Y$ is called the bus admittance matrix. Its diagonal element $Y_{ii}$ is called the self admittance of bus $i$, which equals the sum of all branch admittances connecting to bus $i$. The off-diagonal element of the bus admittance matrix $Y_{ij}$ is the negative of branch admittance between buses $i$ and $j$. If there is no line between buses $i$ and $j$, this term is zero. Obviously, the bus admittance matrix is a sparse matrix.

In addition, the bus current can be represented by bus voltage and power, that is,

$$\dot{I}_i = \frac{\widehat{S}_i}{\widehat{V}_i} = \frac{\widehat{S}_{Gi} - \widehat{S}_{Di}}{\widehat{V}_i} = \frac{(P_{Gi} - P_{Di}) - j(Q_{Gi} - Q_{Di})}{\widehat{V}_i} \tag{2.4}$$

where

S: the complex power injection vector
$P_{Gi}$: the real power output of the generator connecting to bus $i$
$Q_{Gi}$: the reactive power output of the generator connecting to bus $i$
$P_{Di}$: the real power load connecting to bus $i$
$Q_{Di}$: the reactive power load connecting to bus $i$.

Substituting equation (2.4) into equation (2.1), we have

$$\frac{(P_{Gi} - P_{Di}) - j(Q_{Gi} - Q_{Di})}{\widehat{V}_i} = Y_{i1}\dot{V}_1 + Y_{i2}\dot{V}_2 + \cdots + Y_{in}\dot{V}_n, \quad i = 1, 2, \dots, n \tag{2.5}$$

In the power flow problem, the load demands are known variables. We define the following bus power injections as

$$P_i = P_{Gi} - P_{Di} \tag{2.6}$$

$$Q_i = Q_{Gi} - Q_{Di} \tag{2.7}$$

Substituting the above two equations into equation (2.5), we can get the general form of power flow equation as

$$\frac{P_i - jQ_i}{\widehat{V}_i} = \sum_{j=1}^{n} Y_{ij}\dot{V}_j, \quad i = 1, 2, \dots, n \tag{2.8}$$

or

$$P_i + jQ_i = \dot{V}_i \sum_{j=1}^{n} \widehat{Y}_{ij}\widehat{V}_j, \quad i = 1, 2, \dots, n \tag{2.9}$$

If we divide equation (2.9) into real and imaginary parts, we can get two equations for each bus with four variables, that is, bus real power $P$, reactive power $Q$, voltage $V$, and angle $\theta$. To solve the power flow equations, two of these should be known for each bus. According to the practical conditions of the power system operation, as well as known variables of the bus, we can have three bus types as follows:

**(1)** PV bus: For this type of bus, the bus real power $P$ and the magnitude of voltage $V$ are known, and the bus reactive power $Q$ and the angle of voltage $\theta$ are unknown. Generally, the bus connected to the generator is a PV bus.

**(2)** PQ bus: For this type of bus, the bus real power $P$ and reactive power $Q$ are known, and the magnitude and the angle of voltage $(V, \theta)$ are unknown. Generally, the bus connected to load is a PQ bus. However, the power output of some generators is constant or cannot be adjusted under the particular operation conditions. The corresponding bus will also be a PQ bus.

**(3)** Slack bus: The slack bus is also called the swing bus, or the reference bus. Since power loss of the network is unknown during power flow calculation, at least one bus power cannot be given, which will balance the system power. In addition, it is necessary to have a bus with a zero voltage angle as reference for the calculation of the other voltage angles. Generally, the slack bus is a generator-related bus, whose magnitude and angle of voltage $(V, \theta)$ are known. The bus real power $P$ and reactive power $Q$ are unknown variables. Traditionally, there is only one slack bus in the power flow calculation. In practical applications, distributed slack buses are used, so all buses that connect the adjustable generators can be selected as slack buses and used to balance the power mismatch through some rules. One of these rules is that the system power mismatch is balanced by all slacks on the basis of the unit participation factors.

Since the voltage of the slack bus is given, only $n - 1$ bus voltages need to be calculated. Thus, the number of power flow equations is $2(n - 1)$.

## 2.2 NEWTON-RAPHSON METHOD

### 2.2.1 Principle of Newton-Raphson Method

A nonlinear equation in a single variable can be expressed as

$$f(x) = 0 \tag{2.10}$$

For solving this equation, select an initial value $x^0$. The difference between the initial value and the final solution will be $\Delta x^0$. Then $x = x^0 + \Delta x^0$ is the solution of nonlinear equation (2.10), that is,

$$f(x^0 + \Delta x^0) = 0 \tag{2.11}$$

Expanding the above equation with the Taylor series, we get

$$f(x^0 + \Delta x^0) = f(x^0) + f'(x^0)\Delta x^0 + f''(x^0)\frac{(\Delta x^0)^2}{2!} + \cdots$$
$$+ f^{(n)}(x^0)\frac{(\Delta x^0)^n}{n!} + \cdots = 0 \tag{2.12}$$

where $f'(x^0), \ldots, f^{(n)}(x^0)$ are the derivatives of the function $f(x)$.

If the difference $\Delta x^0$ is very small (meaning that the initial value $x^0$ is close to the solution of the function), the terms of the second and higher derivatives can be neglected. Thus equation (2.12) becomes a linear equation as below:

$$f(x^0 + \Delta x^0) = f(x^0) + f'(x^0)\Delta x^0 = 0 \tag{2.13}$$

Then we get

$$\Delta x^0 = -\frac{f(x^0)}{f'(x^0)} \tag{2.14}$$

The new solution will be

$$x^1 = x^0 + \Delta x^0 = x^0 - \frac{f(x^0)}{f'(x^0)} \tag{2.15}$$

Since equation (2.13) is an approximate equation, the value of $\Delta x^0$ is also an approximation. Thus the solution $x$ is not a real solution. Further iterations are needed. The iteration equation is

$$x^{k+1} = x^k + \Delta x^k = x^k - \frac{f(x^k)}{f'(x^k)} \tag{2.16}$$

The iteration can be stopped if one of the following conditions is met:

$$|\Delta x^k| < \varepsilon_1$$

or

$$|f(x^k)| < \varepsilon_2 \tag{2.17}$$

where $\varepsilon_1, \varepsilon_2$, which are the permitted convergence precisions, are small positive numbers.

The Newton method can also be expanded to a nonlinear equation with $n$ variables.

$$\left. \begin{aligned} f_1(x_1, x_2, \ldots, x_n) &= 0 \\ f_2(x_1, x_2, \ldots, x_n) &= 0 \\ \cdots \\ f_n(x_1, x_2, \ldots, x_n) &= 0 \end{aligned} \right\} \tag{2.18}$$

For a given set of initial values $x_1^0, x_2^0, \ldots, x_n^0$, we have the corrected values $\Delta x_1^0, \Delta x_2^0, \ldots, \Delta x_n^0$. Then equation (2.18) becomes

$$\left. \begin{aligned} f_1(x_1^0 + \Delta x_1^0, x_2^0 + \Delta x_2^0, \ldots, x_n^0 + \Delta x_n^0) &= 0 \\ f_2(x_1^0 + \Delta x_1^0, x_2^0 + \Delta x_2^0, \ldots, x_n^0 + \Delta x_n^0) &= 0 \\ \cdots \\ f_n(x_1^0 + \Delta x_1^0, x_2^0 + \Delta x_2^0, \ldots, x_n^0 + \Delta x_n^0) &= 0 \end{aligned} \right\} \tag{2.19}$$

Similarly, expanding equation (2.19) and neglecting the terms of second and higher derivatives, we get

$$
\left.\begin{aligned}
f_1(x_1^0, x_2^0, \ldots, x_n^0) + \left.\frac{\partial f_1}{\partial x_1}\right|_{x_1^0} \Delta x_1^0 + \left.\frac{\partial f_1}{\partial x_2}\right|_{x_2^0} \Delta x_2^0 + \cdots + \left.\frac{\partial f_1}{\partial x_n}\right|_{x_n^0} \Delta x_n^0 = 0 \\
f_2(x_1^0, x_2^0, \ldots, x_n^0) + \left.\frac{\partial f_2}{\partial x_1}\right|_{x_1^0} \Delta x_1^0 + \left.\frac{\partial f_2}{\partial x_2}\right|_{x_2^0} \Delta x_2^0 + \cdots + \left.\frac{\partial f_2}{\partial x_n}\right|_{x_n^0} \Delta x_n^0 = 0 \\
\cdots \\
f_n(x_1^0, x_2^0, \ldots, x_n^0) + \left.\frac{\partial f_n}{\partial x_1}\right|_{x_1^0} \Delta x_1^0 + \left.\frac{\partial f_n}{\partial x_2}\right|_{x_2^0} \Delta x_2^0 + \cdots + \left.\frac{\partial f_n}{\partial x_n}\right|_{x_n^0} \Delta x_n^0 = 0
\end{aligned}\right\} \tag{2.20}
$$

Equation (2.20) can also be written as matrix form as

$$
\begin{bmatrix} f_1(x_1^0, x_2^0, \ldots, x_n^0) \\ f_2(x_1^0, x_2^0, \ldots, x_n^0) \\ \cdots \\ f_n(x_1^0, x_2^0, \ldots, x_n^0) \end{bmatrix} = - \begin{bmatrix} \left.\frac{\partial f_1}{\partial x_1}\right|_{x_1^0} & \left.\frac{\partial f_1}{\partial x_2}\right|_{x_2^0} & \cdots & \left.\frac{\partial f_1}{\partial x_n}\right|_{x_n^0} \\ \left.\frac{\partial f_2}{\partial x_1}\right|_{x_1^0} & \left.\frac{\partial f_2}{\partial x_2}\right|_{x_2^0} & \cdots & \left.\frac{\partial f_2}{\partial x_n}\right|_{x_n^0} \\ \vdots & \vdots & & \vdots \\ \left.\frac{\partial f_n}{\partial x_1}\right|_{x_1^0} & \left.\frac{\partial f_n}{\partial x_2}\right|_{x_2^0} & \cdots & \left.\frac{\partial f_n}{\partial x_n}\right|_{x_n^0} \end{bmatrix} \begin{bmatrix} \Delta x_1^0 \\ \Delta x_2^0 \\ \vdots \\ \Delta x_n^0 \end{bmatrix} \tag{2.21}
$$

From equation (2.21), we can get $\Delta x_1^0, \Delta x_2^0, \ldots, \Delta x_n^0$. Then the new solution can be obtained. The iteration equation can be written as follows:

$$
\begin{bmatrix} f_1(x_1^k, x_2^k, \ldots, x_n^k) \\ f_2(x_1^k, x_2^k, \ldots, x_n^k) \\ \cdots \\ f_n(x_1^k, x_2^k, \ldots, x_n^k) \end{bmatrix} = - \begin{bmatrix} \left.\frac{\partial f_1}{\partial x_1}\right|_{x_1^k} & \left.\frac{\partial f_1}{\partial x_2}\right|_{x_2^k} & \cdots & \left.\frac{\partial f_1}{\partial x_n}\right|_{x_n^k} \\ \left.\frac{\partial f_2}{\partial x_1}\right|_{x_1^k} & \left.\frac{\partial f_2}{\partial x_2}\right|_{x_2^k} & \cdots & \left.\frac{\partial f_2}{\partial x_n}\right|_{x_n^k} \\ \vdots & \vdots & & \vdots \\ \left.\frac{\partial f_n}{\partial x_1}\right|_{x_1^k} & \left.\frac{\partial f_n}{\partial x_2}\right|_{x_2^k} & \cdots & \left.\frac{\partial f_n}{\partial x_n}\right|_{x_n^k} \end{bmatrix} \begin{bmatrix} \Delta x_1^k \\ \Delta x_2^k \\ \vdots \\ \Delta x_n^k \end{bmatrix} \tag{2.22}
$$

$$
x_i^{k+1} = x_i^k + \Delta x_i^k \quad i = 1, 2, \ldots, n \tag{2.23}
$$

Equations (2.22) and (2.23) can be expressed as

$$
F(X^k) = -J^k \Delta X^k \tag{2.24}
$$

$$
X^{k+1} = X^k + \Delta X^k \tag{2.25}
$$

where $J$ is an $n \times n$ matrix called a *Jacobian matrix*.

## 2.2.2   Power Flow Solution with Polar Coordinate System

If the bus voltage in equation (2.9) is expressed using the polar coordinate system, the complex voltage and real and reactive powers can be written as

$$\dot{V_i} = V_i(\cos\theta_i + j\sin\theta_i) \tag{2.26}$$

$$P_i = V_i \sum_{j=1}^{n} V_j(G_{ij}\cos\theta_{ij} + B_{ij}\sin\theta_{ij}) \tag{2.27}$$

$$Q_i = V_i \sum_{j=1}^{n} V_j(G_{ij}\sin\theta_{ij} - B_{ij}\cos\theta_{ij}) \tag{2.28}$$

where $\theta_{ij} = \theta_i - \theta_j$, which is the angle difference between buses $i$ and $j$.

Assuming that buses $1 \sim m$ are PQ buses, buses $(m + 1) \sim (n-1)$ are PV buses, and $n$th bus is the slack bus. $V_n$, $\theta_n$ are given, and the magnitudes of the PV buses $V_{m+1} \sim V_{n-1}$ are also given. Then, $n - 1$ bus voltage angles are unknown, and $m$ magnitudes of voltage are unknown. For each PV or PQ bus, we have the following real power mismatch equation:

$$\Delta P_i = P_{is} - P_i = P_{is} - V_i \sum_{j=1}^{n} V_j(G_{ij}\cos\theta_{ij} + B_{ij}\sin\theta_{ij}) = 0 \tag{2.29}$$

For each PQ bus, we also have the following reactive power equation:

$$\Delta Q_{is} = Q_{is} - Q_i = Q_{is} - V_i \sum_{j=1}^{n} V_j(G_{ij}\sin\theta_{ij} - B_{ij}\cos\theta_{ij}) = 0 \tag{2.30}$$

where $P_{is}$, $Q_{is}$ are the calculated bus real and reactive power injections, respectively.

According to the Newton method, the power flow equations (2.29) and (2.30) can be expanded into Taylor series and the following first-order approximation can be obtained:

$$\begin{bmatrix} \Delta P \\ \Delta Q \end{bmatrix} = -J \begin{bmatrix} \Delta\theta \\ \Delta V/V \end{bmatrix}$$

or

$$\begin{bmatrix} \Delta P \\ \Delta Q \end{bmatrix} = - \begin{bmatrix} H & N \\ K & L \end{bmatrix} \begin{bmatrix} \Delta\theta \\ V_D^{-1}\Delta V \end{bmatrix} \tag{2.31}$$

where

$$\Delta P = \begin{bmatrix} \Delta P_1 \\ \Delta P_2 \\ \vdots \\ \Delta P_{n-1} \end{bmatrix} \tag{2.32}$$

$$\Delta Q = \begin{bmatrix} \Delta Q_1 \\ \Delta Q_2 \\ \vdots \\ \Delta Q_m \end{bmatrix} \tag{2.33}$$

$$\Delta\theta = \begin{bmatrix} \Delta\theta_1 \\ \Delta\theta_2 \\ \vdots \\ \Delta\theta_{n-1} \end{bmatrix} \tag{2.34}$$

$$\Delta V = \begin{bmatrix} \Delta V_1 \\ \Delta V_2 \\ \vdots \\ \Delta V_m \end{bmatrix} \tag{2.35}$$

$$V_D = \begin{bmatrix} V_1 & & & \\ & V_2 & & \\ & & \ddots & \\ & & & V_m \end{bmatrix} \tag{2.36}$$

$H$ is an $(n-1)\times(n-1)$ matrix, and its element is $H_{ij} = \frac{\partial\Delta P_i}{\partial\theta_j}$.

$N$ is an $(n-1)\times m$ matrix, and its element is $N_{ij} = V_j\frac{\partial\Delta P_i}{\partial V_j}$.

$K$ is an $m\times(n-1)$ matrix, and its element is $K_{ij} = \frac{\partial\Delta Q_i}{\partial\theta_j}$.

$L$ is an $m\times m$ matrix, and its element is $L_{ij} = V_j\frac{\partial\Delta Q_i}{\partial V_j}$.

If $i \neq j$, the expressions for the elements in Jacobian matrix are as follows:

$$H_{ij} = -V_iV_j(G_{ij}\sin\theta_{ij} - B_{ij}\cos\theta_{ij}) \tag{2.37}$$

$$N_{ij} = -V_iV_j(G_{ij}\cos\theta_{ij} - B_{ij}\sin\theta_{ij}) \tag{2.38}$$

$$K_{ij} = V_iV_j(G_{ij}\cos\theta_{ij} - B_{ij}\sin\theta_{ij}) \tag{2.39}$$

$$L_{ij} = -V_iV_j(G_{ij}\sin\theta_{ij} - B_{ij}\cos\theta_{ij}) \tag{2.40}$$

If $i = j$, the expressions for the elements in the Jacobian matrix are as follows:

$$H_{ii} = V_i^2 B_{ii} + Q_i \tag{2.41}$$

$$N_{ii} = -V_i^2 G_{ii} - P_i \tag{2.42}$$

$$K_{ii} = V_i^2 G_{ii} - P_i \tag{2.43}$$

$$L_{ii} = V_i^2 B_{ii} - Q_i \tag{2.44}$$

The steps for calculation of the Newton power flow solution are as follows [1,2]:

Step (1):  Given input data.

Step (2):  Form bus admittance matrix.

Step (3):   Assume the initial values of bus voltage.

Step (4):   Compute the power mismatch according to equations (2.29) and (2.30). Check whether the convergence conditions are satisfied.

$$\max|\Delta P_i^k| < \varepsilon_1 \tag{2.45}$$

$$\max|\Delta Q_i^k| < \varepsilon_2 \tag{2.46}$$

If equations (2.45) and (2.46) are met, stop the iteration, and calculate the line flows and real and reactive powers of the slack bus. If not, go to next step.

Step (5):   Compute the elements in the Jacobian matrix (2.37)–(2.44).

Step (6):   Compute the corrected values of the bus voltage using equation (2.31). Then compute the bus voltage.

$$V_i^{k+1} = V_i^k + \Delta V_i^k \tag{2.47}$$

$$\theta_i^{k+1} = \theta_i^k + \Delta\theta_i^k \tag{2.48}$$

Step (7):   Return to Step (4) with new values of the bus voltage.

***Example 2.1:***   The test example for power flow calculation, which is shown in Figure 2.1, is taken from [2].

The parameters of the branches are as follows:

$$z_{12} = 0.10 + j0.40$$

$$y_{120} = y_{210} = j0.01528$$

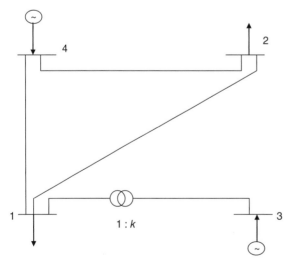

Figure 2.1   Four buses power system.

$$z_{13} = j0.30, \ k = 1.1$$

$$z_{14} = 0.12 + j0.50$$

$$y_{140} = y_{410} = j0.01920$$

$$z_{24} = 0.08 + j0.40$$

$$y_{240} = y_{420} = j0.01413$$

Buses 1 and 2 are PQ buses, bus 3 is a PV bus, and bus 4 is a slack bus. The given data are

$$P_1 + jQ_1 = -0.30 - j0.18$$

$$P_2 + jQ_2 = -0.55 - j0.13$$

$$P_3 = 0.5; \ V_3 = 1.1;$$

$$V_4 = 1.05; \ \theta_4 = 0$$

First, we form the bus admittance matrix as follows:

$$Y = \begin{bmatrix} 1.0421 - j8.2429 & -0.5882 + j2.3529 & j3.6666 & -0.4539 + j1.8911 \\ -0.5882 + j2.3529 & 1.0690 - j4.7274 & 0 & -0.4808 + j2.4038 \\ j3.6666 & 0 & -j3.3333 & 0 \\ -0.4539 + j1.8911 & -0.4808 + j2.4038 & 0 & 0.9346 - j4.2616 \end{bmatrix}$$

Given the initial bus voltage,

$$\dot{V}_1^0 = \dot{V}_2^0 = 1.0 \angle 0^0, \quad \dot{V}_3^0 = 1.1 \angle 0^0$$

Computing the bus power mismatch using equations (2.29) and (2.30), we get

$$\Delta P_1^0 = P_{1s} - P_1^0 = -0.30 - (-0.02269) = -0.27731$$

$$\Delta P_2^0 = P_{2s} - P_2^0 = -0.55 - (-0.02404) = -0.52596$$

$$\Delta P_3^0 = P_{3s} - P_3^0 = 0.5$$

$$\Delta Q_1^0 = Q_{1s} - Q_1^0 = -0.18 - (-0.12903) = -0.05097$$

$$\Delta Q_2^0 = Q_{2s} - Q_2^0 = -0.13 - (-0.14960) = 0.0196$$

Then computing the bus voltage correction using equation (2.31),

$$\Delta \theta_1^0 = -0.5059^0, \quad \Delta \theta_2^0 = -6.1776^0, \quad \Delta \theta_3^0 = 6.5970^0$$

$$\Delta V_1^0 = -0.0065, \quad \Delta V_2^0 = -0.0237$$

The new bus voltage will be

$$\theta_1^1 = \theta_1^0 + \Delta\theta_1^0 = -0.5059^0$$

$$\theta_2^1 = \theta_2^0 + \Delta\theta_2^0 = -6.1776^0$$

$$\theta_3^1 = \theta_3^0 + \Delta\theta_3^0 = 6.5970^0$$

$$V_1^1 = V_1^0 + \Delta V_1^0 = 0.9935$$

$$V_2^1 = V_2^0 + \Delta V_2^0 = 0.9763$$

Conduct the second iteration using the new voltage values. If the convergence tolerance is $\varepsilon = 10^{-5}$, the power flow will be converged after three iterations; these are shown in Tables 2.1 and 2.2.

In the final step, we compute the power of the slack bus and the power flows for all branches:

For the slack bus,

$$P_4 + jQ_4 = 0.36788 + j0.26470$$

For the branches,

$$P_{12} + jQ_{12} = 0.24624 - j0.01465$$

$$P_{13} + jQ_{13} = -0.50000 - j0.02926$$

**TABLE 2.1    Bus Power Mismatch Change**

| Iteration $k$ | $\Delta P_1$ | $\Delta P_2$ | $\Delta P_3$ | $\Delta Q_1$ | $\Delta Q_2$ |
|---|---|---|---|---|---|
| 0 | $-0.27731$ | $-0.52596$ | $0.5$ | $-0.05097$ | $0.01960$ |
| 1 | $-4.0 \times 10^{-3}$ | $-2.047 \times 10^{-2}$ | $4.51 \times 10^{-3}$ | $-4.380 \times 10^{-2}$ | $-2.454 \times 10^{-2}$ |
| 2 | $1.0 \times 10^{-4}$ | $-4.2 \times 10^{-4}$ | $8.0 \times 10^{-5}$ | $-4.5 \times 10^{-4}$ | $-3.2 \times 10^{-4}$ |
| 3 | $<10^{-5}$ | $<10^{-5}$ | $<10^{-5}$ | $<10^{-5}$ | $<10^{-5}$ |

**TABLE 2.2    Bus Voltage Change**

| Iteration $k$ | $\theta_1$ | $\theta_2$ | $\theta_3$ | $V_1$ | $V_2$ |
|---|---|---|---|---|---|
| 1 | $-0.5059^0$ | $-6.1776^0$ | $6.5970^0$ | $0.9935$ | $0.9763$ |
| 2 | $-0.5008^0$ | $-6.4452^0$ | $6.7300^0$ | $0.9848$ | $0.9650$ |
| 3 | $-0.5002^0$ | $-6.4504^0$ | $6.7323^0$ | $0.9847$ | $0.9648$ |

$$P_{14} + jQ_{14} = -0.04624 - j0.13609$$

$$P_{21} + jQ_{21} = -0.23999 + j0.01063$$

$$P_{24} + jQ_{24} = -0.31001 - j0.14063$$

$$P_{31} + jQ_{31} = 0.50000 + j0.09341$$

$$P_{41} + jQ_{41} = 0.04822 + j0.10452$$

$$P_{42} + jQ_{42} = 0.31967 + j0.16018$$

### 2.2.3 Power Flow Solution with Rectangular Coordinate System

***Newton Method*** If the bus voltage in equation (2.9) is expressed using a rectangular coordinate system, the complex voltage and real and reactive powers can be written as

$$\dot{V}_i = e_i + jf_i \tag{2.49}$$

$$P_i = e_i \sum_{j=1}^{n} (G_{ij}e_j - B_{ij}f_j) + f_i \sum_{j=1}^{n} (G_{ij}f_j + B_{ij}e_j) \tag{2.50}$$

$$Q_i = f_i \sum_{j=1}^{n} (G_{ij}e_j - B_{ij}f_j) - e_i \sum_{j=1}^{n} (G_{ij}f_j + B_{ij}e_j) \tag{2.51}$$

For each PQ bus, we have the following power mismatch equations:

$$\Delta P_i = P_{is} - P_i = P_{is} - e_i \sum_{j=1}^{n} (G_{ij}e_j - B_{ij}f_j) - f_i \sum_{j=1}^{n} (G_{ij}f_j + B_{ij}e_j) = 0 \tag{2.52}$$

$$\Delta Q_i = Q_{si} - Q_i = Q_{si} - f_i \sum_{j=1}^{n} (G_{ij}e_j - B_{ij}f_j) + e_i \sum_{j=1}^{n} (G_{ij}f_j + B_{ij}e_j) = 0 \tag{2.53}$$

For each PV bus, we have the following equations:

$$\Delta P_i = P_{is} - P_i = P_{is} - e_i \sum_{j=1}^{n} (G_{ij}e_j - B_{ij}f_j) - f_i \sum_{j=1}^{n} (G_{ij}f_j + B_{ij}e_j) = 0 \tag{2.54}$$

$$\Delta V_i^2 = V_{is}^2 - V_i^2 = V_{is}^2 - (e_i^2 + f_i^2) = 0 \tag{2.55}$$

There are $2(n - 1)$ equations in equations (2.52)–(2.55). According to the Newton method, we have the following correction equation:

$$\Delta F = -J\Delta V \tag{2.56}$$

where

$$\Delta F = \begin{bmatrix} \Delta P_1 \\ \Delta Q_1 \\ \vdots \\ \Delta P_m \\ \Delta Q_m \\ \Delta P_{m+1} \\ \Delta V^2_{m+1} \\ \vdots \\ \Delta P_{n-1} \\ \Delta V^2_{n-1} \end{bmatrix} \qquad (2.57)$$

$$\Delta V = \begin{bmatrix} \Delta e_1 \\ \Delta f_1 \\ \vdots \\ \Delta e_m \\ \Delta f_m \\ \Delta e_{m+1} \\ \Delta f_{m+1} \\ \vdots \\ \Delta e_{n-1} \\ \Delta f_{n-1} \end{bmatrix} \qquad (2.58)$$

$$J =$$

$$\begin{bmatrix}
\dfrac{\partial \Delta P_1}{\partial e_1} & \dfrac{\partial \Delta P_1}{\partial f_1} & \cdots & \dfrac{\partial \Delta P_1}{\partial e_m} & \dfrac{\partial \Delta P_1}{\partial f_m} & \dfrac{\partial \Delta P_1}{\partial e_{m+1}} & \dfrac{\partial \Delta P_1}{\partial f_{m+1}} & \cdots & \dfrac{\partial \Delta P_1}{\partial e_{n-1}} & \dfrac{\partial \Delta P_1}{\partial f_{n-1}} \\[2mm]
\dfrac{\partial \Delta Q_1}{\partial e_1} & \dfrac{\partial \Delta Q_1}{\partial f_1} & \cdots & \dfrac{\partial \Delta Q_1}{\partial e_m} & \dfrac{\partial \Delta Q_1}{\partial f_m} & \dfrac{\partial \Delta Q_1}{\partial e_{m+1}} & \dfrac{\partial \Delta Q_1}{\partial f_{m+1}} & \cdots & \dfrac{\partial \Delta Q_1}{\partial e_{n-1}} & \dfrac{\partial \Delta Q_1}{\partial f_{n-1}} \\[2mm]
\vdots & \vdots & & \vdots & \vdots & \vdots & \vdots & & \vdots & \vdots \\[2mm]
\dfrac{\partial \Delta P_m}{\partial e_1} & \dfrac{\partial \Delta P_m}{\partial f_1} & \cdots & \dfrac{\partial \Delta P_m}{\partial e_m} & \dfrac{\partial \Delta P_m}{\partial f_m} & \dfrac{\partial \Delta P_m}{\partial e_{m+1}} & \dfrac{\partial \Delta P_m}{\partial f_{m+1}} & \cdots & \dfrac{\partial \Delta P_m}{\partial e_{n-1}} & \dfrac{\partial \Delta P_m}{\partial f_{n-1}} \\[2mm]
\dfrac{\partial \Delta Q_m}{\partial e_1} & \dfrac{\partial \Delta Q_m}{\partial f_1} & \cdots & \dfrac{\partial \Delta Q_m}{\partial e_m} & \dfrac{\partial \Delta Q_m}{\partial f_m} & \dfrac{\partial \Delta Q_m}{\partial e_{m+1}} & \dfrac{\partial \Delta Q_m}{\partial f_{m+1}} & \cdots & \dfrac{\partial \Delta Q_m}{\partial e_{n-1}} & \dfrac{\partial \Delta Q_m}{\partial f_{n-1}} \\[2mm]
\dfrac{\partial \Delta P_{m+1}}{\partial e_1} & \dfrac{\partial \Delta P_{m+1}}{\partial f_1} & \cdots & \dfrac{\partial \Delta P_{m+1}}{\partial e_m} & \dfrac{\partial \Delta P_{m+1}}{\partial f_m} & \dfrac{\partial \Delta P_{m+1}}{\partial e_{m+1}} & \dfrac{\partial \Delta P_{m+1}}{\partial f_{m+1}} & \cdots & \dfrac{\partial \Delta P_{m+1}}{\partial e_{n-1}} & \dfrac{\partial \Delta P_{m+1}}{\partial f_{n-1}} \\[2mm]
\dfrac{\partial \Delta V^2_{m+1}}{\partial e_1} & \dfrac{\partial \Delta V^2_{m+1}}{\partial f_1} & \cdots & \dfrac{\partial \Delta V^2_{m+1}}{\partial e_m} & \dfrac{\partial \Delta V^2_{m+1}}{\partial f_m} & \dfrac{\partial \Delta V^2_{m+1}}{\partial e_{m+1}} & \dfrac{\partial \Delta V^2_{m+1}}{\partial f_{m+1}} & \cdots & \dfrac{\partial \Delta V^2_{m+1}}{\partial e_{n-1}} & \dfrac{\partial \Delta V^2_{m+1}}{\partial f_{n-1}} \\[2mm]
\vdots & \vdots & & \vdots & \vdots & \vdots & \vdots & & \vdots & \vdots \\[2mm]
\dfrac{\partial \Delta P_{n-1}}{\partial e_1} & \dfrac{\partial \Delta P_{n-1}}{\partial f_1} & \cdots & \dfrac{\partial \Delta P_{n-1}}{\partial e_m} & \dfrac{\partial \Delta P_{n-1}}{\partial f_m} & \dfrac{\partial \Delta P_{n-1}}{\partial e_{m+1}} & \dfrac{\partial \Delta P_{n-1}}{\partial f_{m+1}} & \cdots & \dfrac{\partial \Delta P_{n-1}}{\partial e_{n-1}} & \dfrac{\partial \Delta P_{n-1}}{\partial f_{n-1}} \\[2mm]
\dfrac{\partial \Delta V^2_{n-1}}{\partial e_1} & \dfrac{\partial \Delta V^2_{n-1}}{\partial f_1} & \cdots & \dfrac{\partial \Delta V^2_{n-1}}{\partial e_m} & \dfrac{\partial \Delta V^2_{n-1}}{\partial f_m} & \dfrac{\partial \Delta V^2_{n-1}}{\partial e_{m+1}} & \dfrac{\partial \Delta V^2_{n-1}}{\partial f_{m+1}} & \cdots & \dfrac{\partial \Delta V^2_{n-1}}{\partial e_{n-1}} & \dfrac{\partial \Delta V^2_{n-1}}{\partial f_{n-1}}
\end{bmatrix}$$

$$(2.59)$$

If $i \neq j$, the expressions for the elements in the Jacobian matrix are as follows:

$$\frac{\partial \Delta P_i}{\partial e_i} = -\frac{\partial \Delta Q_i}{\partial f_i} = -(G_{ij}e_i + B_{ij}f_i) \tag{2.60}$$

$$\frac{\partial \Delta P_i}{\partial f_i} = -\frac{\partial \Delta Q_i}{\partial e_i} = -(G_{ij}f_i - B_{ij}e_i) \tag{2.61}$$

$$\frac{\partial \Delta V_i^2}{\partial e_i} = -\frac{\partial \Delta V_i^2}{\partial f_i} = 0 \tag{2.62}$$

If $i = j$, the expressions for the elements in the Jacobian matrix are as follows:

$$\frac{\partial \Delta P_i}{\partial e_i} = -\sum_{j=1}^{n}(G_{ij}e_j - B_{ij}f_j) - G_{ii}e_i - B_{ii}f_i \tag{2.63}$$

$$\frac{\partial \Delta P_i}{\partial f_i} = -\sum_{j=1}^{n}(G_{ij}f_j + B_{ij}e_j) - G_{ii}f_i + B_{ii}e_i \tag{2.64}$$

$$\frac{\partial \Delta Q_i}{\partial e_i} = \sum_{j=1}^{n}(G_{ij}f_j + B_{ij}e_j) - G_{ii}f_i + B_{ii}e_i \tag{2.65}$$

$$\frac{\partial \Delta Q_i}{\partial f_i} = -\sum_{j=1}^{n}(G_{ij}e_j - B_{ij}f_j) + G_{ii}e_i + B_{ii}f_i \tag{2.66}$$

$$\frac{\partial \Delta V_i^2}{\partial e_i} = -2e_i \tag{2.67}$$

$$\frac{\partial \Delta V_i^2}{\partial f_i} = -2f_i \tag{2.68}$$

Equation (2.56) can be written as matrix form as

$$\begin{bmatrix} \Delta F_1 \\ \Delta F_2 \\ \cdots \\ \Delta F_{n-1} \end{bmatrix} = - \begin{bmatrix} J_{11} & J_{12} & \cdots & J_{1,n-1} \\ J_{21} & J_{22} & \cdots & J_{2,n-1} \\ \vdots & \vdots & & \vdots \\ J_{n-1,1} & J_{n-1,2} & \cdots & J_{n-1,n-1} \end{bmatrix} \begin{bmatrix} \Delta V_1 \\ \Delta V_2 \\ \vdots \\ \Delta V_{n-1} \end{bmatrix} \tag{2.69}$$

where $\Delta F_i$ and $\Delta V_i$ are two-dimensional vectors. $J_{ij}$ is a $2 \times 2$ matrix.

$$\Delta V_i = \begin{bmatrix} \Delta e_i \\ \Delta f_i \end{bmatrix} \qquad (2.70)$$

For the PQ bus, we have

$$\Delta F_i = \begin{bmatrix} \Delta P_i \\ \Delta Q_i \end{bmatrix} \qquad (2.71)$$

$$J_{ij} = \begin{bmatrix} \dfrac{\partial \Delta P_i}{\partial e_j} & \dfrac{\partial \Delta P_i}{\partial f_j} \\ \dfrac{\partial \Delta Q_i}{\partial e_j} & \dfrac{\partial \Delta Q_i}{\partial f_j} \end{bmatrix} \qquad (2.72)$$

For the PV bus, we have

$$\Delta F_i = \begin{bmatrix} \Delta P_i \\ \Delta V_i^2 \end{bmatrix} \qquad (2.73)$$

$$J_{ij} = \begin{bmatrix} \dfrac{\partial \Delta P_i}{\partial e_j} & \dfrac{\partial \Delta P_i}{\partial f_j} \\ \dfrac{\partial \Delta V_i^2}{\partial e_j} & \dfrac{\partial \Delta V_i^2}{\partial f_j} \end{bmatrix} \qquad (2.74)$$

It can be observed from equations (2.60)–(2.68) that the elements of the Jacobian matrix are functions of the bus voltage, which are updated through iterations. The element of the submatrix $J_{ij}$ of the Jacobian matrix in equation (2.69) is related to the corresponding element in the bus admittance matrix $Y_{ij}$. If $Y_{ij} = 0$, then $J_{ij} = 0$. Therefore, the Jacobian matrix in equation (2.69) is also a sparse matrix that is the same as the bus admittance matrix.

The steps of the rectangular coordinate system-based Newton power flow solution are similar to those in the polar coordinate system-based algorithm, which was described in Section 2.2.2.

***Example 2.2:*** For the same system in Example 2.1, the Newton method with the rectangular coordinate system is used to solve power flow.

The bus admittance matrix is the same as in Example 2.1. Given the initial values of the bus voltages,

$$e_1^0 = e_2^0 = e_3^0 = 1.0,$$

$$f_1^0 = f_2^0 = f_3^0 = 0.0,$$

$$e_4^0 = 1.05, \ f_4^0 = 0.0$$

Computing the bus power mismatch and $\Delta V_i^2$ with equations (2.52) and (2.55), we get

$$\Delta P_1^0 = P_{1s} - P_1^0 = -0.30 - (-0.02269) = -0.2773$$

$$\Delta P_2^0 = P_{2s} - P_2^0 = -0.55 - (-0.02404) = -0.5260$$

$$\Delta P_3^0 = P_{3s} - P_3^0 = 0.500$$

$$\Delta Q_1^0 = Q_{1s} - Q_1^0 = -0.18 - 0.23767 = -0.4176$$

$$\Delta Q_2^0 = Q_{2s} - Q_2^0 = -0.13 - (-0.14960) = 0.0196$$

$$\Delta V_3^{2(0)} = |V_{3s}|^2 - |V_3^0|^2 = 0.210$$

Computing the elements of the Jacobian matrix with equations (2.60) and (2.68), we get the following correction equation:

$$-\begin{bmatrix} -1.01936 & -8.00523 & 0.58823 & 2.35294 & 0.00000 & 3.66666 \\ -8.48049 & 1.06478 & 2.35294 & -0.58823 & 3.66666 & 0.00000 \\ 0.58823 & 2.35294 & -1.04496 & -4.87698 & 0.00000 & 0.00000 \\ 2.35294 & -0.58823 & -4.57777 & 1.09304 & 0.00000 & 0.00000 \\ 0.00000 & 3.66666 & 0.00000 & 0.00000 & 0.00000 & -3.66666 \\ 0.00000 & 0.00000 & 0.00000 & 0.00000 & -2.00000 & 0.00000 \end{bmatrix} \begin{bmatrix} \Delta e_1^0 \\ \Delta f_1^0 \\ \Delta e_2^0 \\ \Delta f_2^0 \\ \Delta e_3^0 \\ \Delta f_3^0 \end{bmatrix} = \begin{bmatrix} \Delta P_1^0 \\ \Delta Q_1^0 \\ \Delta P_2^0 \\ \Delta Q_2^0 \\ \Delta P_3^0 \\ \Delta Q_3^0 \end{bmatrix}$$

It can be observed from the above equation that most of the elements in the Jacobian matrix that have the maximal absolute values are not on the diagonals, which easily cause a calculation error. To avoid this, we switch rows 1 and 2, rows 3 and 4, rows 5 and 6, when we get,

$$-\begin{bmatrix} -8.48049 & 1.06478 & 2.35294 & -0.58823 & 3.66666 & 0.00000 \\ -1.01936 & -8.00523 & 0.58823 & 2.35294 & 0.00000 & 3.66666 \\ 2.35294 & -0.58823 & -4.57777 & 1.09304 & 0.00000 & 0.00000 \\ 0.58823 & 2.35294 & -1.04496 & -4.87698 & 0.00000 & 0.00000 \\ 0.00000 & 0.00000 & 0.00000 & 0.00000 & -2.00000 & 0.00000 \\ 0.00000 & 3.66666 & 0.00000 & 0.00000 & 0.00000 & -3.66666 \end{bmatrix} \begin{bmatrix} \Delta e_1^0 \\ \Delta f_1^0 \\ \Delta e_2^0 \\ \Delta f_2^0 \\ \Delta e_3^0 \\ \Delta f_3^0 \end{bmatrix} = \begin{bmatrix} \Delta Q_1^0 \\ \Delta P_1^0 \\ \Delta Q_2^0 \\ \Delta P_2^0 \\ \Delta Q_3^0 \\ \Delta P_3^0 \end{bmatrix}$$

Solving the above correction equation, we get

$$
\begin{bmatrix}
\Delta e_1^0 \\
\Delta f_1^0 \\
\Delta e_2^0 \\
\Delta f_2^0 \\
\Delta e_3^0 \\
\Delta f_3^0
\end{bmatrix}
=
\begin{bmatrix}
-0.0037 \\
-0.0094 \\
-0.0222 \\
-0.1081 \\
0.1050 \\
0.1269
\end{bmatrix}
$$

The new bus voltage will be

$$e_1^1 = e_1^0 + \Delta e_1^0 = 0.9963$$

$$f_1^1 = f_1^0 + \Delta f_1^0 = -0.0094$$

$$e_2^1 = e_2^0 + \Delta e_2^0 = 0.9778$$

$$f_2^1 = f_2^0 + \Delta f_2^0 = -0.1081$$

$$e_3^1 = e_3^0 + \Delta e_3^0 = 1.1050$$

$$f_3^1 = f_3^0 + \Delta f_3^0 = 0.1269$$

We then conduct the second iteration, using the new voltage values. If the convergence tolerance is $\varepsilon = 10^{-5}$, the power flow will be converged after three iterations, which are shown in Tables 2.3 and 2.4.

The final bus voltages are expressed in the polar coordinate system as

$$\dot{V}_1 = 0.9847\angle - 0.500^0$$

$$\dot{V}_2 = 0.9648\angle - 6.450^0$$

$$\dot{V}_3 = 1.1\angle 6.732^0$$

Finally, we compute the power of the slack bus as

$$P_4 + jQ_4 = 0.36788 + j0.26469$$

**TABLE 2.3   The Change in Bus Mismatches**

| Iteration $k$ | $\Delta P_1$ | $\Delta Q_1$ | $\Delta P_2$ | $\Delta Q_2$ | $\Delta P_3$ | $\Delta V_3^2$ |
|---|---|---|---|---|---|---|
| 0 | $-0.2773$ | $-0.4176$ | $-0.5260$ | $0.0196$ | $0.500$ | $0.210$ |
| 1 | $2.90 \times 10^{-3}$ | $-4.18 \times 10^{-3}$ | $-1.28 \times 10^{-2}$ | $-5.50 \times 10^{-2}$ | $-1.91 \times 10^{-3}$ | $-2.71 \times 10^{-2}$ |
| 2 | $-1.29 \times 10^{-5}$ | $-6.74 \times 10^{-5}$ | $-2.86 \times 10^{-4}$ | $-1.07 \times 10^{-3}$ | $4.58 \times 10^{-5}$ | $-1.60 \times 10^{-4}$ |
| 3 | $<10^{-5}$ | $<10^{-5}$ | $<10^{-5}$ | $<10^{-5}$ | $<10^{-5}$ | $<10^{-5}$ |

**TABLE 2.4   The Change in Bus Voltages**

| Iteration $k$ | $e_1 + jf_1$ | $e_2 + jf_2$ | $e_3 + jf_3$ |
|---|---|---|---|
| 1 | $0.9963 - j0.0094$ | $0.9778 - j0.1081$ | $1.1050 + j0.1269$ |
| 2 | $0.9848 - j0.0086$ | $0.9590 - j0.1084$ | $1.0925 + j0.1289$ |
| 3 | $0.9846 - j0.0086$ | $0.9587 - j0.1084$ | $1.0924 + j0.1290$ |

Compared with Example 2.1, the same power flow solution is obtained.

***Second-Order Power Flow Method***   It is noted that equations (2.50) and (2.51) are a second-order equations on voltage. They can be expanded into Taylor series without approximation [3], that is,

$$P_{iSP} = P_{is} + \frac{\partial P_i}{\partial e^T}\Delta e + \frac{\partial P_i}{\partial f^T}\Delta f$$

$$+ \frac{1}{2}\left[\Delta e^T \frac{\partial^2 P_i}{\partial e \partial e^T}\Delta e + \Delta e^T \frac{\partial^2 P_i}{\partial e \partial f^T}\Delta f + \Delta f^T \frac{\partial^2 P_i}{\partial f \partial e^T}\Delta e + \Delta f^T \frac{\partial^2 P_i}{\partial f \partial f^T}\Delta f\right] \tag{2.75}$$

$$Q_{iSP} = Q_{is} + \frac{\partial Q_i}{\partial e^T}\Delta e + \frac{\partial Q_i}{\partial f^T}\Delta f$$

$$+ \frac{1}{2}\left[\Delta e^T \frac{\partial^2 Q_i}{\partial e \partial e^T}\Delta e + \Delta e^T \frac{\partial^2 Q_i}{\partial e \partial f^T}\Delta f + \Delta f^T \frac{\partial^2 Q_i}{\partial f \partial e^T}\Delta e + \Delta f^T \frac{\partial^2 Q_i}{\partial f \partial f^T}\Delta f\right] \tag{2.76}$$

The matrix form is

$$\begin{bmatrix} \Delta P \\ \Delta Q \end{bmatrix} = J\begin{bmatrix} \Delta e \\ \Delta f \end{bmatrix} + \begin{bmatrix} SP \\ SQ \end{bmatrix} \tag{2.77}$$

where $J$ is the Jacobian matrix

$$J = \begin{bmatrix} \dfrac{\partial P_i}{\partial e^T} & \dfrac{\partial P_i}{\partial f^T} \\ \dfrac{\partial Q_i}{\partial e^T} & \dfrac{\partial Q_i}{\partial f^T} \end{bmatrix} \tag{2.78}$$

SP and SQ are the second-order term vectors and can be simplified as [3]

$$SP = P_{is}(\Delta e, \Delta f) \tag{2.79}$$

$$SQ = Q_{is}(\Delta e, \Delta f) \tag{2.80}$$

There are no third- or higher-order terms in equation (2.77). If we ignore the second-order term, it will be similar to the Newton algorithm we just discussed in this section. Here, we keep the second-order term, and estimate the values based on the previous iteration values of voltage components. Thus equation (2.77) can be written as

$$\begin{bmatrix} \Delta P - SP \\ \Delta Q - SQ \end{bmatrix} = J \begin{bmatrix} \Delta e \\ \Delta f \end{bmatrix} \tag{2.81}$$

From the above, we obtain the increment voltage components:

$$\begin{bmatrix} \Delta e \\ \Delta f \end{bmatrix} = J^{-1} \begin{bmatrix} \Delta P - SP \\ \Delta Q - SQ \end{bmatrix} \tag{2.82}$$

For a PV bus, the voltage magnitude is fixed, thus the increment voltage components must satisfy the following equation:

$$e_i \Delta e_i + f_i \Delta f_i = V_i \Delta V_i \tag{2.83}$$

Therefore, the reactive power equation in (2.77) for a PV bus will be replaced by the above equation.

The second-order power flow algorithm is summarized below.

**(1)** Given the input data, initialize all the arrays.

**(2)** Set SP and SQ vectors equal to zero.

**(3)** Compute the $P_{is}, Q_{is}$ vectors.

**(4)** Compute the power mismatches $\Delta P$ and $\Delta Q$. Check whether the convergence conditions are satisfied.

$$\max_i |\Delta P_i^k| < \varepsilon_1 \tag{2.84}$$

$$\max_i |\Delta Q_i^k| < \varepsilon_2 \tag{2.85}$$

If equations (2.84) and (2.85) are met, stop the iteration, and calculate the line flows and real and reactive powers of the slack bus. If not, go to the next step.

**(5)** Compute the Jacobian matrix.

**(6)** Compute the $\Delta e, \Delta f$ using equation (2.82).

**(7)** Update the voltages

$$e^{k+1} = e^k + \Delta e \tag{2.86}$$

$$f^{k+1} = f^k + \Delta f \tag{2.87}$$

**(8)** Compute the second-order terms SP and SQ using $\Delta e, \Delta f$. Then go back to step (3).

## 2.3    GAUSS-SEIDEL METHOD

For a nonlinear equation with $n$ variables (2.18), we can obtain the solutions as

$$\left.\begin{array}{l} x_1 = g_1\left(x_1, x_2, \ldots, x_n\right) \\ x_2 = g_2(x_1, x_2, \ldots, x_n) \\ \cdots \\ x_n = g_n(x_1, x_2, \ldots, x_n) \end{array}\right\} \tag{2.88}$$

If the values of the variables at the $k$th iteration are obtained, substituting them into the right side of the above equation, we can get the new values of these variables as follows:

$$\left.\begin{array}{l} x_1^{k+1} = g_1\left(x_1^k, x_2^k, \ldots, x_n^k\right) \\ x_2^{k+1} = g_2(x_1^k, x_2^k, \ldots, x_n^k) \\ \cdots \\ x_n^{k+1} = g_n(x_1^k, x_2^k, \ldots, x_n^k) \end{array}\right\} \tag{2.89}$$

or

$$x_i^{k+1} = g_i(x_1^k, x_2^k, \ldots, x_n^k), \quad i = 1, 2, \ldots, n \tag{2.90}$$

The iteration will be stopped if the following convergence conditions are satisfied for all variables:

$$|x_i^{k+1} - x_i^k| < \varepsilon \tag{2.91}$$

The Newton method that is described in Section 2.2 is based on this iteration calculation. To speed up the convergence, the formula of the iteration calculation is modified as follows:

$$\left.\begin{array}{l} x_1^{k+1} = g_1\left(x_1^k, x_2^k, \ldots, x_n^k\right) \\ x_2^{k+1} = g_2(x_1^{k+1}, x_2^k, \ldots, x_n^k) \\ \cdots \\ x_n^{k+1} = g_n(x_1^{k+1}, x_2^{k+1}, \ldots, x_{n-1}^{k+1}, x_n^k) \end{array}\right\} \tag{2.92}$$

or

$$x_i^{k+1} = g_i\left(x_1^{k+1}, x_2^{k+1}, \ldots, x_{n-1}^{k+1}, x_n^k\right), \quad i = 1, 2, \ldots, n \tag{2.93}$$

The main idea of the approach is to substitute the new values of variables in the calculation of the next variable immediately, rather than waiting until the next iteration.

This iteration method is called the Gauss-Seidel method. It can be also used to solve the power flow equations.

Assuming the system consists of $n$ buses. Buses 1-$m$ are PQ buses, buses $(m + 1)$-$(n - 1)$ are PV buses, and $n$th bus is the slack bus. The iteration calculation does not include the slack bus.

From equation (2.8), we get

$$\dot{V}_i = \frac{1}{Y_{ii}} \left[ \frac{P_i - jQ_i}{\widehat{V}_i} - \sum_{\substack{j=1 \\ j \neq i}}^{n} Y_{ij} \dot{V}_j \right] \tag{2.94}$$

According to the Gauss-Seidel method, the iteration formula of equation (2.94) can be written as

$$\dot{V}_i^{k+1} = \frac{1}{Y_{ii}} \left[ \frac{P_i - jQ_i}{\widehat{V}_i^k} - \sum_{j=1}^{i-1} Y_{ij} \dot{V}_j^{k+1} - \sum_{j=i+1}^{n} Y_{ij} \dot{V}_j^k \right] \tag{2.95}$$

For the PQ bus, the real and reactive powers are known. Thus, if the initial bus voltage $\dot{V}_i^0$ is given, we can use equation (2.95) to perform the iteration calculation.

For the PV bus, the bus real power and the magnitude of the voltage are known. It is necessary to give the initial value for bus reactive power. The bus reactive power will then be computed by iterative calculation. That is

$$Q_i^k = \text{Im}\left[ \dot{V}_i^k \widehat{I}_i^k \right] = \text{Im}\left[ \dot{V}_i^k \left( \sum_{j=1}^{i-1} \widehat{Y}_{ij} \widehat{V}_j^{k+1} + \sum_{j=i}^{n} \widehat{Y}_{ij} \widehat{V}_j^k \right) \right] \tag{2.96}$$

After the iteration is over, all bus real and reactive powers, as well as the voltages, are obtained. The power of the slack bus can be obtained by solving the following equation:

$$P_n + jQ_n = \dot{V}_n \sum_{j=1}^{n} \widehat{Y}_{nj} \widehat{V}_j \tag{2.97}$$

The line power flow can also be obtained as follows:

$$S_{ij} = P_{ij} + jQ_{ij} = \dot{V}_i \widehat{I}_{ij} = \dot{V}_i^2 y_{i0} + \dot{V}_i (\widehat{V}_i - \widehat{V}_j) \widehat{y}_{ij} \tag{2.98}$$

where $y_{ij}$ is the admittance of the branch $ij$ and $y_{i0}$ is the admittance of the ground branch at the end $i$.

## 2.4    P-Q DECOUPLING METHOD

### 2.4.1    Fast Decoupled Power Flow

According to Section 2.2.2, the updated equation in the Newton power flow method
is as follows:

$$\begin{bmatrix} \Delta P \\ \Delta Q \end{bmatrix} = - \begin{bmatrix} H & N \\ K & L \end{bmatrix} \begin{bmatrix} \Delta \theta \\ V_D^{-1} \Delta V \end{bmatrix} \tag{2.99}$$

The Newton power flow is a robust power flow algorithm. It is also called full
AC power flow as there is no simplification in the calculation. However, the disadvantage of the Newton power flow is that the terms in the Jacobian matrix must be
recalculated in each iteration. Actually, the reactance of the branch is generally far
greater than the resistance of the branch in a practical power system. Thus there exists
a strong relationship between the real power and voltage angle, but weak coupling
between the real power and the magnitude of voltage. This means the real power is
hardly influenced by changes in voltage magnitude, that is,

$$\frac{\partial \Delta P_i}{\partial V_j} \approx 0 \tag{2.100}$$

While there is a strong coupling relationship between the reactive power and
magnitude of voltage, coupling between the reactive power and voltage angle is weak.
This means that the reactive power is hardly influenced by changes in voltage angle,
that is,

$$\frac{\partial \Delta Q_i}{\partial \theta_j} \approx 0 \tag{2.101}$$

Therefore, the values of the elements in the submatrices $N$ and $K$ in equation (2.99)
are very small, that is

$$N_{ij} = V_j \frac{\partial \Delta P_i}{\partial V_j} \approx 0 \tag{2.102}$$

$$K_{ij} = \frac{\partial \Delta Q_i}{\partial \theta_j} \approx 0 \tag{2.103}$$

Equation (2.99) becomes

$$\begin{bmatrix} \Delta P \\ \Delta Q \end{bmatrix} = - \begin{bmatrix} H & 0 \\ 0 & L \end{bmatrix} \begin{bmatrix} \Delta \theta \\ V_D^{-1} \Delta V \end{bmatrix} \tag{2.104}$$

or

$$\Delta P = - H \Delta \theta \tag{2.105}$$

$$\Delta Q = - L V_D^{-1} \Delta V = -L(\Delta V / V_D) \tag{2.106}$$

The simplified equations (2.105) and (2.106) make power flow iteration very easy. The bus real power mismatch is only used to revise the voltage angle, and the bus reactive power mismatch is only used to revise the voltage magnitude. These two equations are iteratively calculated, respectively, until the convergence conditions are satisfied. This method is called the real and reactive power decoupling method.

Actually, equations (2.105) and (2.106) can be further simplified. Since the difference of the voltage angles of two ends in the line $ij$ is small (generally less than $10^0 - 20^0$), $\sin(\theta_i - \theta_j)$ is also small. Thus we have

$$\cos \theta_{ij} = \cos(\theta_i - \theta_j) \cong 1$$
$$G_{ij} \sin \theta_{ij} \ll B_{ij}$$

Assume that

$$Q_i \ll V_i^2 B_{ii}$$

Then the elements of the matrix $H$ and $L$ can be expressed as

$$H_{ij} = V_i V_j B_{ij} \quad i,j = 1, 2, \dots, n-1 \tag{2.107}$$

$$L_{ij} = V_i V_j B_{ij} \quad i,j = 1, 2, \dots, m \tag{2.108}$$

or we have the following derivatives

$$\frac{\partial P_i}{\partial \theta_j} = -V_i V_j B_{ij} \quad i,j = 1, 2, \dots, n-1 \tag{2.109}$$

$$\frac{\partial Q_i}{\left(\dfrac{\partial V_j}{V_j}\right)} = -V_i V_j B_{ij} \quad i,j = 1, 2, \dots, m \tag{2.110}$$

Therefore, the matrices $H$ and $L$ can be written as

$$
H = \begin{bmatrix}
V_1 B_{11} V_1 & V_1 B_{12} V_2 & \cdots & V_1 B_{1,n-1} V_{n-1} \\
V_2 B_{21} V_1 & V_2 B_{22} V_2 & \cdots & V_2 B_{2,n-1} V_{n-1} \\
\vdots & \vdots & & \vdots \\
V_{n-1} B_{n-1,1} V_1 & V_{n-1} B_{n-1,2} V_2 & \cdots & V_{n-1} B_{n-1,n-1} V_{n-1}
\end{bmatrix}
$$

$$
= \begin{bmatrix}
V_1 & & & \\
& V_2 & & \\
& & \ddots & \\
& & & V_{n-1}
\end{bmatrix}
\begin{bmatrix}
B_{11} & B_{12} & \cdots & B_{1,n-1} \\
B_{21} & B_{22} & \cdots & B_{2,n-1} \\
\vdots & \vdots & & \vdots \\
B_{n-1,1} & B_{n-1,2} & \cdots & B_{n-1,n-1}
\end{bmatrix}
$$

$$
\times \begin{bmatrix}
V_1 & & & \\
& V_2 & & \\
& & \ddots & \\
& & & V_{n-1}
\end{bmatrix} = V_{D1} B' V_{D1} \tag{2.111}
$$

$$L = \begin{bmatrix} V_1 B_{11} V_1 & V_1 B_{12} V_2 & \cdots & V_1 B_{1m} V_m \\ V_2 B_{21} V_1 & V_2 B_{22} V_2 & \cdots & V_2 B_{2m} V_m \\ \vdots & \vdots & & \vdots \\ V_m B_{m1} V_1 & V_m B_{m2} V_2 & \cdots & V_m B_{mm} V_m \end{bmatrix}$$

$$= \begin{bmatrix} V_1 & & & \\ & V_2 & & \\ & & \ddots & \\ & & & V_m \end{bmatrix} \begin{bmatrix} B_{11} & B_{12} & \cdots & B_{1m} \\ B_{21} & B_{22} & \cdots & B_{2m} \\ \vdots & \vdots & & \vdots \\ B_{m1} & B_{m2} & \cdots & B_{mm} \end{bmatrix}$$

$$\times \begin{bmatrix} V_1 & & & \\ & V_2 & & \\ & & \ddots & \\ & & & V_m \end{bmatrix} = V_{D2} B'' V_{D2} \tag{2.112}$$

Substitute equations (2.111) and (2.112) into equations (2.105) and (2.106), we have

$$\Delta P = V_{D1} B' V_{D1} \Delta \theta \tag{2.113}$$

$$\Delta Q = V_{D2} B'' \Delta V \tag{2.114}$$

Rewrite equations (2.113) and (2.114) as follows

$$\frac{\Delta P}{V_{D1}} = B' V_{D1} \Delta \theta \tag{2.115}$$

$$\frac{\Delta Q}{V_{D2}} = B'' \Delta V \tag{2.116}$$

where

$$B' = - \begin{bmatrix} B_{11} & B_{12} & \cdots & B_{1,n-1} \\ B_{21} & B_{22} & \cdots & B_{2,n-1} \\ \vdots & \vdots & & \vdots \\ B_{n-1,1} & B_{n-1,2} & \cdots & B_{n-1,n-1} \end{bmatrix} = \begin{bmatrix} -B_{11} & -B_{12} & \cdots & -B_{1,n-1} \\ -B_{21} & -B_{22} & \cdots & -B_{2,n-1} \\ \vdots & \vdots & & \vdots \\ -B_{n-1,1} & -B_{n-1,2} & \cdots & -B_{n-1,n-1} \end{bmatrix}$$

$$B'' = - \begin{bmatrix} B_{11} & B_{12} & \cdots & B_{1m} \\ B_{21} & B_{22} & \cdots & B_{2m} \\ \vdots & \vdots & & \vdots \\ B_{m1} & B_{m2} & \cdots & B_{mm} \end{bmatrix} = \begin{bmatrix} -B_{11} & -B_{12} & \cdots & -B_{1m} \\ -B_{21} & -B_{22} & \cdots & -B_{2m} \\ \vdots & \vdots & & \vdots \\ -B_{m1} & -B_{m2} & \cdots & -B_{mm} \end{bmatrix}$$

Equations (2.113) and (2.114) are the simplified power flow adjustment equations, which can be written in matrix form.

$$
\begin{bmatrix}
\dfrac{\Delta P_1}{V_1} \\[2mm]
\dfrac{\Delta P_2}{V_2} \\[2mm]
\vdots \\[2mm]
\dfrac{\Delta P_{n-1}}{V_{n-1}}
\end{bmatrix}
=
\begin{bmatrix}
-B_{11} & -B_{12} & \cdots & -B_{1,n-1} \\
-B_{21} & -B_{22} & \cdots & -B_{2,n-1} \\
\vdots & \vdots & & \vdots \\
-B_{n-1,1} & -B_{n-1,2} & \cdots & -B_{n-1,n-1}
\end{bmatrix}
\begin{bmatrix}
V_1 \Delta \theta_1 \\
V_2 \Delta \theta_2 \\
\vdots \\
V_{n-1} \Delta \theta_{n-1}
\end{bmatrix}
\tag{2.117}
$$

$$
\begin{bmatrix}
\dfrac{\Delta Q_1}{V_1} \\[2mm]
\dfrac{\Delta Q_2}{V_2} \\[2mm]
\vdots \\[2mm]
\dfrac{\Delta Q_m}{V_m}
\end{bmatrix}
=
\begin{bmatrix}
-B_{11} & -B_{12} & \cdots & -B_{1m} \\
-B_{21} & -B_{22} & \cdots & -B_{2m} \\
\vdots & \vdots & & \vdots \\
-B_{m1} & -B_{m2} & \cdots & -B_{mm}
\end{bmatrix}
\begin{bmatrix}
\Delta V_1 \\
\Delta V_2 \\
\vdots \\
\Delta V_m
\end{bmatrix}
\tag{2.118}
$$

In equations (2.117) and (2.118), matrices $B'$ and $B''$ only contain the imaginary part of the bus admittance matrix. Thus they are constant symmetrical matrices and need to be triangularized once only at the beginning of the analysis. Therefore, equations (2.117) and (2.118) are termed the *fast decoupled power flow model* [4–6].

In practical application, the voltage magnitudes of the right side in equations (2.115) and (2.117) are assumed to be 1.0. In this way, the real power adjustment equation in the fast decoupled power flow model can be further simplified as

$$
\frac{\Delta P}{V} = B' \Delta \theta
\tag{2.119}
$$

$$
\begin{bmatrix}
\dfrac{\Delta P_1}{V_1} \\[2mm]
\dfrac{\Delta P_2}{V_2} \\[2mm]
\vdots \\[2mm]
\dfrac{\Delta P_{n-1}}{V_{n-1}}
\end{bmatrix}
=
\begin{bmatrix}
-B_{11} & -B_{12} & \cdots & -B_{1,n-1} \\
-B_{21} & -B_{22} & \cdots & -B_{2,n-1} \\
\vdots & \vdots & & \vdots \\
-B_{n-1,1} & -B_{n-1,2} & \cdots & -B_{n-1,n-1}
\end{bmatrix}
\begin{bmatrix}
\Delta \theta_1 \\
\Delta \theta_2 \\
\vdots \\
\Delta \theta_{n-1}
\end{bmatrix}
\tag{2.120}
$$

In addition, there are two fast decoupled power flow versions according to a different handling of the constant matrices $B', B''$. These are the BX and XB versions.

For the XB version, the resistance is ignored during the calculation of $B'$. The elements of $B', B''$ are computed as

$$B'_{ij} = B_{ij} \tag{2.121}$$

$$B'_{ii} = -\sum_{j \neq i} B'_{ij} \tag{2.122}$$

$$B'_{ij} = \frac{B^2_{ij} + G^2_{ij}}{B_{ij}} \tag{2.123}$$

$$B''_{ii} = -2B_{i0} - \sum_{j \neq i} B''_{ij} \tag{2.124}$$

where $B_{i0}$ is the shunt reactance to ground.

In the practical calculation, the following assumptions are also adopted in the XB version of the fast decoupled power flow model:

- Assume $r_{ij} \ll x_{ij}$, which leads to $B_{ij} = -\frac{1}{x_{ij}}$.
- Eliminate all shunt reactance to ground.
- Omit all effects from phase shift transformers.

The XB version of the fast decoupled power flow model can then be expressed as

$$B'_{ij} = -\frac{1}{x_{ij}} \tag{2.125}$$

$$B'_{ii} = \sum_{j \neq i} \frac{1}{x_{ij}} \tag{2.126}$$

$$B''_{ij} = -\frac{x_{ij}}{r^2_{ij} + x^2_{ij}} \tag{2.127}$$

$$B''_{ii} = -\sum_{j \neq i} B''_{ij} \tag{2.128}$$

where $r_{ij}, x_{ij}$ are the resistance and reactance of the branch $ij$, respectively.

For the BX version, the resistance is ignored during the calculation of $B''$. The elements of $B', B''$ are computed as

$$B'_{ij} = \frac{B^2_{ij} + G^2_{ij}}{B_{ij}} \tag{2.129}$$

$$B'_{ii} = -\sum_{j \neq i} B'_{ij} \tag{2.130}$$

$$B''_{ij} = B_{ij} \tag{2.131}$$

$$B_{ii}'' = -2B_{i0} - \sum_{j \neq i} B_{ij}'' \tag{2.132}$$

Similarly, the BX version of the fast decoupled power flow model can also be simplified as

$$B_{ij}' = -\frac{x_{ij}}{r_{ij}^2 + x_{ij}^2} \tag{2.133}$$

$$B_{ii}' = \sum_{j \neq i} \frac{x_{ij}}{r_{ij}^2 + x_{ij}^2} \tag{2.134}$$

$$B_{ij}'' = -\frac{1}{x_{ij}} \tag{2.135}$$

$$B_{ii}'' = -\sum_{j \neq i} B_{ij}'' \tag{2.136}$$

It is noted that the fast decoupled power flow algorithm may fail to converge when some of the major assumptions such as $r_{ij} \ll x_{ij}$ do not hold. In such cases, the Newton power flow or decoupled power flow without major approximation is recommended.

***Example 2.3:***   In this example, we solve the system in Example 2.1 using the decoupled PQ method.

First form the $B', B''$ matrices as follows:

$$B' = \begin{bmatrix} -8.2429 & 2.3529 & 3.6666 \\ 2.3529 & -4.7274 & 0.0000 \\ 3.6666 & 0.0000 & -3.3333 \end{bmatrix}$$

$$B'' = \begin{bmatrix} -8.2429 & 2.3529 \\ 2.3539 & -4.7274 \end{bmatrix}$$

On conducting the triangular decomposition of B and $B'$, we obtain Tables 2.5 and 2.6.

Given that the initial bus voltage is

$$\dot{V}_1^0 = \dot{V}_2^0 = 1.0\angle 0^0, \quad \dot{V}_3^0 = 1.1\angle 0^0, \quad \dot{V}_4^0 = 1.05\angle 0^0$$

we compute the bus real power mismatch with equation (2.29), to get

$$\Delta P_1^0 = P_{1s} - P_1^0 = -0.30 - (-0.02269) = -0.27731$$

**TABLE 2.5   Result of Triangular Decomposition of $B'$**

| | | |
|---|---|---|
| −0.121317 | −0.285452 | −0.444829 |
| | −0.246565 | −0.258069 |
| | | −0.698234 |

**TABLE 2.6  Result of Triangular Decomposition of $B''$**

| | |
|---|---|
| $-0.121317$ | $-0.285452$ |
| | $-0.246565$ |

$$\Delta P_2^0 = P_{2s} - P_2^0 = -0.55 - (-0.02404) = -0.52596$$

$$\Delta P_3^0 = P_{3s} - P_3^0 = 0.5$$

$$\frac{\Delta P_1^0}{V_1^0} = -0.27731$$

$$\frac{\Delta P_2^0}{V_2^0} = -0.52596$$

$$\frac{\Delta P_3^0}{V_3^0} = 0.45455$$

Computing the voltage angle by solving the correction equation (2.117), we have

$$\Delta \theta_1^0 = -0.737^0, \quad \Delta \theta_2^0 = -6.742^0, \quad \Delta \theta_3^0 = 6.366^0$$

$$\theta_1^1 = \theta_1^0 + \Delta \theta_1^0 = -0.737^0$$

$$\theta_2^1 = \theta_2^0 + \Delta \theta_2^0 = -6.742^0$$

$$\theta_3^1 = \theta_3^0 + \Delta \theta_3^0 = 6.366^0$$

Then we perform the reactive power iteration. Computing the bus real power mismatch with equation (2.30), we get

$$\Delta Q_1^0 = Q_{1s} - Q_1^0 = -0.18 - (-0.14041) = -3.95903 \times 10^{-2}$$

$$\Delta Q_2^0 = Q_{2s} - Q_2^0 = -0.13 - (-0.00155) = -0.13155$$

$$\frac{\Delta Q_1^0}{V_1^0} = -0.03959$$

$$\frac{\Delta Q_2^0}{V_2^0} = -0.13155$$

Computing voltage magnitude by solving correction equation (2.118),

$$\Delta V_1^0 = -0.0149, \ \Delta V_2^0 = -0.0352$$

$$V_1^1 = V_1^0 + \Delta V_1^0 = 0.9851$$

$$V_2^1 = V_2^0 + \Delta V_2^0 = 0.9648$$

**TABLE 2.7 Bus Power Mismatch Change**

| Iteration $k$ | $\Delta P_1$ | $\Delta P_2$ | $\Delta P_3$ | $\Delta Q_1$ | $\Delta Q_2$ |
|---|---|---|---|---|---|
| 0 | $-0.27731$ | $-0.52596$ | $0.5$ | $-3.95903 \times 10^{-2}$ | $-0.13155$ |
| 1 | $4.051 \times 10^{-3}$ | $1.444 \times 10^{-2}$ | $8.691 \times 10^{-3}$ | $-2.037 \times 10^{-3}$ | $1.568 \times 10^{-3}$ |
| 2 | $-6.603 \times 10^{-3}$ | $-3.488 \times 10^{-3}$ | $6.826 \times 10^{-4}$ | $-1.537 \times 10^{-3}$ | $-1.123 \times 10^{-3}$ |
| 3 | $-1.227 \times 10^{-3}$ | $2.148 \times 10^{-3}$ | $-4.967 \times 10^{-5}$ | $-2.694 \times 10^{-4}$ | $7.3477 \times 10^{-4}$ |
| 4 | $9.798 \times 10^{-5}$ | $-1.552 \times 10^{-4}$ | $-1.140 \times 10^{-5}$ | $2.513 \times 10^{-5}$ | $-3.277 \times 10^{-5}$ |
| 5 | $<10^{-5}$ | $<10^{-5}$ | $<10^{-5}$ | $<10^{-5}$ | $<10^{-5}$ |

**TABLE 2.8 Bus Voltage Change**

| Iteration $k$ | $\theta_1$ | $\theta_2$ | $\theta_3$ | $V_1$ | $V_2$ |
|---|---|---|---|---|---|
| 1 | $-0.737^0$ | $-6.742^0$ | $6.366^0$ | $0.9851$ | $0.9648$ |
| 2 | $-0.349^0$ | $-6.356^0$ | $6.871^0$ | $0.9850$ | $0.9650$ |
| 3 | $-0.497^0$ | $-6.475^0$ | $6.737^0$ | $0.9847$ | $0.9646$ |
| 4 | $-0.500^0$ | $-6.448^0$ | $6.732^0$ | $0.9847$ | $0.9648$ |
| 5 | $-0.500^0$ | $-6.450^0$ | $6.732^0$ | $0.9847$ | $0.9648$ |

We now conduct the second iteration, using new voltage values. If the convergence tolerance is $\varepsilon = 10^{-5}$, the power flow will be converged after five iterations, which are shown in Tables 2.7 and 2.8.

Compared with the Newton method, the decoupled PQ method gave almost the same results.

## 2.4.2 Decoupled Power Flow without Major Approximation

Assuming the voltage magnitude in the Newton power flow model (2.99) to be 1.0, we have

$$\begin{bmatrix} \Delta P \\ \Delta Q \end{bmatrix} = - \begin{bmatrix} H & N \\ K & L \end{bmatrix} \begin{bmatrix} \Delta \theta \\ \Delta V \end{bmatrix} \tag{2.137}$$

Premultiplying the $\Delta P$ equations by $KH^{-1}$ and adding the resulting equations to the $\Delta Q$ equations leads to the system of equations

$$\begin{bmatrix} \Delta P \\ \Delta Q - KH^{-1}\Delta P \end{bmatrix} = - \begin{bmatrix} H & N \\ 0 & L - KH^{-1}N \end{bmatrix} \begin{bmatrix} \Delta \theta \\ \Delta V \end{bmatrix} \tag{2.138}$$

Premultiplying the $\Delta Q$ equations by $NL^{-1}$ and adding the resulting equations to the $\Delta P$ equations leads to the system of equations

$$\begin{bmatrix} \Delta P - NL^{-1}\Delta Q \\ \Delta Q \end{bmatrix} = - \begin{bmatrix} H - NL^{-1}K & 0 \\ K & L \end{bmatrix} \begin{bmatrix} \Delta \theta \\ \Delta V \end{bmatrix} \tag{2.139}$$

By combining the operations performed to obtain equations (2.138) and (2.139), we get

$$\begin{bmatrix} \Delta P - NL^{-1}\Delta Q \\ \Delta Q - KH^{-1}\Delta P \end{bmatrix} = -\begin{bmatrix} H - NL^{-1}K & 0 \\ 0 & L - KH^{-1}N \end{bmatrix} \begin{bmatrix} \Delta \theta \\ \Delta V \end{bmatrix} \qquad (2.140)$$

or

$$\begin{bmatrix} \Delta P - NL^{-1}\Delta Q \\ \Delta Q - KH^{-1}\Delta P \end{bmatrix} = -\begin{bmatrix} H_{eq} & 0 \\ 0 & L_{eq} \end{bmatrix} \begin{bmatrix} \Delta \theta \\ \Delta V \end{bmatrix} \qquad (2.141)$$

where the equivalent matrices $H_{eq}$ and $L_{eq}$ are defined as

$$H_{eq} = H - NL^{-1}K \qquad (2.142)$$

$$L_{eq} = L - KH^{-1}N \qquad (2.143)$$

It can be observed that equation (2.140) or (2.141) is equivalent to the original system (2.137) but has the decoupled solution structure in which $\Delta \theta$ and $\Delta V$ are calculated separately. This decoupled procedure is not an approximation method that ignores the off-diagonal submatrices $N$ and $K$, which was adopted in the fast decoupled power flow method in Section 2.4.1. Thus the solution will be close to the Newton power flow solution. However, the solution procedures are different from those in the Newton method, where the differences $\Delta \theta$ and $\Delta V$ are not computed simultaneously but separately.

The following decoupled algorithm can be used to solve equation (2.138) for $\Delta \theta$ and $\Delta V$ [6]:

Step (1):   Compute the temporary angle corrections:

$$\Delta \theta_H = -H^{-1}\Delta P(V, \theta) \qquad (2.144)$$

Step (2):   Compute the voltage corrections:

$$\Delta V = -L_{eq}^{-1}\Delta Q(V, \theta + \Delta \theta_H) \qquad (2.145)$$

Step (3):   Compute the additional angle corrections:

$$\Delta \theta_N = -H^{-1}N\Delta V \qquad (2.146)$$

It can be verified that $\Delta V$ and $\Delta \theta = \Delta \theta_H + \Delta \theta_N$ are the solution vectors of equation (2.138). This algorithm considers the coupling effect represented by $K$.

For equation (2.139), we have the dual algorithm:

Step (1):   Compute the temporary voltage corrections:

$$\Delta V_L = -L^{-1}\Delta Q(V, \theta) \qquad (2.147)$$

Step (2):   Compute the angle corrections:

$$\Delta\theta = -H_{eq}^{-1}\Delta P(V + \Delta V_L, \theta) \tag{2.148}$$

Step (3):   Compute the additional voltage corrections:

$$\Delta V_K = -L^{-1}K\Delta\theta \tag{2.149}$$

where $\Delta V = \Delta V_L + \Delta V_K$

Although the above iteration algorithms (2.144)–(2.146) and (2.147)–(2.149) yield the correct solutions for the power flow model (2.137), they are not suited for practical implementation [6], for the following reasons:

- In the first algorithm, the angle corrections $\Delta\theta$ are computed in two steps ($\Delta\theta_H$ and $\Delta\theta_N$), while in the second algorithm, the voltage magnitude corrections $\Delta V$ are computed in two steps ($\Delta V_L$ and $\Delta V_K$).
- The matrices $H_{eq}$ and $L_{eq}$ may be full.

The following iteration algorithm is suggested because of the above two difficulties. For solving equations (2.144)–(2.146), the iteration steps for the suggested algorithm are described in the following.

$$\Delta\theta_H^k = -H^{-1}\Delta P\left(V^k, \theta^k\right) \tag{2.150}$$

$$\Delta\theta_{temp}^{k+1} = \left(\theta^k + \Delta\theta_H^k\right) \tag{2.151}$$

$$\Delta V^k = -L_{eq}^{-1}\Delta Q\left(V^k, \theta_{temp}^{k+1}\right) \tag{2.152}$$

$$\Delta V^{k+1} = V^k + \Delta V^k \tag{2.153}$$

$$\Delta\theta_N^k = -H^{-1}N\Delta V^k \tag{2.154}$$

$$\theta^{k+1} = \left(\Delta\theta_{temp}^{k+1} + \Delta\theta_N^k\right) \tag{2.155}$$

Then compute the temporary angle vector of the next iteration:

$$\Delta\theta_H^{k+1} = -H^{-1}\Delta P\left(V^{k+1}, \theta^{k+1}\right) \tag{2.156}$$

$$\Delta\theta_{temp}^{k+2} = \left(\theta^{k+1} + \Delta\theta_H^{k+1}\right) \tag{2.157}$$

By adding the two successive angle corrections, we get

$$\Delta\theta_N^k + \Delta\theta_H^{k+1} = -H^{-1}\left[\Delta P\left(V^{k+1}, \theta^{k+1}\right) - N\Delta V^k\right]$$

$$\approx -H^{-1}\left[\Delta P\left(V^{k+1}, \theta_{temp}^{k+1}\right) - H\Delta\theta_N^k - N\Delta V^k\right] \tag{2.158}$$

$$\approx -H^{-1}\Delta P\left(V^{k+1}, \theta_{temp}^{k+1}\right)$$

The above combined angle correction can be obtained by a single forward/backward solution using the active mismatches computed at $V^{k+1}$ and $\theta_{temp}^{k+1}$. Similar iteration steps can be obtained for the algorithm (2.147)–(2.149).

## 2.5 DC POWER FLOW

AC power flow algorithms have high calculation precision, but do not have high speed. In real power dispatch or power market analysis, the requirement for calculation precision is not very high, but the requirement for calculation speed is of most concern, especially for a large-scale power system. A larger number of simplification power flow algorithms than fast decoupled power flow algorithms are used. One algorithm is called "MW Only." In this method, the $Q - V$ equation in the fast decoupled power flow model is completed dropped. Only the following $P - \theta$ equation is used to correct the angle according to the real power mismatch.

$$
\begin{bmatrix}
\dfrac{\Delta P_1}{V_1} \\[2mm]
\dfrac{\Delta P_2}{V_2} \\[2mm]
\vdots \\[2mm]
\dfrac{\Delta P_{n-1}}{V_{n-1}}
\end{bmatrix}
=
\begin{bmatrix}
-B_{11} & -B_{12} & \cdots & -B_{1,n-1} \\
-B_{21} & -B_{22} & \cdots & -B_{2,n-1} \\
\vdots & \vdots & & \vdots \\
-B_{n-1,1} & -B_{n-1,2} & \cdots & -B_{n-1,n-1}
\end{bmatrix}
\begin{bmatrix}
\Delta\theta_1 \\
\Delta\theta_2 \\
\vdots \\
\Delta\theta_{n-1}
\end{bmatrix}
\tag{2.159}
$$

In the MW-only power flow calculation, the voltage magnitude can be handled either as constant or 1.0 during each $P - \theta$ iteration. For the convergence, only real power mismatch is checked no matter what the reactive power mismatch is.

Another most simplified power flow algorithm is the DC power flow algorithm. It is also an MW only method but makes the following assumptions:

**(1)** All the voltage magnitudes are equal to 1.0.

**(2)** The resistance of the branch is ignored, that is, the susceptance of the branch is

$$
B_{ij} = -\frac{1}{x_{ij}}
\tag{2.160}
$$

**(3)** The angle difference on the two ends of the branch is very small, so that

$$
\sin\theta_{ij} = \theta_i - \theta_j
\tag{2.161}
$$

$$
\cos\theta_{ij} = 1
\tag{2.162}
$$

**(4)** All ground branches are ignored, that is,

$$
B_{i0} = B_{j0} = 0
\tag{2.163}
$$

Therefore, the DC power flow model will be

$$
\begin{bmatrix} \Delta P_1 \\ \Delta P_2 \\ \vdots \\ \Delta P_{n-1} \end{bmatrix} = [B'] \begin{bmatrix} \Delta \theta_1 \\ \Delta \theta_2 \\ \vdots \\ \Delta \theta_{n-1} \end{bmatrix} \tag{2.164}
$$

or

$$
[\Delta P] = [B'][\Delta \theta] \tag{2.165}
$$

where the elements of the matrix $B'$ are the same as those in the XB version of fast decoupled power flow but we ignore the matrix $B''$, that is,

$$
B'_{ij} = -\frac{1}{x_{ij}} \tag{2.166}
$$

$$
B'_{ii} = -\sum_{j \neq i} B'_{ij} \tag{2.167}
$$

The DC power flow is a purely linear equation, so only one iteration calculation is needed to obtain the power flow solution. However, it is only good for calculating real power flows through transmission lines and transformers. The power flowing through each line using the DC power flow is then

$$
P_{ij} = -B_{ij}(\theta_i - \theta_j) = \frac{\theta_i - \theta_j}{x_{ij}} \tag{2.168}
$$

## 2.6 STATE ESTIMATION

Power system state estimation derives a real-time model through the received data from a redundant measurement set. Different kinds of methods about state estimation are introduced in [7]. Among them, the weighted least squares (WLS) state estimation methods are widely used. WLS state estimation minimizes the weighted sum of squares of the residuals, which will be introduced in this section.

### 2.6.1 State Estimation Model

Consider an $N$ bus power network for which $m$ measurements are taken. Assuming a nonlinear model for the electrical network, the relationships between measured quantities and state variables can be expressed as

$$
z = \begin{bmatrix} z_1 \\ z_2 \\ \vdots \\ z_m \end{bmatrix} = \begin{bmatrix} h_1(x_1, x_2, \dots, x_n) \\ h_2(x_1, x_2, \dots, x_n) \\ \vdots \\ h_m(x_1, x_2, \dots, x_n) \end{bmatrix} + \begin{bmatrix} e_1 \\ e_2 \\ \vdots \\ e_m \end{bmatrix} = h(x) = e \tag{2.169}
$$

where

$z$: the measurement vector

$x$: the system state vector

$e$: the vector of measurement errors

$h$: the nonlinear function relating measurement $i$ to the state vector $x$.

There are three most commonly used measurement types in power system state estimation. They are the bus power injections, the line power flows, and the bus voltage magnitudes. These measurement equations can be expressed using the state variables, which are given below from the power flow equations mentioned in Section 2.2:

1. Real and reactive power injection at bus $i$:

$$P_i = V_i \sum_{j=1}^{n} V_j (G_{ij} \cos \theta_{ij} + B_{ij} \sin \theta_{ij}) \qquad (2.170)$$

$$Q_i = V_i \sum_{j=1}^{n} V_j (G_{ij} \sin \theta_{ij} - B_{ij} \cos \theta_{ij}) \qquad (2.171)$$

2. Real and reactive power flow from bus $i$ to bus $j$:

$$P_{ij} = V_i^2 (G_{si} + G_{ij}) - V_i V_j (G_{ij} \cos \theta_{ij} + B_{ij} \sin \theta_{ij}) \qquad (2.172)$$

$$Q_{ij} = - V_i^2 (B_{si} + B_{ij}) - V_i V_j (G_{ij} \sin \theta_{ij} - B_{ij} \cos \theta_{ij}) \qquad (2.173)$$

where

$V_i$: the voltage magnitude at bus $i$

$\theta_i$: the voltage angle at bus $i$

$P_i$: the real power injection at bus $i$

$Q_i$: the reactive power injection at bus $i$

$\theta_{ij}$: the voltage angle different between bus $i$ and $j$

$P_{ij}$: the real power flow from bus $i$ to bus $j$

$Q_{ij}$: the reactive power flow from bus $i$ to bus $j$

$G_{ij}$: the conductance of branch $ij$

$B_{ij}$: the susceptance of branch $ij$.

Consider a system having $N$ buses; the state vector will have $(2N - 1)$ elements, $N$ bus voltage magnitudes, and $(N - 1)$ phase angles. The state vector $x$ will have the following form assuming bus 1 is selected as the reference:

$$x^T = [\theta_2 \theta_3 \ \dots \ \theta_N V_2 V_3 \ \dots \ V_N]$$

Let $E(e)$ denote the expected value of $e$, with the following assumptions:

$$E(e_i) = 0, \quad i = 1, \ldots, m \tag{2.174}$$

$$E(e_i e_j) = 0 \tag{2.175}$$

The measurement errors are assumed to be independent and their covariance matrix is given by a diagonal matrix $R$:

$$\text{Cov}(e) = E[e \cdot e^T] = R = \text{diag}\{\sigma_1^2, \sigma_2^2, \ldots, \sigma_m^2\} \tag{2.176}$$

The standard deviation $\sigma_i$ of each measurement $i$ is computed to reflect the expected accuracy of the corresponding meter used.

## 2.6.2  WLS Algorithm for State Estimation

Power system state estimation is formulated by use of the WLS criterion, which is a function of the estimation residuals. The WLS estimator will minimize the following objective function [7,8]:

$$J(x) = \sum_{i=1}^{m} \frac{(z_i - h_i(x))^2}{R_{ii}} = [z - h(x)]^T R^{-1} [z - h(x)] \tag{2.177}$$

At the minimum value of the objective function, the first-order optimality conditions have to be satisfied. These can be expressed in compact form as follows:

$$g(x) = \frac{\partial J(x)}{\partial x} = -H^T(x) R^{-1}[z - h(x)] = 0 \tag{2.178}$$

where

$$H(x) = \left[ \frac{\partial h(x)}{\partial x} \right] = \begin{bmatrix} \dfrac{\partial P_i}{\partial \theta} & \dfrac{\partial P_i}{\partial V} \\[2mm] \dfrac{\partial P_{ij}}{\partial \theta} & \dfrac{\partial P_{ij}}{\partial V} \\[2mm] \dfrac{\partial Q_i}{\partial \theta} & \dfrac{\partial Q_i}{\partial V} \\[2mm] \dfrac{\partial Q_{ij}}{\partial \theta} & \dfrac{\partial Q_{ij}}{\partial V} \\[2mm] \dfrac{\partial I_{ij}}{\partial \theta} & \dfrac{\partial I_{ij}}{\partial V} \\[2mm] 0 & \dfrac{\partial V_{ij}}{\partial V} \end{bmatrix} \tag{2.179}$$

is the measurement Jacobian matrix. The expressions of each partition can be computed using equations (2.170)–(2.173).

The nonlinear function $g(x)$ can be expanded into its Taylor series around the state vector $x^k$, that is,

$$g(x) = g(x^k) + G(x^k)(x - x^k) + \cdots = 0 \tag{2.180}$$

Neglecting the higher-order terms in the above expression, an iterative solution scheme known as the *Gauss-Newton* method is used to solve the following equation:

$$x^{k+1} = x^k - [G(x^k)]^{-1} \cdot g(x^k) \tag{2.181}$$

where

$k$: the iteration index

$x^k$: the solution vector at iteration $k$

$G(x)$: the gain matrix, which is expressed as follows.

$$G(x^k) = \frac{\partial g(x^k)}{\partial x} = H^T(x^k)R^{-1}H(x^k) \tag{2.182}$$

$$g(x^k) = - H^T(x^k)R^{-1}(z - h(x^k)) \tag{2.183}$$

Generally, the gain matrix $G(x)$ is sparse, positive definite, and symmetric, provided that the system is fully observable. It can be decomposed into its triangular factors. At each iteration $k$, the following sparse linear set of equations are solved using WLS algorithm.

$$\left[G\left(x^k\right)\right] \Delta x^{k+1} = H^T(x^k)R^{-1}\left[z - h\left(x^k\right)\right]$$

$$\Delta x^k = x^{k+1} - x^k \tag{2.184}$$

Equation (2.184) is called a normal equation. WLS state estimation uses the iterative solution of the normal equation. Iterations start at an initial guess $x^0$ which is typically chosen as the flat start, that is, all bus voltages are assumed to be 1.0 per unit and in phase with each other. The iterative solution algorithm for WLS state estimation can be summarized as follows:

(1) Initially set the iteration counter $k = 0$, and set the maximum iteration number $k_{max}$.

(2) If $k > k_{max}$, then terminate the iterations.

(3) Calculate the measurement function $h(x^k)$, the measurement Jacobian $H(x^k)$, and the gain matrix $G(x^k)$.

(4) Solve equation (2.181) to get $\Delta x^k$.

(5) Check for convergence, that is, $\max|\Delta x^k| \leq \varepsilon$. If yes, stop. Otherwise, go to the next step.

(6) Update $x^{k+1} = x^k + \Delta x^k$,   $k \Leftarrow k + 1$, and go to step 2.

# PROBLEMS AND EXERCISES

1. What is the PV bus?

2. What is the PQ bus?

3. What is the slack bus?

4. Can a PV bus become a PQ bus? Why?

5. State the principle of the Newton-Raphson method.

6. What are the differences between the XB and BX versions in the fast decoupled power flow method?

7. Describe the advantages and disadvantages of the major power flow calculation methods (Newton-Raphson method, PQ decoupled method, and DC power flow method)

8. What is the Jacobian matrix?

9. What is the "MW Only" power flow method?

10. State "True" or "False"

   10.1  The slack bus is also called reference bus.

   10.2  Generally, a bus connected to a load is a PQ bus.

   10.3  A bus connected to a generator must be a PV bus.

   10.4  In the PQ decoupled power flow, real power and voltage have a strong coupling relationship.

   10.5  In the PQ decoupled power flow, reactive power and voltage angle have a weak coupling relationship.

   10.6  There is no iteration in DC power flow calculation.

   10.7  The DC power flow method has higher precision than the "MW Only" power flow method.

   10.8  The fast decoupled power flow method is faster than the DC power flow method.

   10.9  A flat voltage of 1.0 per unit is used in the DC power flow method.

   10.10  Only the single slack bus can be used in power flow calculation.

11. A power system is shown in Figure 2.1. The parameters of the branches are as follows:

$$z_{12} = 0.10 + j0.30$$

$$y_{120} = y_{210} = j0.015$$

$$z_{13} = j0.30, \ k = 1.1$$

$$z_{14} = 0.10 + j0.50$$

$$y_{140} = y_{410} = j0.019$$

$$z_{24} = 0.12 + j0.50$$

$$y_{240} = y_{420} = j0.014$$

Buses 1 and 2 are PQ buses, bus 3 is a PV bus, and bus 4 is a slack bus. The given data are

$$P_1 + jQ_1 = -0.3 - j0.15$$

$$P_2 + jQ_2 = -0.6 - j0.10$$

$$P_3 = 0.5; \ V_3 = 1.1;$$

$$V_4 = 1.05; \ \theta_4 = 0$$

**(1)** Use the Newton-Raphson method with the polar coordinate system to solve the power flow.

**(2)** Use Newton-Raphson method with rectangular coordinate system to solve the power flow.

**(3)** Use the PQ decoupled method to solve the power flow.

# REFERENCES

1. Zhu JZ. *Power System Optimal Operation*. Tutorial of Chongqing University; 1990.
2. He Y, Wen ZY, Wang FY, Zhou QH. *Power Systems Analysis*. Huazhong Polytechnic University Press; 1985.
3. Keyhani A, Abur A, Hao S. Evaluation of power flow techniques for personal computers. IEEE Trans. on Power Syst. 1989;4(2):817–826.
4. Alsac O, Sttot B. Fast decoupled power flow. IEEE Trans. on Power Syst. 1974;93:859–869.
5. Van Amerongen RAM, "A general purpose version of the fast decoupled power flow," IEEE Summer Meeting, 1988.
6. Monticelli A, Garcia A, Saavedra OR. Decoupled power flow: hypothesis, derivations, and testing. IEEE Trans. on Power Syst. 1990;5(4):1425–1431.
7. Abur A, Expósito AG. *Power System State Estimation Theory and Implementation*. New York: Wiley-IEEE Press; 2004.
8. Zhu JZ. Power system state estimation. In: Robinson OE, editor. *Electric Power Systems in Transition*. New York: Nova Science Publishers, Inc.; 2010.

# CHAPTER 3

# SENSITIVITY CALCULATION

Currently, sensitivity analysis is becoming more and more important in practical power system operations and also in power market operations. This chapter analyzes and discusses all kinds of sensitivity factors such as the loss sensitivity factor, generator shift factor, pricing node shift factor, constraint shift factor, line outage distribution factor (LODF), outage transfer distribution factor (OTDF), response factor for the transfer path, and voltage sensitivity factor. It also addresses the practical application of these sensitivity factors, including a practical method to convert the sensitivities with different references.

## 3.1 INTRODUCTION

This chapter focuses on the analysis and implementation details of the calculations of several sensitivities such as loss sensitivity, voltage sensitivity, generator constraint shift factor, and area-based constraint shift factor in the practical transmission network and energy markets. The power operator uses these to study and monitor market and system behavior and detect possible problems in the operation. These sensitivities' calculations are also used to determine whether the on-line capacity as indicated in the resource plan is located in the right place in the network to serve the forecast demand. If there is congestion or violation, the generation scheduling based on the sensitivities' calculations can determine whether a different allocation of the available resources could resolve the congestion or violation problem.

In the early energy market, transmission losses were neglected for reasons of computational simplicity, but they are addressed in the standard market design (SMD) [1–4]. Loss calculation is considered for the dispatch functions of SMD such as location-based marginal prices (LMPs). Loss allocation does not affect generation levels or power flows; however, it does modify the value of LMP [5]. The early and classic loss calculation approach is the loss formula—B coefficient method [6], which has been replaced by the more accurate inverse Jacobian transpose method [7]. Numerous loss calculation methods have been proposed in the literature and these can be categorized as pro rata [8], incremental [9], proportional sharing [10], and Z-bus loss allocation [11].

*Optimization of Power System Operation*, Second Edition. Jizhong Zhu.
© 2015 The Institute of Electrical and Electronics Engineers, Inc. Published 2015 by John Wiley & Sons, Inc.

Calculation of loss sensitivity is based on the distributed slack buses in the energy control center [6,11–13]. In real-time energy markets, LMP or economic dispatch is implemented on the basis of market-based reference, which is an arbitrary slack bus, instead of the distributed slack buses in the traditional energy management system. Meanwhile, the existing loss calculation methods in traditional EMS systems are generally based on the generator slacks or references. Since the units with automatic generation control (AGC) are selected as the distributed slacks, and the patterns or status of AGC units are variable for different time periods in the real-time energy market, the sensitivity values will keep changing, which complicates the issue. This chapter presents a fast and useful formula to calculate loss sensitivity for any slack bus [14,15].

The simultaneous feasibility test (SFT) performs the network sensitivity analysis in the base case and in contingency cases in the power system. The base case and postcontingency MW flows are compared against their respective limits to generate the set of critical constraints. For each critical constraint, SFT calculates constraint coefficients (shift factors) that represent linearized sensitivity factors between the constrained quantity (e.g., MW branch flow) and MW injections at network buses. The B-matrix used to calculate the shift factors is constructed to reflect proper network topology [16–18].

The objective of SFT is to identify whether network operation is feasible for a real power injection scenario. If operational limits are violated, generic constraints are generated that can be used to prevent the violation if presented with the same network conditions [16].

In the energy market systems, the trade is often considered between the source and the sink (i.e., the point of delivery, POD, and point of receipt, POR). The source and the sink may be an area or any bus group. Therefore, area-based sensitivities are needed, which can be computed through the constraint shift factors within the area.

Another type of sensitivity that is frequently used is related to voltage stability, especially static voltage stability, which investigates the stability of an operating point and applies a linearized model. Static voltage instability is mainly associated with reactive power imbalance. This imbalance mainly occurs on a local network or a specified bus in a system, which is called the weak bus. Therefore, the reactive power supports have to be locally adequate.

Voltage sensitivity analysis can detect the weak buses/nodes in the power system where the voltage is low. It can be used to select the optimal locations of VAR support service [19–25]. According to the sensitivity values, the voltage benefit factor (VBF) and loss benefit factor (LBF), a ranking of VAR support sites can also be obtained.

## 3.2 LOSS SENSITIVITY CALCULATION

This section presents a fast and useful formula to calculate the loss sensitivity for any slack bus. The formula is based on the loss sensitivity results from distributed slacks without computing a new set of sensitivity factors through traditional power flow calculation. In particular, loads are selected as distributed slacks rather than the usual generator slacks. The loss sensitivity values will be unchanged for the same network topology no matter how the status of the AGC units changes.

In the energy market, the formulation of the optimum economic dispatch can be represented as follows:

$$\min \ F = \sum_j C_j P_{Gj}, \quad j \in NG \tag{3.1}$$

such that

$$\sum P_D + P_L = \sum_j P_{Gj}, \quad j \in NG \tag{3.2}$$

$$\sum_j S_{ij} P_{Gj} \le P_{i\max} \quad j \in NG, \ i \in K_{\max} \tag{3.3}$$

$$P_{Gj\min} \le P_{Gj} \le P_{Gj\max}, \quad j \in NG \tag{3.4}$$

where

$P_D$: the real power load;
$P_{i\max}$: the maximum requirement of power supply at the active constraint $i$;
$P_{Gj}$: the real power output at generator bus $j$;
$P_{Gj\min}$: the minimal real power output at generator $j$;
$P_{Gj\max}$: the maximal real power output at generator $j$;
$P_L$: the network losses;
$S_{ij}$: the sensitivity (shift factor) for resource or unit $j$ and active constraint $i$ with respect to the market-based reference;
$C_j$: the real-time price for the resource (or unit) $j$;
$K_{\max}$: the maximum number of active constraints;
$NG$: the number of units.

The Lagrangian function is obtained from equations (3.1) and (3.2).

$$F_L = \sum_i f_i(P_i) + \lambda \left( \sum_i P_{Di} + P_L - \sum_j P_{Gj} \right) \tag{3.5}$$

Traditionally, generation reference (single or distributed slack) is used in the calculation of loss allocation. This works, but may be inconvenient or confusing for users who frequently use loss factors. The reason is that the AGC status or patterns of units are variable in real-time EMS or energy markets. Loss sensitivity values based on distributed unit references will keep changing because of the change in unit AGC status. Thus the distributed load slack or reference is used here.

The optimality criteria of the Lagrange function (3.5) are written as follows:

$$\frac{\partial F_L}{\partial P_{Di}} = \frac{df_i}{dP_{Di}} + \lambda \left( 1 + \frac{\partial P_L}{\partial P_{Di}} \right) = 0 \quad i \in ND \tag{3.6}$$

$$\frac{\partial F_L}{\partial P_{Gj}} = \frac{df_i}{dP_{Gj}} + \lambda\left(\frac{\partial P_L}{\partial P_{Gj}} - 1\right) = 0 \quad j \in NG \tag{3.7}$$

$$\frac{df_i}{dP_{Di}} L_{Di} = \lambda \quad i \in ND \tag{3.8}$$

$$L_{Di} = -\frac{1}{1 + \frac{\partial P_L}{\partial P_{Di}}} \quad i \in ND \tag{3.9}$$

$$\frac{df_i}{dP_{Gj}} L_{Gj} = \lambda \quad j \in NG \tag{3.10}$$

$$L_{Gj} = \frac{1}{1 - \frac{\partial P_L}{\partial P_{Gj}}} \quad j \in NG \tag{3.11}$$

where

$\lambda$: the Lagrangian multiplier;

$\dfrac{\partial P_L}{\partial P_{Di}}$: the loss sensitivity with respect to load at bus $i$;.

$\dfrac{\partial P_L}{\partial P_{Gj}}$: the loss sensitivity with respect to unit at bus $j$.

We use $\frac{\partial P_L}{\partial P_i}$, which is the loss sensitivity with respect to an injection at bus $i$, to stand for both $\frac{\partial P_L}{\partial P_{Di}}$ and $\frac{\partial P_L}{\partial P_{Gj}}$. Since distributed slack buses are used here, all loss sensitivity factors are nonzero.

If an arbitrary slack bus, $k$, is selected, then $P_k$ is the function of the other injections, that is,

$$P_k = f(P_i) \quad i \in n, \ i \neq k \tag{3.12}$$

where $n$ is the total number of buses in the system and $P_i$ is the power injection at bus $i$, which includes the load $P_{Di}$ and generation $P_{Gj}$. Actually, the load can be treated as a negative generation. Then equations (3.9) and (3.11) can be changed to equation (3.13), and equations (3.8) and (3.10) can be changed to equation (3.14).

$$L_i = \frac{1}{1 - \frac{\partial P_L}{\partial P_i}} \quad i \in n \tag{3.13}$$

$$\frac{df_i}{dP_i} L_i = \lambda \quad i \in n \tag{3.14}$$

Equation (3.2) will be rewritten as

$$P_L = P_k + \sum_{i \neq k} P_i \quad i \in n \tag{3.15}$$

The new Lagrangian function can be obtained from equations (3.1) and (3.15).

$$F_L^* = \sum_i f_i(P_i) + \lambda \left( P_L - P_k - \sum_{i \neq n} P_i \right) \tag{3.16}$$

The optimality criteria can be obtained from the Lagrangian function (3.16).

$$\frac{\partial F_L^*}{\partial P_i} = \frac{df_i}{dP_i} + \frac{df_k}{dP_k} \frac{\partial P_k}{\partial P_i} + \lambda \left( \frac{\partial P_L}{\partial P_i} - \frac{\partial P_k}{\partial P_i} - 1 \right) = 0 \quad i \in n, \quad i \neq k \tag{3.17}$$

From equation (3.15), we get

$$\frac{\partial P_L}{\partial P_i} = 1 + \frac{\partial P_k}{\partial P_i} \tag{3.18}$$

From equations (3.17) and (3.18), we get

$$\frac{df_i}{dP_i} L_i^* = \frac{df_k}{dP_k} \tag{3.19}$$

$$L_i^* = \frac{1}{1 - \frac{\partial P_L}{\partial P_i}} \quad i \in n, \, i \neq k \tag{3.20}$$

It is noted that $L_i$ and $L_i^*$ are similar, but they have different meanings [14]. The former is based on the distributed slack buses, and the latter is based on an arbitrary slack bus $k$. Similarly, the loss sensitivity in $L_i$ is based on the distributed slack, that is, $\left. \frac{\partial P_L}{\partial P_i} \right|_{DS}$ (the subscript DS means distributed slack); the loss sensitivity in $L_i^*$ is based on an arbitrary single slack bus $k$, that is, $\left. \frac{\partial P_L}{\partial P_i} \right|_k$. Note that the $k$th loss sensitivity, with bus $k$ as the slack bus, is zero.

From equations (3.14) and (3.19), we have the following equation:

$$L_i^* = \frac{L_i}{L_k}, \quad L_k^* = 1 \tag{3.21}$$

From equations (3.13), (3.20), and (3.21), we get

$$\frac{1}{1 - \left. \frac{\partial P_L}{\partial P_i} \right|_k} = \frac{1 - \left. \frac{\partial P_L}{\partial P_k} \right|_{DS}}{1 - \left. \frac{\partial P_L}{\partial P_i} \right|_{DS}} \tag{3.22}$$

$$1 - \left. \frac{\partial P_L}{\partial P_i} \right|_k = \frac{1 - \left. \frac{\partial P_L}{\partial P_i} \right|_{DS}}{1 - \left. \frac{\partial P_L}{\partial P_k} \right|_{DS}} \tag{3.23}$$

Hence, with one set of incremental transmission loss coefficients for the distributed slack buses, the loss sensitivity for an arbitrary slack bus can be calculated from the following formula:

$$\left.\frac{\partial P_L}{\partial P_i}\right|_k = \frac{\left.\frac{\partial P_L}{\partial P_i}\right|_{DS} - \left.\frac{\partial P_L}{\partial P_k}\right|_{DS}}{1 - \left.\frac{\partial P_L}{\partial P_k}\right|_{DS}} \tag{3.24}$$

The formula for loss sensitivity calculation is very simple, but it is accurate and efficient for real-time energy markets. It will avoid computing a new set of the loss sensitivity factors whenever the slack bus $k$ changes. Consequently, it means huge time savings. In addition, the loss factors based on the distributed load reference will not be changed no matter how the AGC statuses of units vary, as long as the network topology is the same as before.

## 3.3   CALCULATION OF CONSTRAINED SHIFT SENSITIVITY FACTORS

### 3.3.1   Definition of Constraint Shift Factors

The objective of SFT is to identify whether or not network operation is feasible for a real power injection scenario. If operational limits are violated, generic constraints and corresponding sensitivities (the shift factors) are generated, which can be used to prevent violation if presented with the same network conditions. Meanwhile, the shift factors can also be used in generation scheduling or economic dispatch to alleviate the overload of transmission lines.

The SFT calculations include contingency analysis (CA), in which decoupled power flow (DPF) or DC power flow is used. The set of component changes that can be analyzed include transmission line, transformer, circuit breaker, load demand, and generator outages. SFT informs the users about the contingencies that could cause conditions violating operating limits. These limits include branch overloads, abnormal voltages, and voltage angle differences across specified parts of the network. SFT reports the sensitivity (shift factor) of the constraint with respect to the controls. These controls include unit MW control, phase shifter, and load MW control.

***Unit MW Control***   The unit MW control is the most efficient and least expensive among the available controls. The formulation of sensitivity for a unit can be written as follows:

$$S_{kj} = \frac{\partial P_k}{\partial P_{Gj}} \quad k = 1, \dots, K_{max}, \quad j = 1, \dots PG_{max} \tag{3.25}$$

where

$S_{kj}$: the sensitivity of the power change on constraint $k$ with respect to power change on the unit MW control $j$;

$P_k$: the MW power on the constraint $k$;

$P_{Gj}$: the MW power on generating unit control $j$;

$K_{max}$: the maximum number of constraints;

$PG_{max}$: The maximum number of generator unit MW controls.

According to KCL , it is impossible that power change on the branch constraint will be greater than 1 MW if the generator control has only 1 MW power change. Thus the maximum value of the sensitivity of the branch constraint with respect to the unit MW control is 1.0 (generally, less than 1.0).

***Phase Shifter Control***    The phase shifter is another efficient control among the available controls. There are some assumptions for the phase shifter in the SFT design. The phase shifter control variable is a tap number (e.g., phase shifter angle). Normally a tap number is an integer, but it can be handled as a real number in practical SFT calculation. In addition, all opened phase shifters will be skipped over, that is, the sensitivity for the phase shifter that is open at any end will not be calculated.

The step on the tap type is the sensitivity of angle with respect to the tap number. Thus the sensitivity of the constraint to the phase shifter relates to the power change on the constraint to the angle change of the phase shifter. The angle unit may be in degrees or radians. Since the value of sensitivity may be very small if the angle unit is in degrees, the radian is adopted in practical calculations. The formulation of sensitivity for phase shifter can be written as follows:

$$S_{kjp} = \frac{\partial P_k}{\partial \phi_{jp}^{ps}} \quad k = 1, \ldots, K_{max}, \quad jp = 1, \ldots PS_{max} \qquad (3.26)$$

where

$S_{kjp}$: the sensitivity of the constraint $k$ to the phase shifter control $jp$;

$\phi_{jp}^{ps}$: the phase shifter angle of the phase shifter control $jp$;

$K_{max}$: the maximum number of constraints;

$PS_{max}$: the maximum number of phase shifter controls.

It is noted that there is a special "branch in constraint" logic that must be implemented when the phase shifter branch itself is in the constraint. Basically, the artificial flow through the transformer branch must be subtracted from the constraint flow.

In addition, the sensitivity of the constraint to the phase shifter control is different from the sensitivity of the constraint to the generator control or other bus injection type controls. The value of latter cannot be greater than 1.0, but the former does not have this constraint.

**Load MW Control**   The load MW control should be the last control when other controls are not available. The formulation of sensitivity for load MW control can be written as follows:

$$S_{kjd} = -\frac{\partial P_k}{\partial P_{jd}} \qquad k = 1, \ldots , K_{max}, \qquad jd = 1, \ldots LD_{max} \qquad (3.27)$$

where
   $S_{kjd}$: the sensitivity of the constraint $k$ to the load MW control $jd$;
   $P_{jd}$: the MW power on load control $jd$;
   $K_{max}$: the maximum number of constraints;
$LD_{max}$: the maximum number of load MW controls in whole system.

It is noted that the sensitivity sign for load MW control is negative. The reason is that increasing load will cause more serious constraint violation rather than reduce the constraint violation. According to the sensitivity relationship between the constraint and the load MW control, it is needed to reduce/shed load for alleviating or deleting the constraint violation.

In the market application, the sensitivity of the pricing node is of interest. The pricing node does not have the generator or load connected to it. Thus the above sensitivity calculation of unit/load control can be expanded to any bus injection, that is,

$$S_{k\,bs} = -\frac{\partial P_k}{\partial P_{bs}} \quad k = 1, \ldots , K_{max}, \qquad bs = 1, \ldots NB_{max}$$

   $S_{k\,bs}$: the sensitivity of the constraint $k$ to the bus injection on bus bs;
   $P_{bs}$: the MW power injection on bus bs;
$NB_{max}$: the maximum number of buses in the whole system.

**Constraint Value**   For each constraint, the constraint value (DC value) is computed from the control values multiplied by sensitivities. The formulation can be written as follows:

$$DCVAL_k = \sum_{j=1}^{Umax} VAL\_U_j^* S_{kj} \qquad (3.28)$$

where

$DCVAL_k$: the constraint value for the constraint $k$;
   $VAL\_U_j$: the value of control $j$; here, the controls include unit MW control, phase shifter, and load MW control;
      $S_{kj}$: the sensitivity or shift factor of the constraint $k$ to the control $j$;
   $U_{max}$: the maximum number of controls.

### 3.3.2 Computation of Constraint Shift Factors

**Constraint Shift Factors without Line Outage**   Constraint shift factors without line outage are also called generation shift factors.

From the DC power flow algorithm, we have the following equation:

$$
\begin{bmatrix} \Delta P_1 \\ \Delta P_2 \\ \vdots \\ \Delta P_n \end{bmatrix} = [B'] \begin{bmatrix} \Delta \theta_1 \\ \Delta \theta_2 \\ \vdots \\ \Delta \theta_n \end{bmatrix}
\tag{3.29}
$$

Then the standard matrix calculation of the DC power flow can be written as follows;

$$
\boldsymbol{\theta} = [X]\mathbf{P}
\tag{3.30}
$$

Since the DC power flow model is a linear model, we may calculate the perturbations about a given set of system conditions by using the same model. Thus, we can compute the changes in bus phase angles $\Delta\boldsymbol{\theta}$ for a given set of changes in the bus power injections $\Delta\mathbf{P}$:

$$
\Delta\boldsymbol{\theta} = [X]\Delta\mathbf{P}
\tag{3.31}
$$

where the net perturbation of the reference bus equals the sum of the perturbations of all the other buses.

Now we compute the generation shift factors for the generator on bus $i$. To do this, we will set the perturbation on bus $i$ to $+1$ pu and the perturbation on all the other buses to zero. Then we can solve for the change in bus phase angles using the following matrix calculation:

$$
\Delta\boldsymbol{\theta} = [X] \begin{bmatrix} +1 \\ -1 \end{bmatrix} \begin{matrix} \leftarrow \text{ row } i \\ \leftarrow \text{ ref row} \end{matrix}
\tag{3.32}
$$

The vector of bus power injection perturbations in equation 3.32 represents the situation when a 1-pu power increase is made at bus $i$ and is compensated by a 1-pu decrease in power at the reference bus. The $\Delta\boldsymbol{\theta}$ values in equation are thus equal to the derivative of the bus angles with respect to a change in power injection at bus $i$.

Thus the constraint shift factors $S_{ki}$ without considering the line outage can be derived as follows.

Let $p$ and $q$ be the two ends of the constraint $k$; the power flowing on the constraint line $k$ using DC power flow is

$$
P_k = \frac{1}{x_k}(\theta_p - \theta_q)
\tag{3.33}
$$

The generation shift factors are defined as

$$
S_{ki} = \frac{dP_k}{dP_i} = \frac{d}{dP_i}\left[\frac{1}{x_k}(\theta_p - \theta_q)\right]
$$

$$= \frac{1}{x_k}\left(\frac{d\theta_p}{dP_i} - \frac{d\theta_q}{dP_i}\right) = \frac{1}{x_k}(X_{pi} - X_{qi}) \tag{3.34}$$

In practical applications, the generation shift factors of the network can be directly obtained from $[B']$ through forward and back calculation.

Assume a branch $k$ that is from $p$ to $q$ with the reactance $x_k$.

From $[B'][\theta] = [P]$, we get

$$[B'][\theta] = \begin{bmatrix} 0 \\ \vdots \\ 0 \\ +\dfrac{1}{x_k} \\ 0 \\ \vdots \\ 0 \\ -\dfrac{1}{x_k} \\ 0 \\ \vdots \\ 0 \end{bmatrix} \begin{matrix} \\ \\ \\ \leftarrow \text{row } p \\ \\ \\ \\ \leftarrow \text{row } q \\ \\ \\ \end{matrix} \tag{3.35}$$

Through implying forward and back calculation to the above equation, the solution will be the generation shift factors for all buses with respect to the constraint line $k$.

If a constraint consists of multiple lines (branches), the superposition theory can be applied. For example, a constraint contains two lines $k$ (from $p$ to $q$) and $t$ (from $i$ to $j$) with reactance $x_k$ and $x_t$, respectively. We get following relationship:

$$[B'][\theta] = \begin{bmatrix} 0 \\ \vdots \\ 0 \\ +\dfrac{1}{x_k} \\ 0 \\ \vdots \\ 0 \\ -\dfrac{1}{x_k} \\ 0 \\ \vdots \\ 0 \\ +\dfrac{1}{x_t} \\ 0 \\ \vdots \\ 0 \\ -\dfrac{1}{x_t} \\ 0 \\ \vdots \\ 0 \end{bmatrix} \begin{matrix} \\ \\ \\ \leftarrow \text{row } p \\ \\ \\ \\ \leftarrow \text{row } q \\ \\ \\ \\ \leftarrow \text{row } i \\ \\ \\ \\ \leftarrow \text{row } j \\ \\ \end{matrix}$$

Through implying forward and back calculation to the above equation, the solution will be the generation shift factors for all buses with respect to lines $k$ and $t$.

***Line Outage Distribution Factors (LODF) [17]***   The simulation of line outage is shown in Figure 3.1. Figure 3.1(a) is a network without line outage.

Suppose line $l$ from bus $m$ to bus $n$ were opened by circuit breakers as shown in Figure 3.1(b). A line outage may be modeled by adding two power injections to a system, one at each end of the line to be dropped, which is shown in Figure 3.1(c). The line is actually left in the system and the effects of its being dropped are modeled by injections. Note that when the circuit breakers are opened, no current flows through them and the line is completely isolated from the remainder of the network. In Figure 3.1, the breakers are still closed but injections $\Delta P_m$ and $\Delta P_n$ have been added to bus $m$ and bus $n$, respectively. If $\Delta P_m = P_{mn}$, and $\Delta P_n = -P_{mn}$ where $P_{mn}$ is equal to the power flowing over the line, we will still have no current flowing through the circuit breakers even though they are closed. As far as the remainder of the network is concerned, the line is disconnected.

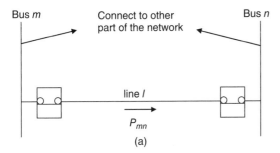

(a)

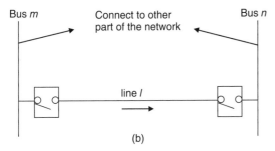

(b)

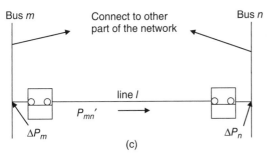

(c)

Figure 3.1   Network for simulating line outage (a) network before line $l$ outage; (b) network after line $l$ outage; and (c) modeling line $l$ outage using injections.

Using equation relating to $\Delta\theta$ and $\Delta P$, we have

$$\Delta\theta = [X]\Delta P \tag{3.36}$$

Since only power injections at buses $m$ and $n$ have been changed after line outage by adding two power injections to a system,

$$\Delta P = \begin{bmatrix} 0 \\ \vdots \\ 0 \\ \Delta P_m \\ 0 \\ \vdots \\ 0 \\ \Delta P_n \\ 0 \\ \vdots \\ 0 \end{bmatrix} \tag{3.37}$$

Thus we can get the incremental changes of the phase angle at buses $m$ and $n$ of the line $l$ from the outage

$$\Delta\theta_m = X_{mn}\Delta P_n + X_{mm}\Delta P_m \tag{3.38}$$

$$\Delta\theta_n = X_{nn}\Delta P_n + X_{nm}\Delta P_m \tag{3.39}$$

where

$\theta_m$: the phase angle at bus $m$ of the line $l$ before the outage;

$\theta_n$: the phase angle at bus $n$ of the line $l$ before the outage;

$P_{mn}$: the power flow on line $l$ from bus $m$ to bus $n$ before the outage;

$\Delta\theta_m$: the incremental changes of the phase angle at bus $m$ of the line $l$ from the outage;

$\Delta\theta_n$: the incremental changes of the phase angle at bus $n$ of the line $l$ from the outage;

$\Delta P_{mn}$: the incremental changes of the power flow in line $l$ after the outage;

$P'_{mn}$: the power flow on line $l$ from bus $m$ to bus $n$ after the outage.

The outage modeling criteria requires that the incremental injections $\Delta P_n$ and $\Delta P_m$ equal the power flowing over the outaged line *after* the injections are imposed. Then, if we let the line reactance be $x_l$,

$$P'_{mn} = \Delta P_m = -\Delta P_n \tag{3.40}$$

$$\Delta P_{mn} = \frac{1}{x_l}(\Delta\theta_m - \Delta\theta_n) \tag{3.41}$$

Since $\Delta P_n = -\Delta P_m$, equations (3.38) and (3.39) can be written as

$$\Delta\theta_m = X_{mn}\Delta P_n + X_{mm}\Delta P_m = X_{mn}(-\Delta P_m) + X_{mm}\Delta P_m$$

$$= (X_{mm} - X_{mn})\Delta P_m \tag{3.42}$$

$$\Delta\theta_n = X_{nn}\Delta P_n + X_{nm}\Delta P_m = X_{nn}(-\Delta P_m) + X_{nm}\Delta P_m$$

$$= (X_{nm} - X_{nn})\Delta P_m \tag{3.43}$$

where

$$X_{mn} = X_{nm} \tag{3.44}$$

Thus,

$$\Delta P_{mn} = \frac{1}{x_l}(\Delta\theta_m - \Delta\theta_n)$$

$$= \frac{1}{x_l}[(X_{mm} - X_{mn})\Delta P_m - (X_{nm} - X_{nn})\Delta P_m]$$

$$= \frac{1}{x_l}(X_{mm} + X_{nn} - 2X_{mn})\Delta P_m \tag{3.45}$$

The power flow on line $l$ from bus $m$ to bus $n$ after the outage $P'_{mn}$ is computed as follows:

$$P'_{mn} = P_{mn} + \Delta P_{mn}$$

$$= P_{mn} + \frac{1}{x_l}(X_{mm} + X_{nn} - 2X_{mn})\Delta P_m \tag{3.46}$$

From equations (3.40) and (3.46), we get

$$\Delta P_m = P_{mn} + \frac{1}{x_l}(X_{mm} + X_{nn} - 2X_{mn})\Delta P_m \tag{3.47}$$

that is,

$$\Delta P_m = \frac{P_{mn}}{1 - \frac{1}{x_l}(X_{mm} + X_{nn} - 2X_{mn})} \tag{3.48}$$

Since there are only two nonzero elements at buses $m$ and $n$ in the power injection vector, the incremental change of phase angle at any bus $i$ can be computed as follows:

$$\Delta\theta_i = X_{in}\Delta P_n + X_{im}\Delta P_m$$

$$= (X_{im} - X_{in})\Delta P_m$$

$$= (X_{im} - X_{in}) \times \frac{P_{mn}}{1 - \frac{1}{x_l}(X_{mm} + X_{nn} - 2X_{mn})}$$

$$= \frac{x_l(X_{im} - X_{in})P_{mn}}{x_l - (X_{mm} + X_{nn} - 2X_{mn})} = S_{i,l}P_{mn} \tag{3.49}$$

where

$$S_{i,l} = \frac{\Delta\theta_i}{\Delta P_l} = \frac{x_l(X_{im} - X_{in})}{x_l - (X_{mm} + X_{nn} - 2X_{mn})} \tag{3.50}$$

which is the sensitivity factor of the change in the phase angle of bus $i$ with respect to power flow in line $l$ before the outage.

For computing the effect of line $l$ outage on the other line $k$, the LODF is defined as follows:

$$\text{LODF}_{k,l} = \frac{\Delta P_k}{\Delta P_l} = \frac{\frac{1}{x_k}(\Delta\theta_p - \Delta\theta_q)}{\Delta P_l}$$

$$= \frac{1}{x_k}\left(\frac{\Delta\theta_p}{\Delta P_l} - \frac{\Delta\theta_q}{\Delta P_l}\right)$$

$$= \frac{1}{x_k}(S_{p,l} - S_{q,l}) \tag{3.51}$$

From equation (3.50), $S_{p,l}$, $S_{q,l}$ can be written as

$$S_{p,l} = \frac{\Delta\theta_p}{\Delta P_l} = \frac{x_l(X_{pm} - X_{pn})}{x_l - (X_{mm} + X_{nn} - 2X_{mn})} \tag{3.52}$$

$$S_{q,l} = \frac{\Delta\theta_q}{\Delta P_l} = \frac{x_l(X_{qm} - X_{qn})}{x_l - (X_{mm} + X_{nn} - 2X_{mn})} \tag{3.53}$$

Thus,

$$\text{LODF}_{k,l} = \frac{1}{x_k}(S_{p,l} - S_{q,l})$$

$$= \frac{1}{x_k}\left(\frac{x_l(X_{pm} - X_{pn})}{x_l - (X_{mm} + X_{nn} - 2X_{mn})} - \frac{x_l(X_{qm} - X_{qn})}{x_l - (X_{mm} + X_{nn} - 2X_{mn})}\right)$$

$$= \frac{1}{x_k}\left(\frac{x_l(X_{pm} - X_{pn}) - x_l(X_{qm} - X_{qn})}{x_l - (X_{mm} + X_{nn} - 2X_{mn})}\right)$$

$$= \frac{1}{x_k} \left( \frac{x_l \left( X_{pm} - X_{qm} - X_{pn} + X_{qn} \right)}{x_l - (X_{mm} + X_{nn} - 2X_{mn})} \right)$$

$$= \frac{\frac{x_l}{x_k}(X_{pm} - X_{qm} - X_{pn} + X_{qn})}{x_l - (X_{mm} + X_{nn} - 2X_{mn})} \tag{3.54}$$

***Outage Transfer Distribution Factors (OTDF)*** Because we know that the generation shift factors and LODFs are linear models, we can use superposition to extend them to compute the network constraint sensitivity factors after a branch has been lost. They are also called the OTDFs. Let us compute the sensitivity factor OTDF between line $k$ and generator bus $j$ when line $l$ is opened. This is calculated by first assuming that the change in generation on bus $j$, $\Delta P_j$, has a direct effect on line $k$ and an indirect effect through its influence on the power flowing in line $l$, which, in turn, influences line $k$ when line $l$ is in outage. Then

$$\Delta P_k = S_{kj}\Delta P_j + \text{LODF}_{k,l}\Delta P_l$$

$$= S_{kj}\Delta P_j + \text{LODF}_{k,l}(S_{lj}\Delta P_j)$$

$$= (S_{kj} + \text{LODF}_{k,l}S_{lj})\Delta P_j \tag{3.55}$$

Therefore, the sensitivity OTDF after line $l$ outage can be defined as

$$\text{OTDF}_{k,j} = \frac{\Delta P_k}{\Delta P_j} = (S_{kj} + \text{LODF}_{k,l}S_{lj}) \tag{3.56}$$

where

$\text{OTDF}_{k,j}$: the sensitivity factor between line $k$ and generator bus $j$ when line $l$ was opened.

### 3.3.3 Constraint Shift Factors with Different References

The shift factors computed in SFT is based on the reference bus in energy management system (EMS) topology, but it can be easily converted to any market-based reference.

Let $y$ be the market-based reference unit, and the shift factor of the constraint $k$ with respect to any unit $j$ that is obtained on the basis of the EMS reference bus be $S_{kj}$. For unit $y$, the shift factor of the constraint $k$ is $S_{ky}$. Then, the shift factors after converting to the market-based reference unit $y$ can be computed as follows.

$$S'_{ky} = 0 \quad k = 1, \ldots, K_{\max} \tag{3.57}$$

$$S'_{kj} = S_{kj} - S_{ky} \quad k = 1, \ldots, K_{\max}, \quad j \neq y \tag{3.58}$$

where

$S_{kj}$: the shift factor of the constraint $k$ with respect to unit $j$ that is based on the EMS reference;

$S_{ky}$: the shift factor of the constraint $k$ with respect to unit $y$ that is based on the EMS reference;

$S'_{kj}$: the shift factor of the constraint $k$ with respect to unit $j$ that is based on the market-based reference $y$;

$S'_{ky}$: the shift factor of the constraint $k$ with respect to unit $y$ that is based on the market-based reference $y$.

We know that the shift factor of the constraint is related to the selected reference, that is, the value of the shift factor will be different if the reference is different even though the system topology and conditions are the same. Sometimes the system operators would like to have stable shift factor values without concern about the selection of the reference bus/unit. Thus the distributed load reference will be used to get the unique constraint shift factors if the system topology and conditions are unchanged.

Let $S_{k\,\mathrm{ldref}}$ be the sensitivity of load distribution reference for the constraint $k$, and the shift factor of the constraint $k$ with respect to any control $j$ that is obtained on the basis of EMS reference bus be $S_{kj}$. Then the shift factors based on the load distribution reference LDREF can be computed as follows.

$$S'_{kj} = S_{kj} - S_{k\,\mathrm{ldref}} \quad k = 1, \dots, K_{\max} \tag{3.59}$$

where

$S_{k\,\mathrm{ldref}}$: the sensitivity of load distribution reference for the constraint $k$, that is,

$$S_{k\,\mathrm{ldref}} = \frac{\sum_{jd=1}^{\mathrm{LD}_{\max}} (S_{kjd} * \mathrm{LD}_{jd})}{\sum_{jd=1}^{\mathrm{LD}_{\max}} \mathrm{LD}_{jd}} \quad k = 1, \dots, K_{\max} \tag{3.60}$$

where

$S_{kjd}$: the sensitivity of load $jd$ with respect to the constraint $k$;
$\mathrm{LD}_{jd}$: the load demand at load bus $jd$.

In practical energy markets such as the independent system operator (ISO), the system consists of many areas, but there is one major area in the ISO system that is called the internal area, while the others are called external areas. If the internal area is of major concern during the price calculation in this market system, the load distribution reference can be selected on the basis of the internal area alone. Similarly, Let $\mathrm{LDA}_{\max}$ be the total number of load controls in the internal area of ISO system, which is less than the total number of load controls in whole ISO system, $\mathrm{LD}_{\max}$. The shift factors based on the area load distribution reference LDAREF can be computed as follows:

$$S'_{kj} = S_{kj} - S_{k\,\text{ldaref}} \quad k = 1, \dots, K_{\max} \tag{3.61}$$

where

$S_{k\,\text{ldaref}}$: the sensitivity of load distribution reference in area $A$ for the constraint $k$, that is,

$$S_{k\,\text{ldaref}} = \frac{\displaystyle\sum_{jd=1}^{\text{LDA}_{\max}} (S_{kjd} * \text{LD}_{jd})}{\displaystyle\sum_{jd=1}^{\text{LDA}_{\max}} \text{LD}_{jd}} \quad k = 1, \dots, K_{\max} \tag{3.62}$$

$$\text{LDA}_{\max} \in \text{LD}_{\max}$$

where

$\text{LDA}_{\max}$: the maximum number of load MW controls in area $A$.

### 3.3.4  Sensitivities for the Transfer Path

A transfer path is an energy transfer channel between a point of delivery (POD) and point of receipt (POR). The POD is the point of interconnection on the transmission provider's transmission system where capacity and/or energy transmitted by the transmission provider will be made available to the receiving party. The POR is the point of interconnection on the transmission provider's transmission system where capacity and/or energy transmitted will be made available to the transmission provider by the delivering party.

This pair POD and POR defines a path and the direction of flow in that path. For internal paths, this would be a specific location in the area. For an external path, this may be an area-to-area interface. Similar to the concept of POD/POR, a transfer path can also be defined as one from the source to sink.

If POD/POR (or source/sink) is a single unit or single injection node, the sensitivity of POR or POD is the same as the constrained shift factor, which is mentioned in Sections 3.2 and 3.3. If POD/POR (or source/sink) is an area, the sensitivity of POR or POD can be computed as follows.

Let $PF_j$ be the participation factor of unit $j$, and the shift factor of the constraint $k$ with respect to any unit $j$ be $S_{kj}$. The area-based shift factor of the constraint $k$ is $S_{kA}$, which can be computed as follows:

$$S_{kA} = \frac{\displaystyle\sum_{j \in A} (PF_j \times S_{kj})}{\displaystyle\sum_{j \in A} PF_j} \quad k = 1, \dots, K_{\max}, j \in A \tag{3.63}$$

where

$S_{kA}$: the area based shift factor of the constraint $k$;

$\mathrm{PF}_j$: the participation factor of the unit $j$.

Similarly, if we consider the effect of the outage, the area-based shift factor of constraint $k$ can be computed as follows:

$$S_{kA} = \frac{\sum\limits_{j \in A}(\mathrm{PF}_j \times \mathrm{OTDF}_{kj})}{\sum\limits_{j \in A} \mathrm{PF}_j} \qquad k = 1, \ldots, K_{\max}, \ j \in A \qquad (3.64)$$

If a transfer path is from area $A$ to area $B$, the sensitivity of the transfer path will be computed as

$$S_{\mathrm{TP}}(A \rightarrow B) = S_{kA} - S_{kB} \qquad (3.65)$$

If a transfer path is from an injection node $i$ to another injection node $j$, the sensitivity of the transfer path will be computed as

$$S_{\mathrm{TP}}(I \rightarrow J) = \mathrm{OTDF}_{ki} - \mathrm{OTDF}_{kj} \qquad (3.66)$$

If a transfer path is from an injection node $i$ to area $A$, or from an area $A$ to an injection node $i$, the corresponding sensitivities of the transfer path will be computed as

$$S_{\mathrm{TP}}(I \rightarrow A) = \mathrm{OTDF}_{ki} - S_{kA} \qquad (3.67)$$

$$S_{\mathrm{TP}}(A \rightarrow I) = S_{kA} - \mathrm{OTDF}_{ki} \qquad (3.68)$$

## 3.4 PERTURBATION METHOD FOR SENSITIVITY ANALYSIS

So far, the sensitivity analysis methods described in this chapter have been based on the matrix (either B′ matrix or the Jacobian matrix). The sensitivity values that are computed on the basis of partial differential terms will be stable, or will not change as long as the system topology remains the same.

Sometimes, the perturbation method is also used in sensitivity calculation.

### 3.4.1 Loss Sensitivity

The perturbation method for loss sensitivity calculation is described in the following.

1. Perform power flow calculation, and obtain the initial system power loss $P_{L0}$.
2. Simulate the calculation of the loss sensitivity with respect to the generator $i$. Increase the power output of generator $i$ for $\Delta P_{Gi}$ (if computing the loss sensitivity of load $k$, reduce the power demand of load k for $\Delta P_{Dk}$), and the slack unit will absorb the same amount of $\Delta P_{Gi}$.

3. Run the power flow again, and get the new system power loss $P_L$.
4. Compute the loss sensitivity as below.

(i) For unit loss sensitivity:

$$LS_{Gi} = \frac{P_L - P_{L0}}{\Delta P_{Gi}} \quad i \in NG \tag{3.69}$$

(ii) For load loss sensitivity:

$$LS_{Dk} = \frac{P_L - P_{L0}}{\Delta P_{Dk}} \quad i \in ND \tag{3.70}$$

where $LS_{Gi}$, and $LS_{Dk}$ are the loss sensitivity values with respect to the unit $i$ and load $k$, respectively.

## 3.4.2 Generator Shift Factor Sensitivity

The perturbation method for generator shift factor sensitivity calculation is .

1. Chose a unit $i$ and a branch constraint $j$.
2. Perform power flow calculation, and obtain the initial power flow $P_{j0}$ for branch $j$.
3. Simulate the calculation of the generator shift factor sensitivity of the branch $j$ with respect to the generator $i$. Increase the power output of generator $i$ for $\Delta P_{Gi}$; the slack unit will absorb the same amount of $\Delta P_{Gi}$.
4. Run power flow again, and get the new power flow $P_j$ for the branch $j$.
5. Compute the generator shift factor sensitivity as follows:

$$GSF_{j,i} = \frac{P_j - P_{j0}}{\Delta P_{Gi}} \quad i \in NG \tag{3.71}$$

where $GSF_{j,i}$ is the generator shift factor sensitivity of the branch $j$ with respect to the unit $i$.

The calculation of the load shift factor sensitivity is similar to the generator shift factor sensitivity, considering the load as the negative generation.

## 3.4.3 Shift Factor Sensitivity for the Phase Shifter

The perturbation method for the phase shifter shift factor sensitivity calculation is shown as follows.

1. Choose a phase shifter $t$ and a branch constraint $j$.
2. Perform power flow calculation and obtain the initial power flow $P_{j0}$ for the branch $j$.

3. Simulate the calculation of the shift factor sensitivity of the branch $j$ with respect to the phase shifter $t$. Increase the taps of the phase shifter $i$ for $\Delta T_t$ (or angle change $\Delta \theta_t$), which can be simulated by changing the suceptance of the phase shifter.

4. Run power flow again and get the new power flow $P_j$ for branch $j$.

5. Compute the phase shifter shift factor sensitivity as follows.

$$SF_{j,t} = \frac{P_j - P_{j0}}{\Delta T_t}$$

or $\qquad SF_{j,t} = \frac{P_j - P_{j0}}{\Delta \theta_t}$ $\qquad\qquad$ (3.72)

where $SF_{j,t}$ is the shift factor sensitivity of the branch $j$ with respect to the phase shifter $t$.

### 3.4.4   Line Outage Distribution Factor (LODF)

The perturbation method for the LODF calculation is described in the following.

1. Choose a branch $l$ that will be simulated as outage and a branch constraint $j$.

2. Perform power flow calculation before the branch $l$ is open, and obtain the initial power flow $P_{j0}$ for the branch $j$, and $P_{l0}$ for the branch $l$.

3. Simulate the calculation of LODF. Open the branch $l$ while the unit power and load power remain unchanged.

4. Run power flow again, and get the new power flow $P_j$ for branch j. The power flow $P_l$ for the branch $l$ will be zero because branch $l$ is in outage.

5. Compute the LODF of branch $j$ as the branch $l$ is in outage as follows:

$$LODF_{j,l} = \frac{P_j - P_{j0}}{P_{l0}}$$ $\qquad\qquad$ (3.73)

where $LODF_{j,l}$ is the LODF of branch $j$ with respect to outage branch $l$.

### 3.4.5   Outage Transfer Distribution Factor (OTDF)

The perturbation method for the OTDF calculation is described in the following.

1. Choose a unit $i$, a branch $l$ that will be simulated as outage, and a branch constraint $j$.

2. Perform power flow calculation before the branch $l$ is open and obtain the initial power flow $P_{j0}$ for the branch $j$ and $P_{l0}$ for the branch $l$.

3. First of all, simulate the calculation of the generator shift factor sensitivity of the branches $j$ and $l$ with respect to the generator $i$. Increase the power output of generator $i$ for $\Delta P_{Gi}$; the slack generator will absorb the same amount of $\Delta P_{Gi}$.

4. Conduct a power flow calculation, and get the new power flows $P_j$ for the branch $j$ and $P_l$ for the branch $l$.

5. Compute the generator shift factor sensitivity for the branches $j$ and $l$, respectively.

$$\text{GSF}_{j,i} = \frac{P_j - P_{j0}}{\Delta P_{Gi}} \quad i \in NG \tag{3.74}$$

$$\text{GSF}_{l,i} = \frac{P_l - P_{l0}}{\Delta P_{Gi}} \quad i \in NG \tag{3.75}$$

6. Then simulate the calculation of LODF for branch $j$ with respective to the outage branch $l$. Open branch $l$ while the unit power and load power remain unchanged.

7. Once again run power flow, and get the new power flow $P'_j$ for branch j. The power flow $P'_l$ for branch $l$ will be zero because branch $l$ is in outage.

8. Compute the LODF of branch $j$ as branch $l$ is in outage as follows:

$$\text{LODF}_{j,l} = \frac{P'_j - P_j}{P_l} \tag{3.76}$$

Finally, the sensitivity OTDF of branch $j$ after line $l$ outage can be obtained as follows:

$$\text{OTDF}_{j,i} = \text{GSF}_{j,i} + \text{LODF}_{j,l}\text{GSF}_{l,i} \tag{3.77}$$

where $\text{OTDF}_{j,i}$ is the sensitivity factor between line $j$ and generator bus $i$ when line $l$ is opened.

It is noted that the perturbation method for sensitivity calculation is very straightforward, but there is a disadvantage, namely, the values of sensitivity depend highly on the solution in addition to the topology. Even if the system topology is not changed, the values of the sensitivity may be a little different for different initial points. Thus, to obtain the accurate sensitivity results, the approach based on a matrix is recommended. If the perturbation method is used, the amount of the perturbation should be small so that the solution is close to the initial operation points.

## 3.5 VOLTAGE SENSITIVITY ANALYSIS

Before we do voltage sensitivity analysis, we need to understand the concept and importance of voltage stability. Voltage stability is the ability of a power system to maintain adequate voltage magnitude so that when the system nominal load is increased, the actual power transferred to that load will increase. The main cause of voltage instability is the inability of the power system to meet the demand for reactive power. The voltage stability problem consists of two aspects: a large disturbance aspect and a small disturbance one. The former is called dynamic stability, and the latter is called static stability. The large disturbance involves short circuit and addresses

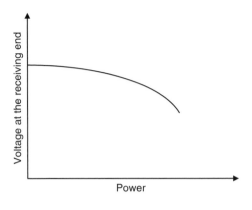

Voltage at the receiving end

Power

Figure 3.2   A plot of power versus voltage.

postcontingency system response. The small disturbance investigates the stability of an operating point and applies a linearized model. The voltage sensitivity analysis herein is used for static voltage stability.

Static voltage instability is mainly associated with reactive power imbalance. This imbalance mainly occurs in a local network or a specified bus in a system. Therefore, the reactive power supports have to be locally adequate. With static voltage stability, slowly developing changes in the power system occur that eventually lead to a shortage of reactive power and declining voltage. This phenomenon can be seen in Figure 3.2, a plot of power transferred versus voltage at the receiving end.

These kinds of plots are generally called $P - V$ curves or "nose" curves. As power transfer increases, the voltage at the receiving end decreases. Eventually, a critical (nose) point, the point at which the system reactive power is out of usage, is reached where any further increase in active power transfer will lead to very rapid decrease in voltage magnitude. Before reaching the critical point, a large voltage drop due to heavy reactive power losses is observed. The only way to save the system from voltage collapse is to reduce the reactive power load or add additional reactive power before reaching the point of voltage collapse.

The purpose of the voltage sensitivity analysis is to improve the voltage profile and to minimize system real power losses through optimal reactive power controls (i.e., by adding VAR supports). These goals are achieved by proper adjustments of VAR variables in power networks through seeking the weak buses in the system. Therefore, if the voltage magnitude at generator buses, VAR compensation (VAR support), and transformer tap position are chosen as the control variables, the optimal VAR control model can be represented as

$$\min \ P_L(Q_S, V_G, T) \tag{3.78}$$

such that

$$Q(Q_S, V_G, T, V_D) = 0 \tag{3.79}$$

$$Q_{G\min} \leq Q_G(Q_S, V_G, T) \leq Q_{G\max} \tag{3.80}$$

$$V_{D\min} \leq V_D(Q_S, V_G, T) \leq V_{D\max} \tag{3.81}$$

$$Q_{S\min} \leq Q_S \leq Q_{S\max} \tag{3.82}$$

$$V_{G\min} \leq V_G \leq V_{G\max} \tag{3.83}$$

$$T_{\min} \leq T \leq T_{\max} \tag{3.84}$$

where

$P_L$: the system real power loss;
$V_G$: the voltage magnitude at generator buses;
$Q_S$: the VAR support in the system;
$Q_G$: the VAR generation in the system;
$T$: the tap position of the transformer;
$V_D$: the voltage magnitude at load buses.

The subscripts "min" and "max" represent the lower and upper limits of the constraint, respectively.

Two kinds of sensitivity-related factors can be computed through equations (3.78)–(3.84). Here they are called voltage benefit factors (VBFs) and loss benefit factors (LBFs), which are expressed as follows.

$$\text{LBF}_i = \frac{\sum\limits_i (P_{L0} - P_L(Q_{si}))}{Q_{si}} \times 100\% \quad i \in ND \tag{3.85}$$

$$\text{VBF}_i = \frac{\sum\limits_i (V_i(Q_{si}) - V_{i0})}{Q_{si}} \times 100\% \quad i \in ND \tag{3.86}$$

where

$Q_{si}$: the amount of VAR support at the load bus $i$;
$\text{LBF}_i$: the loss benefit factors from the VAR compensation $Q_{si}$;
$\text{VBF}_i$: the voltage benefit factors from the VAR compensation $Q_{si}$;
$P_{L0}$: power transmission losses in the system without VAR compensation;
$P_L(Q_{si})$: the power transmission losses in the system with VAR compensation $Q_{si}$;
$V_{i0}$: the voltage magnitude at load bus $i$ without VAR compensation.
$V_i(Q_{si})$: the voltage magnitude at load bus $i$ with VAR compensation $Q_{si}$;
$ND$: the number of load buses.

## 3.6 REAL-TIME APPLICATION OF THE SENSITIVITY FACTORS

In the EMS system and energy markets, the loss sensitivity factors and constraint shift factors are applied for LMP and/or alleviating overload (AOL) calculation. The above-mentioned loss sensitivities, constraint shift factors, and the corresponding constraint elements (transmission lines or transformers) will be passed to the constraint logger and then passed to the LMP calculator. The practical constraints can be divided into the following types:

**(1)** Automatic constraints

All branches (lines, transformers, and interfaces) with violations from EMS real-time contingency analysis (RTCA) calculation.

**(2)** Watch list constraints

The branches without violation in EMS RTCA calculation but with the branch flows that are close to their limits.

**(3)** Active constraints

The constraints from the LMP calculator that are needed to recompute the constraint shift factors.

**(4)** Flowgate constraints

The constraints from the marketing system that are needed to compute the shift factors with respect to the flowgate constraint. The term "flowgate" refers to a single-grid facility or a set of facilities.

**(5)** Quick selection constraints

Any branches (lines, transformers and interfaces) for which the operators want to know the shift factors and monitor the branch flows.

Sensitivity analysis and LMP calculation process is shown in Figure 3.3. The market will require that the LMP be determined on a periodic basis. To support this calculation, the network topology and data including loss sensitivities, network constraints, and their shift factors gathered in real time can be transferred to the LMP automatically through SE (state estimator), RTCA and SFT applications. If the results of the LMP calculator meet the constraints described in equations (3.3) and (3.4), the LMP calculation is deemed successful and the LMP results may be recorded and recommended. If the LMP calculation results in any constraint violation, the violated constraint will be sent back to AOL, and the LMP recalculation will be performed until all constraints are met.

## 3.7   SIMULATION RESULTS

The calculation results of the several sensitivities are illustrated with the IEEE 14-bus system and ALSTOM Grid 60-bus system. The one-line diagram of the ALSTOM Grid 60-bus system is shown in Figure 3.4. The 60-bus system, which has three areas, consists of 24 generation units (15 units are available in the tests), 32 loads, 43 transmission lines, and 54 transformers.

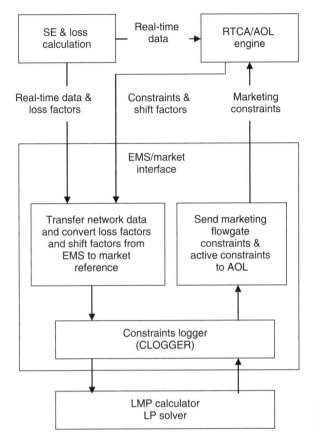

Figure 3.3    Application of the sensitivity factors.

## 3.7.1    Sample Computation for Loss Sensitivity Factors

The following test cases are used to analyze the loss sensitivity in this chapter:

Case 1    Calculate loss sensitivities using the distributed generation slack and load slack, respectively. All units are AGC units (i.e., the status of unit AGC is ON).

Case 2    Calculate loss sensitivities using the distributed generation slack and load slack, respectively. All units are AGC units except the units under station Douglas in Area 1.

Case 3    Calculate loss sensitivities using the distributed generation slack and load slack, respectively. All units are AGC units except the units under station HEARN in Area 1.

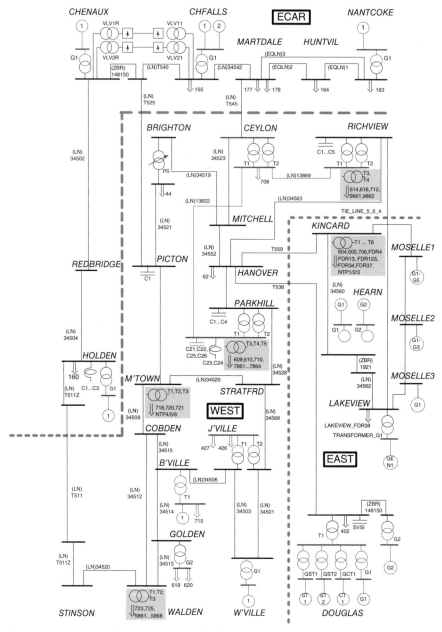

Figure 3.4   One-line diagram of ALSTOM Grid 60 bus system (Area 1-EAST, Area 2-WEST, Area 3-ECAR).

Case 4    Calculate loss sensitivities using the distributed generation slack and load slack, respectively. All units are AGC units except the units in Area 2.

Case 5    Calculate loss sensitivities using the distributed generation slack and load slack, respectively. All units are AGC units except the units under station HOLDEN in Area 3.

Case 6    Calculate loss sensitivities for the selected single slack based on the loss factors under the distributed slack.

The simulation results are shown in Tables 3.1–3.6. All loss sensitivity factors for units and loads are computed. For the purpose of the simplification, only loss sensitivities of generators are listed in Tables 3.1–3.6, in which column 1 is the name of station and units. Column 2 is the area number that the unit belongs to. Column 3 is the AGC status of the unit.

Tables 3.1–3.5 are the test results and comparison of loss sensitivity calculation based on the distributed generation reference and distributed load reference, respectively. The loss factors computed from the distributed unit reference are listed in column 4 of Tables 3.1–3.5. The loss factors computed from the distributed load reference are listed in column 5 of Tables 3.1–3.5.

Generally, the values of loss sensitivities based on the generation reference are different from those based on the load reference, because the distribution of the units is not exactly the same as the distribution of loads in the power system. The loss factors will be close or equal if the units are close to the load locations. This

**TABLE 3.1    Test Results and Comparison of Loss Sensitivity Calculation (Case 1: All Units on AGC)**

| Station, Generator | Area No. | AGC Unit | Loss Sensitivity Distributed Generation Slack | Loss Sensitivity Distributed Load Slack |
|---|---|---|---|---|
| DOUGLAS, G2 | 1 | YES | 0.0151 | 0.0170 |
| DOUGLAS, G1 | 1 | YES | 0.0121 | 0.0140 |
| DOUGLAS, CT1 | 1 | YES | 0.0099 | 0.0118 |
| DOUGLAS, CT2 | 1 | YES | 0.0099 | 0.0118 |
| DOUGLAS, ST | 1 | YES | 0.0097 | 0.0116 |
| HEARN, G1 | 1 | YES | −0.0165 | −0.0146 |
| HEARN, G2 | 1 | YES | −0.0165 | −0.0146 |
| LAKEVIEW, G1 | 1 | YES | −0.0188 | −0.0170 |
| BVILLE, 1 | 2 | YES | −0.0010 | −0.0042 |
| WVILLE, 1 | 2 | YES | 0.0007 | −0.0025 |
| CHENAUX, 1 | 3 | YES | −0.0089 | −0.0089 |
| CHEALLS, 1 | 3 | YES | 0.0212 | 0.0212 |
| CHEALLS, 2 | 3 | YES | 0.0212 | 0.0212 |
| HOLDEN, 1 | 3 | YES | 0.0010 | 0.0010 |
| NANTCOKE, 1 | 3 | YES | −0.0122 | −0.0122 |

**TABLE 3.2  Test Results and Comparison of Loss Sensitivity Calculation (Case 2: All Units on AGC Except the Units Under Station Douglas in Area 1)**

| Station, Generator | Area No. | AGC Unit | Loss Sensitivity Distributed Generation Slack | Loss Sensitivity Distributed Load Slack |
|---|---|---|---|---|
| DOUGLAS, G2 | 1 | NO | 0.0328 | 0.0170 |
| DOUGLAS, G1 | 1 | NO | 0.0299 | 0.0140 |
| DOUGLAS, CT1 | 1 | NO | 0.0278 | 0.0118 |
| DOUGLAS, CT2 | 1 | NO | 0.0278 | 0.0118 |
| DOUGLAS, ST | 1 | NO | 0.0276 | 0.0116 |
| HEARN, G1 | 1 | YES | 0.0015 | −0.0146 |
| HEARN, G2 | 1 | YES | 0.0015 | −0.0146 |
| LAKEVIEW, G1 | 1 | YES | −0.0008 | −0.0170 |
| BVILLE, 1 | 2 | YES | −0.0010 | −0.0042 |
| WVILLE, 1 | 2 | YES | 0.0007 | −0.0025 |
| CHENAUX, 1 | 3 | YES | −0.0089 | −0.0089 |
| CHEALLS, 1 | 3 | YES | 0.0212 | 0.0212 |
| CHEALLS, 2 | 3 | YES | 0.0212 | 0.0212 |
| HOLDEN, 1 | 3 | YES | 0.0010 | 0.0010 |
| NANTCOKE, 1 | 3 | YES | −0.0122 | −0.0122 |

can be observed from Table 3.1, where all units are on AGC status. For the 60-bus system, each load in area 3 has at least one unit connected, so the loss factors in area 3 are the same for both the distributed generation slack and distributed load slack.

It is noted that from Tables 3.1–3.5 that the loss sensitivity factors based on the distributed load slack are the same whether the status of the units is changed or not. But the loss factors based on the distributed generation references are changed as the AGC status of the units are different.

Generally, the change of AGC status of the units only affects the loss sensitivities in the same area that these units belong to.

It can be seen from Tables 3.2 and 3.3 that, when AGC status of the units in area 1 changes, only the loss factors in area 1 is affected. The loss factors in the other areas are unchanged. For Table 3.5, when AGC status of the units in area 3 changes, only the loss factors in area 3 are affected. The loss factors in the other areas are unchanged. But for Table 3.4, there is no AGC unit in area 2; it means that there is no unit reference in area 2. Then the AGC units in the other areas will pick up the power mismatch (i.e. area 1 in this case). Thus, the loss factors in areas 1 and 2 are changed. The loss factors in the other areas are unchanged.

Through the above comparisons, it can be observed that the method of the distributed load references for loss sensitivity calculation is superior to the method of the distributed generation references in the real-time energy markets, as the AGC status of the units are changeable in the real-time system.

TABLE 3.3  Test Results and Comparison of Loss Sensitivity Calculation (Case 3: Only Units Under HEARN in Area 1 Not on AGC)

| Station, Generator | Area No. | AGC Unit | Loss Sensitivity Distributed Generation Slack | Loss Sensitivity Distributed Load Slack |
|---|---|---|---|---|
| DOUGLAS, G2 | 1 | YES | 0.0126 | 0.0170 |
| DOUGLAS, G1 | 1 | YES | 0.0096 | 0.0140 |
| DOUGLAS, CT1 | 1 | YES | 0.0074 | 0.0118 |
| DOUGLAS, CT2 | 1 | YES | 0.0074 | 0.0118 |
| DOUGLAS, ST | 1 | YES | 0.0072 | 0.0116 |
| HEARN, G1 | 1 | NO | −0.0190 | −0.0146 |
| HEARN, G2 | 1 | NO | −0.0190 | −0.0146 |
| LAKEVIEW, G1 | 1 | YES | −0.0213 | −0.0170 |
| BVILLE, 1 | 2 | YES | −0.0010 | −0.0042 |
| WVILLE, 1 | 2 | YES | 0.0007 | −0.0025 |
| CHENAUX, 1 | 3 | YES | −0.0089 | −0.0089 |
| CHEALLS, 1 | 3 | YES | 0.0212 | 0.0212 |
| CHEALLS, 2 | 3 | YES | 0.0212 | 0.0212 |
| HOLDEN, 1 | 3 | YES | 0.0010 | 0.0010 |
| NANTCOKE, 1 | 3 | YES | −0.0122 | −0.0122 |

The results of loss sensitivity calculation for a single slack, which are computed from the proposed formula (3.24), are shown in Table 3.6. Column 3 in Table 3.6 is the set of the loss sensitivity coefficients for the distributed slack buses. Column 4 in Table 3.6 is the set of loss sensitivity factors with a single slack bus at the location of HOLDEN 1. Column 5 in Table 3.6 is the set of loss sensitivity factors with a single slack bus at the location of Douglas.

It is noted that all the loss sensitivities are nonzero if distributed slacks are selected. If a single slack is selected, the loss sensitivity of the slack equals zero.

Since the loss sensitivity values based on the distributed slacks from EMS are unchanged as long as the system topology is the same, the loss sensitivities for any market-based single slack can be easily and quickly acquired by use of the loss sensitivity formula (3.24). Therefore, a large amount of the computations are avoided whenever the loss sensitivities for a market-based reference are needed in the real-time energy markets.

Since the loss sensitivity values based on the distributed load slacks are unchanged as long as the system topology is the same, we can easily and quickly get the loss factors for any single slack by use of the proposed loss sensitivity formula. Therefore, a large amount of the computations are avoided whenever the loss factors are needed for a single slack in the real-time energy markets. For example, a practical system with 25,000 buses, the CPU time of computing loss factors using the traditional power flow calculation is about 60 seconds, but less than 0.1 second

TABLE 3.4 Test Results and Comparison of Loss Sensitivity Calculation (Case 4: All Units on AGC Except the Units in Area 2)

| Station, Generator | Area No. | AGC Unit | Loss Sensitivity Distributed Generation Slack | Loss Sensitivity Distributed Load Slack |
|---|---|---|---|---|
| DOUGLAS, G2 | 1 | YES | 0.0152 | 0.0170 |
| DOUGLAS, G1 | 1 | YES | 0.0122 | 0.0140 |
| DOUGLAS, CT1 | 1 | YES | 0.0100 | 0.0118 |
| DOUGLAS, CT2 | 1 | YES | 0.0100 | 0.0118 |
| DOUGLAS, ST | 1 | YES | 0.0099 | 0.0116 |
| HEARN, G1 | 1 | YES | −0.0167 | −0.0146 |
| HEARN, G2 | 1 | YES | −0.0167 | −0.0146 |
| LAKEVIEW, G1 | 1 | YES | −0.0191 | −0.0170 |
| BVILLE, 1 | 2 | NO | −0.0210 | −0.0042 |
| WVILLE, 1 | 2 | NO | −0.0193 | −0.0025 |
| CHENAUX, 1 | 3 | YES | −0.0089 | −0.0089 |
| CHEALLS, 1 | 3 | YES | 0.0212 | 0.0212 |
| CHEALLS, 2 | 3 | YES | 0.0212 | 0.0212 |
| HOLDEN, 1 | 3 | YES | 0.0010 | 0.0010 |
| NANTCOKE, 1 | 3 | YES | −0.0122 | −0.0122 |

if the proposed method is used. This is a huge time saving in the real-time energy markets.

In order to verify the correctness of the loss sensitivity equation (3.24), the loss factors are computed and compared using the traditional power flow calculation. The results and comparison are shown in Figures 3.5 and 3.6 as well as Tables 3.7 and 3.8, in which column 3 is the set of results from the power flow calculation, and column 4 is the set of results from equation (3.24). Table 3.7 shows the comparison of loss factor results for single slack bus at HOLDEN-1. Table 3.8 shows the comparison of loss factor results for single slack bus at DOUGLAS-ST.

The difference or error of the results between the proposed method and power flow method is obtained from the following equation.

$$|\text{Error\%}| = \left| \frac{\text{LF}_{\text{PM}}(i) - \text{LF}_{\text{PF}}(i)}{\text{LF}_{\text{PF}}(i)} \times 100\% \right| \quad i \in n \qquad (3.87)$$

where

Error %: the percentage of the computation error for the proposed formula.

$\text{LF}_{\text{PM}}$: the loss factor computed from the proposed method.

$\text{LF}_{\text{PF}}$: the loss factor obtained using the traditional power flow calculation.

It can be seen from Tables 3.7 and 3.8 that the loss sensitivity results from the two methods are very close. The maximum error is less than 0.6%.

**TABLE 3.5    Test Results and Comparison of Loss Sensitivity Calculation (Case 5: All Units on AGC Except Unit 3 Under Station HOLDEN in Area 3)**

| Station, Generator | Area No. | AGC Unit | Loss Sensitivity Distributed Generation Slack | Loss Sensitivity Distributed Load Slack |
|---|---|---|---|---|
| DOUGLAS, G2 | 1 | YES | 0.0151 | 0.0170 |
| DOUGLAS, G1 | 1 | YES | 0.0121 | 0.0140 |
| DOUGLAS, CT1 | 1 | YES | 0.0099 | 0.0118 |
| DOUGLAS, CT2 | 1 | YES | 0.0099 | 0.0118 |
| DOUGLAS, ST | 1 | YES | 0.0097 | 0.0116 |
| HEARN, G1 | 1 | YES | −0.0165 | −0.0146 |
| HEARN, G2 | 1 | YES | −0.0165 | −0.0146 |
| LAKEVIEW, G1 | 1 | YES | −0.0188 | −0.0170 |
| BVILLE, 1 | 2 | YES | −0.0010 | −0.0042 |
| WVILLE, 1 | 2 | YES | 0.0007 | −0.0025 |
| CHENAUX, 1 | 3 | YES | −0.0085 | −0.0089 |
| CHEALLS, 1 | 3 | YES | 0.0216 | 0.0212 |
| CHEALLS, 2 | 3 | YES | 0.0216 | 0.0212 |
| HOLDEN, 1 | 3 | NO | 0.0014 | 0.0010 |
| NANTCOKE, 1 | 3 | YES | −0.0118 | −0.0122 |

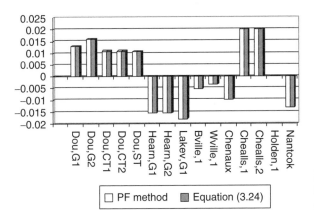

Figure 3.5    Comparison of loss factor results for single slack bus at HOLDEN-1.

## 3.7.2    Sample Computation for Constrained Shift Factors

Tables 3.9–3.12 are the results of the detected constraint and the corresponding shift factors. The results for the constraint branch T525 at Station CHENAUX are listed in Table 3.9.

In Table 3.10, column 1 is the name of station and units. Column 2 is the area number that the unit belongs to. Column 3 is the AGC status of the unit. Column 4 is

**TABLE 3.6   Test Results of Loss Sensitivity Calculation (Distributed Slack Vs Single Slack)**

| Station, Generator | AGC Unit | Loss Sensitivity Distributed Slack | Loss Sensitivity Single Slack, HOLDEN 1 | Loss Sensitivity Single Slack, Douglas ST |
|---|---|---|---|---|
| DOUGLAS, G2 | YES | 0.017000 | 0.016016 | 0.005463 |
| DOUGLAS, G1 | YES | 0.014000 | 0.013013 | 0.002428 |
| DOUGLAS, CT1 | YES | 0.011800 | 0.010811 | 0.000202 |
| DOUGLAS, CT2 | YES | 0.011800 | 0.010811 | 0.000202 |
| DOUGLAS, ST | YES | 0.011600 | 0.010611 | 0.000000 |
| HEARN, G1 | YES | −0.014600 | −0.015616 | −0.026507 |
| HEARN, G2 | YES | −0.014600 | −0.015616 | −0.026507 |
| LAKEVIEW, G1 | YES | −0.017000 | −0.018018 | −0.028936 |
| BVILLE, 1 | YES | −0.004200 | −0.005205 | −0.015985 |
| WVILLE, 1 | YES | −0.002500 | −0.003504 | −0.014265 |
| CHENAUX, 1 | YES | −0.008900 | −0.009910 | −0.020741 |
| CHEALLS, 1 | YES | 0.021200 | 0.020220 | 0.009713 |
| CHEALLS, 2 | YES | 0.021200 | 0.020220 | 0.009713 |
| HOLDEN, 1 | YES | 0.001000 | 0.000000 | −0.010724 |
| NANTCOKE, 1 | YES | −0.012200 | −0.013213 | −0.024079 |

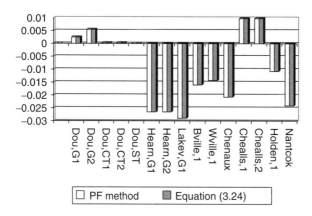

Figure 3.6   Comparison of loss factor results for single slack bus at DOUGLAS-ST.

the unit participation factors. Column 5 is the set of the shift factors of the constraint T525 with respect to the units for the EMS-based reference at station DOUGLAS.

It is noted that all the shift factors are zero for the units in area 1 for the EMS-based reference as the reference is located in area 1 and all units in area 1 are close to the reference unit. If the market-based slack is selected, the shift factors for the market-based reference can be easily obtained from equations (3.57) and (3.58).

**TABLE 3.7    Comparison of Loss Sensitivity Calculation Results for Single Slack Bus at HOLDEN-1 (The Proposed Method Vs Power Flow Method)**

| Station, Generator | AGC Unit | Loss Sensitivity, HOLDEN 1-PF Method | Loss Sensitivity, HOLDEN 1-Equation (3.24) | \|Error %\| |
|---|---|---|---|---|
| DOUGLAS, G2 | YES | 0.016029 | 0.016016 | 0.08110 |
| DOUGLAS, G1 | YES | 0.013053 | 0.013013 | 0.30644 |
| DOUGLAS, CT1 | YES | 0.010817 | 0.010811 | 0.05547 |
| DOUGLAS, CT2 | YES | 0.010817 | 0.010811 | 0.05547 |
| DOUGLAS, ST | YES | 0.010621 | 0.010611 | 0.09415 |
| HEARN, G1 | YES | −0.015630 | −0.015616 | 0.08957 |
| HEARN, G2 | YES | −0.015630 | −0.015616 | 0.08957 |
| LAKEVIEW, G1 | YES | −0.018110 | −0.018018 | 0.50801 |
| BVILLE, 1 | YES | −0.005220 | −0.005205 | 0.23002 |
| WVILLE, 1 | YES | −0.003500 | −0.003504 | 0.02855 |
| CHENAUX, 1 | YES | −0.009920 | −0.009910 | 0.11088 |
| CHEALLS, 1 | YES | 0.020247 | 0.020220 | 0.13335 |
| CHEALLS, 2 | YES | 0.020247 | 0.020220 | 0.13335 |
| HOLDEN, 1 | YES | 0.000000 | 0.000000 | 0.00000 |
| NANTCOKE, 1 | YES | −0.013240 | −0.013213 | 0.20393 |

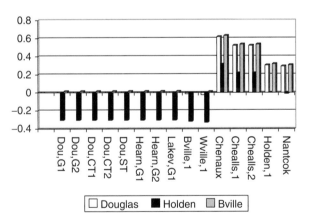

Figure 3.7    The shift factors with different references.

Table 3.11 shows the shift factors of the constraint T525 with respect to the units for the market-based reference at the location of HOLDEN 1 and BVILLE, respectively. The relationships of the shift factors to different references are also shown in Figure 3.7.

**TABLE 3.8 Comparison of Loss Sensitivity Calculation Results for Single Slack Bus at Douglas-ST (The Proposed Method Vs Power Flow Method)**

| Station, Generator | AGC Unit | Loss Sensitivity, Douglas ST-PF Method | Loss Sensitivity, Douglas ST-Equation (3.24) | \|Error %\| |
|---|---|---|---|---|
| DOUGLAS, G2 | YES | 0.005467 | 0.005463 | 0.07317 |
| DOUGLAS, G1 | YES | 0.002421 | 0.002428 | 0.28914 |
| DOUGLAS, CT1 | YES | 0.000202 | 0.000202 | 0.14829 |
| DOUGLAS, CT2 | YES | 0.000202 | 0.000202 | 0.14829 |
| DOUGLAS, ST | YES | 0.000000 | 0.000000 | 0.00000 |
| HEARN, G1 | YES | −0.026530 | −0.026507 | 0.08669 |
| HEARN, G2 | YES | −0.026530 | −0.026507 | 0.08669 |
| LAKEVIEW, G1 | YES | −0.028950 | −0.028936 | 0.04836 |
| BVILLE, 1 | YES | −0.016000 | −0.015985 | 0.09999 |
| WVILLE, 1 | YES | −0.014280 | −0.014265 | 0.10504 |
| CHENAUX, 1 | YES | −0.020770 | −0.020741 | 0.13962 |
| CHEALLS, 1 | YES | 0.009714 | 0.009713 | 0.01029 |
| CHEALLS, 2 | YES | 0.009714 | 0.009713 | 0.01029 |
| HOLDEN, 1 | YES | −0.010730 | −0.010724 | 0.07454 |
| NANTCOKE, 1 | YES | −0.024090 | −0.024079 | 0.02491 |

**TABLE 3.9 Example of the Active Constraint (Branch T525 at Station CHENAUX)**

| Constraint Name | Rating (MVA) | Actual Flow (MVA) | Constraint Deviation | Percent of Violation |
|---|---|---|---|---|
| Branch T525 | 1171.4 | 1542.7 | 371.3 | 131.7 |

Table 3.12 shows the area-based shift sensitivity factors of the constraint T525, which are computed on the basis of unit shift factors and participation factors within the area. If the unit participation factors change, the value of the area based sensitivity change.

Table 3.13 shows the sensitivity factors of the transfer path with respect to the constraint T525. There are four types transfer paths:

(1) Transfer type 1—Area-Area: Both POR and POD (or SOURCE and SINK) are areas.

(2) Transfer type 2—Single point: Both POR and POD (or SOURCE and SINK) are single injection nodes.

(3) Transfer type 3—Point-Area: The POR (SOURCE) is a single injection node and POD (SINK) is an area.

**TABLE 3.10  Test Results of Shift Factors for the Active Constraint T525 at EMS Reference (Station Douglas)**

| Station, Generator | Area No. | Unit in Serve | Unit Participation Factor | Shift Factors on EMS Reference at Station DOUGLAS |
|---|---|---|---|---|
| DOUGLAS, G2 | 1 | YES | 1.5 | 0.000000 |
| DOUGLAS, G1 | 1 | YES | 1.8 | 0.000000 |
| DOUGLAS, CT1 | 1 | YES | 1.2 | 0.000000 |
| DOUGLAS, CT2 | 1 | YES | 1.6 | 0.000000 |
| DOUGLAS, ST | 1 | YES | 0.9 | 0.000000 |
| HEARN, G1 | 1 | YES | 0.5 | 0.000000 |
| HEARN, G2 | 1 | YES | 0.8 | 0.000000 |
| LAKEVIEW, G1 | 1 | YES | 1.1 | 0.000000 |
| BVILLE, 1 | 2 | YES | 1.2 | −0.013650 |
| WVILLE, 1 | 2 | YES | 1.3 | −0.024336 |
| CHENAUX, 1 | 3 | YES | 1.7 | 0.617887 |
| CHEALLS, 1 | 3 | YES | 0.6 | 0.521795 |
| CHEALLS, 2 | 3 | YES | 1.9 | 0.521795 |
| HOLDEN, 1 | 3 | YES | 2.2 | 0.304269 |
| NANTCOKE, 1 | 3 | YES | 0.7 | 0.291815 |

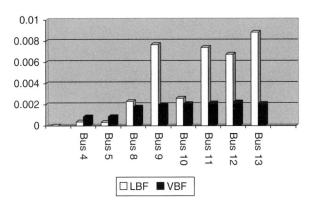

Figure 3.8  Voltage sensitivity analysis of 14-bus system.

**(4)** Transfer type 4—Area-Point: The POR (SOURCE) is an area and POD (SINK) is a single injection node.

It is noted from Table 3.13 that the sensitivity of the transfer path will be the same no matter which reference is used.

### 3.7.3  Sample Computation for Voltage Sensitivity Analysis

Table 3.14 and Figure 3.8 show the major VAR support sites as well as the corresponding benefit factors *LBF* and *VBF* for the IEEE 14-bus system. It can be observed from

**TABLE 3.11 Test Results of Shift Factors for the Active Constraint T525 at Different Market References**

| Station, Generator | Area No. | Unit in Serve | Shift Factors on Market Reference at Station HOLDEN | Shift Factors on Market Reference at Station BVILLE |
|---|---|---|---|---|
| DOUGLAS, G2 | 1 | YES | −0.304269 | 0.013650 |
| DOUGLAS, G1 | 1 | YES | −0.304269 | 0.013650 |
| DOUGLAS, CT1 | 1 | YES | −0.304269 | 0.013650 |
| DOUGLAS, CT2 | 1 | YES | −0.304269 | 0.013650 |
| DOUGLAS, ST | 1 | YES | −0.304269 | 0.013650 |
| HEARN, G1 | 1 | YES | −0.304269 | 0.013650 |
| HEARN, G2 | 1 | YES | −0.304269 | 0.013650 |
| LAKEVIEW, G1 | 1 | YES | −0.304269 | 0.013650 |
| BVILLE, 1 | 2 | YES | −0.317919 | 0.000000 |
| WVILLE, 1 | 2 | YES | −0.328605 | 0.010686 |
| CHENAUX, 1 | 3 | YES | 0.313618 | 0.631537 |
| CHEALLS, 1 | 3 | YES | 0.217526 | 0.535445 |
| CHEALLS, 2 | 3 | YES | 0.217526 | 0.535445 |
| HOLDEN, 1 | 3 | YES | 0.000000 | 0.317946 |
| NANTCOKE, 1 | 3 | YES | −0.012454 | 0.305465 |

Figure 3.8 that buses 9, 11, 12, and 13 have relatively big sensitivity values. The VAR supports at these locations will have bigger benefits than other locations in the IEEE 14-bus system.

## 3.8 CONCLUSION

This chapter introduces several approaches to compute the sensitivities in the practical transmission network and energy markets. The analysis and implementation details of load sensitivity, voltage sensitivity, generator constraint shift factor, and area-based constraint shift factor are presented. The chapter also comprehensively discusses how to compute the sensitivities under the different references, as well as how to convert the sensitivities based on the EMS system reference into the ones based on the market system reference. These sensitivities' calculations can be used to determine whether the on-line capacity as indicated in the resource plan is located in the right place on the network to serve the forecast demand. This chapter will be especially useful for power engineers because sensitivity analysis has already become daily routine in the power industry. The researchers, students and power engineers will also have the big picture on power system sensitivity analysis.

**TABLE 3.12  Test Results of Area Based Sensitivity for the Active Constraint T525 at Different References**

| Area Name | Area No. | Sensitivities on EMS Reference at Station DOUGLAS | Sensitivities on Market Reference at Station HOLDEN | Sensitivities on Market Reference at Station BVILLE |
|---|---|---|---|---|
| EAST | 1 | 0.000000 | −0.304269 | 0.013650 |
| WEST | 2 | −0.019207 | −0.323499 | −0.005557 |
| ECAR | 3 | 0.454726 | 0.150458 | 0.468385 |

**TABLE 3.13  Test Results of Sensitivity for Transfer Path for the Active Constraint T525 at Different References**

| Transfer Path | Path Type | Sensitivities on EMS Reference at Station DOUGLAS | Sensitivities on Market Reference at Station HOLDEN | Sensitivities on Market Reference at Station BVILLE |
|---|---|---|---|---|
| ECAR-WEST | Area-area | 0.473933 | 0.473950 | 0.473940 |
| WEST-EAST | Area-area | −0.019207 | −0.019230 | −0.019207 |
| ECAR-EAST | Area-area | 0.454726 | 0.454727 | 0.454735 |
| BV1-DOUGG1 | Single point | −0.013650 | −0.013650 | −0.013650 |
| WV1-DOUGG1 | Single point | −0.024336 | −0.024336 | −0.024336 |
| CX1-DOUGG1 | Single point | 0.617887 | 0.617887 | 0.617887 |
| CS1-DOUGG1 | Single point | 0.521795 | 0.521795 | 0.521795 |
| CS2-DOUGG1 | Single point | 0.521795 | 0.521795 | 0.521795 |
| HD1-DOUGG1 | Single point | 0.304269 | 0.304269 | 0.304269 |
| NK1-DOUGG1 | Single point | 0.291815 | 0.291815 | 0.291815 |
| BV1-WV1 | Single point | 0.010686 | 0.010686 | 0.010686 |
| CX1-CS1 | Single point | 0.096092 | 0.096092 | 0.096092 |
| HD1-NK1 | Single point | 0.012454 | 0.012454 | 0.012454 |
| HD1-BV1 | Single point | 0.317919 | 0.317919 | 0.317919 |
| HD1-WV1 | Single point | 0.328605 | 0.328605 | 0.328605 |
| BV1-EAST | Point-area | −0.013650 | −0.013650 | −0.013650 |
| HD1-EAST | Point-area | 0.304269 | 0.304269 | 0.304269 |
| HD1-WEST | Point-area | 0.323476 | 0.323476 | 0.323476 |
| WV1-ECAR | Point-area | −0.479062 | −0.479062 | −0.479062 |
| EAST-WV1 | Area-point | 0.024336 | 0.024336 | 0.024336 |
| ECAR-BV1 | Area-point | 0.468376 | 0.468376 | 0.468376 |
| WEST-DOUGG1 | Area-point | −0.019207 | −0.019207 | −0.019207 |

TABLE 3.14 Voltage Sensitivity Analysis Results for IEEE 14 Bus Systems

| VAR Support Site | $\text{LBF}_i$ | $\text{VBF}_i$ |
|---|---|---|
| Bus 4 | 0.000376 | 0.000855 |
| Bus 5 | 0.000337 | 0.000884 |
| Bus 8 | 0.002309 | 0.001775 |
| Bus 9 | 0.007674 | 0.001989 |
| Bus 10 | 0.002618 | 0.002097 |
| Bus 11 | 0.007407 | 0.002175 |
| Bus 12 | 0.006757 | 0.002268 |
| Bus 13 | 0.008840 | 0.002122 |

# PROBLEMS AND EXERCISES

1. What is the LODF?

2. What is the OTDF?

3. What does loss sensitivity mean?

4. What is the constraint shift factor?

5. What is the load distribution reference?

6. In practical application, why is load distribution reference generally used, rather than generation distribution reference?

7. What are VBF and LBF?

8. How are the sensitivities for a given transfer path computed?

9. State the role of SFT in the energy market.

10. State "True" or "False"

   10.1 The change of a unit power output will change the value of the sensitivities.

   10.2 The matrix B' is used to compute constraint shift factor sensitivities.

   10.3 The matrix B'' is used to compute loss sensitivities.

   10.4 The values of the sensitivities will be the same for different references if the network topology is unchanged.

   10.5 The constraint shift factor of the slack bus is zero if a single slack bus is selected.

   10.6 All sensitivities with respect to bus injections are not greater than 1.0

   10.7 A source/sink may be a single unit, single load, an area, or group of nodes.

# REFERENCES

1. Dy-Liyacco TE. Control centers are here to stay. IEEE Comput. Appl. Pow 2002;15(4):18–23.
2. Winser N. FERC's standard market design: the ITC perspective, 2002 IEEE PES summer meeting, Chicago, IL. July 22–26, 2002.

3. Ott A. Experience with PJM market operation, system design, and implementation. IEEE Trans. on Power Syst. 2003;18(2):528–534.

4. Kathan D. FERC's standard market design proposal, 2003 ACEEE/CEE National Symposium on Market Transformation, Washington, DC, April 15, 2003.

5. Zhu JZ, Hwang D, Sadjadpour A. An approach of generation scheduling in energy markets, POWER-CON 2006, Chongqing, October 22–28, 2006.

6. Kirchamayer LK. *Economic Operation of Power Systems*. New York: Wiley; 1958.

7. Dommel HW, Tinney WF. Optimal power flow solutions. IEEE Trans. on PAS 1968;PAS-87(10): 1866–1876.

8. Ilic M, Galiana FD, Fink L. *Power Systems Restructuring: Engineering and Economics*. Norwell, MA: Kluwer; 1998.

9. Kirschen D, Allan R, Strbac G. Contributions of individual generators to loads and flows. IEEE Trans. Power Syst. 1997;12(1):52–60.

10. Schweppe F, Caramanis M, Tabors R, Bohn R. *Spot Pricing of Electricity*. Norwell, MA: Kluwer; 1988.

11. Conejo AJ, Galiana FD, Kochar I. Z-Bus loss allocation. IEEE Trans. Power Syst. 2001;16(1): 105–110.

12. Galiana FD, Conjeo AJ, Korkar I. Incremental transmission loss allocation under pool dispatch. IEEE Trans. Power Syst. 2002;17(1):26–33.

13. Elgerd OI. *Electric Energy Systems Theory: An Introduction*. New York: McGraw-Hill; 1982.

14. Zhu JZ, Hwang D, Sadjadpour A. Loss sensitivity calculation and analysis, in Proceeding 2003 IEEE General Meeting, Toronto, July 13–18, 2003.

15. Zhu JZ, Hwang D, Sadjadpour A. Real time loss sensitivity calculation in power systems operation. Electr. Pow. Syst. Res. 2005;73(1):53–60.

16. Zhu JZ, Hwang D, Sadjadpour A. The implementation of alleviating overload in energy markets, IEEE/PES 2007 general meeting, June 24–28, 2007.

17. Zhu JZ, Hwang D, Sadjadpour A. Calculation of several sensitivity in real time transmission network and energy markets, Power-Grid Europe 2007, Spain, June 23–26, 2007.

18. Wood AJ, Wollenberg BF. *Power Generation, Operation, and Control*. New York: 2nd ed.; 1996.

19. Zhu JZ, Irving MR. Combined active and reactive dispatch with multiple objectives using an analytic hierarchical process. IEE Proc. C, 1996;143(4):344–352.

20. Zhu JZ, Momoh JA. Optimal VAR pricing and VAR placement using analytic hierarchy process. Electr. Pow. Syst. Res. 1998;48(1):11–17.

21. Mansour MO, Abdel-Rahman TM. Non-linear VAR optimization using decomposition and coordination. IEEE Trans. PAS, 1984;103:246–255.

22. Dandachi NH, Rawlins MJ, Alsac O, Stott B. OPF for reactive pricing studies on the NGC system. IEEE Power Industry Computer Applications Conference, PICA'95, Utah, May 1995, pp. 11–17.

23. Alsac O, Stott B. Optimal power flow with steady-state security. IEEE Trans., PAS, 1974;93:745–751.

24. Momoh JA, Zhu JZ. Improved interior point method for OPF problems. IEEE Trans. on Power Syst. 1999;14(3):1114–1120.

25. Begovic M, Phadke AG. Control of voltage stability using sensitivity analysis. IEEE Trans. on Power Syst. 1992;7:114–123.

# CHAPTER 4

# CLASSIC ECONOMIC DISPATCH

This chapter first introduces the input–output characteristic of a power-generating unit as well as the corresponding practical calculation method, and then presents several well-known optimization methods to solve the classic economic dispatch problem. Finally, the applications of the latest methods such as neural network and genetic algorithm to classic economic dispatch (ED) are analyzed.

## 4.1 INTRODUCTION

The aim of real power economic dispatch (ED) is to make the generator's fuel consumption or the operating cost of the whole system minimal by determining the power output of each generating unit under the constraint condition of the system load demands. This is also called the classic economic dispatch, in which line security constraints are neglected [1]. The fundamental of the ED problem is the set of input–output characteristics of a power generating unit.

## 4.2 INPUT–OUTPUT CHARACTERISTICS OF GENERATOR UNITS

### 4.2.1 Input–Output Characteristic of Thermal Units

For thermal units, we call the input–output characteristic the *generating unit fuel consumption function*, or *operating cost function*. The unit of the generator fuel consumption function is Btu per hour heat input to the unit (or MBtu/h). The fuel cost rate times Btu/h is the $ per hour ($/h) input to the unit for fuel. The output of the generating unit will be denoted by $P_G$, the megawatt net power output of the unit.

In addition to fuel consumption cost, the operating cost of a unit includes labor cost, maintenance cost, and fuel transportation cost. It is difficult to express these costs directly as a function of the output of the unit, so these costs are included as a fixed portion of the operating cost.

The thermal unit system generally consists of the boiler, the steam turbine, and the generator. The input of the boiler is fuel and the output is the volume of steam. The relationship between the input and output can be expressed as a convex curve.

*Optimization of Power System Operation*, Second Edition. Jizhong Zhu.
© 2015 The Institute of Electrical and Electronics Engineers, Inc. Published 2015 by John Wiley & Sons, Inc.

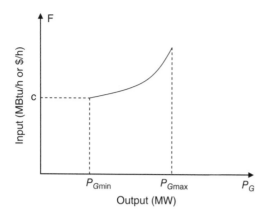

Figure 4.1 Input-output characteristic of the generating unit.

The input of the turbine-generator unit is the volume of steam and the output is the electrical power. A typical boiler–turbine-generator unit consists of a single boiler that generates steam to drive a single turbine-generator set. The input–output characteristic of the whole generating unit system can be obtained by combining directly the input–output characteristic of the boiler and the input–output characteristic of the turbine-generator unit. It is a convex curve, which is shown in Figure 4.1.

It can be observed from the input–output characteristic of the generating unit that the power output is limited by the minimal and maximal capacities of the generating unit, that is,

$$P_{Gmin} \leq P_G \leq P_{Gmax} \qquad (4.1)$$

The minimal power output is determined by the technical condition or other factors of the boiler or turbine. Generally, the minimum load at which a unit can operate is influenced more by the steam generator and the regenerative cycle than by the turbine. The only critical parameters for the turbine are the shell and rotor metal differential temperatures, exhaust hood temperature, and rotor and shell expansion. Minimum load limitations of the boiler are generally caused by fuel combustion stability, and the values, which will differ with different types of boiler and fuel, are about 25–50% of the design capacity. Minimum load limitations of the turbine–generator unit are caused by inherent steam generator design constraints, which are generally about 10–15%. The maximal power output of the generating unit is determined by the design capacity or rate capacity of the boiler, turbine, or generator.

Generally, the input–output characteristic of the generating unit is nonlinear. The widely used input–output characteristic of the generating unit is a quadratic function, that is,

$$F = aP_G^2 + bP_G + c \qquad (4.2)$$

where $a$, $b$, and $c$ are the coefficients of the input–output characteristic. The constant $c$ is equivalent to the fuel consumption of the generating unit operation without power output, which is shown in Figure 4.1.

## 4.2.2 Calculation of Input–Output Characteristic Parameters

The parameters of the input–output characteristic of the generating unit may be determined by the following approaches [2]:

1. based on the experiments of the generating unit efficiency;
2. based on the historic records of the generating unit operation;
3. based on the design data of the generating unit provided by manufacturer.

In the practical power systems, we can easily obtain the fuel statistical data and power output statistic data. Through analyzing and computing some data set $(F_k, P_k)$, we can determine the shape of the input–output characteristic and the corresponding parameters. For example, if the quadratic curve is the best match according to the statistical data, we can use the least square method to compute the parameters. The calculation procedures are as follows.

Let $(F_k, P_k)$ be obtained from the statistical data, where $k = 1, 2, \ldots \ldots n$, and the fuel curve is a quadratic function. To determine the coefficients $a$, $b$, and $c$, compute the following error for each data pair $(F_k, P_k)$:

$$\Delta F_k = (aP_k^2 + bP_k + c) - F_k \tag{4.3}$$

According to the principle of least squares, we form the following objective function and make it minimal, that is,

$$J = (\Delta F_k)^2 = \sum_{k=1}^{n} (aP_k^2 + bP_k + c - F_k)^2 \tag{4.4}$$

We will get the necessary conditions for an extreme value of the objective function when we take the first derivative of the above function $J$ with respect to each of the independent variables $a$, $b$, and $c$, and set the derivatives equal to zero:

$$\frac{\partial J}{\partial a} = \sum_{k=1}^{n} 2P_k^2(aP_k^2 + bP_k + c - F_k) = 0 \tag{4.5}$$

$$\frac{\partial J}{\partial b} = \sum_{k=1}^{n} 2P_k(aP_k^2 + bP_k + c - F_k) = 0 \tag{4.6}$$

$$\frac{\partial J}{\partial c} = \sum_{k=1}^{n} 2(aP_k^2 + bP_k + c - F_k) = 0 \tag{4.7}$$

From equations (4.5)–(4.7), we get

$$\left( \sum_{k=1}^{n} P_k^2 \right) a + \left( \sum_{k=1}^{n} P_k \right) b + nc = \sum_{k=1}^{n} F_k \tag{4.8}$$

$$\left(\sum_{k=1}^{n} P_k^3\right) a + \left(\sum_{k=1}^{n} P_k^2\right) b + \left(\sum_{k=1}^{n} P_k\right) c = \sum_{k=1}^{n} (F_k P_k) \qquad (4.9)$$

$$\left(\sum_{k=1}^{n} P_k^4\right) a + \left(\sum_{k=1}^{n} P_k^3\right) b + \left(\sum_{k=1}^{n} P_k^2\right) c = \sum_{k=1}^{n} (F_k P_k^2) \qquad (4.10)$$

The coefficients $a$, $b$, and $c$ can be obtained by solving the equations (4.8)–(4.10).

***Example 4.1:*** We collected some statistical data for a generating unit in one power plant. The capacity limits of the generator were

$$150 \leq P_G \leq 200$$

Four sample data of unit consume fuel were selected, namely, 0.405, 0.379, 0.368, and 0.399 Btu/MW·h, which correspond to power output 150, 170, 185, and 200 MW, respectively (Figure 4.2). The corresponding fuel consumptions are computed and listed in Table 4.1.

From Table 4.1, we get

$$\sum_{k=1}^{n} P_k = 150 + 170 + 185 + 200 = 705$$

$$\sum_{k=1}^{n} P_k^2 = 150^2 + 170^2 + 185^2 + 200^2 = 1.256 \times 10^5$$

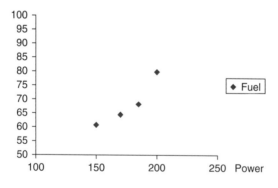

Figure 4.2 Four statistic data points.

TABLE 4.1 Four Sample Data for a Generating Unit

| Sample Data | $K = 1$ | $K = 2$ | $K = 3$ | $K = 4$ |
|---|---|---|---|---|
| Unit consume fuel (Btu/MW·h) | 0.405 | 0.379 | 0.368 | 0.399 |
| Power output (MW) | 150.0 | 170.0 | 185.0 | 200.0 |
| Consume fuel (Btu/h) | 60.75 | 64.43 | 68.08 | 79.80 |

$$\sum_{k=1}^{n} P_k{}^3 = 150^3 + 170^3 + 185^3 + 200^3 = 2.2619 \times 10^7$$

$$\sum_{k=1}^{n} P_k{}^4 = 150^4 + 170^4 + 185^4 + 200^4 = 4.112 \times 10^9$$

$$\sum_{k=1}^{n} F_k = 60.75 + 64.43 + 68.08 + 79.80 = 273.06$$

$$\sum_{k=1}^{n} F_k P_k = 60.75 \times 150 + 64.43 \times 170 + 68.08 \times 185 + 79.80 \times 200$$

$$= 4.86 \times 10^4$$

$$\sum_{k=1}^{n} F_k P_k{}^2 = 60.75 \times 150^2 + 64.43 \times 170^2 + 68.08 \times 185^2 + 79.80 \times 200^2$$

$$= 8.75 \times 10^6$$

From equations (4.8)–(4.10), we get

$$1.256 \times 10^5 a + 705b + 4c = 273.06$$

$$2.2619 \times 10^7 a + 1.26 \times 10^5 b + 705c = 4.86 \times 10^4$$

$$4.112 \times 10^9 a + 2.26 \times 10^7 b + 1.26 \times 10^5 c = 8.75 \times 10^6$$

Solving these equations, we get the coefficients of the fuel consumption function of the generating unit:

$$a = 0.0009, b = 0.0457, c = 31.9$$

The obtained quadratic function for fuel consumption is as follows:

$$F = 0.0009 P_G{}^2 + 0.0457 P_G + 31.9$$

The simulated input–output curve is shown in Figure 4.3. It is noted that the accuracy of calculation will be increased if more data samples are used.

### 4.2.3  Input–Output Characteristic of Hydroelectric Units

The input–output characteristic of the hydroelectric unit is similar to that of the thermal unit, but the input, which is in terms of volume of water per unit time, is different. The unit of water volume is in $m^3/h$. The output is the same, that is, electric power. Figure 4.4 shows a typical input–output curve of a hydroelectric unit where the net hydraulic head is constant. This characteristic shows an almost linear curve of input

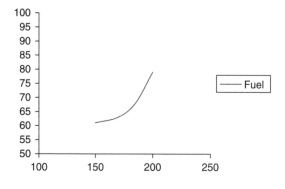

Figure 4.3 Simulated input-output curve.

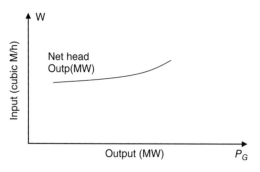

Figure 4.4 Hydroelectric unit input-output curve.

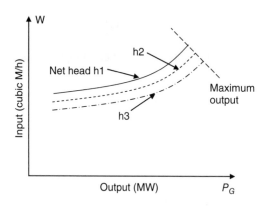

Figure 4.5 Hydroelectric unit input-output curve with variable water head.

water volume requirements per unit time as a function of power output as the power output increases from minimum to rated load. Above this point corresponding to the rated load, the water volume requirements increase as the efficiency of the unit falls off.

Figure 4.5 shows the input–output curve of a hydroelectric plant with variable head. This type of characteristic occurs whenever the variation in the storage pond and/or afterbay elevations is a fairly large percentage of the overall net hydraulic head.

## 4.3 THERMAL SYSTEM ECONOMIC DISPATCH NEGLECTING NETWORK LOSSES

### 4.3.1 Principle of Equal Incremental Rate

Given a system that consists of two generators connected to a single bus serving a received electrical load $P_D$, the input–output characteristic of the two generating units are $F_1(P_{G1})$ and $F_2(P_{G2})$, respectively. The total fuel consumption of the system $F$ is the sum of the fuel consumptions of the two generating units. Assuming there is no power output limitation for both generators, the essential constraint on the operation of this system is that the sum of the output powers must equal the load demand. The economic power dispatch problem of the system, which is to minimize $F$ under the above-mentioned constraint, can be expressed as

$$\min F = F_1(P_{G1}) + F_2(P_{G2}) \tag{4.11}$$

s.t.

$$P_{G1} + P_{G2} = P_D \tag{4.12}$$

According to the principle of equal incremental rate [1], the total fuel consumption $F$ will be minimal if the incremental fuel rates of two generators are equal, that is,

$$\frac{dF_1}{dP_{G1}} = \frac{dF_2}{dP_{G2}} = \lambda \tag{4.13}$$

where $\frac{dF_i}{dP_{Gi}}$ is the incremental fuel rate of generating unit $i$, which corresponds to the slope of the input–output curve of the generating unit.

If the two generators operate under different incremental fuel rates, and

$$\frac{dF_1}{dP_{G1}} > \frac{dF_2}{dP_{G2}} \tag{4.13}$$

the total output power remains the same. If generator 1 reduces output power by $\Delta P$, generator 2 will increase output power by $\Delta P$. Then generator 1 will reduce fuel consumption by $\frac{dF_1}{dP_{G1}} \Delta P$, and generator 2 will increase fuel consumption by $\frac{dF_2}{dP_{G2}} \Delta P$. The total savings in fuel consumption will be

$$\Delta F = \frac{dF_1}{dP_{G1}} \Delta P - \frac{dF_2}{dP_{G2}} \Delta P = \left( \frac{dF_1}{dP_{G1}} - \frac{dF_2}{dP_{G2}} \right) \Delta P > 0 \tag{4.14}$$

It can be observed from equation (4.14) that $\Delta F$ will be zero when $\frac{dF_1}{dP_{G1}} = \frac{dF_2}{dP_{G2}}$, that is, the incremental fuel rates of the two generators are equal.

***Example 4.2:*** The input–output characteristics of two generating units are as follows:

$$F_1 = 0.0008P_{G1}{}^2 + 0.2P_{G1} + 5 \ \text{Btu/h}$$

$$F_2 = 0.0005P_{G2}{}^2 + 0.3P_{G2} + 4 \ \text{Btu/h}$$

We wish to determine the economic operation point for these two units when delivering a total of 500 MW power demand.

First of all, we can obtain the incremental fuel rate of the two generating units as follows:

$$\lambda_1 = \frac{dF_1}{dP_{G1}} = 0.0016P_{G1} + 0.2$$

$$\lambda_1 = \frac{dF_2}{dP_{G2}} = 0.001P_{G2} + 0.3$$

According to the principle of equal incremental rate (4.13), we have

$$\lambda_1 = \lambda_2$$

that is,

$$0.0016P_{G1} + 0.2 = 0.001P_{G2} + 0.3$$

or

$$1.6P_{G1} - P_{G2} = 100$$

Given the system load is 500 MW, then

$$P_{G1} + P_{G2} = 500$$

Solving the above two equations for $P_{G1}, P_{G2}$, we get

$$P_{G1} = 230.77 \ \text{MW}$$

$$P_{G2} = 269.23 \ \text{MW}$$

***Example 4.3:*** Suppose the input–output characteristics of the two generating units are slightly different from that in Example 4.2, given by the following:

$$F_1 = 0.0008P_{G1}{}^2 + 0.02P_{G1} + 5 \ \text{Btu/h}$$

$$F_2 = 0.0005P_{G2}{}^2 + 0.03P_{G2} + 4 \ \text{Btu/h}$$

We still wish to determine the economic operation point for these two units when delivering a total of 500 MW power demand.

First of all, we can obtain the incremental fuel rate of the two generating units as follows:

$$\lambda_1 = \frac{dF_1}{dP_{G1}} = 0.0016P_{G1} + 0.02$$

$$\lambda_2 = \frac{dF_2}{dP_{G2}} = 0.001P_{G2} + 0.03$$

According to the principle of equal incremental rate (4.13), we have

$$\lambda_1 = \lambda_2$$

that is,

$$0.0016P_{G1} + 0.02 = 0.001P_{G2} + 0.03$$

or

$$1.6P_{G1} - P_{G2} = 10$$

Given the system load is 500 MW, then

$$P_{G1} + P_{G2} = 500$$

Solving the above two equations for $P_{G1}, P_{G2}$, we get

$$P_{G1} = 196.15 \ \text{MW}$$

$$P_{G2} = 303.85 \ \text{MW}$$

### 4.3.2  Economic Dispatch without Network Losses

***Neglecting the Constraints of Power Output***   The equal incremental principle can be used for a system with $N$ thermal-generating units. Given that the input–output characteristic of $N$ generating units are $F_1(P_{G1}), F_2(P_{G2}), \ldots, F_n(P_{Gn})$, respectively, and the total system load is $P_D$. The problem is to minimize total fuel consumption $F$ subject to the constraint that the sum of the power generated must equal the received load, that is,

$$\min F = F_1(P_{G1}) + F_2(P_{G2}) + \ldots + F_n(P_{Gn}) = \sum_{i=1}^{N} F_i(P_{Gi}) \qquad (4.15)$$

such that

$$\sum_{i=1}^{N} P_{Gi} = P_D \qquad (4.16)$$

This is a constrained optimization problem, and it can be solved by the Lagrange multiplier method. First of all, the Lagrange function should be formed by adding the constraint function to the objective function after the constraint function has been multiplied by an undetermined multiplier.

$$L = F + \lambda \left( P_D - \sum_{i=1}^{N} P_{Gi} \right) \tag{4.17}$$

where $\lambda$ is the Lagrange multiplier.

The necessary conditions for the extreme value of the Lagrange function are to set the first derivative of the Lagrange function with respect to each of the independent variables equal to zero.

$$\frac{\partial L}{\partial P_{Gi}} = \frac{\partial F}{\partial P_{Gi}} - \lambda = 0, \quad i = 1, 2, \dots N \tag{4.18}$$

or

$$\frac{\partial F}{\partial P_{Gi}} = \lambda, \quad i = 1, 2, \dots, N \tag{4.19}$$

Since the fuel consumption function of each generating unit is only related to its own power output, equation (4.19) can be written as

$$\frac{dF_i}{dP_{Gi}} = \lambda, \quad i = 1, 2, \dots, N \tag{4.20}$$

or

$$\frac{dF_1}{dP_{G1}} = \frac{dF_2}{dP_{G2}} = \dots = \frac{dF_N}{dP_{GN}} = \lambda \tag{4.21}$$

Equation (4.20) is the principle of equal incremental rate of economic power operation for multiple generating units.

***Example 4.4:***   Suppose the input–output characteristics of three generating units are as follows:

$$F_1 = 0.0006P_{G1}^2 + 0.5P_{G1} + 6 \ \text{Btu/h}$$

$$F_2 = 0.0005P_{G2}^2 + 0.6P_{G2} + 5 \ \text{Btu/h}$$

$$F_3 = 0.0007P_{G3}^2 + 0.4P_{G3} + 3 \ \text{Btu/h}$$

We wish to determine the economic operation point for these three units when delivering a total of 500 MW and 800 MW power demand, respectively.

(A) Total load $P_D = 500$ MW

The incremental fuel rates of the three generating units are calculated as follows.

$$\lambda_1 = \frac{dF_1}{dP_{G1}} = 0.0012P_{G1} + 0.5$$

$$\lambda_2 = \frac{dF_2}{dP_{G2}} = 0.001P_{G2} + 0.6$$

$$\lambda_3 = \frac{dF_3}{dP_{G3}} = 0.0014P_{G3} + 0.4$$

According to the principle of equal incremental rate, we have

$$\lambda_1 = \lambda_2 = \lambda_3$$

that is,

$$0.0012P_{G1} + 0.5 = 0.001P_{G2} + 0.6 = 0.0014P_{G3} + 0.4$$

From the above equation, we get

$$1.2P_{G1} - P_{G2} = 100$$

$$1.2P_{G1} - 1.4P_{G3} = -100$$

Given a system load is 500 MW, then

$$P_{G1} + P_{G2} + P_{G3} = 500$$

Solving the above three equations for $P_{G1}, P_{G2}, P_{G3}$, we get

$$P_{G1} = 172.897 \ \text{MW}$$

$$P_{G2} = 107.477 \ \text{MW}$$

$$P_{G3} = 219.626 \ \text{MW}$$

The corresponding system incremental fuel rate under this load level is

$$\lambda = 0.70748$$

**(B)** Total load $P_D = 800$ MW
Similar to (A), we can get the following equations.

$$1.2P_{G1} - P_{G2} = 100$$

$$1.2P_{G1} - 1.4P_{G3} = -100$$

$$P_{G1} + P_{G2} + P_{G3} = 800$$

Solving the above three equations for $P_{G1}, P_{G2}, P_{G3}$, we get

$$P_{G1} = 271.028 \text{ MW}$$

$$P_{G2} = 225.234 \text{ MW}$$

$$P_{G3} = 303.738 \text{ MW}$$

The corresponding system incremental fuel rate under this load level is

$$\lambda = 0.82523$$

**Considering the Constraints of Power Output**   We have discussed the equal incremental principle of economic operation. Thus, we know that the necessary condition for economic operation of a thermal power system is that the incremental fuel rates (or incremental cost rates) of all the units are equal. However, we have not considered the two inequalities, that is, the power output of each unit must be greater than or equal to the minimum power permitted and must also be less than or equal to the maximum power permitted on that particular unit.

Considering the inequality constraints, the problem of ED can be written as follows;

$$\min F = F_1(P_{G1}) + F_2(P_{G2}) + \dots + F_n(P_{Gn}) = \sum_{i=1}^{N} F_i(P_{Gi}) \qquad (4.22)$$

s.t.

$$\sum_{i=1}^{N} P_{Gi} = P_D \qquad (4.23)$$

$$P_{Gimin} \leq P_{Gi} \leq P_{Gimax} \qquad (4.24)$$

The equal incremental principle can be still applied to equations (4.22)–(4.24). The calculation process is as follows:

**(1)** Neglect the inequality equation (4.24). Distribute the power among the units according to the equal incremental principle.

**(2)** Check the power output limits for each unit according to equation (4.24). If the power output is outside the limits, set the power output equal to the corresponding limit, that is,

$$\text{If } P_{Gk} \geq P_{Gkmax}, P_{Gk} = P_{Gkmax} \qquad (4.25)$$

$$\text{If } P_{Gk} \leq P_{Gkmin}, P_{Gk} = P_{Gkmin} \qquad (4.26)$$

**(3)** Handle the violated unit as a negative load, that is,

$$P'_{Dk} = -P_{Gk} \quad \text{for violated units } k(k = 1, \dots nk)$$

**(4)** Recompute the power balance equation as follows;

$$\sum_{\substack{i=1 \\ i \notin nk}}^{N} P_{Gi} = P_D + \sum_{k=1}^{nk} P'_{Dk} \tag{4.27}$$

or

$$\sum_{\substack{i=1 \\ i \notin nk}}^{N} P_{Gi} = P_D - \sum_{k=1}^{nk} P_{Gk} \tag{4.28}$$

**(5)** Go back to step (1) until the inequalities of all the units are met.

***Example 4.5:*** Example 4.3 is used here but considering the inequality constraints of two units, which are given as follows:

$$100 \leq P_{G1} \leq 250 \ \text{MW}$$

$$150 \leq P_{G2} \leq 300 \ \text{MW}$$

From Example 4.3, we know the economic operation point for these two units without inequalities when delivering a total of 500 MW power demand, that is,

$$P_{G1} = 196.15 \ \text{MW}$$

$$P_{G2} = 303.85 \ \text{MW}$$

By checking the inequality constraints of the units, we can see that the power output of unit 2 violated its upper limit. Thus, set the power output of unit 2 to its upper limit.

$$P_{G2} = 303.85 \geq 300(P_{G2max}), P_{G2} = 300 \ \text{MW}$$

So the power dispatch becomes

$$P_{G1} = 200 \ \text{MW}$$

$$P_{G2} = 300 \ \text{MW}$$

***Example 4.6:*** Example 4.4 is used here but considering the inequality constraints of the three units, which are given as follows:

$$100 \leq P_{G1} \leq 250 \ \text{MW}$$

$$100 \leq P_{G2} \leq 250 \ \text{MW}$$

$$150 \leq P_{G3} \leq 350 \ \text{MW}$$

(A) Total load $P_D = 500$ MW

When delivering a total of 500 MW power demand, the dispatch from Example 4.4 is

$$P_{G1} = 172.897 \text{ MW}$$

$$P_{G2} = 107.477 \text{ MW}$$

$$P_{G3} = 219.626 \text{ MW}$$

By checking the inequality constraints of the units, we know that all their power outputs are within the limits. Thus, they are the optimum results and there is no violation of the inequality constraints.

(B) Total load $P_D = 800$ MW

When delivering a total of 800 MW power demand, the dispatch from Example 4.4 is

$$P_{G1} = 271.028 \text{ MW}$$

$$P_{G2} = 225.234 \text{ MW}$$

$$P_{G3} = 303.738 \text{ MW}$$

By checking the inequality constraints of units, we see that the power output of unit 1 violated its upper limit. According to equation (4.25), we get

$$P_{G1} = 250 \text{ MW}$$

According to equation (4.27), we have

$$P'_{D1} = -250 \text{ MW}$$

From equation (4.28), we get the new power balance equation

$$P_{G2} + P_{G3} = 800 - 250 = 550$$

Applying the principle of equal incremental rate for units 2 and 3, we have

$$\lambda_2 = \frac{dF_2}{dP_{G2}} = 0.001P_{G2} + 0.6$$

$$\lambda_3 = \frac{dF_3}{dP_{G3}} = 0.0014P_{G3} + 0.4$$

$$\lambda_2 = \lambda_3$$

that is,

$$0.001P_{G2} + 0.6 = 0.0014P_{G3} + 0.4$$

Then we can get the following two equations

$$P_{G2} - 1.4P_{G3} = -200$$

$$P_{G2} + P_{G3} = 550$$

Solving the above three equations, the power dispatch becomes

$$P_{G1} = 250.0 \text{ MW}$$

$$P_{G2} = 237.5 \text{ MW}$$

$$P_{G3} = 312.5 \text{ MW}$$

## 4.4  CALCULATION OF INCREMENTAL POWER LOSSES

Network losses were neglected in the previous sections on ED. It is much more difficult to solve the ED problem with network losses than the previous cases with no losses. There have been two general approaches to compute network losses and the corresponding incremental power losses. The first is the development of a mathematical expression for the losses in the network solely as a function of the power output of each of the units. This is called the B-coefficient method. The other method is based on power flow equations. The details on how to compute incremental power losses are discussed in Chapter 3. Here, we just describe the simple B-coefficient method.

Let $S_L$ be the plural power losses of the network; the corresponding real and reactive power losses being $P_L$ and $Q_L$. The plural power losses equal the sum of the plural power injections of nodes, which can be expressed as

$$S_L = P_L + jQ_L = \dot{V}^T \overset{*}{I} \tag{4.29}$$

$$\dot{V} = Z\dot{I} \tag{4.30}$$

$$Z = R + jX \tag{4.31}$$

$$\dot{I} = I_P + jI_Q \tag{4.32}$$

where

$V$: the node voltage
$I$: the node current
$I_P$: the node current component corresponding to real power
$I_Q$: the node current component corresponding to reactive power
$Z$: the node impedance matrix.

Substituting equations (4.30)–(4.32) into equation (4.29), and we get the real power losses as follows:

$$P_L = I_P^T R I_P + I_Q^T R I_Q \tag{4.33}$$

The node current can also be expressed as

$$\dot{I}_i = \frac{P_i + jQ_i}{\dot{V}_i} = \frac{P_i + jQ_i}{V_i e^{-j\theta_i}} = \frac{(P_i + jQ_i)e^{j\theta_i}}{V_i} \tag{4.34}$$

Since

$$e^{j\theta_i} = \cos\theta_i + j\ \sin\theta_i \tag{4.35}$$

thus,

$$\dot{I}_i = \frac{(P_i + jQ_i)(\cos\theta_i + j\ \sin\theta_i)}{V_i} \tag{4.36}$$

From equation (4.36), we get

$$I_{Pi} = \frac{(P_i \cos\theta_i + Q_i \sin\theta_i)}{V_i} \tag{4.37}$$

$$I_{qi} = \frac{(P_i \sin\theta_i - Q_i \cos\theta_i)}{V_i} \tag{4.38}$$

Substituting equations (4.37), (4.38) into equation (4.33), we get

$$P_L = [P^T \quad Q^T]\begin{bmatrix} A & -B \\ B & A \end{bmatrix}\begin{bmatrix} P \\ Q \end{bmatrix} \tag{4.39}$$

Where the elements of $A$ and $B$ are

$$A_{ij} = \frac{R_{ij}\cos(\theta_i - \theta_j)}{V_i V_j} \tag{4.40}$$

$$B_{ij} = \frac{R_{ij}\sin(\theta_i - \theta_j)}{V_i V_j} \tag{4.41}$$

Suppose each node power consists of power generation and power demand. Then the node power and matrices $A$ and $B$ can be divided into two parts, namely,

$$P^T = [P_G^T \quad P_D^T] \tag{4.42}$$

$$Q^T = [Q_G^T \quad Q_D^T] \tag{4.43}$$

$$A = \begin{bmatrix} A_{GG} & A_{GD} \\ A_{DG} & A_{DD} \end{bmatrix} \tag{4.44}$$

$$B = \begin{bmatrix} B_{GG} & B_{GD} \\ B_{DG} & B_{DD} \end{bmatrix} \tag{4.45}$$

Substituting equations (4.42)–(4.45) into equation (4.39), we get

$$P_L = [P_G^T \quad Q_G^T] \begin{bmatrix} A_{GG} & -B_{GG} \\ B_{GG} & A_{GG} \end{bmatrix} \begin{bmatrix} P_G \\ Q_G \end{bmatrix} + [C_{GD}^T \quad C_{DG}^T] \begin{bmatrix} P_G \\ Q_G \end{bmatrix} + C \tag{4.46}$$

where

$$C = [P_D^T \quad Q_D^T] \begin{bmatrix} A_{DD} & -B_{DD} \\ B_{DD} & A_{DD} \end{bmatrix} \begin{bmatrix} P_D \\ Q_D \end{bmatrix} \tag{4.47}$$

$$C_{GD} = 2(B_{GD}Q_D - A_{GD}P_D) \tag{4.48}$$

$$C_{DG} = 2(B_{DG}^T P_D - A_{DG}^T Q_D) \tag{4.49}$$

Assuming the relationship between real and reactive power output of the generator is linear, that is,

$$Q_{Gi} = Q_{G0i} - D_i P_{Gi} \tag{4.50}$$

equation (4.46) can be written as

$$P_L = P_G^T B_L P_G + B_{L0}^T P_G + B_0 \tag{4.51}$$

where

$$B_L = FA_{GG}F + A_{GG} + 2FB_{GG} \tag{4.52}$$

$$B_{L0}^T = 2Q_{G0}^T(A_{GG}F + B_{GG}) + C_{DG}^T F + C_{GD}^T \tag{4.53}$$

$$B_0 = Q_{G0}^T A_{GG} Q_{G0} + C_{DG}^T Q_{G0} + C \tag{4.54}$$

Equation (4.51) is the B-coefficient formula for network losses. The incremental power losses can be obtained from equation (4.51):

$$\frac{\partial P_L}{\partial P_G} = 2B_L P_G + B_{L0}^T \tag{4.55}$$

## 4.5 THERMAL SYSTEM ECONOMIC DISPATCH WITH NETWORK LOSSES

Considering the network power losses, the problem of thermal system ED can be written as follows:

$$\min F = F_1(P_{G1}) + F_2(P_{G2}) + \cdots + F_n(P_{Gn}) = \sum_{i=1}^{N} F_i(P_{Gi}) \tag{4.56}$$

such that

$$\sum_{i=1}^{N} P_{Gi} = P_D + P_L \tag{4.57}$$

$$P_{Gimin} \leq P_{Gi} \leq P_{Gimax} \tag{4.58}$$

The Lagrange function is written as

$$L = F + \lambda \left( P_D + P_L - \sum_{i=1}^{N} P_{Gi} \right) \tag{4.59}$$

The necessary conditions for the extreme value of the Lagrange function are to set the first derivative of the Lagrange function with respect to each of the independent variables equal to zero.

$$\frac{\partial L}{\partial P_{Gi}} = \frac{dF_i}{dP_{Gi}} - \lambda \left( 1 - \frac{\partial P_L}{\partial P_{Gi}} \right) = 0, \quad i = 1, 2, \dots, N \tag{4.60}$$

or

$$\frac{\partial F_i}{\partial P_{Gi}} \times \frac{1}{\left( 1 - \frac{\partial P_L}{\partial P_{Gi}} \right)} = \frac{dF_i}{dP_{Gi}} a_i = \lambda, \quad i = 1, 2, \dots, N \tag{4.61}$$

where

$$a_i = \frac{1}{\left( 1 - \frac{\partial P_L}{\partial P_{Gi}} \right)} \tag{4.62}$$

is the correction coefficient for network losses.

Considering the network losses, the equal incremental principle of classic ED can be written as

$$\frac{\partial F_i}{\partial P_{Gi}} a_i = \lambda, \quad i = 1, 2, \dots, N \tag{4.63}$$

or

$$\frac{dF_1}{dP_{G1}} a_1 = \frac{dF_2}{dP_{G2}} a_2 = \dots = \frac{dF_N}{dP_{GN}} a_N = \lambda \tag{4.64}$$

Equation (4.64) is also called the coordination equation of economic power operation.

The solution procedure of thermal system economic power dispatch is as follows:

(1) Pick a set of staring values $P_{G0i}$ that sum to the load.
(2) Calculate the incremental fuel $\frac{dF_i}{dP_{Gi}}$.
(3) Calculate the incremental losses $\frac{\partial P_L}{\partial P_{Gi}}$ as well as the total losses.
(4) Calculate the value of $\lambda$ and $P_{Gi}$ according to the coordination equation (4.64) and power balance equation.

**(5)** Compare the $P_{Gi}$ from step (4) with the starting points $P_{Gi0}$. If there is no significant change in any one of the values, go to step (6), otherwise go back to step (2).

**(6)** Done.

## 4.6 HYDROTHERMAL SYSTEM ECONOMIC DISPATCH

### 4.6.1 Neglecting Network Losses

The hydrothermal system ED is usually more complex than the economic operation of an all-thermal generation system. All hydro-systems are different. The reasons for the differences are the natural differences in the watersheds, the differences in the man-made storage and release elements used to control the water flows, and the very many different types of natural and manmade constraints imposed on the operation of hydroelectric systems. The coordination of the operation of hydroelectric plants involves the scheduling of water release. According to the scheduling period, the hydro-system operation can be divided into long-range hydro-scheduling and short-range hydro-scheduling problems.

The long-range hydro-scheduling problem involves the long-range forecasting of water availability and the scheduling of reservoir water release for an interval of time that depends on the reservoir capacities. Typical long-range scheduling is for anywhere from 1 week to 1 year or several years. For hydro schemes with a capacity of impounding water over several seasons, the long-range problem involves meteorological and statistics analyses. Herein we focus on the short-range hydro-scheduling problem.

Short-range hydro-scheduling refers to a time period from 1 day to 1 week. It involves hour-by-hour scheduling of all generation on a hydrothermal system to achieve minimum production cost (or minimum fuel consumption) for the given time period.

Let $P_T$, $F(P_T)$ be the power output and the input–output characteristic of thermal plant, and let $P_H$, $W(P_H)$ be the power output and input–output characteristic of the hydroelectric plant. The hydrothermal system ED problem can be expressed as

$$\min F_{\Sigma} = \int_0^T F[P_T(t)]dt \tag{4.65}$$

such that

$$P_H(t) + P_T(t) - P_D(t) = 0 \tag{4.66}$$

$$\int_0^T W[P_H(t)]dt - W_{\Sigma} = 0 \tag{4.67}$$

We divide the operation period $T$ into $s$ time stages

$$T = \sum_{k=1}^{s} \Delta t_k \tag{4.68}$$

For any time stage, suppose the power output of the hydro plant and thermal plant as well as load demand are constant. Then, equations (4.66) and (4.67) are changed as

$$P_{Hk} + P_{Tk} - P_{Dk} = 0, \quad k = 1, 2, \dots, s \tag{4.69}$$

$$\sum_{k=1}^{s} W(P_{Hk})\Delta t_k - W_\Sigma = \sum_{k=1}^{s} W_k \Delta t_k - W_\Sigma = 0 \tag{4.70}$$

The objective function (4.65) is also changed as

$$F_\Sigma = \sum_{k=1}^{s} F(P_{Tk})\Delta t_k = \sum_{k=1}^{s} F_k \Delta t_k \tag{4.71}$$

The Lagrange function is written as

$$L = \sum_{k=1}^{s} F_k \Delta t_k - \sum_{k=1}^{s} \lambda_k (P_{Hk} + P_{Tk} - P_{Dk})\Delta t_k + \gamma \left( \sum_{k=1}^{s} W_k \Delta t_k - W_\Sigma \right) \tag{4.72}$$

The necessary conditions for the extreme value of the Lagrange function are

$$\frac{\partial L}{\partial P_{Hk}} = \gamma \frac{dW_k}{dP_{Hk}} \Delta t_k - \lambda_k \Delta t_k = 0 \quad k = 1, 2, \dots, s \tag{4.73}$$

$$\frac{\partial L}{\partial P_{Tk}} = \frac{dF_k}{dP_{Tk}} \Delta t_k - \lambda_k \Delta t_k = 0 \quad k = 1, 2, \dots, s \tag{4.74}$$

$$\frac{\partial L}{\partial \lambda_k} = -(P_{Hk} + P_{Tk} - P_{Dk})\Delta t_k = 0 \quad k = 1, 2, \dots, s \tag{4.75}$$

$$\frac{\partial L}{\partial \gamma} = \sum_{k=1}^{s} W_k \Delta t_k - W_\Sigma = 0 \tag{4.76}$$

From equations (4.73) and (4.74), we get

$$\frac{dF_k}{dP_{Tk}} = \gamma \frac{dW_k}{dP_{Hk}} = \lambda_k \quad k = 1, 2, \dots, s \tag{4.77}$$

If the time stage is very short, equation (4.77) can be expressed as

$$\frac{dF}{dP_T} = \gamma \frac{dW}{dP_H} = \lambda \tag{4.78}$$

Equation (4.78) is the equal incremental principle of the hydrothermal system ED. It means that when the thermal unit increases power output $\Delta P$, the incremental fuel consumption will be

$$\Delta F = \frac{dF}{dP_T} \Delta P \tag{4.79}$$

When the hydro unit increases power output $\Delta P$, the incremental water consumption will be

$$\Delta W = \frac{dW}{dP_H}\Delta P \tag{4.80}$$

From equations (4.78)–(4.80), we obtain

$$\gamma = \frac{\Delta F}{\Delta W} \tag{4.81}$$

where $\gamma$ is the coefficient that converts water consumption to fuel. In other words, the water consumption of a hydro unit multiplied by $\gamma$ is equivalent to the fuel consumption of a thermal unit. Thus, the hydro unit is equivalent to a thermal unit.

Generally, the value of $\gamma$ is related to given water consumption of the hydro unit during a time period (e.g., 1 day). If the given water consumption is very high, the hydro unit can produce a larger power output to meet the load demand. In this case, a smaller value of $\gamma$ will be selected. Otherwise, a bigger value of $\gamma$ will be selected. The calculation procedures of hydrothermal system ED are as follows:

**(1)** Given an initial value $\gamma(0)$. Set the iteration number $k = 0$

**(2)** Compute power distribution for hydrothermal system for all time stages according to equation (4.77).

**(3)** Check if the total water consumption $W(k)$ equals the given water consumption, that is,

$$\left|W(k) - W_\Sigma\right| < \varepsilon \tag{4.82}$$

If this condition is met, stop calculation, otherwise, go to the next step.

**(4)** If $W(k) > W_\Sigma$ it means that the selected $\gamma$ is too small. Make $\gamma(k+1) > \gamma(k)$. If $W(k) < W_\Sigma$ it means that the selected $\gamma$ is too big. Make $\gamma(k+1) < \gamma(k)$. Go back to step (2).

**Example 4.7:** A system has one thermal plant and one hydro plant. The input–output characteristic of the thermal plant is

$$F = 0.00035P_T^2 + 0.4P_T + 3 \text{ Btu/h}$$

The input–output characteristic of the hydro plant is

$$W = 0.0015P_H{}^2 + 0.8P_H + 2 \text{ m}^3/\text{s}$$

The daily water consumption of hydro plant is

$$W_\Sigma = 1.5 \times 10^7 \text{ m}^3$$

The daily load demands of the system are as follows (Figure 4.6):

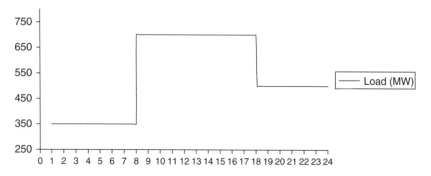

Figure 4.6    Daily load demands for Example 4.7.

The power output limits of the thermal plant is

$$50 \leq P_T \leq 600 \text{ MW}$$

The power output limits of the hydro plant is

$$50 \leq P_H \leq 450 \text{ MW}$$

The problem is to determine the ED for this hydrothermal system.

According to the input–output characteristics of the thermal plant and hydro plant and equation (4.78), we can write the coordination equation as follows:

$$0.0007 \ P_T + 0.4 = \gamma(0.003P_H + 0.8)$$

From the load curve, we can know that there are three time stages. The loads are the same within each time stage. Thus, for each time stage, we get the corresponding power balance equation.

$$P_{Hk} + P_{Tk} = P_{Dk} \quad k = 1, 2, 3$$

From the above two equations, we get

$$P_{Hk} = \frac{0.4 - 0.8\gamma + 0.0007P_{Dk}}{0.003\gamma + 0.0007} \quad k = 1, 2, 3$$

$$P_{Tk} = \frac{-0.4 + 0.8\gamma + 0.003\gamma P_{Dk}}{0.003\gamma + 0.0007} \quad k = 1, 2, 3$$

Select the initial value of $\gamma$ to be 0.5. For the first time stage, the load level is 350 MW and we get

$$P_{H1} = \frac{0.4 - 0.8 \times 0.5 + 0.0007 \times 350}{0.003 \times 0.5 + 0.0007} = 111.36 \text{ MW}$$

$$P_{T1} = \frac{-0.4 + 0.8 \times 0.5 + 0.003 \times 0.5 \times 350}{0.003 \times 0.5 + 0.0007} = 238.64 \text{ MW}$$

For the second time stage, the load level is 700 MW and we get

$$P_{H2} = \frac{0.4 - 0.8 \times 0.5 + 0.0007 \times 700}{0.003 \times 0.5 + 0.0007} = 222.72 \, \text{MW}$$

$$P_{T2} = \frac{-0.4 + 0.8 \times 0.5 + 0.003 \times 0.5 \times 700}{0.003 \times 0.5 + 0.0007} = 477.28 \, \text{MW}$$

For the third time stage, the load level is 500 MW and we get

$$P_{H3} = \frac{0.4 - 0.8 \times 0.5 + 0.0007 \times 500}{0.003 \times 0.5 + 0.0007} = 159.09 \, \text{MW}$$

$$P_{T3} = \frac{-0.4 + 0.8 \times 0.5 + 0.003 \times 0.5 \times 500}{0.003 \times 0.5 + 0.0007} = 340.91 \, \text{MW}$$

According to the power output of the hydro plant and input–output characteristic of the hydro plant, we can compute the daily water consumption.

$$W_{\Sigma} = (0.0015 \times 111.36^2 + 0.8 \times 111.36 + 2) \times 8 \times 3600 +$$

$$(0.0015 \times 222.72^2 + 0.8 \times 222.72 + 2) \times 10 \times 3600 +$$

$$(0.0015 \times 159.09^2 + 0.8 \times 159.09 + 2) \times 6 \times 3600 = 1.5937 \times 10^7 \text{m}^3$$

The water consumption is greater than the daily given amount. So increase the value of $\gamma$, say $\gamma = 0.52$, recompute the power output. For the first time stage, the load level is 350 MW and we get

$$P_{H1} = \frac{0.4 - 0.8 \times 0.52 + 0.0007 \times 350}{0.003 \times 0.5 + 0.0007} = 101.33 \, \text{MW}$$

$$P_{T1} = \frac{-0.4 + 0.8 \times 0.52 + 0.003 \times 0.52 \times 350}{0.003 \times 0.52 + 0.0007} = 248.67 \, \text{MW}$$

For the second time stage, the load level is 700 MW and we get

$$P_{H2} = \frac{0.4 - 0.8 \times 0.52 + 0.0007 \times 700}{0.003 \times 0.52 + 0.0007} = 209.73 \, \text{MW}$$

$$P_{T2} = \frac{-0.4 + 0.8 \times 0.52 + 0.003 \times 0.52 \times 700}{0.003 \times 0.52 + 0.0007} = 490.27 \, \text{MW}$$

For the third time stage, the load level is 500 MW and we get

$$P_{H3} = \frac{0.4 - 0.8 \times 0.52 + 0.0007 \times 500}{0.003 \times 0.52 + 0.0007} = 147.79 \, \text{MW}$$

$$P_{T3} = \frac{-0.4 + 0.8 \times 0.52 + 0.003 \times 0.52 \times 500}{0.003 \times 0.52 + 0.0007} = 352.21 \, \text{MW}$$

**TABLE 4.2    Iteration Process of Example 4.7**

| Iteration | $\gamma$ | $P_{H1}$(MW) | $P_{H1}$(MW) | $P_{H1}$(MW) | $W_{\Sigma}(10^7 \text{ m}^3)$ |
|---|---|---|---|---|---|
| 1 | 0.5000 | 111.360 | 222.720 | 159.090 | 1.5937 |
| 2 | 0.5200 | 101.330 | 209.730 | 147.790 | 1.4628 |
| 3 | 0.5140 | 104.280 | 213.560 | 151.110 | 1.5010 |
| 4 | 0.5145 | 104.207 | 213.463 | 151.031 | 1.5000 |

Then the daily water consumption can be computed as

$$W_{\Sigma} = (0.0015 \times 101.33^2 + 0.8 \times 101.33 + 2) \times 8 \times 3600 +$$

$$(0.0015 \times 209.73^2 + 0.8 \times 209.73 + 2) \times 10 \times 3600 +$$

$$(0.0015 \times 147.79^2 + 0.8 \times 147.79 + 2) \times 6 \times 3600 = 1.4628 \times 10^7 \text{m}^3$$

The water consumption is less than the daily given amount. So reduce the value of $\gamma$, recompute the power output until the water consumption equals the daily given amount, or equation (4.82) is satisfied. The iteration process is listed in Table 4.2.

After fourth iteration, the water consumption almost equals the daily given amount. Stop the calculation.

## 4.6.2 Considering Network Losses

Suppose there are $m$ hydro plants and $n$ thermal plants. The system load is given in the time period. The given water consumption of hydro plant $j$ is $W_{\Sigma j}$. The hydrothermal system ED with network loss can be expressed as follows:

$$\min F_{\Sigma} = \sum_{i=1}^{n} \int_{0}^{T} F_i[P_{Ti}(t)]dt \tag{4.83}$$

such that

$$\sum_{j=1}^{m} P_{Hj}(t) + \sum_{i=1}^{n} P_{Ti}(t) - P_L(t) - P_D(t) = 0 \tag{4.84}$$

$$\int_{0}^{T} W_j[P_{Hj}(t)]dt - W_{\Sigma j} = 0 \tag{4.85}$$

Similarly to Section 4.6.1, we divide the operation period $T$ into $s$ time stages

$$T = \sum_{k=1}^{s} \Delta t_k \tag{4.86}$$

We get

$$F_{\Sigma} = \sum_{i=1}^{n} \sum_{k=1}^{s} F_{ik}(P_{Tik}) \Delta t_k \tag{4.87}$$

$$\sum_{j=1}^{m} P_{Hjk} + \sum_{i=1}^{n} P_{Tik} - P_{Lk} - P_{Dk} = 0 \quad k = 1, 2, \ldots, s \tag{4.88}$$

$$\sum_{k=1}^{s} W_{jk}(P_{Hjk})\Delta t_k - W_{\Sigma j} = 0, \quad j = 1, 2, \ldots, m \tag{4.89}$$

The Lagrange function will be

$$L = \sum_{i=1}^{n} \sum_{k=1}^{s} F_{ik}(P_{Tik})\Delta t_k - \sum_{k=1}^{s} \lambda_k \left( \sum_{i=1}^{m} P_{Hik} + \sum_{i=1}^{n} P_{Tik} - P_{Lk} - P_{Dk} \right) \Delta t_k$$
$$+ \sum_{j=1}^{m} \gamma_j \left( \sum_{k=1}^{s} W_{jk}\left(P_{Hjk}\right) \Delta t_k - W_{\Sigma j} \right) \tag{4.90}$$

The necessary conditions for the extreme value of the Lagrange function are

$$\frac{\partial L}{\partial P_{Hjk}} = \gamma_j \frac{dW_{jk}}{dP_{Hjk}}\Delta t_k - \lambda_k \left( 1 - \frac{\partial P_{Lk}}{\partial P_{Hjk}} \right) \Delta t_k = 0$$
$$j = 1, 2, \ldots, m; \quad k = 1, 2, \ldots, s \tag{4.91}$$

$$\frac{\partial L}{\partial P_{Tik}} = \frac{dF_{ik}}{dP_{Tik}}\Delta t_k - \lambda_k \left( 1 - \frac{\partial P_{Lk}}{\partial P_{Tik}} \right) \Delta t_k = 0$$
$$i = 1, 2, \ldots, n; \quad k = 1, 2, \ldots, s \tag{4.92}$$

$$\frac{\partial L}{\partial \lambda_k} = - \left( \sum_{j=1}^{m} P_{Hjk} + \sum_{i=1}^{n} P_{Tik} - P_{Lk} - P_{Dk} \right) \Delta t_k = 0$$
$$k = 1, 2, \ldots, s \tag{4.93}$$

$$\frac{\partial L}{\partial \gamma_j} = \sum_{k=1}^{s} W_{jk}\Delta t_k - W_{j\Sigma} = 0 \quad j = 1, 2, \ldots m \tag{4.94}$$

From equations (4.91) and (4.92), we get

$$\frac{dF_{ik}}{dP_{Tik}} \times \frac{1}{1 - \frac{\partial P_{Lk}}{\partial P_{Tik}}} = \gamma_j \frac{dW_{jk}}{dP_{Hjk}} \times \frac{1}{1 - \frac{\partial P_{Lk}}{\partial P_{Hjk}}} = \lambda_k$$
$$k = 1, 2, \ldots, s \tag{4.95}$$

Equation (4.95) is true for any time stage, that is,

$$\frac{dF_i}{dP_{Ti}} \times \frac{1}{1 - \frac{\partial P_L}{\partial P_{Ti}}} = \gamma_j \frac{dW_j}{dP_{Hj}} \times \frac{1}{1 - \frac{\partial P_L}{\partial P_{Hj}}} = \lambda \qquad (4.96)$$

Equation (4.96) is the coordination equation of hydrothermal system ED, considering network losses.

## 4.7   ECONOMIC DISPATCH BY GRADIENT METHOD

### 4.7.1   Introduction

We discussed the equal incremental principle for classical ED in the previous sections. Generally, the equal incremental principle is good only if the input–output characteristic of the generation unit is a quadratic function, or the incremental input–output characteristic is a piecewise linear function [2]. But the input–output characteristic of the generating unit may be a cubic function, or more complex. For example,

$$F_{Gi} = A + BP_{Gi} + CP_{Gi}^2 + DP_{Gi}^3 + \ldots\ldots$$

Thus, other methods are needed to get the optimum solution for the above function. We discuss the gradient method in this section.

### 4.7.2   Gradient Search in Economic Dispatch

The principle of the gradient method is that the minimum of a function, $f(x)$, can be found by a series of steps that always go in the *downward* direction. The gradient of the function $f(x)$ can be expressed as follows:

$$\nabla f = \begin{bmatrix} \dfrac{\partial f}{\partial x_1} \\[2mm] \dfrac{\partial f}{\partial x_2} \\[2mm] \vdots \\[2mm] \dfrac{\partial f}{\partial x_n} \end{bmatrix} \qquad (4.97)$$

The gradient $\nabla f$ always points to the direction of maximum ascent. If we want to move in the direction of maximum descent, we negate the gradient. Thus the direction of steepest descent for minimizing a function can be found by use of the direction of the negative gradient. Given any starting point $x^0$, the new point $x^1$ should be obtained as follows:

$$x^1 = x^0 - \varepsilon \nabla f \qquad (4.98)$$

where $\varepsilon$ is a scale that is used to process the convergence of the gradient method.

Applying the gradient method to ED, the objective function will be

$$\min \ F = \sum_{i=1}^{N} f_i(P_{Gi}) \tag{4.99}$$

The constraint is the real power balance equation, that is,

$$\sum_{i=1}^{N} P_{Gi} = P_D \tag{4.100}$$

As mentioned before, to solve this classic ED problem, the Lagrange function should be constructed first, that is,

$$L = F + \lambda \left( P_D - \sum_{i=1}^{N} P_{Gi} \right) = \sum_{i=1}^{N} f_i(P_{Gi}) + \lambda \left( P_D - \sum_{i=1}^{N} P_{Gi} \right) \tag{4.101}$$

The gradient of the Lagrange function is

$$\nabla L = \begin{bmatrix} \dfrac{\partial L}{\partial P_{G1}} \\[2mm] \dfrac{\partial L}{\partial P_{G2}} \\[2mm] \vdots \\[2mm] \dfrac{\partial L}{\partial P_{GN}} \\[2mm] \dfrac{\partial L}{\partial \lambda} \end{bmatrix} = \begin{bmatrix} \dfrac{df_1(P_{G1})}{dP_{G1}} - \lambda \\[2mm] \dfrac{df_2(P_{G2})}{dP_{G2}} - \lambda \\[2mm] \vdots \\[2mm] \dfrac{df_N(P_{GN})}{dP_{GN}} - \lambda \\[2mm] P_D - \sum_{i=1}^{N} P_{Gi} \end{bmatrix} \tag{4.102}$$

To use the gradient $\nabla L$ to solve the ED problem, the starting values $P_{G1}^0, P_{G2}^0, \ldots, P_{GN}^0$, and $\lambda^0$ should be given. Then the new values will be computed by the following equation.

$$x^1 = x^0 - \varepsilon \nabla L \tag{4.103}$$

where the vectors $x^1, x^0$ are

$$x^0 = \begin{bmatrix} P_{G1}^0 \\[2mm] P_{G2}^0 \\[2mm] \vdots \\[2mm] P_{GN}^0 \\[2mm] \lambda^0 \end{bmatrix} \tag{4.104}$$

$$x^1 = \begin{bmatrix} P^1_{G1} \\ P^1_{G2} \\ \vdots \\ P^1_{GN} \\ \lambda^1 \end{bmatrix} \tag{4.105}$$

The more general expression of the gradient search is as follows:

$$x^n = x^{n-1} - \varepsilon \nabla L \tag{4.106}$$

where $n$ is the iteration number.

The calculation steps for applying the gradient method to classic ED are summarized in the following.

Step 1:  Select the starting values $P^0_{G1}, P^0_{G2}, \ldots, P^0_{GN}$, where

$$P^0_{G1}, P^0_{G2} + \cdots + P^0_{GN} = P_D$$

Step 2:  Compute the initial $\lambda^0_i$ for each generator.

$$\lambda^0_i = \left. \frac{df_i(P_{Gi})}{\partial P_{Gi}} \right|_{P^0_{Gi}}, \quad i = 1, \ldots, N$$

Step 3:  Compute the initial average incremental cost $\lambda^0$

$$\lambda^0 = \frac{1}{N} \sum_{i=1}^{N} \lambda^0_i$$

Step 4:  Compute the gradient as follows:

$$\nabla L^1 = \begin{bmatrix} \dfrac{df_1(P^0_{G1})}{dP_{G1}} - \lambda^0 \\ \dfrac{df_2(P^0_{G2})}{dP_{G2}} - \lambda^0 \\ \vdots \\ \dfrac{df_N(P^0_{GN})}{dP_{GN}} - \lambda^0 \\ P_D - \sum_{i=1}^{N} P^0_{Gi} \end{bmatrix}$$

Step 5:  If $\nabla L = 0$, the solution converges. Stop the iteration. Otherwise, go to the next step.

Step 6:  Select a scale $\varepsilon$ for handling the convergence.

Step 7:  Compute the new values $P^1_{G1}, P^1_{G2}, \ldots, P^1_{GN}, \lambda^1$ according to equation (4.106).

Step 8:  Substitute the new values into equation (4.102) in step (4), and recompute the gradient.

**Example 4.8:**  For the same data in Example 4.4, solve for the ED with a total load of 500 MW. The solution is as follows:

Select the starting values $P^0_{G1} = 300$, $P^0_{G2} = 150$, $P^0_{G3} = 250$, and

$$P^0_{G1} + P^0_{G2} + P^0_{G3} = 500$$

Compute the initial $\lambda^0_i$ for each generator.

$$\lambda^0_1 = \frac{df_1(P^0_{G1})}{dP_{G1}} = 0.0012 \times 150 + 0.5 = 0.68$$

$$\lambda^0_2 = \frac{df_2(P^0_{G2})}{dP_{G2}} = 0.001 \times 100 + 0.6 = 0.70$$

$$\lambda^0_3 = \frac{df_3(P^0_{G3})}{dP_{G3}} = 0.0014 \times 250 + 0.4 = 0.75$$

Compute the initial average incremental cost $\lambda^0$

$$\lambda^0 = \frac{1}{3}\sum_{i=1}^{3}\lambda^0_i = \frac{1}{3}(0.68 + 0.7 + 0.75) = 0.71$$

Compute the gradient as follows:

$$\nabla L^1 = \begin{bmatrix} 0.68 - 0.71 \\ 0.70 - 0.71 \\ 0.75 - 0.71 \\ 500 - (150 + 100 + 250) \end{bmatrix} = \begin{bmatrix} -0.03 \\ -0.01 \\ 0.04 \\ 0.00 \end{bmatrix}$$

Select a scale $\varepsilon = 300$ for handling the convergence, and compute the new values $P^1_{G1}, P^1_{G2}, \ldots, P^1_{GN}, \lambda^1$ according to equation (4.106).

$$\begin{bmatrix} P^1_{G1} \\ P^1_{G2} \\ P^1_{G3} \\ \lambda^1 \end{bmatrix} = \begin{bmatrix} 150 \\ 100 \\ 250 \\ 0.71 \end{bmatrix} - 300 \begin{bmatrix} -0.03 \\ -0.01 \\ 0.04 \\ 0.0 \end{bmatrix} = \begin{bmatrix} 159 \\ 103 \\ 238 \\ 0.71 \end{bmatrix}$$

Then compute the new gradient as follows:

$$\nabla L^2 = \begin{bmatrix} (0.0012 \times 159 + 0.5) - 0.71 \\ (0.0010 \times 103 + 0.6) - 0.71 \\ (0.0014 \times 238 + 0.4) - 0.71 \\ 500 - (159 + 103 + 238) \end{bmatrix} = \begin{bmatrix} -0.0192 \\ -0.0070 \\ 0.0232 \\ 0.0000 \end{bmatrix}$$

$$\begin{bmatrix} P^2_{G1} \\ P^2_{G2} \\ P^2_{G3} \\ \lambda^2 \end{bmatrix} = \begin{bmatrix} 159 \\ 103 \\ 238 \\ 0.71 \end{bmatrix} - 300 \begin{bmatrix} -0.0192 \\ -0.0070 \\ 0.0232 \\ 0.0 \end{bmatrix} = \begin{bmatrix} 164.76 \\ 105.10 \\ 231.04 \\ 0.71 \end{bmatrix}$$

Once again compute the new gradient.

$$\nabla L^3 = \begin{bmatrix} (0.0012 \times 164.76 + 0.5) - 0.71 \\ (0.0010 \times 105.10 + 0.6) - 0.71 \\ (0.0014 \times 231.04 + 0.4) - 0.71 \\ 500 - (164.76 + 105.1 + .231.04) \end{bmatrix} = \begin{bmatrix} -0.0123 \\ -0.0049 \\ 0.0135 \\ 0.9000 \end{bmatrix}$$

The gradient $\nabla L^3 \neq 0$, so compute a new solution.

The iterations have led to no solution because the element $\lambda$ in the gradient had a huge jump and could not be converged. To solve this problem, we present three methods in the following.

**Gradient Method 1** In the calculation of the gradient, the element $\lambda$ will be removed, that is,

$$\nabla L = \begin{bmatrix} \dfrac{\partial L}{\partial P_{G1}} \\ \dfrac{\partial L}{\partial P_{G2}} \\ \vdots \\ \dfrac{\partial L}{\partial P_{GN}} \end{bmatrix} = \begin{bmatrix} \dfrac{df_1(P_{G1})}{dP_{G1}} - \lambda \\ \dfrac{df_2(P_{G2})}{dP_{G2}} - \lambda \\ \vdots \\ \dfrac{df_N(P_{GN})}{dP_{GN}} - \lambda \end{bmatrix} \tag{4.107}$$

We always set the value of $\lambda$ equal to the average of the incremental cost of the generators at the iterated generation values, that is,

$$\lambda^k = \frac{1}{N} \sum_{i=1}^{N} \left[ \frac{df_i(P^k_{Gi})}{dP_{Gi}} \right] \tag{4.108}$$

**Example 4.9:** Reworking example 4.8 using gradient method 1, the results are shown in Table 4.3

**TABLE 4.3    Gradient Method 1 Results ($\varepsilon = 300$)**

| Iteration | $P_{G1}$ | $P_{G2}$ | $P_{G3}$ | $\lambda$ |
|---|---|---|---|---|
| 0 | 150 | 100 | 250 | 0.71 |
| 1 | 159 | 103 | 238 | 0.709 |
| 2 | 164.46 | 104.8 | 230.74 | 0.7084 |
| 3 | 169.7388 | 105.5388 | 226.348 | 0.7086 |
| 4 | 171.21 | 106.4688 | 223.888 | 0.7085 |
| 5 | 172.11 | 107.0688 | 222.418 | 0.7083 |
| 6 | 172.65 | 107.4288 | 221.518 | 0.7082 |

This solution is much more stable and converges to the optimum solution. However, gradient method 1 cannot guarantee that the total outputs of the generators meet the total load demand.

**Gradient Method 2**    This method is modified from method 1, but we need to check the power balance equation each time when we finish the iteration of gradient calculation. The method is described in the following.

If $\sum_{i=1}^{N}(P_{Gi}^{k}) > P_D$, select the unit with the maximal incremental generation cost to pick up the power difference.

$$P_{GS'}^{k}|_{\lambda_{\max}} = P_{GS}^{k} - \left( \sum_{i=1}^{N} (P_{Gi}^{k}) - P_D \right) \qquad (4.109)$$

If $\sum_{i=1}^{N}(P_{Gi}^{k}) < P_D$, select the unit with the minimal incremental generation cost to pick up the power difference.

$$P_{GS'}^{k}|_{\lambda_{\max}} = P_{GS}^{k} + \left( P_D - \sum_{i=1}^{N} (P_{Gi}^{k}) \right) \qquad (4.110)$$

Then, recompute the average incremental generation cost, and conduct a new iteration.

***Example 4.10:***    Reworking Example 4.9 using gradient method 2, the results are shown in Table 4.4.

This solution is much more stable and converges to the optimum solution. Obviously, gradient method 2 can guarantee that the total outputs of generators meet the total load.

**TABLE 4.4   Gradient Method 2 Results ($\varepsilon = 300$)**

| Iteration | $P_{G1}$ | $P_{G2}$ | $P_{G3}$ | $P_{total}$ | $\lambda$ |
|-----------|----------|----------|----------|-------------|-----------|
| 0 | 150 | 100 | 250 | 500 | 0.71 |
| 1 | 159 | 103 | 238 | 500 | 0.709 |
| 2 | 164.46 | 104.8 | 230.74 | 500 | 0.7084 |
| 3 | 169.7388 | 105.5388 | 224.7224* | 500 | 0.7079 |
| 4 | 171.0108* | 106.2678 | 222.7214 | 500 | 0.7078 |

*The corresponding unit is selected to balance the total generation and total load.

***Gradient Method 3***   This method is similar to method 2 but with some simplification. One fixed unit is selected as the slack machine. For example, selecting the last unit as the slack generator, we get

$$P_{GN} = P_D - \sum_{i=1}^{N-1} (P_{Gi}) \tag{4.111}$$

The objective function becomes

$$F = f_1(P_{G1}) + f_2(P_{G2}) +, \dots , f_N(P_{GN})$$

$$= f_1(P_{G1}) + f_2(P_{G2}) +, \dots , f_N\left(P_D - \sum_{i=1}^{N-1} (P_{Gi})\right) \tag{4.112}$$

The gradient will become

$$\nabla F = \begin{bmatrix} \dfrac{dF}{dP_{G1}} \\[2mm] \dfrac{dF}{dP_{G2}} \\[2mm] \vdots \\[2mm] \dfrac{dF}{dP_{G(N-1)}} \end{bmatrix} = \begin{bmatrix} \dfrac{df_1(P_{G1})}{dP_{G1}} - \dfrac{df_N(P_{GN})}{dP_{GN}} \\[2mm] \dfrac{df_2(P_{G2})}{dP_{G2}} - \dfrac{df_N(P_{GN})}{dP_{GN}} \\[2mm] \vdots \\[2mm] \dfrac{df_{(N-1)}(P_{G(N-1)})}{dP_{G(N-1)}} - \dfrac{df_N(P_{GN})}{dP_{GN}} \end{bmatrix} \tag{4.113}$$

The gradient iteration will be the same as before.

$$x^n = x^{n-1} - \varepsilon \nabla F \tag{4.114}$$

and

$$x = \begin{bmatrix} P_{G1} \\ P_{G2} \\ \vdots \\ P_{G(N-1)} \end{bmatrix} \tag{4.115}$$

**TABLE 4.5 Gradient Method 3 Results ($\varepsilon = 300$)**

| Iteration | $P_{G1}$ | $P_{G2}$ | $P_{G3}$ | $P_{total}$ |
|-----------|----------|----------|----------|-------------|
| 0 | 150 | 100 | 250 | 500 |
| 1 | 171 | 115 | 214 | 500 |
| 2 | 169.32 | 110.38 | 220.3 | 500 |
| 3 | 170.8908 | 109.792 | 219.317 | 500 |
| 4 | 171.4728 | 108.937 | 219.590 | 500 |

*Example 4.11:* Reworking Example 4.8 using gradient method 3, the results are shown in Table 4.5.

This solution is also stable and converges to the optimum solution, which is similar to method 2. Obviously, gradient method 3 can also guarantee that the total outputs of generators meet the total load.

## 4.8 CLASSIC ECONOMIC DISPATCH BY GENETIC ALGORITHM

### 4.8.1 Introduction

Another type of method that is used to solve classic ED problem is the genetic algorithm (GA) [3–5]. The theoretical foundation for GA was first described by Holland [18] and was extended by Goldberg [19]. GA provides a solution to a problem by working with a population of individuals each representing a possible solution. Each possible solution is termed a "chromosome." New points of the search space are generated through GA operations, known as reproduction, crossover, and mutation. These operations consistently produce fitter offspring through successive generations, which rapidly lead the search toward global optima. The features of GA are different from other search techniques in the following aspects:

(1) The algorithm is a multipath that searches many peaks in parallel, hence reducing the possibility of local minimum trapping.

(2) GA works with a bit string encoding instead of the real parameters. The coding of parameters will help the genetic operator to evolve the current state into the next state with minimum number of computations.

(3) Instead of the optimization function, GA evaluates the fitness of each string to guide its search. The genetic algorithm only needs to evaluate objective function (fitness) to guide its search. There is no requirement for the operation of derivatives.

(4) GA explores the search space where the probability of finding improved performance is high.

The main operators of GA used are the following:

- The *crossover operator* is applied with a certain probability. The parent generations are combined (exchange bits) to form two new generations that inherit solution characteristics from both parents. Crossover, although being the primary search operator, cannot produce information that does not already exist within the population.

- The *mutation operator* is also applied with a small probability. Randomly chosen bits of the offspring genotype flip from 0 to 1 and *vice versa* to give characteristics that do not exist in the parent population. Generally, mutation is considered as a secondary but not useless operator that gives a nonzero probability to every solution to be considered and evaluated.

- *Elitism* is implemented so that the best solution of every generation is copied to the next so that the possibility of its destruction through a genetic operator is eliminated.

- *Fitness Scaling* refers to a nonlinear transformation of genotype fitness in order to emphasize small differences between near-optimal qualities in a converged population.

The GA-type algorithms are actually of unconstrained optimization; all information must be expressed in a fitness function. As mentioned at the beginning of this chapter, the classic ED problem neglectsnetwork losses and network constraints. Thus the fitness function for classic ED can be easily formed.

## 4.8.2 GA-Based ED Solution

According to Section 4.3, the classic ED problem can be stated as follows:

$$\min F = \sum_{i=1}^{N} F_i(P_{Gi}) \tag{4.116}$$

such that

$$\sum_{i=1}^{N} P_{Gi} = P_D \tag{4.117}$$

In the application of GA to ED, the outputs of the $N - 1$ "free generators" can be chosen arbitrarily within limits while the output of the "reference generator" (or slack bus generator) is constrained by the power balance. It is assumed that the $N$th generator is the reference generator. GAs do not work on the real generator outputs themselves, but on bit string encoding of these outputs. The output of the free generators is encoded in strings. For example, an 8-bit string (an unsigned 8-bit integer) that gives a resolution of $2^8$ discrete power values in the range $(P_{Gmin}, P_{Gmax})$. These $(N - 1)$ strings are concatenated to form a consolidated solution bit string of $8*(N - 1)$ bits called a genotype. A population of $m$ genotypes must be initially generated at random. Each genotype is decoded to a power output vector. The output of the reference unit is

$$P_{GN} = P_D - \sum_{i=1}^{N-1} P_{Gi} \qquad (4.118)$$

Adding penalty factors $h_1, h_2$ to the violation of power output of the slack bus unit, we can combine equations (4.117) and (4.118) as follows:

$$F_A = \sum_{i=1}^{N} F_i(P_{Gi}) + h_1(P_{GN} - P_{GNmax})^2 + h_2(P_{GNmin} - P_{GN})^2 \qquad (4.119)$$

where, $P_{GNmin}, P_{GNmax}$ are respectively the lower and upper limits of the power output of the slack bus unit. The value of the penalty factors should be large so that there is no violation for unit output at the final solution. Since GA is designed for the solution of the maximization problem, the GA fitness function is defined as the inverse of equation (4.119).

$$F_{fitness} = \frac{1}{F_A} \qquad (4.120a)$$

In the ED problem, the problem variables correspond to the power generation of the units. Each string represents a possible solution and is made of substrings, each corresponding to a generating unit. The length of each substring is decided on the basis of the maximum/minimum limits on the power generation of the corresponding unit and the solution accuracy desired. The string length, which depends upon the length of each substring, is chosen on the basis of a trade-off between solution accuracy and solution time. Longer strings may provide better accuracy, but result in more solution time. Thus, the step size of a unit can be computed as follows:

$$\varepsilon_i = \frac{P_{Gimax} - P_{Gimin}}{2^n - 1} \qquad (4.120b)$$

where $n$ is the length of substring in binary codes corresponding to a unit.

For example, there are six units in a system, and the sixth unit is selected as the slack bus unit. The power output limits of the five free units are

$$20 \leq P_{G1} \leq 100 (MW)$$

$$10 \leq P_{G2} \leq 100 (MW)$$

$$50 \leq P_{G3} \leq 200 (MW)$$

$$20 \leq P_{G4} \leq 120 (MW)$$

$$50 \leq P_{G5} \leq 250 (MW)$$

If the length of substring in binary codes is selected as 4, the step size of each unit will be

$$\varepsilon_1 = \frac{P_{G1\,max} - P_{G1\,min}}{2^4 - 1} = \frac{100 - 20}{15} = 5.33 \text{ MW}$$

$$\varepsilon_2 = \frac{P_{G2\,max} - P_{G2\,min}}{2^4 - 1} = \frac{100 - 10}{15} = 6.00 \text{ MW}$$

$$\varepsilon_3 = \frac{P_{G3\,max} - P_{G3\,min}}{2^4 - 1} = \frac{200 - 50}{15} = 10.00 \text{ MW}$$

$$\varepsilon_4 = \frac{P_{G4\,max} - P_{G4\,min}}{2^4 - 1} = \frac{120 - 20}{15} = 6.67 \text{ MW}$$

$$\varepsilon_5 = \frac{P_{G5\,max} - P_{G5\,min}}{2^4 - 1} = \frac{250 - 50}{15} = 13.33 \text{ MW}$$

If the length of substring in binary codes is selected as 5, the step size of each unit will be

$$\varepsilon_1 = \frac{P_{G1\,max} - P_{G1\,min}}{2^5 - 1} = \frac{100 - 20}{31} = 2.58 \text{ MW}$$

$$\varepsilon_2 = \frac{P_{G2\,max} - P_{G2\,min}}{2^5 - 1} = \frac{100 - 10}{31} = 2.90 \text{ MW}$$

$$\varepsilon_3 = \frac{P_{G3\,max} - P_{G3\,min}}{2^5 - 1} = \frac{200 - 50}{31} = 4.84 \text{ MW}$$

$$\varepsilon_4 = \frac{P_{G4\,max} - P_{G4\,min}}{2^5 - 1} = \frac{120 - 20}{31} = 3.23 \text{ MW}$$

$$\varepsilon_5 = \frac{P_{G5\,max} - P_{G5\,min}}{2^5 - 1} = \frac{250 - 50}{31} = 6.45 \text{ MW}$$

It can be observed that the long string has smaller step size, which verifies that the length of the substring in binary codes affects the solution accuracy and solution speed.

In standard GAs, all the strings in the population are reformed during a generation. Parents are crossed on the basis of their performance in comparison to the average fitness of the population and mutation is allowed to occur on the offspring. Selective pressure is provided by the fitness measure; the differential need not be great to achieve good results. Both selective pressure and initial population sizes may be tuned to match the problem space. The type of crossover and rate of mutation needs to be selected on the basis of the problem type. For a large scale of power system, there are many generators. If the standard GA is used in ED, it appears to increase performance. A little improvement on the GA operator is needed, that is, we do not replace the entire population with each generation. Instead GA operator probabilistically chooses two parents to reform into two offspring. Recombination and mutation occur, and then one of the offspring is discarded randomly. The remaining offspring is placed in the population according to its fitness in relation to the rest of the strings. The lowest-valued string is discarded. This keeps high-valued strings within the population, directly accumulating high-performance hyperplanes. It also bases the reproductive opportunity upon rank with the population, not upon a string's fitness value in comparison with the average of the population, reducing the impact of selective pressure fluctuation. It also reduces the importance of choosing a proper evaluation function for fitness in that the difference in the fitness function between two adjacent strings is irrelevant.

To use GA programming to solve classic ED, the following parameters are needed for data input.

- Number of chromosomes (that consist a generation)
- Bit resolution per generator
- Number of cross-points
- Number of generations
- Initial crossover probability (%)
- Initial mutation probability (%)
- Minimal power output of each unit
- Maximal power output of each unit
- Status of the unit
- The coefficient of unit cost function
- Total load demand.

*Example 4.12:*   For Example 4.6, using genetic algorithm to distribute the 500 MW load to three units. The GA parameters are selected as follows:

- Number of chromosomes $= 100$
- Bit resolution per generator $= 8$
- Number of cross-points $= 2$
- Number of generations $= 9000$
- Initial crossover probability $= 92\%$
- Initial mutation probability $= 0.1\%$

For the total load of 500 MW, the output results are as follows:

$$P_{G1} = 172.897 \text{ MW}$$

$$P_{G2} = 107.477 \text{ MW}$$

$$P_{G3} = 219.626 \text{ MW}$$

## 4.9 CLASSIC ECONOMIC DISPATCH BY HOPFIELD NEURAL NETWORK

Since Hopfield introduced neural networks in the early 1980s [6], the Hopfield neural networks (HNNs) have been used in many different applications. This section presents the application of the HNN to the classic ED problem [7–10].

### 4.9.1 Hopfield Neural Network Model

Let $u_i$ be $i$th neuron input, and $V_i$ be its output. Suppose there are $N$ neurons that are connected together, the nonlinear differential equations of the HNN are described as follows:

$$\begin{cases} C_i \dfrac{du_i}{dt} = \displaystyle\sum_{j=1}^{N} T_{ij}V_j + \dfrac{u_i}{R_i} + I_i \\ V_i = g\left(u_i\right) \qquad i = 1, 2, \dots, N \end{cases} \tag{4.121}$$

where

$$\frac{1}{R_i} = \theta_i + \sum_{j=1}^{N} T_{ij}$$

$$V_i = g(u_i) \tag{4.122}$$

are the nonlinear characteristics of the neuron.

For a very high gain parameter $\lambda$ of the neuron, the output equation can be defined as

$$V_i = g(\lambda u_i) = g\left(\frac{u_i}{u_0}\right) = \frac{1}{1 + \exp\left(-\frac{u_i + \theta_i}{u_0}\right)} \tag{4.123}$$

where $\theta_i$ is the threshold bias.

The energy function of the system (4.121) is defined as

$$E = -\frac{1}{2}\sum_{i=1}^{N}\sum_{j=1}^{N} T_{ij}V_iV_j - \sum_{i=1}^{N} V_iI_i + \sum_{i=1}^{N}\frac{1}{R_i}\int_{0}^{V_i} g^{-1}(V)dV \tag{4.124}$$

From equation (4.124), we get

$$\frac{dE}{dt} = \sum_i \frac{\partial E}{\partial V_i}\frac{dV_i}{dt} \tag{4.125}$$

where

$$\frac{\partial E}{\partial V_i} = -\frac{1}{2}\sum_j T_{ij}V_j - \frac{1}{2}\sum_j T_{ji}V_j + \frac{u_i}{R_i} - I_i$$

$$= -\frac{1}{2}\sum_j (T_{ji} - T_{ij})V_j - \left(\sum_j T_{ij}V_j - \frac{u_i}{R_i} + I_i\right)$$

$$= -\frac{1}{2}\sum_j (T_{ji} - T_{ij})V_j - C_i\frac{du_i}{dt}$$

$$= -\frac{1}{2}\sum_j (T_{ji} - T_{ij})V_j - C_i[g^{-1}(V_i)]'\frac{dV_i}{dt} \qquad (4.126)$$

Substituting equation (4.126) in equation (4.125), we get

$$\frac{dE}{dt} == -\frac{1}{2}\sum_j (T_{ji} - T_{ij})V_j\frac{dV_i}{dt} - C_i[g^{-1}(V_i)]'\left(\frac{dV_i}{dt}\right)^2 \qquad (4.127)$$

Since the weight parameter matrix $T$ in equation (4.121) is symmetric, we have

$$T_{ji} = T_{ij} \qquad (4.128)$$

Substituting equation (4.128) into equation (4.127), we get

$$\frac{dE}{dt} == -C_i[g^{-1}(V_i)]'\left(\frac{dV_i}{dt}\right)^2 \qquad (4.129)$$

Since $g^{-1}$ is a monotone increasing function, and $C_i > 0$,

$$\frac{dE}{dt} == -C_i[g^{-1}(V_i)]'\left(\frac{dV_i}{dt}\right)^2 \leq 0 \qquad (4.130)$$

This shows that the time evolution of the system is a motion in state space that seeks out minima in $E$ and comes to a stop at such points.

### 4.9.2 Mapping of Economic Dispatch to HNN

As discussed above, the classic ED problem without line security can be written as

$$\min F = F_1(P_{G1}) + F_2(P_{G2}) + \cdots + F_n(P_{Gn}) = \sum_{i=1}^{N} F_i(P_{Gi}) \qquad (4.131)$$

such that

$$\sum_{i=1}^{N} P_{Gi} = P_D + P_L \qquad (4.132)$$

$$P_{Gi\min} \leq P_{Gi} \leq P_{Gi\max} \qquad (4.133)$$

Assuming that the generator cost function is a quadratic function, that is,

$$F_i(P_{Gi}) = a_i P_{Gi}^2 + b_i P_{Gi} + c_i \tag{4.134}$$

and the network loss can be represented by the B-coefficient,

$$P_L = \sum_{i=1}^{N} \sum_{j=1}^{N} P_{Gi} B_{ij} P_{Gj} \tag{4.135}$$

To apply HNN to solve the above classic ED problem, the following energy function is defined by augmenting the objective function (4.131) with the constraint (4.132):

$$E = \frac{1}{2} A \left( P_D + P_L - \sum_i P_{Gi} \right)^2 + \frac{1}{2} B \sum_i (a_i P_{Gi}^2 + b_i P_{Gi} + c_i) \tag{4.136}$$

By comparing equation (4.136) with equation (4.224), whose threshold is assumed to be zero, the weight parameters and external input of neuron $i$ in the network [7] are given by

$$T_{ii} = -A - B c_i \tag{4.137}$$

$$T_{ij} = -A \tag{4.138}$$

$$I_i = A(P_D + P_L) - \frac{B b_i}{2} \tag{4.139}$$

where the diagonal weights are nonzero.

The sigmoid function (4.223) can be modified to meet the power limit constraint as follows [7].

$$V_i(k+1) = (P_{i\max} - P_{i\min}) \frac{1}{1 + \exp\left(-\frac{u_i(k) + \theta_i}{u_0}\right)} + P_{i\min} \tag{4.140}$$

In order to speed up convergence of the ED problem solved by HNN, two adjustment methods can be used [9].

**Slope Adjustment Method** Since energy is to be minimized and its convergence depends on the gain parameter $u_0$, the gradient descent method can be applied to adjust the gain parameters.

$$u_0(k+1) = u_0(k) - \eta_s \frac{\partial E}{\partial u_0} \tag{4.141}$$

Where $\eta_s$ is a learning rate.

From equations (4.136) and (4.140), the gradient of energy with respect to the gain parameter can be computed as

$$\frac{\partial E}{\partial u_0} = \sum_i \frac{\partial E}{\partial P_i} \frac{\partial P_i}{\partial u_0} \tag{4.142}$$

The update rule of equation (4.141) needs a suitable choice of the learning rate $\eta_s$. For a small value of $\eta_s$, convergence is guaranteed but speed is too slow. On the other hand, if the learning rate is too high, the algorithm becomes unstable. The suggested learning rate will be

$$0 < \eta_s < \frac{2}{g_{s,\max}^2} \tag{4.143}$$

where

$$g_{s,\max} = \max \|g_s(k)\|$$

$$g_s(k) = \frac{\partial E(k)}{\partial u_0} \tag{4.144}$$

Moreover, the optimal convergence corresponds to

$$\eta_s^* = \frac{1}{g_{s,\max}^2} \tag{4.145}$$

**Bias Adjustment Method**   There is a limitation in the slope adjustment method, in which the slopes are small near the saturation region of the sigmoid function. If every input can use the same maximum possible slope, convergence will be much faster. This can be achieved by changing the bias to shift the input close to the center of the sigmoid function, that is

$$\theta_i(k+1) = \theta_i(k) - \eta_b \frac{\partial E}{\partial \theta_i} \tag{4.146}$$

Where $\eta_b$ is a learning rate.

The bias can be applied to every neuron as in equation (4.223). Thus, from equations (4.136) and (4.140), the derivate of energy with respect to a bias can be computed as

$$\frac{\partial E}{\partial \theta_i} = \frac{\partial E}{\partial P_i} \frac{\partial P_i}{\partial \theta_i} \tag{4.147}$$

The suggested learning rate will be

$$0 < \eta_b < -\frac{2}{g_b(k)} \tag{4.148}$$

where

$$g_b(k) = \sum_i \sum_j T_{ij} \frac{\partial V_i}{\partial \theta} \frac{\partial V_j}{\partial \theta} \qquad (4.149)$$

Moreover, the optimal convergence corresponds to

$$\eta_b = -\frac{1}{g_b(k)} \qquad (4.150)$$

### 4.9.3   Simulation Results

The test example and results of applying HNN to ED are taken from reference [9]. The system data are shown in Table 4.6. Each generator has three types of fuels. There are four values of load demand, that is, 2400, 2500, 2600 and 2700 MW.

The ED results based on the slope adjustment method are shown in Table 4.7. Compared with the conventional Hopfield network, the number of iterations is reduced to about one half, and oscillation is drastically reduced from about 40,000 to less than 100 iterations. In addition, the degree of freedom of the system increases from 1, which is $u_0$, to 2. It can be observed that the final results of the adaptive learning rate are close to those of the fixed learning rate.

The ED results based on the bias adjustment method are shown in Table 4.8, which are similar to those based on the slope adjustment method. For the adaptive learning rate, the number of iterations is reduced and the final results of the adaptive learning rate are better than those of the fixed learning rate.

## APPENDIX A: OPTIMIZATION METHODS USED IN ECONOMIC OPERATION

Herein, we introduce several methods [10–17] that are used for economic power operation of power systems.

Although a wide spectrum of methods exists for optimization, methods can be broadly categorized in terms of the derivative information that is, or is not, used. Search methods that use only function evaluations are most suitable for problems that are very nonlinear or have a number of discontinuities. Gradient methods are generally more efficient when the function to be minimized is continuous in its first derivative. Higher-order methods, such as Newton's method, are only really suitable when the second-order information is readily and easily calculated, because calculation of second-order information using numerical differentiation is computationally expensive.

### A.1   Gradient Method

Gradient methods use information about the slope of the function to dictate a direction of search where the minimum is thought to lie. The simplest of these is the method

**TABLE 4.6    Cost Coefficients for Piecewise Quadratic Cost Function**

| Unit | Generation min $P1 \cdot P2$ max $F1 \cdot F2 \cdot F3$ | $F$ | $C$ | $B$ | $a$ |
|------|------|-----|-----|-----|-----|
| 1 | 100 196 250 250 1 2 2 | 1 | 0.2697e2 | −0.3975e0 | 0.2176e−2 |
|   |   | 2 | 0.2113e2 | −0.3059e0 | 0.1861e−2 |
|   |   | 2 | 0.2113e2 | −0.3059e0 | 0.1861e−2 |
| 2 | 50 114 157 230 2 3 1 | 1 | 0.1184e3 | −0.1269e1 | 0.4194e−2 |
|   |   | 2 | 0.1865e1 | −0.3988e − 1 | 0.1138e−2 |
|   |   | 3 | 0.1365e2 | −0.1980e − 1 | 0.1620e−2 |
| 3 | 200 332 388 500 1 2 3 | 1 | 0.3979e2 | −0.3116e0 | 0.1457e−2 |
|   |   | 2 | −0.5914e2 | 0.4864e0 | 0.1176e−4 |
|   |   | 3 | −0.2876e1 | 0.3389e1 | 0.8035e−3 |
| 4 | 99 138 200 265 1 2 3 | 1 | 0.1983e1 | −0.3114e − 1 | 0.1049e−2 |
|   |   | 2 | 0.5285e2 | −0.6348e0 | 0.2758e−2 |
|   |   | 3 | 0.2668e3 | −0.2338e1 | 0.5935e−2 |
| 5 | 190 338 407 490 1 2 3 | 1 | 0.1392e2 | −0.8733e − 1 | 0.1066e−2 |
|   |   | 2 | 0.9976e2 | −0.5206e0 | 0.1597e−2 |
|   |   | 3 | 0.5399e2 | 0.4462e0 | 0.1498e−3 |
| 6 | 85 138 200 265 2 1 3 | 1 | 0.5285e2 | −0.6348e0 | 0.2758e−2 |
|   |   | 2 | 0.1983e1 | −0.3114e − 1 | 0.1049e−2 |
|   |   | 3 | 0.2668e3 | −0.2338e1 | 0.5935e−2 |
| 7 | 200 331 391 500 1 2 3 | 1 | 0.1893e2 | −0.1325e0 | 0.1107e−2 |
|   |   | 2 | 0.4377e2 | −0.2267e0 | 0.1165e−2 |
|   |   | 3 | −0.4335e2 | 0.3559e0 | 0.2454e−3 |
| 8 | 99 138 200 265 1 2 3 | 1 | 0.1983e1 | −0.3114e − 1 | 0.1049e−2 |
|   |   | 2 | 0.5285e2 | −0.6348e0 | 0.2758e−2 |
|   |   | 3 | 0.2668e3 | −0.2338e1 | 0.5935e−2 |
| 9 | 130 213 370 440 3 1 2 | 1 | 0.8853e2 | −0.5675e0 | 0.1554e−2 |
|   |   | 2 | 0.1530e2 | −0.4514e − 1 | 0.7033e−2 |
|   |   | 3 | 0.1423e2 | −0.1817e − 1 | 0.6121e−3 |
| 10 | 200 362 407 490 1 3 2 | 1 | 0.1397e2 | −0.9938e − 1 | 0.1102e−2 |
|   |   | 2 | −0.6113e2 | 0.5084e0 | 0.4164e−4 |
|   |   | 3 | 0.4671e2 | −0.2024e0 | 0.1137e−2 |

of steepest descent in which a search is performed in a particular direction.

$$S^k = -\nabla f(x^k) \tag{4A.1}$$

where $\nabla f(x^k)$ is the gradient of the objective function.

**TABLE 4.7** Results for the Slope Adjustment Method with Fixed Learning Rate, 1.0 (A) and Adaptive Learning Rate (B)

| Unit | 2400 | MW | 2500 | MW | 2600 | MW | 2700 | MW |
| | A | B | A | B | A | B | A | B |
|---|---|---|---|---|---|---|---|---|
| 1 | 196.8 | 189.9 | 205.6 | 205.1 | 215.7 | 214.5 | 223.2 | 224.6 |
| 2 | 202.7 | 202.9 | 206.7 | 206.5 | 211.1 | 211.4 | 216.1 | 215.7 |
| 3 | 251.2 | 252.1 | 265.3 | 266.4 | 278.9 | 278.8 | 292.5 | 291.9 |
| 4 | 232.5 | 232.9 | 236.0 | 235.8 | 239.2 | 239.3 | 242.6 | 242.6 |
| 5 | 240.4 | 241.7 | 257.9 | 256.8 | 276.1 | 276.1 | 294.1 | 293.6 |
| 6 | 232.5 | 232.9 | 236.0 | 235.9 | 239.2 | 239.1 | 242.4 | 242.5 |
| 7 | 252.5 | 253.4 | 269.5 | 269.3 | 286.0 | 286.7 | 303.5 | 303.0 |
| 8 | 232.5 | 232.9 | 236.0 | 235.8 | 239.2 | 239.3 | 242.7 | 242.6 |
| 9 | 320.2 | 321.0 | 331.8 | 334.0 | 343.4 | 343.6 | 355.8 | 355.7 |
| 10 | 238.9 | 240.4 | 255.5 | 254.4 | 271.2 | 271.2 | 287.3 | 287.8 |
| Total $P$ | 2400.0 | 2400.0 | 2500.0 | 2500.0 | 2600.0 | 2600.0 | 2700.0 | 2700.0 |
| Cost | 481.83 | 481.71 | 526.23 | 526.23 | 574.36 | 574.37 | 626.27 | 626.24 |
| Iters | 99,992 | 84,791 | 80,156 | 86,081 | 72,993 | 79,495 | 99,948 | 99,811 |
| $u_0$ | 95.0 | 110.0 | 120.0 | 100.0 | 130.0 | 120.0 | 160.0 | 120.0 |
| $n$ | 1.5 | 1.0E−04 | 1.0 | 1.0E−04 | 1.0 | 1.0E−04 | 1.0 | 1.0E−04 |

**TABLE 4.8** Results for the Bias Adjustment Method with Fixed Learning Rate, 1.0 (A) and Adaptive Learning Rate (B)

| Unit | 2400 | MW | 2500 | MW | 2600 | MW | 2700 | MW |
| | A | B | A | B | A | B | A | B |
|---|---|---|---|---|---|---|---|---|
| 1 | 197.6 | 189.4 | 208.3 | 206.7 | 212.4 | 217.9 | 221.4 | 228.8 |
| 2 | 201.6 | 201.8 | 206.2 | 205.8 | 209.6 | 210.5 | 213.8 | 214.1 |
| 3 | 252.3 | 253.5 | 265.2 | 265.6 | 280.0 | 278.8 | 293.3 | 292.0 |
| 4 | 232.7 | 232.9 | 235.9 | 235.8 | 238.8 | 239.0 | 242.1 | 242.2 |
| 5 | 239.9 | 242.1 | 257.1 | 258.2 | 277.9 | 275.8 | 295.4 | 293.6 |
| 6 | 232.7 | 232.9 | 235.9 | 235.8 | 238.6 | 239.0 | 242.0 | 242.1 |
| 7 | 251.5 | 253.8 | 268.3 | 269.4 | 288.1 | 285.5 | 305.3 | 302.6 |
| 8 | 232.7 | 232.9 | 235.8 | 235.8 | 238.8 | 239.0 | 242.1 | 242.1 |
| 9 | 318.8 | 319.3 | 330.9 | 330.1 | 341.9 | 342.1 | 345.2 | 352.3 |
| 10 | 240.3 | 241.6 | 256.4 | 256.9 | 274.0 | 272.3 | 290.4 | 290.1 |
| Total $P$ | 2400.0 | 2400.0 | 2500.0 | 2500.0 | 2600.0 | 2600.0 | 2700.0 | 2700.0 |
| Cost | 481.83 | 481.72 | 526.24 | 526.23 | 574.43 | 574.37 | 626.32 | 626.27 |
| Iters | 99,960 | 99,904 | 99,987 | 88,776 | 99,981 | 99,337 | 99,972 | 73,250 |
| $u_0$ | 100.0 | 100.0 | 100.0 | 100.0 | 100.0 | 100.0 | 100.0 | 100.0 |
| theta | 0.0 | 50.0 | 0.0 | 50.0 | 0.0 | 50.0 | 0.0 | 100.0 |
| $n$ | 1.0 | 1.0 | 1.0 | 5.0 | 1.0 | 5.0 | 1.0 | 5.0 |

The optimum search step can be computed as follows.

$$\varepsilon^{*k} = \frac{[\nabla f(x^k)]^T \nabla f(x^k)}{[\nabla f(x^k)]^T H(x^k) \nabla f(x^k)} \tag{4A.2}$$

where $H(x^k)$ is the Hessian matrix of the objective function.

The gradient method based on equation (4A.2) is also called the optimum gradient method. However, this method is very inefficient when the function to be minimized has long narrow valleys.

## A.2   Line Search

*Line search* is a search method that is used as part of a larger optimization algorithm. At each step of the main algorithm, the line-search method searches along the line containing the current point, $x^k$, parallel to the *search direction*, which is a vector determined by the main algorithm, that is, the iteration form of the method can be expressed as

$$x^{k+1} = x^k + \varepsilon d^k \tag{4A.3}$$

where $x^k$ denotes the current iterate, $d^k$ is the search direction, and $\varepsilon$ is a scalar step length parameter.

The line-search method attempts to decrease the objective function along the line $x^k + \varepsilon \ d^k$ by repeatedly minimizing polynomial interpolation models of the objective function. The line-search procedure has two main steps:

- The *bracketing* phase determines the range of points on the line $x^{k+1} = x^k + \varepsilon \ d^k$ to be searched. The *bracket* corresponds to an interval specifying the range of values of $\varepsilon$.

- The *sectioning* step divides the bracket into subintervals, on which the minimum of the objective function is approximated by polynomial interpolation.

The resulting step length $\varepsilon$ satisfies the Wolfe conditions:

$$f(x^k + \varepsilon \ d^k) \leq f(x^k) + \alpha_1 \varepsilon (\nabla f^k)^T d^k \tag{4A.4}$$

$$\nabla f(x^k + \varepsilon \ d^k)^T d_k \geq \alpha_2 \varepsilon (\nabla f^k)^T d^k \tag{4A.5}$$

where $\alpha_1$ and $\alpha_2$ are constants with $0 < \alpha_1 < \alpha_2 < 1$.

The first condition (4A.4) requires that $\varepsilon$ sufficiently decreases the objective function. The second condition (4A.5) ensures that the step length is not too small. Points that satisfy both conditions (4A.4) and (4A.5) are called acceptable points.

## A.3   Newton-Raphson Optimization

The Newton–Raphson optimization is also called the Newton method or Hessian matrix method.

The objective function can be approximately expressed by use of the second-order Taylor series expansion at the point $x^k$, that is,

$$f(x) \approx f(x^k) + [\nabla f(x^k)]^T \Delta x + \frac{1}{2} \Delta x^T H(x^k) \Delta x \qquad (4A.6)$$

The necessary condition that a quadratic function achieves the minimum value is its gradient equals zero.

$$\nabla f(x) = \nabla f(x^k) + H(x^k) \Delta x = 0 \qquad (4A.7)$$

Thus, the general iteration expression is as follows:

$$x^{k+1} = x^k - [H(x^k)]^{-1} \nabla f(x^k) \qquad (4A.8)$$

It is noted that the Hessian matrix $H(x)$ will be constant if the original nonlinear objective function is a quadratic function. In this case, the minimum value of the function will be obtained through one iteration only. Otherwise, the Hessian matrix $H(x)$ will not be constant, and multiple iterations are needed to obtain the minimum of the function. The formula for the search direction is

$$S^k = -[H(x^k)]^{-1} \nabla f(x^k) \qquad (4A.9)$$

The advantage of the Hessian matrix method is fast convergence. The disadvantage is that it needs to compute the inverse of the Hessian matrix, which leads to expensive memory and calculation burden.

## A.4 Trust-Region Optimization

The convergence of the Newton optimization method can be made more robust by using trust regions (TR) [11]. TR-based methods generate steps based on a quadratic model of the objective function. A region around the current solution is defined, within which the model is supposed to be an adequate representation of the objective function. Then a step is selected to minimize this approximate model in the trust region. Both the direction and the length of the step are chosen simultaneously. If a step is not acceptable, the size of the region is reduced and a new solution is found. In general, the step direction changes whenever the size of the trust region is altered [11].

Since the trust-region method uses the gradient $g(x^k)$ and Hessian matrix $H(x^k)$, it requires that the objective function $f(x)$ have continuous first- and second-order derivatives inside the feasible region. The general trust-region problem is expressed as

$$\min f = g^T(x^k) \Delta x + \frac{1}{2} \Delta x^T H(x^k) \Delta x \qquad (4A.10)$$

such that

$$\|\Delta x\| \le \delta \qquad (4A.11)$$

Where $\delta$ is the trust region radius.

The general idea of the trust region is to solve the subproblem represented by equations (4A.10), (4A.11) to obtain a point $y^k$. Then the value of the true objective function is calculated at $y^k$ and compared to the value predicted by the quadratic model, to verify if the point located in the trust region represents an effective progress toward the optimal solution. For this purpose, the size of the trust region is critical to the effectiveness of each step.

In practice, the size of the region is determined according to the evolution of the iterative process. If the model is sufficiently accurate, the size of the trust region is steadily increased to allow bigger steps. Otherwise, the quadratic model is inadequate, so the size of the trust region must be reduced. In order to establish an algorithm to control the trust region radius, define the *reduction ratio* evaluated at the $k$th iteration

$$\rho^k = \frac{J(x^k) - J(x^{k+1})}{Q(x^k) - Q(x^{k+1})} \tag{4A.12}$$

Where $J(x^k)$ and $Q(x^k)$ are the values of the summation of the weighted squared residuals for the actual objective function and the corresponding approximated quadratic model, respectively, evaluated at the $k$th iteration.

## A.5   Newton–Raphson Optimization with Line Search

This technique uses the gradient $g(x^k)$ and Hessian matrix $H(x^k)$ and thus requires that the objective function have continuous first- and second-order derivatives inside the feasible region. If second-order derivatives are computed efficiently and precisely, the method may perform well for medium-sized to large problems, and it does not need many functions, gradients, and Hessian calls.

This algorithm uses a pure Newton step when the Hessian is positive definite and when the Newton step reduces the value of the objective function successfully. Otherwise, a combination of ridging and line search is done to compute successful steps. If the Hessian is not positive definite, a multiple of the identity matrix is added to the Hessian matrix to make it positive definite. In each iteration, a line search is done along the search direction to find an approximate optimum of the objective function. The default line-search method uses quadratic interpolation and cubic extrapolation.

## A.6   Quasi-Newton Optimization

The (dual) quasi-Newton method uses the gradient $g(x^k)$ and does not need to compute second-order derivatives because they are approximated. It works well for medium to moderately large optimization problems where the objective function and the gradient are much faster to compute than the Hessian.

The method builds up curvature information at each iteration to formulate a quadratic model problem of the form

$$\min f(x) = b + c^T x + \frac{1}{2} x^T H x \tag{4A.13}$$

where the Hessian matrix, $H$, is a positive definite symmetric matrix, $c$ is a constant vector, and $b$ is a constant. The optimal solution for this problem occurs when the partial derivatives of $x$ go to zero, that is,

$$\nabla f(x^*) = Hx^* + c = 0 \tag{4A.14}$$

The optimal solution point, $x^*$, can be written as

$$x^* = -H^{-1}c \tag{4A.15}$$

Newton-type methods (as opposed to quasi-Newton methods) calculate $H$ directly and proceed in a direction of descent to locate the minimum after a number of iterations. Calculating $H$ numerically involves a large amount of computation. Quasi-Newton methods avoid this by using the observed behavior of $f(x)$ and $\nabla f(x)$ to build up curvature information to make an approximation to $H$ using an appropriate updating technique.

A large number of Hessian updating methods have been developed. However, the formula of Broyden, Fletcher, Goldfarb, and Shanno (BFGS) is thought to be the most effective for use in a general purpose method [12–17].

The formula given by BFGS is

$$H^{k+1} = H^k + \frac{q^k(q^k)^T}{(q^k)^T S^k} - \frac{(H^k)^T(S^k)^T S^k H^k}{(S^k)^T H^k S^k} \tag{4A.16}$$

where

$$S^k = x^{k+1} - x^k \tag{4A.17}$$

$$q^k = \nabla f(x^{k+1}) - \nabla(x^k) \tag{4A.18}$$

As a starting point, $H^0$ can be set to any symmetric positive definite matrix, for example, the identity matrix $I$. To avoid the inversion of the Hessian $H$, we can derive an updating method that avoids the direct inversion of $H$ by using a formula that makes an approximation of the inverse Hessian $H^{-1}$ at each update. A well-known procedure is the DFP formula of Davidon, Fletcher, and Powell. This uses the same formula as the BFGS method (4A.16) except that $q^k$ is substituted for $S^k$.

The gradient information is either supplied through analytically calculated gradients or derived by partial derivatives using a numerical differentiation method via finite differences. This involves perturbing each of the design variables, $x$, in turn and calculating the rate of change in the objective function.

At each major iteration, $k$, a line search is performed in the direction

$$d = -(H^k)^{-1}\nabla f(x^k) \tag{4A.19}$$

## A.7  Double Dogleg Optimization

The double dogleg optimization method combines the ideas of quasi-Newton and trust region methods. The double dogleg algorithm computes in each iteration the step $S^k$ as the linear combination of the steepest descent or ascent search direction $S_1{}^k$ and a quasi-Newton search direction $S_2{}^k$,

$$S^k = \alpha_1 S_1^k + \alpha_2 S_2^k \tag{4A.20}$$

The step is requested to remain within a prespecified trust region radius. The double dogleg optimization technique works well for medium to moderately large optimization problems where the objective function and the gradient are much faster to compute than the Hessian.

## A.8  Conjugate Gradient Optimization

Second-order derivatives are not used by conjugate gradient optimization. As already discussed, t the method of steepest descent (or gradient method) converges slowly. The method of conjugate gradients is an attempt to mend this problem. "Conjugacy" means that two unequal vectors, $S_i$ and $S_j$, are orthogonal with respect to any symmetric positive definite matrix, for example $Q$, that is,

$$S_i^T Q S_j = 0 \tag{4A.21}$$

This can be looked upon as a generalization of orthogonality, for which $Q$ is the unity matrix. The idea is to let each search direction $S_i$ be dependent on all the other directions searched to locate the minimum of $f(x)$ through equation (4A.21). A set of such search directions is referred to as a Q-orthogonal set, or conjugate set, and it will take a positive definite $n$-dimensional quadratic function to its minimum point in, at most, $n$ exact linear searches. This method is often referred to as conjugate directions, and a short description follows.

The conjugate gradients method is a special case of the method of conjugate directions, where the conjugate set is generated by the gradient vectors. This seems to be a sensible choice as the gradient vectors have proved their applicability in the steepest descent method, and they are orthogonal to the previous search direction.

Subsequently, mutually conjugate directions are chosen so that

$$S^{k+1} = -\nabla f(x^{k+1}) + \beta^k S^k \tag{4A.22}$$

where the coefficient $\beta^k$ is given by, for example, the so called Fletcher–Reeves formula:

$$\beta^k = \frac{[\nabla f(x^{k+1})]^T \nabla f(x^{k+1})}{[\nabla f(x^k)]^T \nabla f(x^k)} \tag{4A.23}$$

The optimum search step can be computed as follows.

$$\varepsilon^{*k} = -\frac{[\nabla f(x^k)]^T S^k}{(S^k)^T H(x^k) S^k} \tag{4A.24}$$

During $n$ successive iterations, uninterrupted by restarts or changes in the working set, the conjugate gradient algorithm computes a cycle of $n$ conjugate search directions. In each iteration, a line search is done along the search direction to find an approximate optimum of the objective function. The default line-search method uses quadratic interpolation and cubic extrapolation to obtain a step size $\varepsilon$ satisfying the Goldstein conditions. One of the Goldstein conditions can be violated if the feasible region defines an upper limit for the step size.

## A.9    Lagrange Multipliers Method

Suppose there are $M$ constraints to be met, then optimization problem can be written as below.

$$\min \ f(x_i), \quad i = 1, 2, \ldots, N \tag{4A.25}$$

such that

$$h_1(x_i) = 0, \quad i = 1, 2, \ldots, N \tag{4A.26}$$

$$h_2(x_i) = 0, \quad i = 1, 2, \ldots, N \tag{4A.27}$$

$$h_M(x_i) = 0, \quad i = 1, 2, \ldots, N \tag{4A.28}$$

The optimum point would possess the property that the gradient of $f(x)$ and the gradient of $h_1$, $h_2$, and $h_M$ are linear dependent, that is,

$$\nabla f + \lambda_1 \nabla h_1 + \lambda_2 \nabla h_2 \cdots + \lambda_M \nabla h_M = 0 \tag{4A.29}$$

The scaling variable $\lambda$ is called a Lagrange multiplier.

In addition, we can write the Lagrange equation according to equations (4A.25)–(4A.28).

$$L(x_i, \lambda_M) = f(x_i) + \lambda_1 h_1(x_i) + \lambda_2 h_2(x_i) \cdots + \lambda_M h_M(x_i) \quad i = 1, 2, \ldots, N \tag{4A.30}$$

To meet the conditions stated in equation (4A.29), we simply require that the partial derivative of the Lagrange function with respect to each of the unknown variables, $x_1, x_2, \ldots, x_N$ and $\lambda_1, \lambda_2, \ldots, \lambda_M$, be equal to zero. That is,

$$\frac{\partial L}{\partial x_1} = 0$$

$$\frac{\partial L}{\partial x_2} = 0$$

$$\vdots$$

$$\frac{\partial L}{\partial x_N} = 0$$

$$\frac{\partial L}{\partial \lambda_1} = 0$$

$$\frac{\partial L}{\partial \lambda_2} = 0$$

$$\vdots$$

$$\frac{\partial L}{\partial \lambda_M} = 0 \tag{4A.31}$$

## A.10  Kuhn–Tucker Conditions

If inequality constraints are involved in the optimization problem, the optimum is reached if the Kuhn–Tucker conditions are met. These can be stated as below.

$$\min f(x_i), \quad i = 1, 2, \ldots, N \tag{4A.32}$$

such that

$$h_j(x_i) = 0, \quad j = 1, 2, \ldots, M_h \tag{4A.33}$$

$$g_j(x_i) \leq 0, \quad j = 1, 2, \ldots, M_g \tag{4A.34}$$

The Lagrange function can be formed on the basis of equations (4A.32)–(4A.34).

$$L(x, \lambda, \mu) = f(x) + \sum_{j=1}^{M_h} \lambda_j h_j(x) + \sum_{j=1}^{M_g} \mu_j g_j(x) \tag{4A.35}$$

The Kuhn–Tucker conditions for the optimum for the points $x^*, \lambda^*, \mu^*$ are

1. $\dfrac{\partial L}{\partial x_i}(x^*, \lambda^*, \mu^*) = 0, \quad i = 1, 2, \ldots, N$
2. $h_j(x^*) = 0, \quad j = 1, 2, \ldots, M_h$
3. $g_j(x^*) \leq 0, \quad j = 1, 2, \ldots, M_g$
4. $\mu_j^* g_j(x^*) = 0, \quad \mu_j^* \geq 0, \quad j = 1, 2, \ldots, M_g$

The first condition is the set of partial derivatives of the Lagrange function that must equal zero at the optimum. The second and third expressions are a restatement of the constraint conditions on the problem. The fourth is the complementary slackness condition. Since the product $\mu_j^* g_j(x^*)$ equals zero, either $\mu_j^*$ equals to zero or $g_j(x^*)$ equals zero, or both equal zero. If $\mu_j^*$ equals zero, $g_j(x^*)$ is free to be nonbinding; if $\mu_j^*$ is positive, $g_j(x^*)$ must be zero. Thus we can know if the inequality constraint is binding or not by looking at the value of $\mu_j^*$.

# PROBLEMS AND EXERCISES

1. What is the principle of equal incremental rate?

2. What is the B-coefficient formula?

3. What is the correction coefficient of network losses?

4. What is the coordination equation of hydrothermal system economic dispatch?

5. State the advantages and limitations of GA-based economic dispatch.

6. The input–output characteristics of two generating units are as follows:

$$F_1 = 0.0012P_{G1}^2 + 0.3P_{G1} + 2 \ \text{Btu/h}$$

$$F_2 = 0.0009P_{G2}^2 + 0.5P_{G2} + 1 \ \text{Btu/h}$$

Determine the economic operation point for these two units when delivering a total of 600 MW power demand.

7. Suppose the input–output characteristics of three generating units are as follows:

$$F_1 = 0.0005P_{G1}^2 + 0.8P_{G1} + 9 \ \text{Btu/h}$$

$$F_2 = 0.0009P_{G2}^2 + 0.5P_{G2} + 6 \ \text{Btu/h}$$

$$F_3 = 0.0006P_{G3}^2 + 0.7P_{G3} + 8 \ \text{Btu/h}$$

Determine the economic operation point for these three units when delivering a total of 600 MW and 800 MW power demand, respectively.

8. The input–output characteristics of two generating units are as follows:

$$F_1 = 0.001P_{G1}^2 + 0.5P_{G1} + 3 \ \text{Btu/h}$$

$$F_2 = 0.002P_{G2}^2 + 0.3P_{G2} + 5 \ \text{Btu/h}$$

The power output limits of the two units are

$$100 \le P_{G1} \le 280 \ \text{MW}$$

$$150 \le P_{G2} \le 300 \ \text{MW}$$

Determine the economic operation point for these two units when delivering a total of 500 MW power demand.

9. Suppose the input–output characteristics of three generating units are as follows:

$$F_1 = 0.0005P_{G1}^2 + 0.6P_{G1} + 9 \ \text{Btu/h}$$

$$F_2 = 0.0013P_{G2}^2 + 0.5P_{G2} + 6 \ \text{Btu/h}$$

$$F_3 = 0.0008P_{G3}^2 + 0.7P_{G3} + 5 \ \text{Btu/h}$$

The power output limits of the three units are

$$100 \leq P_{G1} \leq 200 \ \text{MW}$$

$$150 \leq P_{G2} \leq 300 \ \text{MW}$$

$$150 \leq P_{G3} \leq 300 \ \text{MW}$$

Determine the economic operation point for these three units when delivering a total of 400 MW and 700 MW power demand, respectively.

10. The input–output characteristics of three generating units are as follows.

$$F_1 = 0.0005 P_{G1}{}^2 + 0.8 P_{G1} + 9 \ \text{Btu/h}$$

$$F_2 = 0.0009 P_{G2}{}^2 + 0.5 P_{G2} + 6 \ \text{Btu/h}$$

$$F_3 = 0.0006 P_{G3}{}^2 + 0.7 P_{G3} + 8 \ \text{Btu/h}$$

(1) Use the gradient method to solve the economic dispatch with a total load of 600 MW.

(2) Use gradient method 1 to solve the economic dispatch with a total load of 600 MW.

(3) Use gradient method 2 to solve the economic dispatch with a total load of 600 MW.

(4) Use gradient method 3 to solve the economic dispatch with a total load of 600 MW.

# REFERENCES

1. Kirchamayer LK. *Economic Operation of Power Systems*. New York: Wiley; 1958.
2. Zhu JZ. *Power System Optimal Operation*. Tutorial of Chongqing University, 1990.
3. Nanda J, Narayanan RB. Application of genetic algorithm to economic load dispatch with Line-flow constraints. Electr. Power and Energ. Syst. 2002;24:723–729.
4. Walters DC, Sheble GB. Genetic algorithm solution of economic dispatch with valve point loading. IEEE Trans. on Power Syst. 1993;8(3).
5. Sheble GB, Brittig K. Refined genetic algorithm-economic dispatch example. IEEE Trans. on Power Syst. 1995;10(1).
6. Hopfield JI. Neural networks and physical systems with emergent collective computational abilities. Proc. Natl. Acad. Sci. USA, 1982;79:2554–2558.
7. Park JH, Kim YS, Eom IK, Lee KY. Economic load dispatch for piecewise quadratic cost function using Hopfield neural networks. IEEE Trans on Power Syst. 1993;8(3):1030–1038.
8. King TD, El-Hawary ME, El-Hawary F. Optimal environmental dispatching of electric power system via an improved Hopfield neural network model. IEEE Trans on Power Syst. 1995;10(3):1559–1565.
9. Lee KY, Yome AS, Park JH. Adaptive neural networks for economic load dispatch. IEEE Trans on Power Syst. 1998;13(2):519–526.
10. Wong KP, Fung CC. Simulated-annealing-based economic dispatch algorithm. IEE Proc. Part C, 1993140(6):509–515.
11. Nocedal J, Wright SJ. *Numerical Optimization*. Springer; 1999.
12. Fletcher, R. *Practical Methods of Optimization*. John Wiley and Sons, 1987.
13. Broyden, CG. The convergence of a class of double-rank minimization algorithms. J. Inst. Maths. Applics. 1970;6:76–90.
14. Fletcher, R. A new approach to variable metric algorithms. Comput. J. 1970;13:317–322.

15. Goldfarb D. A family of variable metric updates derived by variational means. Math. Comput. 1970;24:23–26.
16. Shanno DF. Conditioning of quasi-Newton methods for function minimization. Math. Comput. 1970;24:647–656.
17. Fletcher R, Powell MJD. A rapidly convergent descent method for minimization. Comput. J. 1963;6:163–168.
18. Holland JH. *Adaptation in Nature and Artificial Systems*. The University of Michigan Press; 1975.
19. Goldberg DE. *Genetic Algorithms in Search, Optimization and Machine Learning*. Reading, MA: Addision-Wesley; 1989.

# CHAPTER 5

# *SECURITY-CONSTRAINED ECONOMIC DISPATCH*

Security-constrained economic dispatch (SCED) is a simplified optimal power flow (OPF) problem. It is widely used in the power industry. This chapter first introduces several major approaches to solve the SCED problem, such as linear programming (LP), network flow programming (NFP), and quadratic programming (QP). Then, nonlinear convex network flow programming (NLCNFP) and the genetic algorithm (GA) are added to tackle the SCED problem. Implementation details of these methods and a number of numerical examples are provided in this chapter.

## 5.1 INTRODUCTION

Chapter 4 analyzes the model and algorithm of the classic economic dispatch (ED), where network security constraints are neglected. In practical power systems, it is very important to solve ED with network security constraints. Mathematical optimization methods such as LP, QP, and NFP as well as GSs are applied to solve this problem [1–19].

## 5.2 LINEAR PROGRAMMING METHOD

### 5.2.1 Mathematical Model of Economic Dispatch with Security

The mathematical model of real power ED with security constraints can be written as follows (model M-1):

$$\min \ F = \sum_{i \in NG} f_i(P_{Gi}) \tag{5.1}$$

such that

$$\text{s.t.} \sum_{i \in NG} P_{Gi} = \sum_{k \in ND} P_{Dk} + P_L \tag{5.2}$$

$$|P_{ij}| \le P_{ij\max} \quad ij \in NT \tag{5.3}$$

*Optimization of Power System Operation*, Second Edition. Jizhong Zhu.
© 2015 The Institute of Electrical and Electronics Engineers, Inc. Published 2015 by John Wiley & Sons, Inc.

$$P_{Gimin} \leq P_{Gi} \leq P_{Gimax} \quad i \in NG \tag{5.4}$$

where

$P_D$: the real power load

$P_{ij}$: the power flow of transmission line $ij$

$P_{ij\,max}$: the power limits of transmission line $ij$

$P_{Gi}$: the real power output at generator bus $i$

$P_{Gi\,min}$: the minimal real power output at generator $i$

$P_{Gi\,max}$: the maximal real power output at generator $i$

$P_L$: the network losses

$f_i$: the cost function of the generator $i$

$NT$: the number of transmission lines

$NG$: the number of generators.

Since the input–output characteristic of generator units and system power losses are nonlinear functions, the real power ED model is a nonlinear model. An LP method to solve SCED needs to linearize the objective function and constraints in the model.

## 5.2.2 Linearization of ED Model

***Linearization of Objective Function*** Let the initial operation point of generator $i$ be $P_{Gi}^0$. The nonlinear objective function can be expressed by using Taylor series expansion, with only first two terms being considered, that is,

$$f_i(P_{Gi}) \approx f_i(P_{Gi}^0) + \left. \frac{df_i\left(P_{Gi}\right)}{dP_{Gi}} \right|_{P_{Gi}^0} \Delta P_{Gi} = b\Delta P_{Gi} + c$$

or
$$\tag{5.5}$$

$$f_i(\Delta P_{Gi}) = b\Delta P_{Gi}$$

where

$$b = \left. \frac{df_i\left(P_{Gi}\right)}{dP_{Gi}} \right|_{P_{Gi}^0} \tag{5.6}$$

$$c = f_i(P_{Gi}^0) \tag{5.7}$$

are constant, and
$$\Delta P_{Gi} = P_{Gi} - P_{Gi}^0 \tag{5.8}$$

***Linearization of Power Balance Equation*** Since loads are constant for a given time, we can get the following expression through linearizing the real power balance equation

$$\sum_{i \in NG} \left(1 - \frac{\partial P_L}{\partial P_{Gi}}\right)\Bigg|_{P_{Gi}^0} \Delta P_{Gi} = 0 \tag{5.9}$$

***Linearization of Branch Flow Constraints***  The real power flow equation of a branch can be written as follows.

$$P_{ij} = V_i^2 \, g_{ij} - V_i V_j (g_{ij} \cos \theta_{ij} + b_{ij} \sin \theta_{ij}) \tag{5.10}$$

where

$P_{ij}$: the sending end real power on transmission branch $ij$
$V_i$: the node voltage magnitude of node $i$
$\theta_{ij}$: the difference of node voltage angles between the sending end and receiving end of the line; $ij$
$b_{ij}$: the susceptance of transmission branch $ij$
$g_{ij}$: the conductance of transmission branch $ij$.

Through linearizing equation (5.10), we get the incremental branch power expression as follows;

$$\Delta P_{ij} = -V_i^0 V_j^0 (-g_{ij} \sin \theta_{ij}^0 \Delta \theta_{ij} + b_{ij} \cos \theta_{ij}^0 \Delta \theta_{ij}) \tag{5.11}$$

In a high-voltage power network, the value of $\theta_{ij}$ is very small, and the following approximate equations are easily obtained:

$$\sin \theta_{ij} \cong 0 \tag{5.12}$$

$$\cos \theta_{ij} \cong 1 \tag{5.13}$$

In addition, assume that the magnitudes of all bus voltages are the same and equal to 1.0 p.u. Furthermore, suppose the reactance of the branch is much bigger than the resistance of the branch, so that we can neglect the resistance of the branch. Thus,

$$g_{ij} = \frac{R_{ij}}{R_{ij}^2 + X_{ij}^2} \approx 0 \tag{5.14}$$

$$b_{ij} = -\frac{X_{ij}}{R_{ij}^2 + X_{ij}^2} \approx -\frac{X_{ij}}{X_{ij}^2} \approx -\frac{1}{X_{ij}} \tag{5.15}$$

Substituting equations (5.12)–(5.15) into equation (5.11), we get

$$\Delta P_{ij} = -b_{ij}\Delta \theta_{ij} = -b_{ij}(\Delta \theta_i - \Delta \theta_j) = \frac{\Delta \theta_i - \Delta \theta_j}{X_{ij}} \tag{5.16}$$

The above equation can also be written as matrix form, that is,

$$\Delta P_b = B' \Delta \theta \tag{5.17}$$

where the elements of the susceptance matrix $B'$ are

$$B'_{ij} = b_{ij} = -\frac{1}{X_{ij}} \tag{5.18}$$

$$B'_{ii} = -\sum_{\substack{j=1 \\ j \neq i}}^{n} b_{ij} \tag{5.19}$$

From Chapter 2, the bus power injection equation can be written as

$$P_{Gi} - P_{Di} = V_i \sum_{j=1}^{n} V_j (g_{ij} \cos \theta_{ij} + b_{ij} \sin \theta_{ij}) \tag{5.20}$$

Since the load demand is constant, the linearization expression of equation (5.20) can be written as follows:

$$\Delta P_{Gi} = V_i^0 \sum_{j=1}^{n} V_j^0 (-g_{ij} \sin \theta_{ij}^0 \Delta \theta_{ij} + b_{ij} \cos \theta_{ij}^0 \Delta \theta_{ij})$$

$$= V_i^0 \sum_{j=1}^{n} V_j^0 (-g_{ij} \sin \theta_{ij}^0 + b_{ij} \cos \theta_{ij}^0) \Delta \theta_{ij} \tag{5.21}$$

This equation can also be written in the following matrix form:

$$\Delta P_G = H \Delta \theta \tag{5.22}$$

Equation (5.22) stands for the relationship between the incremental generator output power (except for the generator that is taken as the slack unit) and the incremental bus voltage angle. Matrix $H$ can also be simplified using equations (5.12)–(5.15).

According to equations (5.17) and (5.22), we can get the direct linear relationship between the incremental branch power flow and incremental generator output power, that is,

$$\Delta P_b = B' \Delta \theta = B' H^{-1} \Delta P_G = D \Delta P_G \tag{5.23}$$

where

$$D = B' H^{-1} \tag{5.24}$$

which is also called the linear sensitivity of the branch power flow with respect to the generator power output.

Thus, the linear expression of the branch power flow constraints can be written as

$$|D\Delta P_G| \leq \Delta P_{b\max} \tag{5.25}$$

The element of the matrix $\Delta P_{b\max}$ is the incremental power flow limit $\Delta P_{ij\max}$ of the branch $ij$, that is,

$$\Delta P_{ij\max} = P_{ij\max} - P_{ij}^0 \tag{5.26}$$

If the branch outage is considered in the real power ED, the outage transfer distribution factors (OTDFs) in Chapter 3 will be used. So the sensitivity factor OTDF between branch $ij$ and generator bus $i$ when line $l$ is opened is written as

$$\text{OTDF}_{ij,i} = \frac{\Delta P_{ij}}{\Delta P_{Gi}} = (S_{ij,i} + \text{LODF}_{ij,i}S_{l,i}) \tag{5.27}$$

In this case, the branch power flow can be written as

$$\Delta P_{ij} = (S_{ij,i} + \text{LODF}_{ij,i}S_{l,i})\Delta P_{Gi} \tag{5.28}$$

The matrix form of the equation (5.28) is

$$\Delta P_b = D'\Delta P_G \tag{5.29}$$

The corresponding branch power flow constraints are written as

$$|D'\Delta P_G| \leq \Delta P'_{b\max} \tag{5.30}$$

Comparing with $D$, $\Delta P_{b\max}$ in equation (5.25), $D'$, $\Delta P'_{b\max}$ in equation (5.30) consider the effect of the branch outage. In this case, the real power ED is called the $N-1$ security economic dispatch.

***Generator Output Power Constraint***  The incremental form of the generator output power constraint is

$$P_{Gi\min} - P_{Gi}^0 \leq \Delta P_{Gi} \leq P_{Gi\max} - P_{Gi}^0 \quad i \in NG \tag{5.31}$$

### 5.2.3  Linear Programming Model

The linearized ED model can be written as the standard LP form.

$$\min Z = c_1 x_1 + c_2 x_2 + \cdots + c_N x_N$$

such that

$$a_{11}x_1 + a_{12}x_2 + \ldots + a_{1N}x_N \geq b_1$$

$$a_{21}x_1 + a_{22}x_2 + \ldots + a_{2N}x_N \geq b_2$$

$$\vdots$$

$$a_{N1}x_1 + a_{N2}x_2 + \ldots + a_{NN}x_N \geq b_N$$

$$x_{imin} \leq x_i \leq x_{imax}$$

The basic algorithm for LP can be found in the Appendix in Chapter 9.

### 5.2.4   Implementation

***Solution Steps of ED by LP***   The above-mentioned method for solving ED by LP uses an iterative technique to obtain the optimal solution, so it is also called a successive linear programming (SLP) method. The solution procedures of SLP for ED are summarized in the following steps.

Step 1.   Select the set of initial control variables.

Step 2.   Solve the power flow problem to obtain a feasible solution that satisfies the power balance equality constraint.

Step 3.   Linearize the objective function and inequality constraints around the power flow solution and formulate the LP problem.

Step 4.   Solve the LP problem and obtain optimal incremental control variables $\Delta P_{Gi}$.

Step 5.   Update the control variables: $P_{Gi}^{(k+1)} = P_{Gi}^{(k)} + \Delta P_{Gi}$.

Step 6.   Obtain the power flow solution with updated control variables.

Step 7.   Check the convergence. If $\Delta P_{Gi}$ in step 4 are below the user-defined tolerance, the solution converges. Otherwise, go to step 3.

***Test Results***   The LP-based ED method is tested on IEEE 5-bus and 30-bus systems. The network topologies of the IEEE test systems are shown in Figure 5.1. The corresponding system data and parameters are listed in Tables 5.1–5.3. The data and parameters of the 30-bus system are listed in Tables 5.4–5.6.

The calculation results of ED with $N$ security for the IEEE 5-bus system are shown in Table 5.7. The calculation results of ED with $N$ security for the IEEE 30-bus system are show in Table 5.8, and $N-1$ security ED results are listed in Table 5.9.

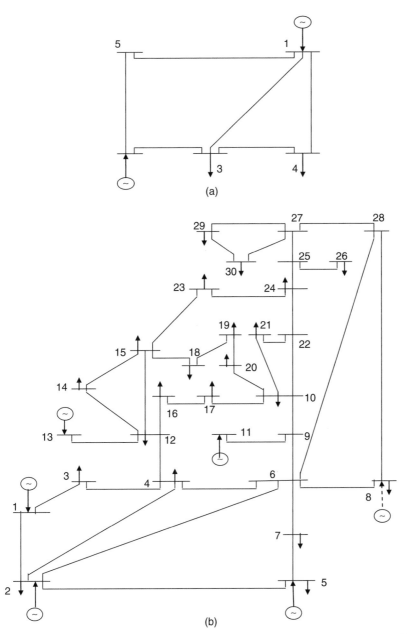

Figure 5.1    IEEE test systems (a) IEEE 5-bus system; (b) one-line diagram of IEEE 30-bus system.

**TABLE 5.1 Generator Data of 5-Bus System**

| Generators | #1 | #2 |
|---|---|---|
| $P_{Gi\,max}$(p.u.) | 1.00 | 1.00 |
| $P_{Gi\,min}$(p.u.) | 0.20 | 0.20 |
| $Q_{Gi\,max}$(p.u.) | 0.80 | 0.80 |
| $Q_{Gi\,min}$(p.u.) | −0.20 | −0.20 |
| Quadratic cost function | | |
| $a_i$ | 50.00 | 50.00 |
| $b_i$ | 351.00 | 389.00 |
| $c_i$ | 44.40 | 40.60 |

**TABLE 5.2 Load Data of 5-Bus System**

| Load Bus | #3 | #4 | #5 |
|---|---|---|---|
| MW load $P_D$(p.u.) | 0.60 | 0.40 | 0.60 |
| MVAR load $Q_D$(p.u.) | 0.30 | 0.10 | 0.20 |

**TABLE 5.3 Line Data of 5-Bus System**

| Line No. | From−to Bus | Resistance | Reactance | Line Charge |
|---|---|---|---|---|
| 1 | 1−3 | 0.10 | 0.40 | 0.00 |
| 2 | 4−1 | 0.15 | 0.60 | 0.00 |
| 3 | 5−1 | 0.05 | 0.20 | 0.00 |
| 4 | 3−2 | 0.05 | 0.20 | 0.00 |
| 5 | 2−5 | 0.05 | 0.20 | 0.00 |
| 6 | 3−4 | 0.10 | 0.40 | 0.00 |

**TABLE 5.4 Generator Data of 30-Bus System**

| Generators | #1 | #2 | #5 | #8 | #11 | #13 |
|---|---|---|---|---|---|---|
| $P_{Gi\,max}$(p.u.) | 2.00 | 0.80 | 0.50 | 0.35 | 0.30 | 0.40 |
| $P_{Gi\,min}$(p.u.) | 0.50 | 0.20 | 0.15 | 0.10 | 0.10 | 0.12 |
| $Q_{Gi\,max}$(p.u.) | 2.50 | 1.00 | 0.80 | 0.60 | 0.50 | 0.60 |
| $Q_{Gi\,min}$(p.u.) | −0.20 | −0.20 | −0.15 | −0.15 | −0.10 | −0.15 |
| Quadratic cost function | | | | | | |
| $a_i$ | 0.00375 | 0.0175 | 0.0625 | 0.0083 | 0.0250 | 0.0250 |
| $b_i$ | 2.00000 | 1.7500 | 1.0000 | 3.2500 | 3.0000 | 3.0000 |
| $c_i$ | 0.00000 | 0.0000 | 0.0000 | 0.0000 | 0.0000 | 0.0000 |

**TABLE 5.5    Load Data of 30-Bus System**

| Bus No. | $P_D$(p.u.) | $Q_D$(p.u.) | Bus No. | $P_D$(p.u.) | $Q_D$(p.u.) |
|---------|-------------|-------------|---------|-------------|-------------|
| 1  | 0.000 | 0.000 | 16 | 0.035 | 0.016 |
| 2  | 0.217 | 0.127 | 17 | 0.090 | 0.058 |
| 3  | 0.024 | 0.012 | 18 | 0.032 | 0.009 |
| 4  | 0.076 | 0.016 | 19 | 0.095 | 0.034 |
| 5  | 0.942 | 0.190 | 20 | 0.022 | 0.007 |
| 6  | 0.000 | 0.000 | 21 | 0.175 | 0.112 |
| 7  | 0.228 | 0.109 | 22 | 0.000 | 0.000 |
| 8  | 0.300 | 0.300 | 23 | 0.032 | 0.016 |
| 9  | 0.000 | 0.000 | 24 | 0.087 | 0.067 |
| 10 | 0.058 | 0.020 | 25 | 0.000 | 0.000 |
| 11 | 0.000 | 0.000 | 26 | 0.035 | 0.023 |
| 12 | 0.112 | 0.075 | 27 | 0.000 | 0.000 |
| 13 | 0.000 | 0.000 | 28 | 0.000 | 0.000 |
| 14 | 0.062 | 0.016 | 29 | 0.024 | 0.009 |
| 15 | 0.082 | 0.025 | 30 | 0.106 | 0.019 |

**TABLE 5.6    Line Data of 30-Bus System**

| Line No. | From–to Bus | Resistance (p.u.) | Reactance (p.u.) | Line Limit (p.u.) |
|----------|-------------|-------------------|------------------|-------------------|
| 1  | 1–2  | 0.0192 | 0.0575 | 1.30 |
| 2  | 1–3  | 0.0452 | 0.1852 | 1.30 |
| 3  | 2–4  | 0.0570 | 0.1737 | 0.65 |
| 4  | 3–4  | 0.0132 | 0.0379 | 1.30 |
| 5  | 2–5  | 0.0472 | 0.1983 | 1.30 |
| 6  | 2–6  | 0.0581 | 0.1763 | 0.65 |
| 7  | 4–6  | 0.0119 | 0.0414 | 0.90 |
| 8  | 5–7  | 0.0460 | 0.1160 | 0.70 |
| 9  | 6–7  | 0.0267 | 0.0820 | 1.30 |
| 10 | 6–8  | 0.0120 | 0.0420 | 0.32 |
| 11 | 6–9  | 0.0000 | 0.2080 | 0.65 |
| 12 | 6–10 | 0.0000 | 0.5560 | 0.32 |
| 13 | 9–10 | 0.0000 | 0.2080 | 0.65 |

(*continued*)

**TABLE 5.6** (*Continued*)

| Line No. | From–to Bus | Resistance (p.u.) | Reactance (p.u.) | Line Limit (p.u.) |
|---|---|---|---|---|
| 14 | 9–11 | 0.0000 | 0.1100 | 0.65 |
| 15 | 4–12 | 0.0000 | 0.2560 | 0.65 |
| 16 | 12–13 | 0.0000 | 0.1400 | 0.65 |
| 17 | 12–14 | 0.1231 | 0.2559 | 0.32 |
| 18 | 12–15 | 0.0662 | 0.1304 | 0.32 |
| 19 | 12–16 | 0.0945 | 0.1987 | 0.32 |
| 20 | 14–15 | 0.2210 | 0.1997 | 0.16 |
| 21 | 16–17 | 0.0824 | 0.1932 | 0.16 |
| 22 | 15–18 | 0.1070 | 0.2185 | 0.16 |
| 23 | 18–19 | 0.0639 | 0.1292 | 0.16 |
| 24 | 19–20 | 0.0340 | 0.0680 | 0.32 |
| 25 | 10–20 | 0.0936 | 0.2090 | 0.32 |
| 26 | 10–17 | 0.0324 | 0.0845 | 0.32 |
| 27 | 10–21 | 0.0348 | 0.0749 | 0.32 |
| 28 | 10–22 | 0.0727 | 0.1499 | 0.32 |
| 29 | 21–22 | 0.0116 | 0.0236 | 0.32 |
| 30 | 15–23 | 0.1000 | 0.2020 | 0.16 |
| 31 | 22–24 | 0.1150 | 0.1790 | 0.16 |
| 32 | 23–24 | 0.1320 | 0.2700 | 0.16 |
| 33 | 24–25 | 0.1885 | 0.3292 | 0.16 |
| 34 | 25–26 | 0.2544 | 0.3800 | 0.16 |
| 35 | 25–27 | 0.1093 | 0.2087 | 0.16 |
| 36 | 28–27 | 0.0000 | 0.3960 | 0.65 |
| 37 | 27–29 | 0.2198 | 0.4153 | 0.16 |
| 38 | 27–30 | 0.3202 | 0.6027 | 0.16 |
| 39 | 29–30 | 0.2399 | 0.4533 | 0.16 |
| 40 | 8–28 | 0.0636 | 0.2000 | 0.32 |
| 41 | 6–28 | 0.0169 | 0.0599 | 0.32 |
| 42 | 10–10 | 0.0000 | −5.2600 | |
| 43 | 24–24 | 0.0000 | −25.0000 | |

**TABLE 5.7    Economic Dispatch Results for 5-Bus System**

| Method | LP | $P_{i\min}$ | $P_{i\max}$ |
|---|---|---|---|
| $P_{G1}$(p.u.) | 0.9786 | 0.2 | 1.0 |
| $P_{G2}$(p.u.) | 0.6662 | 0.2 | 1.0 |
| Total cost ($/hr) | 757.74 | – | – |
| Total loss (p.u.) | 0.0449 | – | – |

**TABLE 5.8    *N* Security Economic Dispatch Results by LP for IEEE 30-Bus System**

| Generation No. | Economic Dispatch | $P_{Gi\min}$ | $P_{Gi\max}$ |
|---|---|---|---|
| $P_{G1}$ | 1.7626 | 0.50 | 2.00 |
| $P_{G2}$ | 0.4884 | 0.20 | 0.80 |
| $P_{G5}$ | 0.2151 | 0.15 | 0.50 |
| $P_{G8}$ | 0.2215 | 0.10 | 0.35 |
| $P_{G11}$ | 0.1214 | 0.10 | 0.30 |
| $P_{G13}$ | 0.1200 | 0.12 | 0.40 |
| Total generation | 2.9290 | – | – |
| Total real power losses | 0.0948 | – | – |
| Total generation cost ($) | 802.4000 | – | – |

**TABLE 5.9    *N* – 1 Security Economic Dispatch Results by LP for IEEE 30-Bus System**

| Generator No. | Economic Dispatch | $P_{Gi\min}$ | $P_{Gi\max}$ |
|---|---|---|---|
| $P_{G1}$(p.u.) | 1.3854 | 0.50 | 2.00 |
| $P_{G2}$(p.u.) | 0.5756 | 0.20 | 0.80 |
| $P_{G5}$(p.u.) | 0.2456 | 0.15 | 0.50 |
| $P_{G8}$(p.u.) | 0.3500 | 0.10 | 0.35 |
| $P_{G11}$(p.u.) | 0.1793 | 0.10 | 0.30 |
| $P_{G13}$(p.u.) | 0.1691 | 0.12 | 0.40 |
| Total generation (p.u.) | 2.9050 | – | – |
| Total Cost ($/hr) | 813.74 | – | – |
| Total loss (p.u.) | 0.0711 | – | – |

## 5.2.5   Piecewise Linear Approach

Assume the objective function is a quadratic characteristic, which can also be linearized by a piecewise linear approach.

If the objective function is divided into $N$ linear segments, the real power variable of each generator will also be divided into $N$ variables. Figure 5.2 is an objective function with three linear segments. The corresponding slopes are $b_1$, $b_2$, and $b_3$, respectively.

From Figure 5.2, the generator power output variables for each segment can be presented as follows:

$$P_{Gimin} \leq P_{Gi1} \leq P_{G1max} \tag{5.32}$$

$$P_{G1max} \leq P_{Gi2} \leq P_{G2max} \tag{5.33}$$

$$P_{G2max} \leq P_{Gi3} \leq P_{Gimax} \tag{5.34}$$

If $P_{Gi\,min}$ is selected as the initial generator output power, the incremental generator power outputs for each segment can be expressed as

$$\Delta P_{Gi1} = P_{Gi1} - P_{Gimin} \tag{5.35}$$

$$\Delta P_{Gi2} = P_{Gi2} - P_{Gi1max} \tag{5.36}$$

$$\Delta P_{Gi3} = P_{Gi3} - P_{Gi2max} \tag{5.37}$$

Thus, the constraint equations (5.32)–(5.34) become

$$0 \leq \Delta P_{Gi1} \leq P_{Gi1max} - P_{Gimin} \tag{5.38}$$

$$0 \leq \Delta P_{Gi2} \leq P_{Gi2max} - P_{Gi1max} \tag{5.39}$$

$$0 \leq \Delta P_{Gi3} \leq P_{Gimax} - P_{Gi2max} \tag{5.40}$$

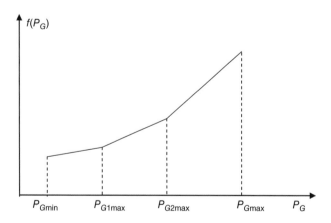

Figure 5.2   Piecewise linear objective function.

The piecewise linear objective function becomes

$$F = \sum_{i=1}^{NG} f_i(P_{Gi}) = \sum_{k=1}^{3} \sum_{i=1}^{NG} b_k \Delta P_{Gik} \tag{5.41}$$

Replacing the incremental generator power output $\Delta P_{Gi}$ in the constraints (5.9) and (5.30) in Section 5.2.2 by $\sum_{k=1}^{3} \Delta P_{Gik}$, we can also obtain the LP model for the ED problem.

## 5.3  QUADRATIC PROGRAMMING METHOD

A QP model contains a quadratic objective function and linear constraints. As mentioned early in this chapter, the ED problem is a nonlinear mathematical model. We discuss the successive LP method for solving the ED problem in Section 5.2. The successive LP method can also be used in the QP model of ED.

### 5.3.1  QP Model of Economic Dispatch

Let the initial operation point of generator $i$ be $P_{Gi}^0$. The nonlinear objective function can be expressed by useing Taylor series expansion, only first three terms being considered, that is,

$$f_i(P_{Gi}) \approx f_i(P_{Gi}^0) + \left. \frac{df_i(P_{Gi})}{dP_{Gi}} \right|_{P_{Gi}^0} \Delta P_{Gi} + \frac{1}{2} \left. \frac{df_i^2(P_{Gi})}{dP_{Gi}^2} \right|_{P_{Gi}^0} \Delta P_{Gi}^2$$

$$= a \Delta P_{Gi}^2 + b \Delta P_{Gi} + c \tag{5.42}$$

or

$$f_i(\Delta P_{Gi}) = a \Delta P_{Gi}^2 + b \Delta P_{Gi} \tag{5.43}$$

where

$$a = \frac{1}{2} \left. \frac{df_i'(P_{Gi})}{dP_{Gi}} \right|_{P_{Gi}^0} \tag{5.44}$$

$$b = f_i'(P_{Gi}) = \left. \frac{df_i(P_{Gi})}{dP_{Gi}} \right|_{P_{Gi}^0} \tag{5.45}$$

$$c = f_i(P_{Gi}^0) \tag{5.46}$$

are constant, and

$$\Delta P_{Gi} = P_{Gi} - P_{Gi}^0 \tag{5.47}$$

Linearizing the constraints using the same approach used in Section 5.2, the QP model of real power ED can be written as follows:

$$\min f_i(\Delta P_{Gi}) = \sum_{i=1}^{N}(a\Delta P_{Gi}^2 + b\Delta P_{Gi}) \tag{5.48}$$

s.t.

$$\sum_{i \in NG}\left(1 - \frac{\partial P_L}{\partial P_{Gi}}\right)\bigg|_{P_{Gi}^0}\Delta P_{Gi} = 0 \tag{5.49}$$

$$P_{Gimin} - P_{Gi}^0 \le \Delta P_{Gi} \le P_{Gimax} - P_{Gi}^0 \quad i \in NG \tag{5.50}$$

$$|D'\Delta P_G| \le \Delta P'_{bmax} \tag{5.51}$$

## 5.3.2 QP Algorithm

The ED model in equations (5.48)–(5.51) can be written as a standard QP model.

$$\min f(X) = CX + X^T QX \tag{5.52}$$

such that

$$AX \le B \tag{5.53}$$

$$X \ge 0 \tag{5.54}$$

where $C$ is an $n$-dimensional row vector describing the coefficients of the linear terms in the objective function, and $Q$ is an $(n \times n)$ symmetric matrix describing the coefficients of the quadratic terms.

As in LP, the decision variables are denoted by the $n$-dimensional column vector $X$, and the constraints are defined by an $(m \times n)$ $A$ matrix and an $m$-dimensional column vector $B$ of right-hand-side coefficients. For the real power ED problem, we know that a feasible solution exists and that the constraint region is bounded.

When the objective function $f(X)$ is strictly convex for all feasible points, the problem has a unique local minimum which is also the global minimum. A sufficient condition to guarantee strictly convexity is for $Q$ to be positive definite. This is generally true for most ED problems.

Equation (5.53) can be expressed as

$$g(X) = (AX - B) \le 0 \tag{5.55}$$

Form the Lagrange function for equations (5.52) and (5.55), that is,

$$L(X, \mu) = CX + X^T QX + \mu g(X) \tag{5.56}$$

where $\mu$ is an $m$-dimensional row vector.

According to the optimization theory, the Kuhn–Tucker (KT) conditions for a local minimum are given as follows.

$$
\begin{cases}
\dfrac{\partial L}{\partial X_j} \geq 0, & j = 1, \ldots, n \\[2mm]
C + 2X^T Q + \mu A \geq 0
\end{cases}
\tag{5.57}
$$

$$
\begin{cases}
\dfrac{\partial L}{\partial \mu_i} \leq 0, & i = 1, \ldots, m \\[2mm]
AX - B \leq 0
\end{cases}
\tag{5.58}
$$

$$
\begin{cases}
X_j \dfrac{\partial L}{\partial X_j} = 0, & j = 1, \ldots, n \\[2mm]
X^T \left( C^T + 2QX + A^T \mu \right) = 0
\end{cases}
\tag{5.59}
$$

$$
\begin{cases}
\mu_i g_i(X) = 0, & i = 1, \ldots, m \\[2mm]
\mu(AX - B) = 0
\end{cases}
\tag{5.60}
$$

$$
\begin{cases}
X \geq 0 \\[1mm]
\mu \geq 0
\end{cases}
\tag{5.61}
$$

If we introduce nonnegative surplus variables **y** to the inequalities in equation (5.57) and nonnegative slack variables **v** to the inequalities in equation (5.58), we get the following equivalent form.

$$
C^T + 2QX + A^T \mu^T - y = 0
\tag{5.62}
$$

$$
AX - B + v = 0
\tag{5.63}
$$

Then, the KT conditions can be written as follows:

$$
2QX + A^T \mu^T - y = -C^T
\tag{5.64}
$$

$$
AX + v = B
\tag{5.65}
$$

$$
X \geq 0, \quad \mu \geq 0, \quad y \geq 0, \quad v \geq 0
\tag{5.66}
$$

$$
y^T X = 0, \quad \mu v = 0
\tag{5.67}
$$

The first two expressions are linear equalities, the third restricts all the variables to be nonnegative, and the fourth is the complementary slackness condition.

Obviously, the KT conditions in equations (5.64)–(5.67) have a linear form with the variables $X$, $\mu$, $y$, and $v$. An approach similar to the modified simplex can be used to solve equations (5.64)–(5.67). The steps of the algorithm are as follows:

(1) Let the structural constraints be equations (5.64) and (5.65) defined by the KT conditions.

(2) If any of the right-hand-side values are negative, multiply the corresponding equation by $-1$.

(3) Add an artificial variable to each equation.

(4) Let the objective function be the sum of the artificial variables.

(5) Put the resultant problem into simplex form.

The goal is to find the solution to the LP problem that minimizes the sum of the artificial variables with the additional requirement that the complementary slackness conditions be satisfied at each iteration. If the sum is zero, the solution will satisfy equations (5.64)–(5.67). To accommodate equation (5.67), the rule for selecting the entering variable must be modified with the following relationships.

$$X_j \text{ and } y_j \text{ are complementary for } j = 1, \dots, n$$

$$\mu_i \text{ and } v_i \text{ are complementary for } i = 1, \dots, m$$

The entering variable will be that whose reduced cost is most negative provided that its complementary variable is not in the basis or would leave the basis on the same iteration. At the conclusion of the algorithm, the vector $x$ defines the optimal solution and the vector $\mu$ defines the optimal dual variables.

This approach has been shown to work well when the objective function is positive definite, and requires computational effort comparable to an LP problem with $m + n$ constraints, where $m$ is the number of constraints and $n$ is the number of variables in the QP. Fortunately, the objective function in economic power dispatch is positive definite. Thus, this approach is very good for solving the QP model of ED.

## 5.3.3 Implementation

The first example is to solve the following QP problem using the algorithm mentioned in Section 5.3.2.

$$\min f(x) = x_1^2 + 4x_2^2 - 8x_1 - 16x_2$$

subject to

$$x_1 + x_2 \leq 5$$

$$x_1 \leq 3$$

$$x_1 \geq 0, \quad x_2 \geq 0$$

*Solution*: Convert the problem into the following QP model.

$$\min f(X) = CX + X^T QX$$

such that

$$AX \leq B$$

$$X \geq 0$$

where

$$C^T = \begin{bmatrix} -8 \\ -16 \end{bmatrix}$$

$$Q = \begin{bmatrix} 1 & 0 \\ 0 & 4 \end{bmatrix}$$

$$A = \begin{bmatrix} 1 & 1 \\ 1 & 0 \end{bmatrix}$$

$$B = \begin{bmatrix} 5 \\ 3 \end{bmatrix}$$

$$X = \begin{bmatrix} x_1 \\ x_2 \end{bmatrix}$$

As can be seen, the **Q** matrix is positive definite so the KT conditions are necessary and sufficient for a global optimum.

Let

$$y = \begin{bmatrix} y_1 \\ y_2 \end{bmatrix}$$

$$v = \begin{bmatrix} v_1 \\ v_2 \end{bmatrix}$$

$$\mu = \begin{bmatrix} \mu_1 \\ \mu_2 \end{bmatrix}$$

According to equations (5.64) and (5.65), we get

$$2x_1 + \mu_1 + \mu_2 - y_1 = 8$$

$$8x_2 + \mu_1 - y_2 = 16$$

$$x_1 + x_2 + v_1 = 5$$

$$x_1 + v_2 = 3$$

To create the appropriate LP problem, we add artificial variables to each constraint and minimize their sum.

$$\min Z = w_1 + w_2 + w_3 + w_4$$

such that

$$2x_1 + \mu_1 + \mu_2 - y_1 + w_1 = 8$$

$$8x_2 + \mu_1 - y_2 + w_2 = 16$$

$$x_1 + x_2 + v_1 + w_3 = 5$$

$$x_1 + v_2 + w_4 = 3$$

$$x_1 \geq 0, x_2 \geq 0, y_1 \geq 0, y_2 \geq 0, v_1 \geq 0, v_2 \geq 0, \mu_1 \geq 0, \mu_2 \geq 0,$$

Applying the presented algorithm to this example, the optimal solution to the original problem is $(x_1^*, x_2^*) = (3, 2)$. Table 5.10 shows the iterations of the solution.

The second example is to apply the presented QP algorithm to solve the real power ED problem. The testing system is the IEEE 30-bus system, the data of which are given in Section 5.2. The following testing cases are conducted.

Case 1: IEEE 30-bus system with the original data.

Case 2: IEEE 30-bus system with the original data, but the limit of the line 1 is reduced to 1.0 p.u.

The security ED results for the two cases are shown in Table 5.11. The results of Case 1 are also compared with those obtained by using LP, which are shown in Table 5.12. It can be observed that the ED results obtained by QP are a little better than those obtained by LP.

## 5.4 NETWORK FLOW PROGRAMMING METHOD

### 5.4.1 Introduction

NFP is a specialized LP. It is characterized by simple manipulation and rapid convergence. NFP models of $N$ security ED have been proposed in recent years.

**TABLE 5.10 Iterations for QP Example**

| Iterations | Basic Variables | Solution | Objective Values | Entering Variable | Leaving Variable |
|---|---|---|---|---|---|
| 1 | $(w_1, w_2, w_3, w_4)$ | (8,16,5,3) | 32 | $x_2$ | $w_2$ |
| 2 | $(w_1, x_2, w_3, w_4)$ | (8,2,3,3) | 14 | $x_1$ | $w_3$ |
| 3 | $(w_1, x_2, x_1, w_4)$ | (2,2,3,0) | 2 | $\mu_1$ | $w_4$ |
| 4 | $(w_1, x_2, x_3, \mu_1)$ | (2,2,3,0) | 2 | $\mu_1$ | $w_1$ |
| 5 | $(\mu_1, x_2, x_3, \mu_1)$ | (2,2,3,0) | 0 | / | / |

**TABLE 5.11   Economic Dispatch Results by QP for IEEE 30-Bus System**

| Generation No. | Case 1 | Case 2 |
|---|---|---|
| $P_{G1}$ | 1.7586 | 1.5174 |
| $P_{G2}$ | 0.4883 | 0.5670 |
| $P_{G5}$ | 0.2151 | 0.2326 |
| $P_{G8}$ | 0.2233 | 0.3045 |
| $P_{G11}$ | 0.1231 | 0.1517 |
| $P_{G13}$ | 0.1200 | 0.1400 |
| Total generation | 2.9285 | 2.9132 |
| Total real power losses | 0.0945 | 0.0792 |
| Total generation cost ($) | 802.3900 | 807.2400 |

**TABLE 5.12   ED Results and Comparison Between QP and LP for IEEE 30-Bus System**

| Generation No. | QP Method | LP Method |
|---|---|---|
| $P_{G1}$ | 1.7586 | 1.7626 |
| $P_{G2}$ | 0.4883 | 0.4884 |
| $P_{G5}$ | 0.2151 | 0.2151 |
| $P_{G8}$ | 0.2233 | 0.2215 |
| $P_{G11}$ | 0.1231 | 0.1214 |
| $P_{G13}$ | 0.1200 | 0.1200 |
| Total generation | 2.9285 | 2.9290 |
| Total real power losses | 0.0945 | 0.0948 |
| Total generation cost ($) | 802.3900 | 802.4000 |

This section first presents a network flow model and uses the out-of-kilter algorithm (OKA) for solving the on-line economic power dispatch with $N$ and $N-1$ security. A fast $N-1$ security analysis method solved by OKA is applied to seek out all the over-constrained cases for all possible single-line outages, and then an "$(N-1)$- constrained zone" is formed that is coordinated with the network flow model. On the basis of the normal operating state, a corrective incremental network flow model for ED is established for the over-constrained cases. Consequently, the calculation burden is reduced significantly and the shortcoming of the NFP imprecision, is mitigated to some extent.

## 5.4.2   Out-of-kilter Algorithm

***OKA Model***   According to graph theory, a network with $n$ nodes and $m$ arcs (branches) can be shown as in Figure 5.3(a). The corresponding minimum cost flow problem can be expressed as follows.

$$\min C = \sum_{ij} C_{ij}f_{ij} \quad ij \in m \tag{5.68}$$

such that

$$\sum_{j\in n}(f_{ij} - f_{ji}) = r_i \quad i \in n \tag{5.69}$$

$$L_{ij} \le f_{ij} \le U_{ij} \quad ij \in m \tag{5.70}$$

where,

$C_{ij}$: the arc cost per unit flow
$f_{ij}$: the flow on the arc $ij$ in the network
$L_{ij}$: the lower bound of the flow on the arc $ij$ in the network
$U_{ij}$: the upper bound of flow on the arc $ij$ in the network
 $n$: the total number of the nodes in the network
 $m$: the total number of the arcs in the network.

According to the "out-of-kilter" algorithm (OKA) of NFP, we can transform the original network into an OKA network by introducing a "return arc" from sink node $t$ to source node $s$, while the internal flows remain unchanged. The return arc flow $f_{ts}$ equals the original network flow $r$. The OKA network model is shown in Figure 5.3(b).

Similarly, if the original network has multiple sources and multiple sinks, which is shown in Figure 5.4(a), the corresponding OKA model can be formed as shown in

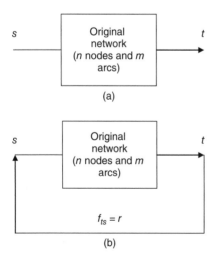

Figure 5.3   (a and b) OKA network model with one source $s$ and one sink $t$.

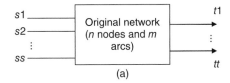

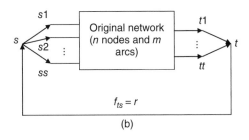

Figure 5.4   (a and b) OKA network model with multiple source $ss$ and multiple sinks $tt$.

Figure 5.4(b), where each source corresponds to a source arc connecting to a total source node $s$ and each sink forms a sink arc connecting to the total sink node $t$.

The corresponding mathematical model for OKA as follows:

$$\min C = \sum_{ij} C_{ij}f_{ij} \quad ij \in (m + ss + tt + 1) \tag{5.71}$$

such that

$$\sum_{j \in n}(f_{ij} - f_{ji}) = 0 \quad i \in n \tag{5.72}$$

$$L_{ij} \le f_{ij} \le U_{ij} \quad ij \in (m + ss + tt + 1) \tag{5.73}$$

where $m$ is the total number of arcs other than the return arc.

**Complementary Slackness Conditions for Optimality of OKA**   The model consisting of equations (5.71)–(5.73) is a specialized LP model. According to the dual theory, the corresponding primary problem and dual problem can be expressed as follows.

*Primary Problem*

$$\max F' = -\sum_{ij} C_{ij}f_{ij} \tag{5.74}$$

such that

$$\sum_{j \in n}(f_{ij} - f_{ji}) = 0 \tag{5.75}$$

$$L_{ij} \le f_{ij} \le U_{ij} \quad i \in n, j \in n, ij \in (m + ss + tt + 1) \tag{5.76}$$

*Dual Problem*

$$\min G = \sum_{ij} U_{ij}\alpha_{ij} - \sum_{ij} L_{ij}\beta_{ij} \tag{5.77}$$

such that

$$C_{ij} + \pi_i - \pi_j + \alpha_{ij} - \beta_{ij} \geq 0 \tag{5.78}$$

$$\alpha_{ij} \geq 0, \quad \beta_{ij} \geq 0 \tag{5.79}$$

$$i \in n, \quad j \in n, \quad ij \in (m + ss + tt + 1)$$

where $\pi$ is the dual variable of the variable $f$ of the primary problem. $\alpha$ and $\beta$ correspond to the dual variables of the upper and lower limits of the primary problem.

When all the variables $f$, $\pi$, $\alpha$, and $\beta$ meet the requirements of the constraints, the following relationship exists between the objective functions of the primary and dual problems.

$$G - F' = \sum_{ij} U_{ij}\alpha_{ij} - \sum_{ij} L_{ij}\beta_{ij} + \sum_{ij} C_{ij}f_{ij}$$

$$= 0 \cdot (\pi_s - \pi_t) + \sum_{ij} U_{ij}\alpha_{ij} - \sum_{ij} L_{ij}\beta_{ij} + \sum_{ij} C_{ij}f_{ij}$$

$$= \sum_j \sum_i \pi_i(f_{ij} - f_{ji}) + \sum_{ij} U_{ij}\alpha_{ij} - \sum_{ij} L_{ij}\beta_{ij} + \sum_{ij} C_{ij}f_{ij} \tag{5.80}$$

$$= \sum_{ij} [\pi_i - \pi_j + \alpha_{ij} - \beta_{ij} + C_{ij}]f_{ij} + \sum_{ij} (U_{ij} - f_{ij})\alpha_{ij}$$

$$+ \sum_{ij} (f_{ij} - L_{ij})\beta_{ij} \geq 0$$

It will be true that $G - F' = 0$ if the solution is optimal. Thus, from equation (5.80) we get

$$(U_{ij} - f_{ij})\alpha_{ij} = 0 \tag{5.81}$$

$$(f_{ij} - L_{ij})\beta_{ij} = 0 \tag{5.82}$$

$$(C_{ij} + \pi_i - \pi_j + \alpha_{ij} - \beta_{ij})f_{ij} = 0 \tag{5.83}$$

that is,

$$(\overline{C}_{ij} + \alpha_{ij} - \beta_{ij})f_{ij} = 0 \tag{5.84}$$

From equations (5.81)–(5.84), we get

Case 1:  $\overline{C}_{ij} > 0$

If $\beta_{ij} = \overline{C}_{ij} + \alpha_{ij}, f_{ij} \neq 0$

Furthermore, if $\alpha_{ij} \geq 0$, $\beta_{ij} \neq 0$, then, from equation (5.82), we can get

$$f_{ij} = L_{ij}$$

Case 2:  $\overline{C}_{ij} < 0$

If $\beta_{ij} = \overline{C}_{ij} + \alpha_{ij}$, then $f_{ij} \neq 0$, and $\alpha_{ij} > \beta_{ij}$

Furthermore, if $\beta_{ij} \geq 0$, $\alpha_{ij} \neq 0$, then, from equation (5.81), we can get

$$f_{ij} = U_{ij}$$

Case 3:  $\overline{C}_{ij} = 0$

From (5.84), we get $(\alpha_{ij} - \beta_{ij})f_{ij} = 0$, which can be analyzed as follows.

(3a): If $f_{ij} = 0$, then $(\alpha_{ij} - \beta_{ij}) \neq 0$

When $\alpha_{ij} > \beta_{ij}$, then $\alpha_{ij} > 0$, we get the following expression from equation (5.81)

$$f_{ij} = U_{ij} \neq 0$$

When $\beta_{ij} > \alpha_{ij}$, then $\beta_{ij} > 0$, we get the following expression from equation (5.82)

$$f_{ij} = L_{ij} \neq 0$$

Both situations are in conflicted with the assumption $f_{ij} = 0$. So we can be sure $f_{ij} \neq 0$ for this case.

(3b): Assuming $\alpha_{ij} = 0$, then $\beta_{ij}f_{ij} = 0$

Since $f_{ij} \neq 0$ from (3a), we have $\beta_{ij} = 0$

Therefore, from equation (5.81) we get

$$f_{ij} \leq U_{ij}$$

From equation (5.82) we get

$$f_{ij} \geq L_{ij}$$

that is, if $\overline{C}_{ij} = 0$, then $L_{ij} \leq f_{ij} \leq U_{ij}$

According to the three cases described above, the complementary slackness conditions for optimality of OKA are summarized as follows:

$$f_{ij} = L_{ij} \quad \text{for} \quad \overline{C}_{ij} > 0 \tag{5.85}$$

$$L_{ij} \leq f_{ij} \leq U_{ij} \quad \text{for} \quad \overline{C}_{ij} = 0 \tag{5.86}$$

$$f_{ij} = U_{ij} \quad \text{for} \quad \overline{C}_{ij} < 0 \tag{5.87}$$

where the relative cost is

$$\overline{C}_{ij} = C_{ij} + \pi_i - \pi_j \tag{5.88}$$

According to equations (5.85)–(5.87) and the labeling technique, the arcs have nine kinds of states, which are shown in Table 5.13.

The complementary slackness conditions for optimality of OKA shown in equations (5.85)–(5.87) correspond to the three "in-kilter" states of the arcs. In addition, there are six "out-of-kilter" states that do not satisfy conditions (5.85)–(5.87). If all the arcs are in kilter, then the optimal solution is obtained. Otherwise, we must vary the relevant arc flows or node potentials (parameter $\pi$) by the labeling technique so that the out-of-kilter states of the arcs come into kilter.

The states of arcs and labeling rules can be explained using Figure 5.5.

**TABLE 5.13  States of OKA Arcs**

| Symbol | $\overline{C}_{ij}$ | $f_{ij}$ | State of Arcs |
|--------|---------------------|----------|---------------|
| $I_1$ | $\overline{C}_{ij} > 0$ | $f_{ij} = L_{ij}$ | In kilter |
| $I_2$ | $\overline{C}_{ij} = 0$ | $L_{ij} < f_{ij} < U_{ij}$ | In kilter |
| | | $f_{ij} = U_{ij}, f_{ij} = L_{ij}$ | In kilter |
| $I_3$ | $\overline{C}_{ij} < 0$ | $f_{ij} = U_{ij}$ | In kilter |
| $II_1$ | $\overline{C}_{ij} > 0$ | $f_{ij} < L_{ij}$ | Out of kilter |
| $II_2$ | $\overline{C}_{ij} = 0$ | $f_{ij} < L_{ij}$ | Out of kilter |
| $II_3$ | $\overline{C}_{ij} < 0$ | $f_{ij} < U_{ij}$ | Out of kilter |
| $III_1$ | $\overline{C}_{ij} > 0$ | $f_{ij} > L_{ij}$ | Out of kilter |
| $III_2$ | $\overline{C}_{ij} = 0$ | $f_{ij} > U_{ij}$ | Out of kilter |
| $III_3$ | $\overline{C}_{ij} < 0$ | $f_{ij} > U_{ij}$ | Out of kilter |

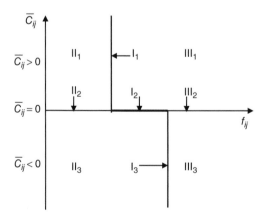

Figure 5.5  States of OKA arcs.

In Figure 5.5, if the arc is in-kilter state, the point $(f_{ij}, \overline{C}_{ij})$ will be located on one of three dark lines $I_1, I_2$, or $I_3$, where the dark line $I_1$ corresponds to the lower bound $L_{ij}$ of flow $f_{ij}$; the dark line $I_3$ corresponds to the upper bound $U_{ij}$ of flow $f_{ij}$; and the dark line $I_2$ corresponds to the flow $f_{ij}$ that is within $L_{ij} < f_{ij} < U_{ij}$.

If the flow of the arc is violated at the upper or lower limits, the point $(f_{ij}, \overline{C}_{ij})$ will be located outside the three dark lines, corresponding to the six "out-of-kilter" states in Figure 5.5. In these situations, the value of the flow of the arc will be either less than its lower limit or higher than its upper limit, that is, $f_{ij} > U_{ij}$ or $f_{ij} < L_{ij}$.

***Labeling Rules and Algorithm of OKA***   According to the labeling technique, the labeling rules of OKA for the forward arc and backward arc under nine OKA states in Table 5.13 are listed in Table 5.14, where the symbol "↑" stands for increase; "↓" stands for decrease; "→" stands for change; "$f_k$" indicates that the flow is outside the feasible region.

According to the labeling rules mentioned above, the OKA is implemented as follows.

*With Incremental Flow Loop*   When an incremental flow loop exists, correct the values of flow for all arcs in the loop. The process is as follows:

**(1)** For forward arcs

    **(a)** If $\overline{C}_{ij} \geq 0, f_{ij} < L_{ij}$, the node $j$ can be labeled. The incremental flow to the node $j$ will be computed as

$$q_j = \min[q_i, L_{ij} - f_{ij}] \tag{5.89}$$

**TABLE 5.14   Labeling Rules for OKA Algorithm**

| Symbol | $f_{ij}$ | Forward Arc $f^+$ Labeling? Why? | Backward Arc $f^-$ Labeling? Why? |
|---|---|---|---|
| $I_1$ | $f_{ij} = L_{ij}$ | No, $f^+ \uparrow \to f_k^+$ | No, $f^- \downarrow \to f_k^-$ |
| $I_2$ | $L_{ij} < f_{ij} < U_{ij}$ | Yes, $f^+ \uparrow \to U$ | Yes, $f^- \downarrow \to L$ |
| | $f_{ij} = U_{ij}$ $f_{ij} = L_{ij}$ | No, $f^+ \uparrow \to f_k^+$ | No, $f^- \downarrow \to f_k^-$ |
| $I_3$ | $f_{ij} = U_{ij}$ | No, $f^+ \uparrow \to f_k^+$ | No, $f^- \downarrow \to f_k^-$ |
| $II_1$ | $f_{ij} < L_{ij}$ | Yes, $f^+ \uparrow \to U$ | No, $f^- \downarrow \to f_k^-$ |
| $II_2$ | $f_{ij} < L_{ij}$ | Yes, $f^+ \uparrow \to U$ | No, $f^- \downarrow \to f_k^-$ |
| $II_3$ | $f_{ij} < U_{ij}$ | Yes, $f^+ \uparrow \to U$ | No, $f^- \downarrow \to f_k^-$ |
| $III_1$ | $f_{ij} > L_{ij}$ | No, $f^+ \uparrow \to f_k^+$ | Yes, $f^- \downarrow \to L$ |
| $III_2$ | $f_{ij} > U_{ij}$ | No, $f^+ \uparrow \to f_k^+$ | Yes, $f^- \downarrow \to L$ |
| $III_3$ | $f_{ij} > U_{ij}$ | No, $f^+ \uparrow \to f_k^+$ | Yes, $f^- \downarrow \to U$ |

**(b)** If $\overline{C}_{ij} \leq 0, f_{ij} < U_{ij}$, the node $j$ can be labeled. The incremental flow to the node $j$ will be computed as

$$q_j = \min[q_i, U_{ij} - f_{ij}] \tag{5.90}$$

**(2)** For backward arcs

**(a)** If $\overline{C}_{ji} \geq 0, f_{ji} > L_{ji}$, the node $j$ is can be labeled. The incremental flow to the node $j$ will be computed as

$$q_j = \min[q_i, f_{ji} - L_{ji}] \tag{5.91}$$

**(b)** If $\overline{C}_{ji} \leq 0, f_{ji} > U_{ji}$, the node $j$ can be labeled. The incremental flow to the node $j$ will be computed as

$$q_j = \min[q_i, f_{ji} - U_{ji}] \tag{5.92}$$

*Without Incremental Flow Loop*  When there an incremental flow loop does not exist, correct the values of the relative cost $\overline{C}_{ij}$, or $\overline{C}_{ji}$ by increasing the cost of the vertex $\pi$. This is because the change in $\overline{C}_{ij}$, or $\overline{C}_{ji}$ causes the change of the path of minimum cost flow. Consequently, a new incremental flow loop will be produced. The process of computing the incremental vertex cost is as follows:

Let $B$ and $\overline{B}$ stand for the set of labeled and unlabeled vertices, respectively. Obviously, the super source $s \in B$ and super sink $t \in \overline{B}$. In addition, define two sets of arcs $A_1$ and $A_2$

$$A_1 = \{ij, \; i \in B, \; j \in \overline{B}, \; \overline{C}_{ij} > 0, \; f_{ij} \leq U_{ij}\} \tag{5.93}$$

$$A_2 = \{ji, \; i \in B, \; j \in \overline{B}, \; \overline{C}_{ji} < 0, \; f_{ji} \geq L_{ij}\} \tag{5.94}$$

The incremental vertex cost is determined as follows:

$$\delta = \min\{\delta_1, \delta_2\} \tag{5.95}$$

where

$$\delta_1 = \min\{|\overline{C}_{ij}|\} > 0 \tag{5.96}$$

$$\delta_2 = \min\{|\overline{C}_{ji}|\} > 0 \tag{5.97}$$

If $A_1$ is an empty set, make $\delta_1 = \infty$; If $A_2$ is an empty set, make $\delta_2 = \infty$. When $\delta = \infty$, it means there is no feasible flow, that is, there is no solution for the given NFP problem. When $\delta < \infty$, update the vertex costs for all unlabeled vertexes, that is,

$$\delta' = \pi_j + \delta \quad j \in \overline{B} \tag{5.98}$$

In this way, the out-of-kilter arc will be changed into an in-kilter arc. When all arcs are in in kilter, the optimum solution is obtained.

The steps of the OKA algorithm are as follows:

Step 1.   Set the initial values of the arc flows. The initial flows are required to satisfy constraint (5.72) only, but not necessarily the constraint (5.73).

Step 2.   Check the state of the arcs. If all arcs are in kilter, then the optimal solution has been found. Terminate the iteration. Otherwise, go to step 3.

Step 3.   Revise the state of the arcs. Arbitrarily choose an arc to be revised from the set of out-of-kilter arcs. Using the labeling technique, when a flow-augmenting loop exists, vary the values of flow $f_{ij}$ for all arcs in this loop. If no flow-augmenting loop is found, adjust the values of $\pi$ at unlabeled nodes, and hence change the relative cost $\overline{C}_{ij}$, or $\overline{C}_{ji}$. Some cross iterations between flow and the relative cost may be needed for the out-of-kilter arc to become in kilter. Once the arc state has been revised, go back to step 2.

It should be noted that the revision process converges after a finite number of iterations.

In comparison with the general algorithm of the minimum cost flow, the following are the main features of the OKA:

**(1)** The nonzero lower bound of flow is allowable.

**(2)** The initial flow does not have to be feasible or zero flow.

**(3)** Nonnegative constraints, $f_{ij} \geq 0$, are released.

**(4)** It is easy to imitate a change in network topology by changing the specified bound values of the flows as the branch outage occurs.

### 5.4.3   *N* Security Economic Dispatch Model

In the normal operating case, the NFP model of real power ED with *N* security can be written as follows.

$$\min F^0 = \sum_{i \in NG} (a_i P_{Gi}^{02} + b_i P_{Gi}^0 + c_i) + h \sum_{j \in NT} R_j P_{Tj}^{02} \tag{5.99}$$

such that

$$\sum_{i(\omega)} P_{Gi}^0 + \sum_{j(\omega)} P_{Tj}^0 + \sum_{k(\omega)} \widehat{P}_{Dk}^0 = 0 \quad \omega \in n \tag{5.100}$$

$$\underline{P}_{Gi} \leq P_{Gi}^0 \leq \overline{P}_{Gi} \tag{5.101}$$

$$\underline{P}_{Tj} \leq P_{Tj}^0 \leq \overline{P}_{Tj} \tag{5.102}$$

$$i \in NG, \quad j \in NT, \quad k \in ND$$

where

$a_i, b_i$ and $c_i$:  the cost coefficients of the $i$th generator

$P_{Gi}^0$:  the real power flow of the generation arc $i$ in the normal operating case

$P_{Tj}^0$:  the real power flow of the transmission arc $j$ in the normal operating case

$P_{Dk}^0$:  the real power flow of the load arc $k$ in the normal operating case

$NG$:  the total number of generation arcs

$NT$:  the total number of transmission arcs

$ND$:  the total number of load arcs

$N$:  the total number of nodes

$R_j$:  the resistance of the transmission arc (line) $j$

$\underline{P}$:  the lower bound of the real power flow through the arc

$\overline{P}$:  the upper bound of the real power flow through the arc.

The positive direction of flow is specified as the flow enters the node and the negative as it leaves the node. The symbol $i(w)$ means that arc $i$ is adjacent to node $w$; so also $j(w)$ and $k(w)$.

The following points should be noted.

**(1)** The second term of the objective,

$$h \sum_{j \in NT} R_j P_{Tj}^{0\,2} \tag{5.103}$$

is the penalty on transmission losses with the system marginal cost $h$ (in $ per MWh). The total transmission loss is represented approximately, but validly, as the sum of the products of the line resistance and the square of the transmitted power on the line. This is obtained from the following real power loss expression of the transmission line:

$$P_{Lj} = \frac{P_{Tj}^2 + Q_{Tj}^2}{V_{Tj}^2} \times R_j \tag{5.104}$$

under the assumptions of 1.0 p.u. flat voltage across the line and local supply of the reactive power.

**(2)** The power loss of an individual line is assumed to be distributed equally to its ends. Thus, the real load $P_{Dk}^0$ in equation (5.100) would involve half the transmission losses on all the lines connected to node $k$, which are estimated preliminarily from the power flow calculation of the normal operation and kept constant, or modified if necessary, that is,

$$\widehat{P}_{Dk}^0 = P_{Dk}^0 + \frac{1}{2} \sum_{j \to k} R_j P_{Tj}^{0\,2} \tag{5.105}$$

The other half of the loss on the line that is not related to load will be added on to the flow of the return arc of the OKA network model.

**(3)** The transmitted real power acts as the independent variable and the line security constraints are introduced into the model straight away. The secure line limit is based on its surge impedance loading (SIL) and its length, and not on the thermal limit.

**(4)** The topology of the power system is preserved as the penalty factors are not calculated in the usual sense. Therefore, the model can be solved easily by NFP as well as the OKA.

Although this model is different from the traditional ED model, it has been verified that they are equivalent [4,10].

The objective function of economic power dispatch in equation (5.99) is a quadratic function. It can be linearized by use of the average cost. From the previous section, we know that the OKA network model of economic power dispatch consists of three types of arcs. They are the generation arc, the transmission arc, and the load arc. Obviously, each generation arc corresponds to a generator, each transmission arc corresponds to a line or transformer, and each load arc corresponds to a real power demand. In addition, there is a return arc. The total arcs in a power network will be $m + 1$, where $m = NG + NT + ND$.

Comparing the ED model shown in equations (5.99)–(5.102) with the OKA model shown in equations (5.71)–(5.73), the average cost and flow limits of each type of arc are

**(1)** The generation arc

$$\overline{C}_{ij} = a_i P_{Gi} + b_i \tag{5.106}$$

$$L_{ij} = \underline{P}_{Gi} \tag{5.107}$$

$$U_{ij} = \overline{P}_{Gi} \tag{5.108}$$

**(2)** The transmission arc

$$\overline{C}_{ij} = h R_j P_{Tj} \tag{5.109}$$

$$L_{ij} = \underline{P}_{Tj} \tag{5.110}$$

$$U_{ij} = \overline{P}_{Tj} \tag{5.111}$$

**(3)** The load arc

$$\overline{C}_{ij} = 0 \tag{5.112}$$

$$L_{ij} = \hat{P}^0_{Dk} \tag{5.113}$$

$$U_{ij} = \hat{P}^0_{Dk} \tag{5.114}$$

**(4)** The return arc

$$\overline{C}_{ij} = 0 \tag{5.115}$$

$$L_{ij} = \sum_{k \in ND} \widehat{P}^0_{Dk} + \frac{1}{2} \sum_{j \in NT} R_j P^{0\,2}_{Tj} \tag{5.116}$$

$$U_{ij} = \sum_{k \in ND} \widehat{P}^0_{Dk} + \frac{1}{2} \sum_{j \in NT} R_j P^{0\,2}_{Tj} \tag{5.117}$$

If the network loss is neglected in the ED OKA model, the cost of the transmission arc will be zero; the load $\widehat{P}_{Dk}$ will be replaced by $P_{Dk}$. Meanwhile, the part of power loss in the return arc will be zero too.

It is noted that the flow $P_{ts}$ on the return arc contains the total loads and network losses, that is,

$$P_{ts} = \sum_{k \in ND} \widehat{P}^0_{Dk} + \frac{1}{2} \sum_{j \in NT} R_j P^{0\,2}_{Tj} \tag{5.118}$$

Substituting equation (5.105) in equation (5.118), we get

$$
\begin{aligned}
P_{ts} &= \sum_{k \in ND} \left( P^0_{Dk} + \frac{1}{2} \sum_{j \to k} R_j P^{0\,2}_{Tj} \right) + \frac{1}{2} \sum_{j \in NT} R_j P^{0\,2}_{Tj} \\
&= \sum_{k \in ND} (P^0_{Dk}) + \frac{1}{2} \sum_{j \in NT} R_j P^{0\,2}_{Tj} + \frac{1}{2} \sum_{j \in NT} R_j P^{0\,2}_{Tj} \\
&= \sum_{k \in ND} (P^0_{Dk}) + \sum_{j \in NT} R_j P^{0\,2}_{Tj} \\
&= P_D + P_L
\end{aligned}
\tag{5.119}
$$

Obviously, the KCL law at the super source node that connects to the return arc will be

$$\sum_{i=1}^{NG} P_{Gi} = P_D + P_L \tag{5.120}$$

This is exactly the real power balance equation in the traditional real power ED model. Thus, it is very easy to compute network losses in the ED OKA model, which involves adjusting the flow in the flow-augmenting loop that contains the return arc.

### 5.4.4 Calculation of $N - 1$ Security Constraints

In the theoretical sense, the total number of $N - 1$ security constraints is very large and equals $n(n - 1)$ for the system with $n$ transmission and transformer branches. In the practical sense, power transmission systems are usually designed well within the capacity of the system load and generation. Only a small proportion of lines may be overloaded, even if a single branch outage occurs. Therefore, it is neither necessary nor reasonable to incorporate all the $N - 1$ security constraints into the calculation model directly. To detect all the possible overconstrained cases, which must be considered, a fast contingency analysis for a single line outage must be performed [20,21].

On the basis of the normal generation schedule obtained from model M-1, the NFP model M-2 of $N - 1$ security analysis is presented as

$$\min F_l = \sum_{j \in NT} R_j P_{Tj}^2(l) \tag{5.121}$$

such that

$$\sum_{i(\omega)} P_{Gi}^0 + \sum_{j(\omega)} P_{Tj}(l) + \sum_{k(\omega)} P_{Dk}^0 = 0 \quad \omega \in n \tag{5.122}$$

$$|P_{Tj}(l)| \leq \gamma \overline{P}_{Tj} \quad l \in NL \tag{5.123}$$

$$P_{Tl} = 0 \tag{5.124}$$

where

$P_{Tl}(l)$: the real power transmitted in line $j$ while line $l$ is in outage

   $NL$: the set of the outage lines

   $\gamma$: a constant greater than unity (say $1 < \gamma < 1.3$).

The following are the differences between the models M-1 and M-2:

**(1)** The generation costs in the objective equation (5.99) and the inequality constraint equation (5.100) vanish as all the generations and loads remain unchanged.

**(2)** Only the transmitted real power $P_{Tl}(l)$ acts as a variable to adjust the power flows. The inequality constraint equation (5.123) has replaced equation (5.102). The constant $\gamma$ is introduced to find the overloaded line when line $l$ appears as an outage.

Once the overconstrained cases have been detected, the maximum value of the violation in line $j$ can be determined by the following equations:

$$\Delta \overline{P}_{Tj} = \max_{l \in NL} \{P_{Tj}(l) - \overline{P}_{Tj}\} \quad j \in NT1 \tag{5.125}$$

$$\Delta \underline{P}_{Tj} = \min_{l \in NL} \{P_{Tj}(l) - \underline{P}_{Tj}\} \quad j \in NT2 \tag{5.126}$$

where $NT1$ and $NT2$ represent the number of lines that violate their upper and lower bounds, respectively, as line $l$ appears as an outage.

### 5.4.5   $N-1$ Security Economic Dispatch

There is no guarantee that the economic schedules with $N$ security in normal operation will not violate the line limits if a single contingency occurs (or multiple contingencies occur). If such a situation does arise, it is necessary to reallocate the generations so that the line constraints are satisfied. An efficient approach to incorporating $N-1$ security constraints as a part of ED is therefore desirable. On the basis of the normal case with consideration of $N$ security and the fast contingency analysis, the network flow model M-3 of $N-1$ security economic power dispatch is presented as follows:

$$\min \Delta F = \sum_{i \in NG} \left( \left. \frac{\partial f_i}{\partial P_{Gi}} \right|_{P^0_{Gi}} \Delta P_{Gi} \right) + h \sum_{j \in NT} \left( \left. \frac{\partial P_{Lj}}{\partial P_{Tj}} \right|_{P^0_{Tj}} \Delta P_{Tj} \right) \tag{5.127}$$

such that

$$\sum_{i(\omega)} \Delta P_{Gi} + \sum_{j(\omega)} \Delta P_{Tj} = 0 \quad \omega \in (NG + NT) \tag{5.128}$$

$$\underline{P}_{Gi} - P^0_{Gi} \le \Delta P_{Gi} \le \overline{P}_{Gi} - P^0_{Gi} \quad i \in NG \tag{5.129}$$

$$|\Delta P_{Gi}| \le \Delta \overline{P}_{Grci} \quad i \in NG \tag{5.130}$$

$$\Delta P_{Tj} = -\Delta \overline{P}_{Tj} \quad j \in NT1 \tag{5.131}$$

$$\Delta P_{Tj} = -\Delta \underline{P}_{Tj} \quad j \in NT2 \tag{5.132}$$

$$\underline{P}_{Tj} - P^0_{Tj} \le \Delta P_{Tj} \le \overline{P}_{Tj} - P^0_{Tj} \quad j \in (NT - NT1 - NT2) \tag{5.133}$$

where $\Delta P_{Gi}$ and $\Delta P_{Tj}$ are the incremental generations and transmissions, respectively. The incremental generation and transmission costs are

$$\left. \frac{\partial f_i}{\partial P_{Gi}} \right|_{P^0_{Gi}} = 2a_i P^0_{Gi} + b_i \tag{5.134}$$

$$\left. \frac{\partial P_{Lj}}{\partial P_{Tj}} \right|_{P^0_{Tj}} = 2R_j P^0_{Tj} \tag{5.135}$$

$\Delta F$ is the objective, that, is the total incremental product cost.

Obviously, M-3 is an incremental optimization model. The following issues should be noted.

**(1)** The objective equation (5.127) and the equality constraint equation (5.128) are obtained under the assumption that the loads are held constant, that is, $\Delta P_{Dk} = 0$. Exceptionally, if there is no feasible solution for problem M-3 in the preventive control, some loads would be curtailed partially or completely, so that the problem becomes solvable. In this case, the incremental loads may act as the variable introduced into M-3 without the cost. The contents of load shedding can be found in Chapter 11.

**(2)** To realize the transition from the $N$ to $N - 1$ security schedule successfully, the limits of the real power generation regulations (regulating speeds), $\Delta \overline{P}_{Grci}$ must be considered. These are determined from the product of the relevant regulating speed and regulating time specified. Thus, the regulating value of the generation is restricted by the two inequalities (5.129) and (5.130), which can be combined into one expression:

$$\max\{-\Delta \overline{P}_{Grci}, \underline{P}_{Gi} - P^0_{Gi}\} \le \Delta P_{Gi} \le \min\{\Delta \overline{P}_{Grci}, \overline{P}_{Gi} - P^0_{Gi}\} \quad i \in NG \tag{5.136}$$

**(3)** If any critical single line outage occurs, then the line security zone will be contracted to some extent. Equations (5.131)–(5.133) reflect the changing number of line security constraints. Recalling equations (5.125) and (5.126), an "$N - 1$ constrained zone", which is in fact formed by the intersection of the secure zones for all single contingencies, can be determined from these equations. This means that an $N - 1$ security problem with the same number of constraints as in the $N$ security problem can be introduced into the network flow model.

Substituting equations (5.125), (5.126), and (5.134)–(5.136) into model M-3, the incremental network flow model of ED with $N - 1$ security, model M-4, becomes

$$\min \Delta F = \sum_{i \in NG} (2a_i P^0_{Gi} + b_i) \Delta P_{Gi} + h \sum_{j \in NT} \left(2R_j P^0_{Tj}\right) \Delta P_{Tj} \tag{5.137}$$

such that

$$\sum_{i(\omega)} \Delta P_{Gi} + \sum_{j(\omega)} \Delta P_{Tj} = 0 \quad \omega \in (NG + NT) \tag{5.138}$$

$$\max\{-\Delta \overline{P}_{Grci}, \underline{P}_{Gi} - P^0_{Gi}\} \le \Delta P_{Gi} \le \min\{\Delta \overline{P}_{Grci}, \overline{P}_{Gi} - P^0_{Gi}\} \tag{5.139}$$

$$\overline{\Delta P}_{Tj} = -\max_{l \in NL} \{P_{Tj}(l) - \overline{P}_{Tj}\} \quad j \in NT1 \tag{5.140}$$

$$\underline{\Delta P}_{Tj} = -\min_{l \in NL} \{P_{Tj}(l) - \overline{P}_{Tj}\} \quad j \in NT2 \tag{5.141}$$

$$\underline{P}_{Tj} - P^0_{Tj} \le \Delta P_{Tj} \le \overline{P}_{Tj} - P^0_{Tj} \quad j \in (NT - NT1 - NT2) \tag{5.142}$$

The linear model M-4 corresponds to the OKA model and it can be solved easily by the OKA.

It is noted that model M-4 can provide the bi-generation schedule, that is, the normal generation schedule from model M-1 is used in the normal operation state, while the post-fault generation schedule from model M-4 is only used in the post-contingency case. Furthermore, it can also be used as a single generation schedule, which is applied both in the normal case and in post-contingency, that is, the unique generation schedule not only guarantees secure operation in the normal case but it also avoids the occurrence of an overload in a possible single contingency. This scheme is easy to implement because no rescheduling is needed. However, because all the $N-1$ line security constraints have to be satisfied, the constraint region is very narrow, and hence the operating cost increases.

### 5.4.6 Implementation

***Major Procedures of the OKA*** The essence of the OKA is to revise the out-of-kilter states of arcs to in-kilter states according to complementary slackness conditions for optimality equations (5.85)–(5.87). It should be noted that the correction process converges after a finite number of iterations. The following is a numerical example, which is taken from reference [2], to illustrate the solution procedure:

The problem is to solve a secure ED of a simple power system shown in Figure 5.6. There are two generators ($P_{G1}$ and $P_{G2}$) and three transmission lines to supply a load $P_D$. The system parameters are as follows.

$$F_1(P_{G1}) = C_1 P_{G1} = 2P_{G1}$$

$$F_2(P_{G2}) = C_2 P_{G2} = 5P_{G2}$$

$$0 \le P_{G1} \le 2$$

$$0 \le P_{G2} \le 2$$

$$P_D = 3$$

$$0 \le P_{l1} \le 1$$

$$0 \le P_{l2} \le 4$$

$$1 \le P_{l3} \le 2$$

where, $l_1$ is the line between the two generators $P_{G1}$ and $P_{G2}$; $l_2$ is the line from the generator $P_{G1}$ to load $P_D$; $l_3$ is the line from the generator $P_{G2}$ to the load $P_D$.

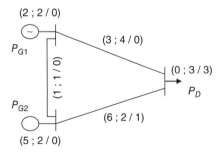

(2 ; 2 / 0)

$P_{G1}$

(3 ; 4 / 0)

(1 ; 1 / 0)

(0 ; 3 / 3)

$P_D$

$P_{G2}$

(6 ; 2 / 1)

(5 ; 2 / 0)

Figure 5.6   A simple power system ($C_{ij}$; $U_{ij}/L_{ij}$).

For simplification, the network loss is neglected. Then the ED model for this example can be written as follows.

$$\min F = 2P_{G1} + 5P_{G2}$$

such that

$$P_{G1} + P_{G2} = 3$$

$$0 \le P_{G1} \le 2$$

$$0 \le P_{G2} \le 2$$

$$0 \le P_{l1} \le 1$$

$$0 \le P_{l2} \le 4$$

$$1 \le P_{l3} \le 2$$

This ED problem can be expressed as the OKA network flow model as already mentioned.

The corresponding network flow model for the OKA is depicted in Figure 5.7. The solution process of the OKA is demonstrated in the following.

(1) Assign the initial values: $f_{13} = f_{32} = f_{24} = f_{41} = 2$, $f_{12} = f_{34} = 0$, and $\pi_1 = \pi_2 = \pi_3 = \pi_4 = 0$. These values and the relevant parameters are given in Figure 5.8(a). Then calculate the relative cost $\overline{C}_{ij}$.

(2) Check the state of the arcs. From Figure 5.8(a) we know that all the arcs are out of kilter except arc 1-2 marked with a star.

(3) Choose an out-of-kilter arc, say arc 4-1. By the labeling technique, no flow-augmenting loop exists because only node 1 can be labeled, but nodes 2-4 cannot. Then change the value of $\pi$ at nodes 2–4 as shown in Figure 5.8(b). In this case, arc 4-1 is still out of kilter, but all the nodes can be labeled. Then, a flow-augmenting loop 1-2-3-4-1 is found and the augmenting value is equal to unity. After the flows in this loop are adjusted, the resultant is shown in Figure 5.8(c). Now, arc 4-1 comes into kilter and so does arc 3-4 at the same time.

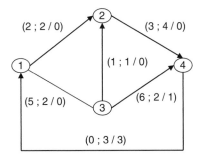

Figure 5.7   Network flow model for the OKA corresponding to Figure 5.6.

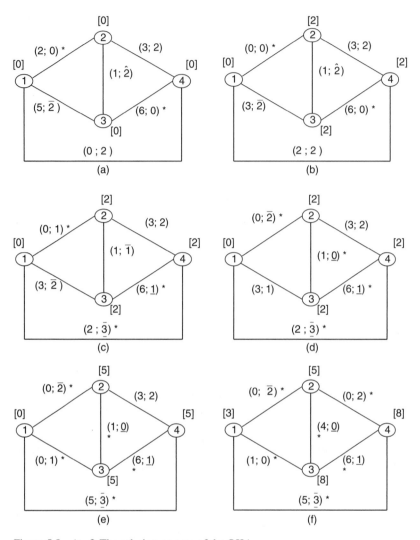

Figure 5.8    (a–f) The solution process of the OKA.

(4) Again check the state of the arcs. We can observe that arcs 1-3, 3-2, and 2-4 are out of kilter.

(5) Choose arc 1-3 to be revised. The flow-augmenting loop 1-2-3-1 is obtained because nodes 1, 2, and 3 can be labeled. Then modify the relevant flows; the results are given in Figure 5.8(d). In this case, arc 1-3 is still out of kilter and the nodes cannot be labeled, except node 1. Through changing the values of $\pi$ and $\overline{C}_{ij}$, arc 1-3 comes into kilter, as shown in Figure 5.8(e).

(6) Check the state of the arcs once more. Only arc 2-4 is in out of kilter state.

(7) Revise the state of arc 2-4. No flow-augmenting loop exists because only node 2 can be labeled. After the values of $\pi$ and $\overline{C}_{ij}$ at nodes 1, 3, and 4 have been changed, arc 2-4 comes into kilter, as shown in Figure 5.8(f).

(8) By checking the state of the arcs, we see that all the arcs are in kilter and all conditions for optimality have been satisfied. This shows that the optimal (minimum cost) power flow of the system is obtained. Stop the iteration.

The optimal results are

(1) The relevant cost

$$\overline{C}_{12} = 0, \overline{C}_{13} = 0, \overline{C}_{23} = 4, \overline{C}_{24} = 0, \overline{C}_{34} = 6, \overline{C}_{41} = 5$$

(2) The vertex cost

$$\pi_1 = 3, \pi_2 = 5, \pi_3 = 8, \pi_4 = 8,$$

(3) Flow on the arcs

$$f_{12} = 2, f_{13} = 1, f_{23} = 0, f_{24} = 2, f_{34} = 1, f_{41} = 3$$

***Numerical Example of Economic Dispatch with N Security*** The proposed model and algorithm have also been tested on the IEEE 5-bus and 30-bus systems. Table 5.15 has the ED results of the 5-bus system obtained by the OKA algorithm, where the total generation costs are 757.50 $/h, and the total system losses are 0.043 p.u. The results are almost the same as those obtained by LP.

The following simulation cases were conducted for the 30-bus system:

Case 1:   the original data including the power limit of the line;

Case 2:   the original data but with the power limit of the lines 2 and 6 reduced to 0.45 and 0.35 p.u., respectively;

Case 3:   the original data but with the power limit of the line 1 reduced to 0.65 p.u;

**TABLE 5.15   Economic Dispatch by OKA (5-Bus System)**

| Generators or Lines | Real Power (p.u.) | Lower Limit (p.u.) | Upper Limit (p.u.) |
|---|---|---|---|
| $P_{G1}$ | 0.9270 | 0.3000 | 1.2000 |
| $P_{G2}$ | 0.7160 | 0.3000 | 1.2000 |
| $P_{13}$ | 0.2160 | 0.0000 | 1.0000 |
| $P_{41}$ | −0.4110 | 0.0000 | 0.5000 |
| $P_{51}$ | −0.3000 | 0.0000 | 0.3000 |
| $P_{32}$ | −0.4000 | 0.0000 | 0.4000 |
| $P_{25}$ | 0.3160 | 0.0000 | 1.0000 |
| $P_{34}$ | 0.0000 | 0.0000 | 0.5000 |

TABLE 5.16 Economic Dispatch by OKA (30-Bus System)

| Case | Case 1 | Case 2 | Case 3 | Case 4 |
|---|---|---|---|---|
| $P_{G1}$(p.u.) | 1.7588 | 1.75000 | 1.34665 | 1.69665 |
| $P_{G2}$(p.u.) | 0.4881 | 0.26236 | 0.64571 | 0.33295 |
| $P_{G5}$(p.u.) | 0.2151 | 0.15000 | 0.15000 | 0.15000 |
| $P_{G8}$(p.u.) | 0.2236 | 0.31270 | 0.31270 | 0.31270 |
| $P_{G11}$(p.u.) | 0.1230 | 0.30000 | 0.30000 | 0.30000 |
| $P_{G13}$(p.u.) | 0.12000 | 0.12000 | 0.12000 | 0.12000 |
| Total cost ($/hr) | 802.51 | 813.75 | 814.24 | 809.68 |
| Total loss (p.u.) | 0.0950 | 0.0782 | 0.0793 | 0.0783 |

TABLE 5.17 Economic Dispatch with Different h by OKA (30-Bus System)

| H | >1600 | 200–1600 | 29–200 | 20–25 |
|---|---|---|---|---|
| $P_{G1}$(p.u.) | 0.56236 | 0.84236 | 1.34665 | 1.34665 |
| $P_{G2}$(p.u.) | 0.80000 | 0.80000 | 0.29571 | 0.64571 |
| $P_{G5}$(p.u.) | 0.50000 | 0.50000 | 0.15000 | 0.15000 |
| $P_{G8}$(p.u.) | 0.31270 | 0.31270 | 0.31270 | 0.31270 |
| $P_{G11}$(p.u.)) | 0.30000 | 0.30000 | 0.30000 | 0.30000 |
| $P_{G13}$(p.u.) | 0.40000 | 0.12000 | 0.12000 | 0.12000 |
| Total cost ($/hr) | 964.86 | 915.21 | 872.52 | 814.24 |
| Total loss (p.u.) | 0.0594 | 0.0620 | 0.0691 | 0.0793 |
| Iteration no. | 1 | 1 | 2 | 3 |

Case 4:   the original data but with the power limit of the line 1 reduced to 1.00 p.u.

The corresponding ED results are shown in Table 5.16.

To analyze the impact of the weighting $h$ on the calculation result, the data of case 3 are used and different values of $h$ are selected. The results are listed in Table 5.17, which show that the optimal results are reached when the weighting $h$ equals 20–25.

**Numerical Example of Economic Dispatch with $N-1$ Security**   The same data of the IEEE 30-bus system are used to compute the ED with $N-1$ security. The results are listed in Tables 5.18, and 5.19.

From Table 5.18, through $N-1$ security analysis and calculation, the $N-1$ security cannot be satisfied as four single-line outages (line number 1, 2, 4, and 5) appear. Thus, these violated constraints need to be introduced in the $N-1$ security

**TABLE 5.18   $N-1$ Security Analysis and Calculation Results (IEEE 30-Bus System)**

| Outage Line Number | Overloaded Lines Caused by Outage |
|---|---|
| 1 | $L_1(1.75662)$, $L_4(1.73162)$, $L_7(-1.08480)$ |
| 2 | $L_1(1.75662)$, $L_{10}(0.56510)$, $L_{12}(-0.39087)$ |
| 4 | $L_1(1.73162)$, $L_{10}(0.56510)$, $L_{12}(0.39087)$ |
| 5 | $L_1(1.73162)$, $L_6(1.30000)$, $L_8(-0.72573)$, $L_{10}(0.56508)$ |

**TABLE 5.19   Results and Comparison of Economic Dispatch with $N-1$ Security (IEEE 30-Bus System)**

| Generator No. | OKA | LP |
|---|---|---|
| $P_{G1}$(p.u.) | 1.40625 | 1.38540 |
| $P_{G2}$(p.u.) | 0.60638 | 0.57560 |
| $P_{G5}$(p.u.) | 0.25513 | 0.24560 |
| $P_{G8}$(p.u.) | 0.30771 | 0.35000 |
| $P_{G11}$(p.u.) | 0.17340 | 0.17930 |
| $P_{G13}$(p.u.) | 0.16154 | 0.16910 |
| Total generation (p.u.) | 2.91041 | 2.90500 |
| Total cost ($/hr) | 813.44 | 813.74 |
| Total loss (p.u.) | 0.07641 | 0.0711 |

ED model to readjust the generators' output until no any violation appears. The final results are shown in Table 5.19.

Through comparison with the conventional LP method that is used to solve ED, the OKA NFP can achieve almost the same results as LP, although sometimes the precision of OKA may be a little lower than that of the LP method; this can be neglected from the viewpoint of the engineering.

It should be noted that the amount of calculation of $N-1$ security EDD is greatly reduced with the presented method because of the use of the "$N-1$-constrained zone," which is formed by the fast $N-1$ security analysis.

## 5.5   NONLINEAR CONVEX NETWORK FLOW PROGRAMMING METHOD

### 5.5.1   Introduction

This section presents a new NLCNFP model of economic dispatch control (EDC), which is solved by a combination approach of QP and NFP. First of all, a new NLC-NFP model of economic power dispatch with security is deduced, based on the load flow equations. Then, a new incremental NLCNFP model of secure and ED can be set up. The new EDC model can be transformed into a QP model, in which the search direction in the space of the flow variables is found. The concept of a maximum basis in the network flow graph is introduced, allowing the constrained QP model to be changed into an unconstrained QP model that is then solved using the reduced gradient method.

## 5.5.2   NLCNFP Model of EDC

**Mathematical Model**   It is well known that the active power flow equations of a transmission line can be written as follows.

$$P_{ij} = V_i^2 g_{ij} - V_i V_j g_{ij} \cos \theta_{ij} - V_i V_j b_{ij} \sin \theta_{ij} \tag{5.143}$$

$$P_{ji} = V_j^2 g_{ij} - V_i V_j (-g_{ij} \cos \theta_{ij} + b_{ij} \sin \theta_{ij}) \tag{5.144}$$

where

$P_{ij}$: the sending end active power on transmission line $ij$
$P_{ji}$: the receiving end active power on transmission line $ij$
$V_i$: the node voltage magnitude of node $i$
$\theta_{ij}$: the difference of node voltage angles between the sending and receiving
      ends of the line $ij$
$b_{ij}$: the susceptance of transmission line $ij$
$g_{ij}$: the conductance of transmission line $ij$.

In a high voltage power network, the value of $\theta_{ij}$ is very small and the following approximate equations are easily obtained.

$$V \cong 1.0 \ p.u. \tag{5.145}$$

$$\sin \theta_{ij} \cong \theta_{ij} \tag{5.146}$$

$$\cos \theta_{ij} \cong 1 - \theta_{ij}^2/2 \tag{5.147}$$

Substituting equations (5.145)–(5.147) in equations (5.143) and (5.144), the active power load flow equations of a line can be simplified and deduced as follows.

$$P_{ij} = P_{ijC} + \frac{1}{2} \left( -\frac{P_{ijC}}{b_{ij}} \right)^2 g_{ij} \tag{5.148}$$

$$P_{ji} = -P_{ijC} + \frac{1}{2} \left( -\frac{P_{ijC}}{b_{ij}} \right)^2 g_{ij} \tag{5.149}$$

where

$$P_{ijC} = -b_{ij}\theta_{ij} \tag{5.150}$$

is called an *equivalent power flow* on transmission line $ij$.

The active power loss on transmission line $ij$ can be obtained according to equations (5.148) and (5.149), that is,

$$P_{Lij} = P_{ij} + P_{ji} = \left(-\frac{P_{ijC}}{b_{ij}}\right)^2 g_{ij}$$

$$= P_{ijC}^2 \frac{\left(R_{ij}^2 + X_{ij}^2\right)}{X_{ij}^2} R_{ij} \tag{5.151}$$

where

$R_{ij}$: the resistance of transmission line $ij$
$X_{ij}$: the reactance of transmission line $ij$.

Let

$$Z_{ijC} = \frac{\left(R_{ij}^2 + X_{ij}^2\right)}{X_{ij}^2} R_{ij} \tag{5.152}$$

The active power loss on the transmission line $ij$ can be expressed as follows.

$$P_{Lij} = P_{ijC}^2 Z_{ijC} \tag{5.153}$$

The traditional NFP model for the ED problem can be written as follows, that is, model M-5.

$$\min F = \sum_{i \in NG} \left(a_i P_{Gi}^2 + b_i P_{Gi} + c_i\right) + h \sum_{ij \in NT} P_{Lij} \tag{5.154}$$

such that

$$P_{Gi} = P_{Di} + \sum_{j \to i} P_{ij} \tag{5.155}$$

$$P_{Gim} \leq P_{Gi} \leq P_{GiM} \quad i \in NG \tag{5.156}$$

$$-P_{ijM} \leq P_{ij} \leq P_{ijM} \quad j \in NT \tag{5.157}$$

where,

$P_{Gi}$: the active power of the generator $i$
$P_{Di}$: the active power demand at load bus $i$
$P_{ij}$: the flow in the line connected to node $i$, which would have a negative value for a line in which the flow is toward node $i$
$a_i, b_i, c_i$: the cost coefficients of the $i$-th generator
$NG$: the number of generators in the power network
$NT$: the number of transmission lines in the power network
$P_{ijM}$: the active power flow constraint on transmission line $ij$
$P_{Lij}$: the active power loss on transmission line $ij$

$h$: the weighting coefficient of the transmission losses

$j \rightarrow i$: represents node $j$ connected to node $i$ through transmission line $ij$

Subscripts $m$ and $M$ represent the lower and upper bounds of the constraint.

The second term of the objective function (equation 5.154) is a penalty on transmission losses based on the system marginal cost $h$ (in $ per MWh). Equation (5.157) is the line security constraint. Equation (5.156) defines the generator power upper and lower limits. Equation (5.155) is Kirchhoff's first law (i.e, the node current law, KCL).

Substituting equation (5.151) or (5.153) in equation (5.154), and substituting equation (5.148) in equation (5.155), the new NLCNFP model M-6 can be written as follows.

$$\min F = \sum_{i \in NG} \left( a_i P_{Gi}^2 + b_i P_{Gi} + c_i \right) + h \sum_{ij \in NT} P_{ijC}^2 Z_{ijC} \qquad (5.158)$$

such that

$$P_{Gi} = P_{Di} + \sum_{j \rightarrow i} \left[ P_{ijC} + \frac{P_{ijC}^2}{2b_{ij}^2} g_{ij} \right] \qquad (5.159)$$

$$P_{Gim} \leq P_{Gi} \leq P_{GiM} \quad i \in NG \qquad (5.156)$$

$$-P_{ijCM} \leq P_{ijC} \leq P_{ijCM} \quad j \in NT \qquad (5.160)$$

where, $Z_{ijC}$ is called an *equivalent impedance* of transmission line $ij$, as shown in equation (5.152).

Obviously, equation (5.159) is equivalent to the general system active balance equation in the traditional EDC model, that is,

$$\sum_{i \in NG} P_{Gi} = \sum_{k \in ND} P_{Dk} + P_L \qquad (5.161)$$

where

$ND$: the number of load nodes

$P_L$: the total system active power losses, which is obtained through the computation of the following equation (5.162), rather than usual power flow calculations.

$$P_L = \sum_{ij \in NT} P_{Lij} = \sum_{ij \in NT} P_{ijC}^2 Z_{ijC} \qquad (5.162)$$

The limiting value of the equivalent line power flow $P_{ijCM}$ in equation (5.160) can be obtained from equation (5.148), that is,

$$P_{ijM} = P_{ijCM} + \frac{1}{2} \left( -\frac{P_{ijCM}}{b_{ij}} \right)^2 g_{ij} \qquad (5.163)$$

According to equation (5.163), we can get the positive limiting value of the equivalent line power flow $P_{ijCM}$ (the negative root of $P_{ijCM}$ is neglected), that is,

$$P_{ijCM} = \frac{\left[ \sqrt{1 + (2g_{ij}P_{ijM}/b_{ij}{}^2)} - 1 \right]}{g_{ij}} \tag{5.164}$$

**Consideration of Kirchhoff's voltage law**   It is well know that Kirchhoff's second law (i.e., the loop voltage law, KVL) has not been considered in the study of secure economic power dispatch using general NFP. This is why there always exists some modeling error when secure economic power dispatch is solved using traditional linear NFP. KVL is considered in this section.

The voltage drop on the transmission line $ij$ can be approximately expressed as

$$V_{ij} = P_{ijC}Z_{ijC} \tag{5.165}$$

In this way, the voltage equation of the $l$th loop can be obtained, that is,

$$\sum_{ij} (P_{ijC}Z_{ijC})\mu_{ij,l} = 0 \quad l = 1, 2, \dots, NM \tag{5.166}$$

where $NM$ is the number of loops in the network and $\mu_{ij,l}$ is the element in the related loop matrix, which takes the value 0 or 1.

Introducing the KVL equation into model M-6, we get the following model M-7, in which the augmented objective function is obtained from the KVL equation (5.166) and objective function (5.158) in the model M-6.

$$\min F_L = \sum_{i \in NG} (a_i P_{Gi}^2 + b_i P_{Gi} + c_i) + h \sum_{ij \in NT} P_{ijC}^2 Z_{ijC}$$

$$- \lambda_l \sum_{ij} (P_{ijC}Z_{ijC})\mu_{ij,l} \quad l = 1, 2, \dots, NM \tag{5.167}$$

subject to constraints in equations (5.156), (5.159), (5.160) where $\lambda_l$ is the Lagrange multiplier, which can be obtained through minimizing equation (5.167) with respect to variable the $P_{ijC}$, that is,

$$2hP_{ijC}Z_{ijC} - \lambda_l \sum_{ij} Z_{ijC}\mu_{ij,l} = 0 \quad l = 1, 2, \dots, NM \tag{5.168}$$

$$\lambda_l = 2hP_{ijC}/\sum_{ij} \mu_{ij,l} \quad l = 1, 2, \dots, NM \tag{5.169}$$

By solving optimization NLCNFP model M-7, the generator power output $P_{Gi}$ and the equivalent line power flow $P_{ijC}$ can be obtained. Therefore, the line power $P_{ij}$, angle $\theta_{ij}$, which is the difference of node voltage angles between the sending and receiving ends of the line, and system active power losses $P_L$ can be computed from equations (5.148), (5.150), and (5.162), respectively, rather than from the usual power flow calculations.

Similarly, the method of handling $N - 1$ security constraints in Section 5.4 is adopted here. Thus, the incremental NLCNFP model of ED with $N - 1$ security, model M-8, becomes

$$\min \Delta F = \sum_{i \in NG} \left(2a_i P_{Gi}^0 + b_i\right) \Delta P_{Gi} + h \sum_{ij \in NT} \left(2Z_{ijC} P_{ijC}^0\right) \Delta P_{ijC} + \lambda_l \sum_{ij} Z_{ijC} \mu_{ij,l} \tag{5.170}$$

such that

$$\Delta P_{Gi} = \sum_{j \to i} \left(1 + \frac{P_{ijC}}{b_{ij}^2} g_{ij}\right) \Delta P_{ijC} \tag{5.171}$$

$$\max\{-\Delta P_{GRCiM}, P_{Gim} - P_{Gi}^0\} \leq \Delta P_{Gi} \leq \min\{\Delta P_{GRCiM}, P_{GiM} - P_{Gi}^0\}, \quad i \in NG \tag{5.172}$$

$$\Delta P_{ijC} = -\max_{l \in NL}\{P_{ijC}(l) - P_{ijCM}\} \quad j \in NT1 \tag{5.173}$$

$$\Delta P_{ijC} = -\min_{l \in NL}\{P_{ijC}(l) + P_{ijCM}\} \quad j \in NT2 \tag{5.174}$$

$$-P_{ijCM} - P_{ijC}^0 \leq \Delta P_{ijC} \leq P_{ijCM} - P_{ijC}^0 \quad j \in (NT - NT1 - NT2) \tag{5.175}$$

It is noted that the $N - 1$ security region may be very narrow because all constraints that are produced by all kinds of single outages are introduced in $N - 1$ security ED. In other words, the feasible range of the generators power output become very small. Consequently, $N - 1$ security is met, but the system economy may not be satisfied. Thus, the idea of multigeneration plans is used. The method is to solve the ED model by considering one single outage only each time. This means that each effective single outage corresponds to one generation plan. Generally, there are not too many effective single outages in a system. Therefore, it will not have many generation plans. The incremental NLCNFP model of multigeneration plans can be written as follows.

$$\min \Delta F = \sum_{i \in NG} \left(2a_i P_{Gi}^0 + b_i\right) \Delta P_{Gi}(l) + h \sum_{ij \in NT} \left(2Z_{ijC} P_{ijC}^0\right) \Delta P_{ijC}(l) + \lambda_l \sum_{ij} Z_{ijC} \mu_{ij,l} \tag{5.176}$$

such that

$$\Delta P_{Gi}(l) = \sum_{j \to i} \left(1 + \frac{P_{ijC}^0}{b_{ij}^2} g_{ij}\right) \Delta P_{ijC}(l) \tag{5.177}$$

$$\max\{-\Delta P_{GRCiM}, P_{Gim} - P_{Gi}^0\} \leq \Delta P_{Gi}(l) \leq \min\{\Delta P_{GRCiM}, P_{GiM} - P_{Gi}^0\}, \quad i \in NG \tag{5.178}$$

$$\Delta P_{ijC}(l) = -(P_{ijC}(l) - P_{ijCM}) \quad j \in NT1, \ l \in NL \qquad (5.179)$$

$$\Delta P_{ijC}(l) = -(P_{ijC}(l) + P_{ijCM}) \quad j \in NT2, \ l \in NL \qquad (5.180)$$

$$-P_{ijCM} - P_{ijC}^0 \le \Delta P_{ijC} \le P_{ijCM} - P_{ijC}^0 \quad j \in (NT - NT1 - NT2) \qquad (5.181)$$

## 5.5.3 Solution Method

Because of the special form of model M-7 or M-8, we introduce the following algorithm for solving it.

Model M-7 or M-8 is easily changed into a standard model of NLCNFP, that is, model M-9:

$$\min C = \sum_{ij} c(f_{ij}) \qquad (5.182)$$

such that

$$\sum_{j \in n} (f_{ij} - f_{ji}) = r_i \quad i \in n \qquad (5.183)$$

$$L_{ij} \le f_{ij} \le U_{ij} \quad ij \in m \qquad (5.184)$$

Equation (5.183) can be written as

$$Af = r \qquad (5.185)$$

where $A$ is an $n \times (n + m)$ matrix in which every column corresponds to an arc in the network and every row corresponds to a node in the network.

Matrix $A$ can be divided into a basic submatrix and nonbasic submatrix, which is similar to the convex simplex method. that is,

$$A = [B, \ S, \ N] \qquad (5.186)$$

where the columns of $B$ form a basis; both $S$ and $N$ correspond to the nonbasic arcs. $S$ corresponds to the nonbasic arcs in which the flows are within the corresponding constraints. $N$ corresponds to the nonbasic arcs in which the flows reach the corresponding bounds.

A similar division can be made for the other variables, that is,

$$f = [f_B, f_S, f_N] \qquad (5.187)$$

$$g(f) = [g_B, g_S, g_N] \qquad (5.188)$$

$$G(f) = \text{diag}[G_B, G_S, G_N] \qquad (5.189)$$

$$D = [D_B, D_S, D_N] \qquad (5.190)$$

where

$g(f)$: the first order gradient of the objective function
$G(f)$: the Hessian matrix of the objective function
  $D$: the search direction in the space of the flow variables.

To solve model M-9, Newton's method can first be used to calculate the search direction in the space of the flow variables. The idea behind Newton's method is that the function being minimized is approximated locally by a quadratic function, and this approximate function is minimized exactly.

Suppose that $f$ is a feasible solution and the search step along the search direction in the space of flow variables $\beta = 1$. Then the new feasible solution can be obtained.

$$f' = f + D \tag{5.191}$$

Substituting equation (5.191) into the equations in the model M-9, the NLCNFP model M-9 can be changed into the following QP model M-10, in which the search direction in the space of the flow variables is to be solved.

$$\min C(D) = \frac{1}{2}D^T G(f)D + g(f)^T D \tag{5.192}$$

such that

$$AD = 0 \tag{5.193}$$

$$D_{ij} \geq 0, \text{ when } f_{ij} = L_{ij} \tag{5.194}$$

$$D_{ij} \leq 0, \text{ when } f_{ij} = U_{ij} \tag{5.195}$$

Model M-10 is a special QP model which has the form of NFP. In order to enhance the calculation speed, we present a new approach, in place of the general QP algorithm, to solve the model M-10. The main calculation steps are described in the following.

**_Neglecting Temporarily Equations (5.194) and (5.195)_** This means that $L_{ij} < f_{ij} < U_{ij}$ in this case. Thus $D_N = 0$ according to the definition of the corresponding nonbasic arc.

From equation (5.193), we know that

$$AD = [B, S, N]\begin{bmatrix} D_B \\ D_S \\ 0 \end{bmatrix} = 0 \tag{5.196}$$

From equation (5.196), we can obtain

$$D_B = -B^{-1}SD_S \tag{5.197}$$

$$D = \begin{bmatrix} -B^{-1}S \\ I \\ 0 \end{bmatrix}D_S = ZD_S \tag{5.198}$$

Substituting equation (5.198) in equation (5.192), we get

$$\min C(D) = \frac{1}{2} D^T G(f) D + g(f)^T D \qquad (5.199)$$

Through minimizing equation (5.199) to variable $D_S$, the model M-10 can be changed into an unconstrained problem, the optimization solution of which can be solved from the following equations.

$$D_N = 0 \qquad (5.200)$$

$$BD_B = -SD_B \qquad (5.201)$$

$$(Z^T G Z) D_S = -Z^T g \qquad (5.202)$$

**Introduction of Equations (5.194) and (5.195)**    According to equations (5.200)–(5.202), $D_S$ can be solved from equation (5.202) and then $D_B$ can be solved from equation (5.201). If $D_B$ violates the constraint equations (5.194) and (5.195), a new basis must be sought to calculate the new search direction in the space of flow variables. This step will not be terminated until $D_B$ satisfies the constraint equations (5.194) and (5.195).

**Introduction of Maximum Basis in Network**    Obviously, the general repeated calculation of $D_B$ and $D_S$, which is similar to that of pivoting in LP, is not only time-consuming but also does not improve the value of the objective function. To speed up the calculation, we adopt a new method to form a basis in advance so that $D_B$ and $D_S$ can satisfy the constraints (5.194) and (5.195). Therefore, the maximum basis in network, which consists of as many free basic arcs as possible, is introduced in this chapter.

The maximum basis in a network can be obtained by solving the following model M-11.

$$\max_B \sum_{ij} d_{ij} A_{ij} \qquad (5.203)$$

where

$$d_{ij} = \begin{cases} 1, & \text{when arc } ij \text{ is a free one, that is, the flow in arc } ij \text{ is within its bounds.} \\ 0, & \text{when arc } ij \text{ is not a free one, that is, the flow in arc } ij \text{ reaches its bounds.} \end{cases}$$

$$A_{ij} = \begin{cases} 1, & \text{when arc } ij \text{ is in the basis } B. \\ 0, & \text{when arc } ij \text{ is not in Basis } B. \end{cases}$$

Suppose basis $B$ is the maximum basis from equation (5.203), only the flows on the free arcs in basis $B$ need to be adjusted in order to satisfy equation (5.203), if the flow on a free nonbasic arc needs to be adjusted [22].

The introduction of the maximum base indicates adjusting the direction of flow, that is, the change of flow is carried out according to the maximum basis.

Through selecting the maximum basis, equations (5.194) and (5.195) in model M-10 can always be satisfied in the calculation of the search direction in the space of the flow variables. Therefore, the QP model M-10 is equivalent to unconstrained problem equations (5.200)–(5.202). To enhance the calculation speed further, equations (5.200)–(5.202) can be solved by the reduced gradient method.

**Reduced Gradient Algorithm with Weight Factor** Equations (5.200)–(5.202) can be written as compact format as follows:

$$(Z^T GZ)D = -Z^T g \tag{5.204}$$

If we use the unit matrix to replace the Hessen matrix $(Z^T GZ)$, we get

$$V = -Z^T g \tag{5.205}$$

$$D = ZV \tag{5.206}$$

where

$V$: the negative reduced gradient
$D$: the direction of the reduced gradient.

The main advantages of the reduced gradient method are (1) the calculation is simple and (2) the required storage space is relatively small. The disadvantage is that it is an approximation. Thus, the reduced gradient algorithm has a linear convergence speed.

To improve the convergence speed of the reduced gradient method, select a positive matrix that is not a unit matrix but can be easily inversed, and use it to replace the Hessian matrix $(Z^T GZ)$. In this way, we get a new reduced gradient with weight, that is,

$$MV = -Z^T g \tag{5.207}$$

where

$M$: the weight of the reduced gradient.

Select the initial value of $Z$ as

$$Z = \begin{bmatrix} -B^{-1}S \\ I \\ 0 \end{bmatrix} \tag{5.208}$$

Substituting equation (5.208) in equation (5.207), we get

$$MV = -Z^T g = -[-S^T(B^T)^{-1}, I, 0] \begin{bmatrix} g_B \\ g_S \\ g_N \end{bmatrix} = S^T(B^T)^{-1}g_B - g_S \tag{5.209}$$

According to equations (5.182) and (5.185), the following Lagrange function can be obtained.

$$L = C(f) - \lambda(Af - r) \tag{5.210}$$

where

$\lambda$: the Lagrange multiplier.

According to the condition of optimization, we have

$$\frac{\partial L}{\partial f} = 0 \tag{5.211}$$

$$\frac{\partial C(f)}{\partial f} - A^T \lambda = 0 \tag{5.212}$$

that is,

$$g(f) = A^T \lambda \tag{5.213}$$

Expanding the above equation, we get

$$\begin{bmatrix} B^T \lambda \\ S^T \lambda \\ N^T \lambda \end{bmatrix} = \begin{bmatrix} g_B \\ g_S \\ g_N \end{bmatrix} \tag{5.214}$$

$$B^T \lambda = g_B \tag{5.215}$$

Substituting equation (5.215) in equation (5.209), we get

$$MV = S^T (B^T)^{-1} B^T \lambda - g_S = S^T \lambda - g_S \tag{5.216}$$

In summary, the calculation steps of a NLCNFP model, which is solved by reduced gradient algorithm with weight, are as follows:

**(1)** Compute $\lambda$ from equation (5.215).

**(2)** Compute $V$ from equation (5.216).

**(3)** Compute $D_S$ from the following expression.

$$D_S = \begin{cases} 0, & \text{when } (f_S)_{ij} = L_{ij}, \text{ and } V_{ij} < 0. \\ 0, & \text{when } (f_S)_{ij} = U_{ij}, \text{ and } V_{ij} > 0. \\ V_{ij}, & \text{Otherwise.} \end{cases} \tag{5.217}$$

**(4)** Compute $D_B$ from equation (5.201).

**(5)** Compute the new value of flow $f' = f + D_B$

In the practical calculation, several parameters related to the algorithm must be addressed.

**(1)** The convergence criteria
The convergence criteria are as follows.

$$\max \left|\left(S^T \lambda - g_S\right)_j\right| \leq \sigma \tag{5.218}$$

where, $\sigma$ is determined according to the required calculation precision.

**(2)** The selection of the weighting matrix $M$
We can select the diagonal matrix of the Hessen matrix $Z^T GZ$ as the weighting matrix M, that is,

$$M = \operatorname{diag}(Z^T GZ) \tag{5.219}$$

**(3)** The selection of the search step
We assume that the search step is along the search direction in the space of flow variables $\beta = 1$. To speed up the convergence, we can use the following approach to compute the optimum search step along the search direction in the space of flow variables. First of all, compute the initial step as follows.

$$\beta^0 = -\frac{g^T D}{D^T GD} \tag{5.220}$$

Then compute the optimum step according to the following equation.

$$\frac{g(f + \beta^* D)^T D}{|g(f)^T D|} \leq \omega, \quad 0 < \omega < 1 \tag{5.221}$$

Meanwhile, the $\beta^*$ must meet the following equation:

$$C(f + \beta^* D) - C(f) \leq \eta, \quad 0 < \eta < 1 \tag{5.222}$$

If the above equation is not satisfied, use half of $\beta^*$ to recompute the flow until the equation is met.

### 5.5.4 Implementation

For examining the NLCNFP model and algorithm, the numerical simulations have been carried out on the IEEE 5-bus and 30-bus systems. The results and comparison of secure EDC are listed on Tables 5.20–5.22. To further raise the precision of EDC and check the operation states of the system, the fast decoupled power flow is also used in the calculation, but only in the first and final stages.

Table 5.20 shows the ED results of the 5-bus system by use of the NLCNFP. The ED results with use of OKA are also listed in Table 5.20 (column 3).

The simulation results of the 30-bus system by NLCNFP are also compared with those obtained by OKA in Section 5.4. The following two cases are used to make the comparison:

**TABLE 5.20    Economic Dispatch Results
Comparison (5-Bus System)**

| Method | OKA | NLCNFP |
|---|---|---|
| $P_{G1}$(p.u.) | 0.92700 | 0.97800 |
| $P_{G2}$(p.u.) | 0.71600 | 0.66670 |
| Total cost ($/hr) | 757.500 | 757.673 |
| Total loss (p.u.) | 0.04300 | 0.04470 |

**TABLE 5.21    ED Results and Comparison Between NLCNFP and OKA for IEEE
30-Bus System**

| Scenario | Scenario 1 | Scenario 1 | Scenario 2 | Scenario 2 |
|---|---|---|---|---|
| Method | NLCNFP | OKA | NLCNFP | OKA |
| $P_{G1}$(p.u.) | 1.7595 | 1.7588 | 1.5018 | 1.69665 |
| $P_{G2}$(p.u.) | 0.4884 | 0.4881 | 0.5645 | 0.33295 |
| $P_{G5}$(p.u.) | 0.2152 | 0.2151 | 0.2321 | 0.15000 |
| $P_{G8}$(p.u.) | 0.2229 | 0.2236 | 0.3207 | 0.31270 |
| $P_{G11}$(p.u.) | 0.1227 | 0.1230 | 0.1518 | 0.30000 |
| $P_{G13}$(p.u.) | 0.1200 | 0.12000 | 0.1413 | 0.12000 |
| Total generation | 2.9286 | 2.9290 | 2.9121 | 2.9151 |
| Total real power losses | 0.0946 | 0.0950 | 0.0781 | 0.0783 |
| Total generation cost ($) | 802.3986 | 802.51 | 807.80 | 809.68 |

**TABLE 5.22    ED Results and Comparison Among NLCNFP, QP and LP for IEEE
30-Bus System**

| Generation No. | NLCNFP Method | QP Method | LP Method |
|---|---|---|---|
| $P_{G1}$ | 1.7595 | 1.7586 | 1.7626 |
| $P_{G2}$ | 0.4884 | 0.4883 | 0.4884 |
| $P_{G5}$ | 0.2152 | 0.2151 | 0.2151 |
| $P_{G8}$ | 0.2229 | 0.2233 | 0.2215 |
| $P_{G11}$ | 0.1227 | 0.1231 | 0.1214 |
| $P_{G13}$ | 0.1200 | 0.1200 | 0.1200 |
| Total generation | 2.9286 | 2.9285 | 2.9290 |
| Total real power losses | 0.0946 | 0.0945 | 0.0948 |
| Total generation cost ($) | 802.3986 | 802.3900 | 802.4000 |

Scenario 1: the original data;

Scenario 2: the original data, but the power limit value of the line 1 is reduced to 1.00 p.u.

The corresponding calculation results and comparison based on two different network flow techniques (NLCNFP and OKA) for these two scenarios are listed in Table 5.21. Obviously, the ED solved by NLCNFP has higher precision than the ED solved by OKA.

Table 5.22 lists the ED results comparison among the NLCNFP method and the conventional LP and QP methods. The agreement between the conventional ED method through power flow calculations and the NLCNFP method can be observed.

According to the $N - 1$ security analysis in Section 5.4, there are four single outages that cause the line violation for the 30-bus system. They are outage lines 1, 2, 4, and 5. Applying the idea of multigeneration plans to the 30-bus system, there will be five generation plans: one for normal operation state and four for the effective single outages, respectively. The detailed results of the multigeneration plans are shown in Table 5.23.

**TABLE 5.23    Multigeneration Plans for IEEE 30-Bus System**

| Generation No. | Normal State | Line 1 Outage | Line 2 Outage | Line 4 Outage | Line 5 Outage |
|---|---|---|---|---|---|
| $P_{G1}$ | 1.7595 | 1.42884 | 1.40919 | 1.41584 | 1.57840 |
| $P_{G2}$ | 0.4884 | 0.55222 | 0.57188 | 0.56521 | 0.38880 |
| $P_{G5}$ | 0.2152 | 0.24135 | 0.24135 | 0.24135 | 0.25512 |
| $P_{G8}$ | 0.2229 | 0.35000 | 0.35000 | 0.35000 | 0.35000 |
| $P_{G11}$ | 0.1227 | 0.17340 | 0.17340 | 0.17340 | 0.17340 |
| $P_{G13}$ | 0.1200 | 0.16154 | 0.16154 | 0.16154 | 0.16154 |
| Total generation | 2.9286 | 2.90735 | 2.90736 | 2.90734 | 2.90726 |
| Total real power losses | 0.0946 | 0.07335 | 0.07336 | 0.07334 | 0.07326 |
| Total generation cost ($) | 802.3986 | 811.36192 | 812.64862 | 812.18859 | 808.30441 |
| N security | Satisfied | – | – | – | – |
| $N - 1$ security | Not satisfied when one of lines #1,2,3,5 is in outage | satisfied | satisfied | satisfied | satisfied |

## 5.6    TWO-STAGE ECONOMIC DISPATCH APPROACH

### 5.6.1    Introduction

This section presents a two-stage ED approach according to the practical operation situation of power systems. The first stage involves the classic economic power dispatch without considering network loss. The initial generation plans of the generator units are determined according to the rank of fuel consumption characteristic of the units or the principle of equal incremental rate. The second stage involves ED considering system power loss and network security constraints. Three objectives can be used for the second stage: (i) minimize the fuel consumption, (ii) minimize system loss, and (iii) minimize the movement of generator output from the initial generation plans.

### 5.6.2    Economic Power Dispatch — Stage One

The equal incremental principle, introduced in Chapter 4, can be used for the first stage of economic power dispatch. Given the input–output characteristic of NG generating units are $F_1(P_{G1})$, $F_2(P_{G2})$, ... , $F_n(P_{Gn})$, respectively, the total system load is $P_D$. The problem is to minimize the total fuel consumption $F$ of the generators, subject to the constraint that the sum of the power generated must equal the received load, that is,

$$\min F = F_1(P_{G1}) + F_2(P_{G2}) + \cdots + F_n(P_{Gn}) = \sum_{i=1}^{NG} F_i(P_{Gi}) \qquad (5.223)$$

such that

$$\sum_{i=1}^{NG} P_{Gi} = P_D \qquad (5.224)$$

This is a constrained optimization problem, and it can be solved by the Lagrange multiplier method. According to Chapter 4, the principle of equal incremental rate of economic power operation for multiple generating units can be obtained as

$$\frac{dF_i}{dP_{Gi}} = \lambda \quad i = 1, 2, \ldots, N \qquad (5.225)$$

or

$$\frac{dF_1}{dP_{G1}} = \frac{dF_2}{dP_{G2}} = \cdots \frac{dF_N}{dP_{GN}} = \lambda \qquad (5.226)$$

The economic operation points $P_{Gi}^0$ of the first stage can be obtained from equations (5.225) or (5.226).

### 5.6.3   Economic Power Dispatch — Stage Two

The second stage of the economic power dispatch includes loss correction and network security constraints. On one hand, the system loss minimization or the fuel consumption minimization can be selected as the objective function. On the other hand, the operators expect the optimal dispatch points close to the economic operation points $P_{Gi}^0$ obtained from the first stage. Thus, the following three objectives may be adopted in the second stage of ED:

**(1)** Minimize the fuel consumption

$$\min F_1 = \sum_{i=1}^{NG} F_i(P_{Gi}) \tag{5.227}$$

**(2)** Minimize the system loss

$$\min F_2 = P_L \tag{5.228}$$

**(3)** Minimize the adjustment of generator output

$$\min F_3 = \sum_{i=1}^{NG} (P_{Gi} - P_{Gi}^0)^2 \tag{5.229}$$

The constraints include real power balance, generator power output limits, and branch power flow constraints, that is,

$$\sum_{i \in NG} P_{Gi} = \sum_{k \in ND} P_{Dk} + P_L \tag{5.230}$$

$$P_{Gimin} \leq P_{Gi} \leq P_{Gimax} \quad i \in NG \tag{5.231}$$

$$|P_{ij}| \leq P_{ijmax} \quad ij \in NT \tag{5.232}$$

where

$P_D$: the real power load
$P_{ij}$: the power flow of transmission line $ij$
$P_{ijmax}$: the power limits of transmission line $ij$
$P_{Gi}$: the real power output at generator bus $i$
$P_{Gimin}$: the minimal real power output at generator $i$
$P_{Gimax}$: the maximal real power output at generator $i$
$P_L$: the network losses
$F_i$: the fuel consumption function of the generator unit $i$
$NT$: the number of transmission lines
$NG$: the number of generators.

It is noted that the two-stage approach for ED can be used for dynamic ED or daily dispatch in the practical operation of the power systems. To actualize the transition from the time point $t$ to $t+1$ schedule successfully, the real power generation regulations constraint, $\Delta P_{GRCimax}$ must be considered, that is,

$$|P_{Gi} - P_{Gi}^0| \le \Delta P_{GRCimax} \quad i \in NG \tag{5.233}$$

or

$$-\Delta P_{GRCimax} + P_{Gi}^0 \le P_{Gi} \le \Delta P_{GRCimax} + P_{Gi}^0 \quad i \in NG \tag{5.234}$$

Thus, the regulating value of the generation is restricted by the two inequality equations (5.231) and (5.234), which can be combined into one expression:

$$\max\{-\Delta P_{GRCimax} + P_{Gi}^0, P_{Gimin}\} \le P_{Gi} \le \min\{\Delta P_{GRCimax} + P_{Gi}^0, P_{Gimax}\} \quad i \in NG \tag{5.235}$$

The ED model for the second stage can be written as

$$\min F = h_1 F_1 + h_2 F_2 + h_3 F_3 \tag{5.236}$$

such that

$$\sum_{i \in NG} P_{Gi} = \sum_{k \in ND} P_{Dk} + P_L \tag{5.237}$$

$$\max\{-\Delta P_{GRCimax} + P_{Gi}^0, P_{Gimin}\} \le P_{Gi} \le \min\{\Delta P_{GRCimax} + P_{Gi}^0, P_{Gimax}\} \quad i \in NG \tag{5.238}$$

$$|P_{ij}| \le P_{ijmax} \quad ij \in NT \tag{5.239}$$

where

$$h_1 + h_2 + h_3 = 1 \tag{5.240}$$

$h_1$: the weighting factor of the fuel consumption objective function
$h_2$: the weighting factor of the loss minimization objective function
$h_3$: the weighting factor of the generator output adjustment objective function.

The weighting factors can be determined according to the practical situation of the specific system. For example, if the network loss is the only concern in a system, we can select $h_2 = 1$ and $h_1 = h_3 = 0$. If the network loss is not a concern, and the economy is the primary concern in a system, we can select $h_1 = 1$ and $h_2 = h_3 = 0$.

The ED model for the second stage can be solved by any algorithm mentioned in the previous sections.

### 5.6.4   Evaluation of System Total Fuel Consumption

In the practical system operation, the system total fuel consumption is the main concern. Generally, the system total fuel consumption includes two parts:

**(1)** the total fuel consumption of the generators;

**(2)** the equivalent fuel consumption of the system power losses.

Generally, the system total fuel consumption before optimization is taken as the reference point. It is expected that the system total fuel consumption obtained after the second stage is less than that in the reference point.

For the reference point, the initial system power losses $P_L^0$ are obtained from a power flow solution. In addition, as the line constraints are not considered before optimization, there may be a branch flow violation. Thus a penalty term for the power violation should be introduced in the calculation of the system total fuel consumption in the reference point. The system total power violation can be computed as follows.

$$\Delta P_{\text{Viol}} = \sum_{ij=1}^{Nl} (P_{ij}^0 - P_{ij\text{max}}) \tag{5.241}$$

where $Nl$ is the set of violated branches.

The equivalent fuel consumption for the power violation is computed as

$$F_{\text{viol}} = \gamma \Delta P_{\text{Viol}} \tag{5.242}$$

Obviously, equivalent fuel consumption for the power violation $F_{\text{viol}}$ will be zero if there is no branch violation (i.e., $Nl$ is empty set).

Thus the system total fuel consumption before optimization will be

$$F_T^1 = \sum_{i=1}^{NG} F_i(P_{Gi}^0) + \gamma P_L^0 + \gamma \Delta P_{\text{Viol}} \tag{5.243}$$

After stage two, the system power losses $P_L$ and the economic operation points are computed by solving the model (5.236)–(5.239) and power flow, that is,

$$F_T^2 = \sum_{i=1}^{NG} F_i(P_{Gi}) + \gamma P_L \tag{5.244}$$

where

$\gamma$: the coefficient for converting the system power loss or branch power violation to the fuel consumption.

The requirement of the two-stage ED will be

$$F_T^2 \le F_T^1 \tag{5.245}$$

where

$F_T^1$: the initial system total fuel consumption.

$F_T^2$: the final system total fuel consumption.

## 5.7 SECURITY CONSTRAINED ECONOMIC DISPATCH BY GENETIC ALGORITHMS

GAs are adaptive search techniques that derive their models from the genetic processes of biological organisms based on evolution theory. In Chapter 4, GAs are applied to solve the classic ED problem, where the network losses and security constraints are neglected.

Considering the network losses $P_L$ and selecting unit $N$ as slack bus unit, the real power balance equation can be written as

$$P_{GN} = P_D + P_L - \sum_{i=1}^{N-1} P_{Gi} \qquad (5.246)$$

The network security constraints can be written as

$$|P_{ij}| \leq P_{ij\text{max}} \quad ij = 1, 2, \dots, NL \qquad (5.247)$$

Adding penalty factors $h_1, h_2$ to the violation of power output of the slack bus unit and $h_3$ to the violation of line power, we can get augmented cost.

$$F_A = \sum_{i=1}^{N} F_i(P_{Gi}) + h_1(P_{GN} - P_{GN\text{max}})^2 + h_2(P_{GN\text{min}} - P_{GN})^2$$

$$+ h_3 \sum_{ij=1}^{NL} (|P_{ij}| - P_{ij\text{max}})^2 \qquad (5.248)$$

GA is designed for the solution of the maximization problem, so the fitness function for solving security ED problem is defined as the inverse of equation (5.248).

$$F_{\text{fitness}} = \frac{1}{F_A} \qquad (5.249)$$

The GA operations are stated in Chapter 4. The calculation steps for solving GA-based ED with line flow constraints are as follows.

**(1)** Select the parameters related to GA such as the population size, number of generations, substring length, and the number of trials.

**(2)** Generate initially random-coded strings as population members in the first generation.

(3) Decode the population to get power generations of the units in the strings.

(4) Perform power flow analysis considering the unit generations in step (3), so that GA is able to evaluate the system transmission loss, slack bus generation, line flows, and hence any violation of the slack bus generation and violation of the line flow limits.

(5) Check whether the number of trials reaches the maximal.
If the number of trials reaches the maximal, and there is no generator power violation and line flow violation, then stop, and output the results.
If the number of trials reaches the maximal, but there exists a generator power violation or line flow violation, then this means that the given trial number is too small. Increase the trial numbers and recompute.
If the number of trials does not reach the maximal, go to the next step.

(6) Evaluate the fitness of the population members (i.e. strings).

(7) Execute a selection of strings based on reproduction, considering the roulette wheel procedure with embedded elitism followed by crossover with embedded mutation to create the new population for the next generation. Go to step (2).

***Example 5.1:*** The method of GAs for solving the security ED problem is tested on the IEEE 30-bus system. The test case is the normal operation state. The parameters related to the GAs are selected as follows.

- Number of chromosomes $= 100$
- Bit resolution per generator $= 8$
- Number of cross-points $= 2$
- Number of generations $= 18000$

**TABLE 5.24 ED Results by Genetic Algorithm and Comparison For IEEE 30-Bus System**

| Generation No. | GA Method | QP Method | LP Method |
|---|---|---|---|
| $P_{G1}$ | 1.7612 | 1.7586 | 1.7626 |
| $P_{G2}$ | 0.4884 | 0.4883 | 0.4884 |
| $P_{G5}$ | 0.2152 | 0.2151 | 0.2151 |
| $P_{G8}$ | 0.2223 | 0.2233 | 0.2215 |
| $P_{G11}$ | 0.1221 | 0.1231 | 0.1214 |
| $P_{G13}$ | 0.1200 | 0.1200 | 0.1200 |
| Total generation | 2.9292 | 2.9285 | 2.9290 |
| Total real power losses | 0.0952 | 0.0945 | 0.0948 |
| Total generation cost ($) | 802.4634 | 802.3900 | 802.4000 |

- Initial crossover probability $= 92\%$
- Initial mutation probability $= 0.1\%$

The total load is 283.4 MW and the output results are listed in Table 5.24. The GA-based ED results are also compared with those obtained by the traditional optimization methods (QP and LP). The same results are obtained.

# APPENDIX A: NETWORK FLOW PROGRAMMING

Network flow programming (NFP) is a special form of linear programming (LP). The algorithms for LP including the simplex method can also be used for the NFP problem. However, as the specialization of NFP, especially when applied to ED problem of power system, some simplified algorithms are more efficient in solving NFP problem. Herein, we only introduce several very important applications of network flow problems that are used in power systems optimal operation [22–27].

## A.1   The Transportation Problem

The transportation problem is to find number of goods to ship from the supply site to the demand site in order to minimize the total transportation cost. As we described in Section 5.4, in the ED of a power system, the supply sites correspond to the generator sources, the demand sites correspond to load demands, and the transportation paths correspond to transmission lines.

In the transportation problem, the supply node is called the *source* and the demand node is called the *sink*. The mathematical representation of the transportation problem is as follows.

$$\min C = \sum_{i=1}^{S} \sum_{j=1}^{D} c_{ij} x_{ij} \tag{5A.1}$$

such that

$$\sum_{j \in D} x_{ij} \le s_i \quad i \in S \tag{5A.2}$$

$$\sum_{i \in S} x_{ij} \ge r_j \quad j \in D \tag{5A.3}$$

$$x_{ij} \ge 0 \quad i \in S, j \in D \tag{5A.4}$$

where

$c_{ij}$: the cost of supply from source $i$ to sink $j$
$x_{ij}$: the supply from source $i$ to sink $j$. It must be nonnegative
$s_i$: the supply from the source

$r_j$: the supply received at the sink
$S$: the total number of source nodes in the network
$D$: the total number of the sink nodes in the network.

Obviously, the transportation problem is not feasible unless the supply is at least as great as the demand.

$$\sum_{i \in S} s_i \geq \sum_{j \in D} r_j \tag{5A.5}$$

If this inequality is satisfied, then the transportation problem is feasible. This is generally true for the ED problem of power systems, in which the total generation equals the total load demand plus the system power loss.

For simplificity, in the transportation problem, it can be assumed that the total demand is equal to the total supply, that is,

$$\sum_{i \in S} s_i = \sum_{j \in D} r_j \tag{5A.6}$$

Under this assumption, the inequalities in constraints (5A.2) and (5A.3) must be satisfied by equalities, that is,

$$\sum_{j \in D} x_{ij} = s_i \quad i \in S \tag{5A.7}$$

$$\sum_{i \in S} x_{ij} = r_j \quad j \in D \tag{5A.8}$$

This corresponds to the ED problem neglecting network loss. We also can use this assumption even for ED with transmission loss as we analyzed in Section 5.4.

This problem can, of course, be solved by the simplex method described in the Appendix of Chapter 9. However, the simplex tableau for this problem involves an $IJ \times (I + J)$ constraint matrix. Instead, we use a more efficient algorithm to solve it. The algorithm consists of four steps.

1. Form a transportation array or table as shown (Table A.1).
2. Find a basic feasible shipping schedule, $x_{ij}$.

**TABLE A.1   Transportation Array**

| | $D_1$ | | $D_2$ | | | $D_D$ | | |
|---|---|---|---|---|---|---|---|---|
| $P_1$ | $c_{11}$ | | $c_{12}$ | | $\cdots$ | $c_{1D}$ | | $s_1$ |
| | | $x_{11}$ | | $x_{12}$ | | | $x_{1D}$ | |
| $P_2$ | $c_{21}$ | | $c_{22}$ | | $\cdots$ | $c_{2D}$ | | $s_2$ |
| | | $x_{21}$ | | $x_{22}$ | | | $x_{2D}$ | |
| | $\vdots$ | | $\vdots$ | | | $\vdots$ | | |
| $P_S$ | $c_{S1}$ | | $c_{S2}$ | | $\cdots$ | $c_{SD}$ | | $s_S$ |
| | | $x_{S1}$ | | $x_{S2}$ | | | $x_{SD}$ | $s_S$ |
| | $r_1$ | | $r_2$ | | | $r_D$ | | |

(a) Choose any available square from the table, say $(i_0, j_0)$, specify $x_{i_0 j_0}$ as large as possible subject to the constraints, and circle this variable.

(b) Delete from consideration whichever row or column has its constraint satisfied, but not both. If there is a choice, do not delete a row (column) if it is the last row (respectively, column) undeleted.

(c) Repeat steps (a) and (b) until the last available square is filled with a circled variable, and then delete from consideration both row and column.

3. Test for optimality.

Given a feasible shipping schedule, $x_{ij}$, we can use the equilibrium theorem to check for optimality. This entails finding feasible $u_i$ and $v_j$ that satisfy the equilibrium conditions

$$v_j - u_i = c_{ij}, \quad \text{for} \quad x_{ij} > 0 \tag{5A.9}$$

where, $u_i$ and $v_j$ are nonnegative dual variables of the primal problem, and satisfy the following constraint.

$$v_j - u_i \leq c_{ij}, \quad \text{for all } i \text{ and } j. \tag{5A.10}$$

Then, the method for checking the optimality as follows:

(a) Set one of the $u_i$ and $v_j$, and use equation (5A.9) for squares containing circled variables to find all the $u_i$ and $v_j$.

(b) Check the feasibility, $v_j - u_i \leq c_{ij}$, for the remaining squares. If feasible, the solution is optimal for the problem and its dual problem.

4. If the test fails, find an improved basic feasible shipping schedule, and repeat step 3.

(a) Choose any square $(i, j)$ with $v_j - u_i > c_{ij}$, set $x_{ij} = \theta$, but keep the constraints satisfied by subtracting and adding $\theta$ to appropriate circled variables.

(b) Choose $\theta$ to be the minimum of the variables in the squares in which $\theta$ is subtracted.

(c) Determine the new variable and remove from the circled variables, one of the variables from which $\theta$ was subtracted that is now zero.

***Example A.1:*** There is a simplified power system that consists of three generators ($G_1 = 6$ p.u., $G_2 = 7$ p.u., and $G_3 = 9$ p.u.) and four load demands ($D_1 = 3$ p.u., $D_2 = 9$ p.u., $D_3 = 4$ p.u., $D_4 = 6$ p.u.). Each generator connects to all loads, respectively. Assume network loss is neglected. To compute the minimal transmission cost flow $P_{ij}$ for this network we can follow the steps described:

1. We can form the transportation table, Table A.2, where the number in the table is the transmission cost for transferring power from the generator to the load.

2. Find an initial power flow $P_{ij}$.

**TABLE A.2  Transportation Array for Example A.1**

| | $D_1$ | $D_2$ | $D_3$ | $D_4$ | |
|---|---|---|---|---|---|
| $P_{G1}$ | 4 | 10 | 12 | 3 | 6 |
| $P_{G2}$ | 8 | 5 | 6 | 4 | 7 |
| $P_{G3}$ | 1 | 3 | 4 | 7 | 9 |
| | 3 | 9 | 4 | 6 | |

Choose any square, say the upper left corner, (1, 1), and make $P_{11}$ as large as possible subject to the constraints. In this case, $P_{11}$ is chosen equal to 3 (we delete the unit for simplification). It means that the supply load $D_1$ from $P_{G1}$. Thus, we get $P_{21} = P_{31} = 0$.

We choose another square, say (1, 2), and make $P_{12}$ as large as possible subject to the constraints. Then $P_{12} = 3$, as there are only three units left at $P_{G1}$. Hence, $P_{13} = P_{14} = 0$. Next, choose square (2, 2), say, and put $P_{22} = 6$, so that load $D_2$ receives all of its demands, 3 units from $P_{G1}$ and 6 units from $P_{G2}$. Hence, $P_{32} = 0$. One continues in this way until all the variables $P_{ij}$ are determined. The results are shown in the Table A.3.

**TABLE A.3  Feasible Flow for Example A.1**

| | $D_1$ | | $D_2$ | | $D_3$ | | $D_4$ | | |
|---|---|---|---|---|---|---|---|---|---|
| $P_{G1}$ | 4 | 3 | 10 | 3 | 12 | | 3 | | 6 |
| $P_{G2}$ | 8 | | 5 | 6 | 6 | 1 | 4 | | 7 |
| $P_{G3}$ | 1 | | 3 | | 4 | 3 | 7 | 6 | 9 |
| | | 3 | | 9 | | 4 | | 6 | |

It is noted that this method of finding the initial feasible solution is simple, but may not be efficient. Here we introduce another approach called the *least cost method*.

We choose a different order for selecting the squares in the example above. We try to find a good initial solution by choosing the squares with the smallest transmission costs first.

It can be observed from the above table that the smallest transmission cost is in the lower left square, which is $c_{31} = 1$. Thus it will be most economical to supply power from generator 3 to load 1. Since the maximal load is 3 for $D_1$, the maximal power flow $P_{31} = 3$ is determined and $D_1$ is satisfied, which can be deleted for the other computation. Of the remaining squares, 3 is the lowest transmission cost (there are two). We might choose the upper right corner next. Thus, $P_{14} = 6$ is determined and we may delete either $P_{G1}$ or $D_4$, but not both, according to rule (2b). Say we delete $P_{G1}$. Next $P_{32} = 6$ is determined and $P_{G3}$ is deleted. Of the generators, only $P_{G2}$ remains, so we can determine $P_{22} = 3$, $P_{23} = 4$ and $P_{24} = 0$. The results are show in Table A.4.

3. Check optimality of the results.

We check the feasible power flow in Table A.4 for optimality. First solve for the $u_i$ and $v_j$. We put $u_2 = 0$ because that allows us to solve quickly for $v_2 = 5$, $v_3 = 6$ , and $v_4 = 4$. (Generally, it is a good idea to start with a $u_i = 0$ (or $v_j = 0$ ) for which there are many determined variables in the corresponding row (column).) Knowing $v_4 = 4$ allows us to solve for $u_1 = 1$. Knowing $v_2 = 5$ allows us to solve for $u_3 = 2$, which allows us to solve for $v_1 = 3$. We write the $v_j$ variables across the top of the array and $u_i$ along the left, as shown in Table A.5.

**TABLE A.4  Feasible Flow Using Least Cost Rule for Example A.1**

|  | $D_1$ | | $D_2$ | | $D_3$ | | $D_4$ | | |
|---|---|---|---|---|---|---|---|---|---|
| $P_{G1}$ | 4 | | 10 | | 12 | | 3 | 6 | 6 |
| $P_{G2}$ | 8 | | 5 | 3 | 6 | 4 | 4 | 0 | 7 |
| $P_{G3}$ | 1 | 3 | 3 | 6 | 4 | | 7 | | 9 |
| | | 3 | | 9 | | 4 | | 6 | |

**TABLE A.5  Optimality Check for Example A.1**

|  | 3 | | 5 | | 6 | | 4 | | |
|---|---|---|---|---|---|---|---|---|---|
| 1 | 4 | | 8 | | 12 | | 3 | 6 | 6 |
| 0 | 8 | | 5 | 3 | 6 | 4 | 4 | 0 | 7 |
| 2 | 1 | 3 | 3 | 6 | 4 | | 7 | | 9 |
| | | 3 | | 9 | | 4 | | 6 | |

Then, check feasibility of the remaining six squares. The upper left square satisfies the constraint $v_j - u_i \leq c_{ij}$, as $3 - 1 = 2 \leq 4$. Similarly, all the squares may be seen to satisfy the constraints, and hence the above gives the solution to both the primal and dual problems. The optimal shipping schedule is as noted, and the value is

$$\sum \sum c_{ij}x_{ij} = 3\cdots 1 + 6\cdots 3 + 3\cdots 5 + 4\cdots 6 + 0\cdots 4 + 6\cdots 3 = 78.$$

We can check if the solution is optimum by computing $\sum v_j r_j - \sum u_i s_i$, which is the objective function of the dual problem. According to Corollary 2 of the duality theorem described in Appendix A, we have

$$\sum \sum c_{ij}x_{ij} = \sum v_j r_j - \sum u_i s_i \tag{5A.11}$$

If both primal and dual problems have the optimal solution,

$$\Sigma v_j r_j - \Sigma u_i s_i = 78$$

Thus, the above solution is optimal.

**Example A.2:** For example A.1 with the following transmission cost (Table A.6).

**TABLE A.6  Transportation Array for Example A.2**

|          | $D_1$ | $D_2$ | $D_3$ | $D_4$ |   |
|----------|-------|-------|-------|-------|---|
| $P_{G1}$ | 4     | 8     | 13    | 3     | 6 |
| $P_{G2}$ | 2     | 5     | 6     | 5     | 7 |
| $P_{G3}$ | 1     | 3     | 4     | 15    | 9 |
|          | 3     | 9     | 4     | 6     |   |

According to least cost rule, we get the feasible flow table (Table A.7).

**TABLE A.7  Least Cost Flow for Example A.2**

|          | $D_1$ | $D_2$ | $D_3$ | $D_4$ |   |
|----------|-------|-------|-------|-------|---|
| $P_{G1}$ | 4     | 8     | 13    | 3   6 | 6 |
| $P_{G2}$ | 2     | 5   3 | 6   4 | 5   | 7 |
| $P_{G3}$ | 1   3 | 3   6 | 4     | 15  | 9 |
|          | 3     | 9     | 4     | 6     |   |

According to the equilibrium condition, we can compute the $u_i$ and $v_j$. The corresponding results are shown in Table A.8.

**TABLE A.8  Optimality Check for Example A.2**

|   | 3 | 5 | 6 | 5 |   |
|---|-------|-------|-------|-------|---|
| 2 | 4     | 8     | 13    | 3   6 | 6 |
| 0 | 2     | 5   3 | 6   4 | 5   | 7 |
| 2 | 1   3 | 3   6 | 4     | 15  | 9 |
|   | 3     | 9     | 4     | 6     |   |

Through checking the optimality in Table A.8, we found the block (2, 1) in Table A.8 cannot satisfy the constraint $v_j - u_i \le c_{ij}$, as $v_1 - u_2 = 3 - 0 = 3 \ge c_{12} = 2$. Thus, the solution in Table A.8 is not optimal. We need to find an improved basic feasible shipping schedule and recheck the optimality.

Choose any square $(i, j)$ with $v_j - u_i > c_{ij}$, set $x_{ij} = \theta$, but keep the constraints satisfied by subtracting and adding $\theta$ to appropriate selected variables. So we would like to add to block $(2, 1)$. This requires subtracting $\theta$ from squares $(3, 1)$ and $(2, 2)$, and adding $\theta$ to square $(3, 2)$, as shown in Table A.9.

**TABLE A.9   Optimality Check for Example A.2**

|   | 3 | 5 | 6 | 5 |   |   |
|---|---|---|---|---|---|---|
| 2 | 4 | 8 | 13 | 3 |   |   |
|   |   |   |   |   | 6 | 6 |
| 0 | 2 | 5 | 6 | 5 |   |   |
|   | $+\theta$ | $-\theta$ 3 | 4 |   |   | 7 |
| 2 | 1 | 3 | 4 | 15 |   |   |
|   | $-\theta$ 3 | $+\theta$ 6 |   |   |   | 9 |
|   | 3 | 9 | 4 | 6 |   |   |

We choose $\theta$ to be the minimum of the $x_{ij}$ in the squares in which we are subtracting $\theta$. In the example, $\theta = 3$. Determine the new variable and remove from the selected variables, one of the variables from which $\theta$ was subtracted and is now zero. Then we get Table A.10. We can check that all the constraints are met, and the optimal solution is 75.

**TABLE A.10   Optimality Check for Example A.2**

|   | 2 | 5 | 6 | 5 |   |   |
|---|---|---|---|---|---|---|
| 2 | 4 | 8 | 13 | 3 |   |   |
|   |   |   |   |   | 6 | 6 |
| 0 | 2 | 5 | 6 | 5 |   |   |
|   | 3 | 0 | 4 |   |   | 7 |
| 2 | 1 | 3 | 4 | 15 |   |   |
|   | 0 | 9 |   |   |   | 9 |
|   | 3 | 9 | 4 | 6 |   |   |

## A.2   Dijkstra Label-Setting Algorithm

Dijkstra's algorithm is a widely used label method for solving network flow problems such as the shortest-path problem. The data structures that are carried from one iteration to the next are a set $F$ of *finished* nodes and two arrays indexed by the nodes of the graph. The first array, $v_j, j \in N$, is just the array of labels. The second array, $h_i$, $i \in N$, indicates the next node to visit from node $i$ in the shortest path. As the algorithm proceeds, the set $F$ contains those nodes for which the shortest path has already been found. This set starts out empty. Each iteration of the algorithm adds one node to it.

The algorithm is called a *label-setting* algorithm because each iteration sets one label to its optimal value. For finished nodes, the labels are fixed at their optimal values. For each unfinished node, the label has a temporary value, which represents the length of the shortest path from that node to the root, subject to the condition that all intermediate nodes on the path must be finished nodes. At those nodes for which no such path exists, the temporary label is set to infinity (or, in practice, a large positive number).

The algorithm is initialized by setting all the labels to infinity except for the root node (or source node), whose label is set to 0. Also, the set of finished nodes is initialized to the empty set. Then, as long as there remain unfinished nodes, the algorithm selects an unfinished node $j$ having the smallest temporary label, adds it to the set of finished nodes, and then updates each unfinished "upstream" neighbor $i$ by setting its label to $c_{ij} + v_j$ if this value is smaller than the current value $v_i$. For each neighbor $i$ whose label gets changed, $h_i$ is set to $j$.

## PROBLEMS AND EXERCISES

**1.** What is SCED?

**2.** What does the economic dispatch with $N - 1$ security mean?

**3.** Compare LP, QP, and NFP that are used for solving SCED.

**4.** State the features of OKA algorithm when it is applied to SCED

**5.** What are the differences between NFP and NLCNFP?

**6.** What is the "$N - 1$ constrained zone"?

**7.** State "True" or "False"

    7.1 SCED considers not only the generator power output limits but also the capacity limits of the transmission lines and transformers.

    7.2 SCED must be linear model

    7.3 SCED does not involve reactive power dispatch.

    7.4 SCED must satisfy the bus voltage constraint.

    7.5 KCL used in NFP is equivalent to the real power balance.

    7.6 NFP is a special LP method.

    7.7 QP has a quadratic objective function as well as quadratic constraints.

    7.8 SCED neglects network losses.

    7.9 Network losses cannot be considered in NFP economic dispatch.

    7.10 NLCNFP can solve the nonlinear SCED problem.

**8.** Solve the following QP problem

$$\min f(x) = \frac{1}{2}(x_1 - 1)^2 + \frac{1}{2}(x_2 - 5)^2$$

subject to

$$-2x_1 + x_2 \leq 2$$

$$-x_1 + x_2 \leq 3$$

$$x_1 \leq 3$$

$$x_1 \geq 0, \quad x_2 \geq 0$$

**9.** A power network, which has two generators $(P_{G1}$ and $P_{G2})$ and three transmission lines to supply a load $P_D$, is shown in Figure 5.6. The system parameters are as follows.

$$F_1(P_{G1}) = C_1 P_{G1} = 3P_{G1}$$

$$F_2(P_{G2}) = C_2 P_{G2} = 5P_{G2}$$

$$0 \leq P_{G1} \leq 4$$

$$0 \leq P_{G2} \leq 3$$

$$P_D = 4$$

$$0 \leq P_{l1} \leq 1$$

$$0 \leq P_{l2} \leq 4$$

$$1 \leq P_{l3} \leq 3$$

**(1)** Use the OKA algorithm to solve this economic dispatch problem.

**(2)** Use the LP method to solve this economic dispatch problem.

**10.** For the same power system and parameters as exercise 8 except for the two generators cost functions, which are quadratic, that is,

$$F_1(P_{G1}) = a_1 P_{G1}^2 + b_1 P_{G1} = P_{G1}^2 + 4P_{G1}$$
$$F_2(P_{G2}) = a_2 P_{G2}^2 + b_2 P_{G2} = 3P_{G2}^2 + 2P_{G2}$$

Use the quadratic programming method to solve this economic dispatch problem.

# REFERENCES

1. Alsac O, Stott B. Optimal load flow with steady-state security. IEEE Trans., PAS 1974;745–751.
2. Zhu JZ, Xu GY. A new economic power dispatch method with security. Electr. Pow. Syst. Res. 1992;25:9–15.
3. Zhu JZ. *Power System Optimal Operation*. Tutorial of Chongqing University; 1990.
4. Irving MR, Sterling MJH. Economic dispatch of active power with constraint relaxation. IEE Proc. C 1983;130:172–177.
5. Lee TH, Thorne DH, Hill EF. A transportation method for economic dispatching—application and comparison. IEEE Trans., PAS 1980;99:2372–2385.
6. Hobson E, Fletcher DL, Stadlin WO. Network flow linear programming techniques and their application to fuel scheduling and contingency analysis. IEEE Trans., PAS 1984;103:1684–1691.
7. Elacqua AJ, Corey SL. Security constrained dispatch at the New York power pool. IEEE Trans., PAS 1982;101:2876–2884.
8. Momoh JA, Brown GF, Adapa R. "Evaluation of interior point methods and their application to power system economic dispatch," Proceedings of the 1993 North American Power Symposium, October 11 and 12, 1993.
9. Zhu JZ, Irving MR. Combined active and reactive dispatch with multiple objectives using an analytic hierarchical process. IEE Proc. C 1996;143(4):344–352.
10. Zhu JZ, Xu GY. Application of out-of-kilter algorithm in network programming technology to real power economic dispatch. J Chongqing Univ 1988;11(2).
11. Zhu JZ, Xu GY. The convex network flow programming model and algorithm of real power economic dispatch with security. Control Decis. 1991;6(1):48–52.
12. Zhu JZ, Chang CS. A new model and algorithm of secure and economic automatic generation control. Electr. Pow. Syst. 1998;45(2):119–127.
13. Zhu JZ. *Application of Network Flow Techniques to Power Systems*. WA: Tianya Press, Technology; 2005.
14. Zhu JZ, Irving MR, Xu GY. A new approach to secure economic power dispatch. Int. J. Elec. Power 1998;20(8):533–538.
15. Zhu JZ, Xu GY. Network flow model of multi-generation plan for on-line economic dispatch with security. Modeling, Simulation & Control, A 1991;32(1):49–55.
16. Zhu JZ, Xu GY. Secure economic power reschedule of power systems. Modeling, Measurement & Control, D 1994;10(2):59–64.
17. Nanda J, Narayanan RB. Application of genetic algorithm to economic load dispatch with Line flow constraints. Int. J. Elec. Power 2002;24:723–729.
18. King TD, El-Hawary ME, El-Hawary F. Optimal environmental dispatching of electric power system via an improved Hopfield neural network model. IEEE Trans on Power Syst. 1995;10(3):1559–1565.
19. Wong KP, Fung CC. Simulated-annealing-based economic dispatch algorithm. IEE Proc. Part C 1993;140(6):509–515.
20. Zhu JZ, Xu GY. Approach to automatic contingency selection by reactive type performance index. IEE Proc. C 1991;138:65–68.
21. Zhu JZ, Xu GY. A unified model and automatic contingency selection algorithm for the P and Q sub-problems. Electr. Pow. Syst. Res. 1995;32:101–105.
22. Smith DK. *Network optimization practice*. Chichester, UK: Ellis Horwood; 1982.

23. Dantzig GB. *Linear Programming and Extensions*. Princeton University Press; 1963.
24. Luenberger DG. *Introduction to Linear and Nonlinear Programming*. USA: Addison-wesley Publishing Company, Inc.; 1973.
25. Hadley G. *Linear Programming*. Reading, MA: Addison—Wesley; 1962.
26. Strayer JK. *Linear Programming and Applications*. Springer-Verlag; 1989.
27. Bazaraa M, Jarvis J, Sherali H. *Linear Programming and Network Flows*. 2 ed. New York: Wiley; 1977.

# CHAPTER 6

# *MULTIAREAS SYSTEM ECONOMIC DISPATCH*

This chapter focuses on the operation of the multiarea system. In addition to the introduction of the wheeling model, multiarea wheeling, the total transfer capability computation in multiareas, this chapter introduces the multiarea economic dispatch (MAED) algorithms based on nonlinear convex network flow programming (NLC-NFP), as well as the nonlinear optimization neural network approach.

## 6.1  INTRODUCTION

Many countries have more than one major generation-transmission utility with local distribution utilities. Because of the recent deregulation of the power industry, the industry structure is important in discussing the interchange of power and energy as the purchase and sale of power and energy is a commercial business in which the parties to any transaction expect to enhance their own economic positions under nonemergency situations. The multiarea system economic dispatch or interconnect systems economic dispatch is for this purpose.

At present, many approaches have been considered for MAED [1–5], which is an extension of economic dispatch. All kinds of optimization algorithms and heuristic approaches have been used in economic dispatch [6–18], which are described in Chapter 5.

## 6.2  ECONOMY OF MULTIAREAS INTERCONNECTION

Electric power systems are interconnected or multiple areas are interconnected to one big system because the interconnected system is more reliable. Here we use the term multiarea system to stand for the interconnected system. In a multiarea system, generations and loads are coordinated with each other through the tie lines among the areas. A load change in any one of areas is taken care of by all generators in all areas. Similarly, if a generator is lost in one control area, governing action from generators in all connected areas will increase generation outputs to make up the mismatch. Another advantage of a multiarea system is that it may be operated at

*Optimization of Power System Operation*, Second Edition. Jizhong Zhu.
© 2015 The Institute of Electrical and Electronics Engineers, Inc. Published 2015 by John Wiley & Sons, Inc.

less cost than if left as separate parts. As described in Chapter 4, it will improve the operating economics if two generators that have different incremental costs are operating together. This concept is also suited for the interconnected multiarea system because the generators' cost functions are different for different areas.

For example, companies that are members of the broker system send hourly buy-and-sell offers for energy to the broker, who matches them according to certain market rules. Hourly, each member transmits an incremental cost and the number of MWh it is willing to sell or its decremental cost and the number of MWh it will buy. The broker sets up the transactions by matching the lowest-cost seller with the highest-cost purchaser, proceeding in this manner until all offers are processed. A common arrangement set up by the broker for the buyers and sellers is to compensate the seller for the incremental generation costs and split the savings of the buyer equally with the seller. The pricing formula for this arrangement is similar to the operation of two generators with different incremental cost rate in a system. But we handle the two generators like two utilities with one selling, the other buying. Then, the transaction's cost rate is computed as below [19].

$$\lambda_c = \lambda_s + \frac{1}{2}(\lambda_b - \lambda_s)$$

$$= \frac{1}{2}(\lambda_b + \lambda_s) \qquad (6.1)$$

where

$\lambda_s$: the incremental cost of the selling utility ($/MWh)
$\lambda_b$: the decremental cost of the buying utility ($/MWh)
$\lambda_c$: the cost rate of the transaction ($/MWh).

***Example 6.1:*** There are four utilities with two selling, and two buying. The related data are listed in Tables 6.1 and 6.2. The maximum pool savings possible is computed as follows.

**TABLE 6.1   Data of Utilities A and B**

| Utilities Selling Energy | Incremental Cost ($/MWh) | MWh for Sale | Seller's Total Increase in Cost($) |
|---|---|---|---|
| A | 20 | 120 | 2400 |
| B | 28 | 80 | 2240 |

**TABLE 6.2   Data of Utilities C and D**

| Utilities Buying Energy | Decremental Cost ($/MWh) | MWh for Purchase | Buyer's Total Decrease in Cost($) |
|---|---|---|---|
| C | 32 | 60 | 1920 |
| D | 46 | 140 | 6440 |

Net pool savings $= (1920 + 6440) - (2440 + 2240) = 3720(\$)$
The broker sets up transactions as shown in the following.

1. Transaction: A sells 120 MWh to D
   The transaction saving $\Delta F_{A-D} = 120 \times (46 - 20) = 3120(\$)$
2. Transaction: B sells 20 MWh to D
   The transaction saving $\Delta F_{B-D} = 20 \times (46 - 28) = 360(\$)$
3. Transaction: B sells 60 MWh to C
   The transaction saving $\Delta F_{B-C} = 60 \times (32 - 28) = 240(\$)$

The total transaction savings are

$$\Delta F_T = 60 \times (32 - 28) = 3120 + 360 + 240 = 3720(\$)$$

Then the rate and payment of each transaction are computed as follows.

1. Transaction: A sells 120 MWh to D
   The rate $\lambda_{A-D} = (46 + 20)/2 = 33(\$/MWh)$
   The payment: $F_{A-D} = 33 \times 120 = 3960(\$)$
2. Transaction: B sells 20 MWh to D
   The rate $\lambda_{A-D} = (46 + 28)/2 = 37 \ (\$/MWh)$
   The payment: $F_{A-D} = 37 \times 20 = 740(\$)$
3. Transaction: B sells 60 MWh to C
   The rate $\lambda_{A-D} = (32 + 28)/2 = 30 \ (\$/MWh)$
   The payment: $F_{A-D} = 30 \times 60 = 1800(\$)$

This means that utility A receives payment \$3960 from utility D, and utility B receives the payment \$2540 from C and D. Then each participant obtains benefit.

$$\Delta F_A = 3960 - 2400 = 1560(\$)$$

$$\Delta F_B = 2540 - 2240 = 300(\$)$$

$$\Delta F_C = 1920 - 1800 = 120(\$)$$

$$\Delta F_D = 6440 - 3960 - 740 = 1740(\$)$$

Obviously, $\Delta F_A + \Delta F_B + \Delta F_C + \Delta F_D = \Delta F_T$.

Therefore, there exist transactions among areas if the areas belong to different companies. One area may have a surplus of power and energy and may wish to sell it to other areas with different companies on a long-term firm supply basis. In excess of this agreed amount, it will be on a "when and if available" basis with different price. Meanwhile, some area may wish to buy energy from the other areas in the connected system. It is possible that the interconnected system will have interchange power being bought and sold simultaneously within several areas. Thus the price for the interchange must be set while taking account of the other transactions. For example, if one area were to sell interchange power to two different areas in sequence, it would probably quote a higher price for the second sale as the first sale would have raised its incremental cost. On the other hand, if the selling utility was a member of a power pool, the sale price might be set by the power and energy pricing portions of the pool agreement to be at a level such that the seller receives the cost of the generation for the sale plus one-half the total savings of all the purchasers. In this case, it is assumed that a pool control center exists, the sale price would be computed by this center, and this would differ from the prices under multiple interchange contracts. In the United States, the independent system operator (ISO) plays this kind of role.

The power pool or ISO is administered from a central location that has responsibility for setting up interchange between members, as well as other administrative tasks. The pool members relinquish certain responsibilities to the pool operating office in return for greater economy in operation. The agreement that the pool members sign is usually very complex. The complexity arises because the members of the pool are attempting to gain greater benefits from the pool operation and to allocate these benefits equitably among the members. In addition to maximizing the economic benefits of interchange between the members, the pool helps member companies by coordinating unit commitment and maintenance scheduling, providing a centralized assessment of system security and reliability, as well as marketing rules, and so on. The increased reliability provided by the pool allows the members to draw energy from the pool transmission network during emergencies as well as covering each others' reserves when generating units are down for maintenance or in outage.

The agreements among the pool members are very important for the operation of a pool system. Obviously, the agreements will become more complicated if the members try to push for maximum economic operation. Nevertheless, the savings obtainable are quite significant and have led many interconnected utility systems (i.e., multiarea systems) throughout the world to form centrally dispatched power pools when feasible. At present, there are several organizations similar to the power pool in the United States. They are MISO, ISONE, CAISO, PJM, NYISO, ERCOT, SPP, Entergy, and so on. These ISOs have SCADA and EMS systems, as well as a market system. They use the real-time data telemetered to central computers that calculate the best economic dispatch for the whole organization (within footprint) and provide signals to the member companies.

***Example 6.2:*** For Example 6.1, assume that four utilities were scheduled to transact energy by a central dispatching scheme, and 12% of the gross system savings was to be set aside to compensate those systems that provided transmission facilities to the pool. The maximum pool savings possible is computed as follows.

The net pool savings without transmission compensation is 3720 ($). Thus the transmission compensation $F_{T\text{comp}} = 3720 \times 12\% = 446.4(\$)$

The weighted average incremental cost for selling can be computed as follows.

$$\overline{\lambda}_s = \frac{\displaystyle\sum_{i=1}^{NS} \lambda_{si} P_{si}}{\displaystyle\sum_{i=1}^{NS} P_{si}} \tag{6.2}$$

where

$\overline{\lambda}_s$: the weighted average incremental cost for selling utilities ($/MWh);
$\lambda_{si}$: the incremental cost for selling utility $i$ ($);
$P_{si}$: the selling power for the selling utility $i$ (MWh);
NS: the number of selling utilities.

The weighted average decremental cost for buying can be computed as follows.

$$\overline{\lambda}_b = \frac{\sum_{j=1}^{NB} \lambda_{bj} P_{bj}}{\sum_{j=1}^{NB} P_{bj}} \tag{6.3}$$

where

$\overline{\lambda}_b$: the weighted average incremental cost for buying utilities ($/MWh);
$\lambda_{bj}$: the decremental cost for buying utility $j$ ($);
$P_{bj}$: the selling power for the buying utility $j$ (MWh);
NB: the number of buying utilities.

For this example, the seller's weighted average incremental cost is

$$\overline{\lambda}_s = \frac{20 \times 120 + 28 \times 80}{120 + 80} = 23.2 (\$/\text{MWh})$$

The buyer's weighted average decremental cost is

$$\overline{\lambda}_b = \frac{32 \times 60 + 46 \times 140}{60 + 140} = 41.8 (\$/\text{MWh})$$

Considering the transmission compensation, the transaction savings for seller and buyer can be computed as below.

$$\Delta F_{si} = (1 - \eta\%) \frac{\overline{\lambda}_b - \lambda_{si}}{2} P_{si} \tag{6.4}$$

$$\Delta F_{bi} = (1 - \eta\%) \frac{\overline{\lambda}_{bi} - \lambda_s}{2} P_{bi} \tag{6.5}$$

where

$\eta\%$: the transmission compensation rate.

For utility A that sells 120 MWh to the pool, the transaction savings are

$$\Delta F_{sA} = (1 - 12\%) \frac{41.8 - 20}{2} \times 120 = 1151.04 (\$)$$

For utility B that sells 80 MWh to the pool, the transaction savings are

$$\Delta F_{sB} = (1 - 12\%)\frac{41.8 - 28}{2} \times 80 = 485.76(\$)$$

For utility C that buys 60 MWh from the pool, the transaction savings are

$$\Delta F_{bC} = (1 - 12\%)\frac{32 - 23.2}{2} \times 60 = 232.32(\$)$$

For utility D that buys 140 MWh from the pool, the transaction savings are

$$\Delta F_{bD} = (1 - 12\%)\frac{46 - 23.2}{2} \times 140 = 1404.48(\$)$$

The total savings are

$$\Delta F_T = \Delta F_{sA} + \Delta F_{sB} + \Delta F_{bC} + \Delta F_{bD}$$

$$= 1151.04 + 485.76 + 232.32 + 1404.48 = 3273.6$$

The practical costs in the transactions for this hour are

A sells 120 MWh and obtains

$$F_A = 120 \times 23.2 + 1151.04 = 3935.04(\$)$$

B sells 80 MWh and obtains

$$F_B = 80 \times 23.2 + 485.76 = 2341.76(\$)$$

C buys 60 MWh with payment

$$F_C = 60 \times 41.8 - 232.32 = 2275.68(\$)$$

D buys 140 MWh with payment

$$F_D = 140 \times 41.8 - 1404.48 = 4447.52(\$)$$

The total payment for this transaction is $F_C + F_D = 2275.68 + 4447.52 = 6723.2$.

The total cost that the sellers obtain is $F_A + F_B = 3935.04 + 2341.76 = 6276.8$

The difference between the total payments and the costs that sellers obtained is 446.4, which equal the transmission charge or compensation.

## 6.3   WHEELING

### 6.3.1   Concept of Wheeling

Wheeling is the heart of the operational and economic issues of an open access transmission. Let use the following example to explain what "wheeling" is. Assume utility

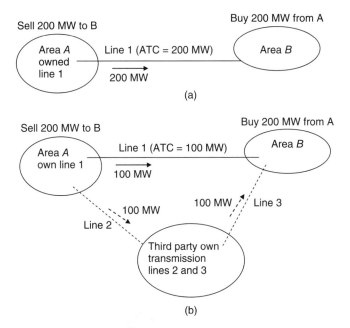

Figure 6.1   (a and b) Explanation of wheeling.

A (e.g., in area A) needs to sell 200 MW to another utility B (e.g., in area B) through its own transmission (line 1) shown in Figure 6.1(a). For simplification of explanation, the network power loss is neglected. If the available transfer capacity (ATC) of line 1 is greater than 200 MW, the transaction is simple and there is no "wheeling." But if the ATC of line 1 is only 100 MW and the same amount of transaction is required, utility A cannot complete the transaction through its own transmission lines in this case. Utility A has to "borrow" the path from the third part that owns transmission lines 2 and 3, which connect to utilities A and B (unless utility A constructs a new transmission line that is an expensive investment). Thus the transaction between utility A and B is completed through the third part, which is shown in Figure 6.1(b). This case involves "wheeling." The corresponding cost or pricing for this transaction is more complicated than that for the case shown in Figure 6.1(a).

Thus we can simply say that "wheeling" is the use of some party's (or parties') transmission system for the benefit of other parties. Each wheeling utility is termed as a wheel. Wheeling occurs on the interconnected areas or systems that contain more than two utilities (or parties) whenever a transaction takes place. When the contracted energy flow enters and leaves the wheeling utility, the flows throughout the wheeling utility's network will change. The transmission losses incurred in the wheeling utility will change. Wheeling rates are the prices it charges for use of its network, which determine payments by the buyers or sellers, or both, to the wheeling utility to compensate it for the generation and network costs incurred.

There are four major types of wheeling depending on the relationships between the wheeling utility and the buyer–seller parties [20].

- *Utility to utility*: this is usually the case of area-to-area wheeling.
- *Utility to private user or requirements customer*: The former is usually the case of area-to-bus wheeling, while the latter is usually the case of area-to-area wheeling, unless the requirements customer is small enough to be fed only at one bus, and thus it becomes area-to-bus wheeling.
- *Private generator to utility*: bus-to-area wheeling.
- *Private generator to private generator*: bus-to-bus wheeling.

Wheeling power may either increase or decrease transmission losses depending on whether the power wheeled flows in the same direction as, or counter to, the native load on the wheeler's lines. Wheeling power on a heavily loaded line causes more energy loss.

The cost of wheeling is a current high-priority problem throughout the power industry for utilities, independent power producers, as well as regulators. The following four factors have led to the importance of the cost of wheeling problem in the United States:

**(1)** enormous growth in transmission facilities at 230 KV and above since the 1960s;

**(2)** cost differentials for electric energy between different but interconnected electric utilities;

**(3)** high cost of new plant construction versus long term, off-system capacity purchase;

**(4)** Dramatic growth in nonutility generation (NUG) capacity, which includes independent power producers (IPPs) and cogenerators, due to the passage of the Public Utility Regulatory Act in 1978 and the subsequent introduction of competitive bidding for generation capacity and energy.

Wheeling is a necessary and important for any NUG, unless the customer of an NUG is the utility itself to which it is directly connected.

It is noted that not all of the transaction flows over the direct interconnections between the two systems. The other systems are all wheeling some amount of the transaction. These are called "parallel path or loop flows" in the United States, where various arrangements have been worked out between the utilities in different regions to facilitate inter-utility transactions that involve wheeling. These past agreements would generally ignore flows over parallel paths where the two systems are contiguous and own sufficient transmission capacity to permit the transfer [19]. In this case, wheeling was not taking place, by mutual agreement. The extension of this agreement to noncontiguous utilities led to the artifice known as the "contract path." To make arrangements for wheeling, the two utilities would rent the capability needed to any path that would interconnect these two utilities.

## 6.3.2   Cost Models of Wheeling

We considered energy transaction prices based on the split-savings concept earlier in this chapter. Both the sellers and wheeling systems would want to recover their cost and would wish to receive a profit by splitting the savings of the purchaser. The transmission services may be offered on the basis of a "cost plus" price. Other pricing schemes have also been used. Most are based upon simplified models that allow such fictions as the "contract path." Some are based on an attempt to mimic a power flow, in that they would base prices on incremental power flows determined in some cases by using DC power flow models. The simplest rate is a charge per MWh transferred, and ignores any path considerations. More complex schemes are based on the marginal cost of transmission that is based on the use of bus incremental costs [19]. The numerical evaluation of bus incremental costs is straightforward for a system in economic dispatch. In that case, the bus penalty factor times the incremental cost of power at the bus is equal to the system cost $\lambda$, except for the generator buses that are at upper or lower limits. This concept is not only for generator buses, but also for load buses, even for any bus that does not have any generator or load connected to it. In the practical marketing system, this kind of bus or node is called the pricing bus or pricing node. It is noted that this method is only good for a small increment of power at a bus, rather than a large increment. If the increment of power is large, the optimal power dispatch must be recalculated and the cost is not equal to the incremental cost. We treat this case in the following sections as well as in Chapter 8 on optimal power flow.

In this section, several cost models of wheeling are discussed.

***Short-Run Marginal Cost Model***  The short-run marginal costs (SRMC) of wheeling are the costs of the last MWh of energy wheeled, which can be computed from the difference in the marginal costs of electricity at the entry and exit buses, that is, the difference in the spot prices of these buses.

Figure 6.2 gives a wheeling example with system A selling $\Delta P_W$ MW to system C and system B wheeling that amount. As we mentioned above, if the operators were to purchase the block of wheeled power at bus $i$ at the incremental cost and sell it to system C at the incremental cost of power at bus $j$, the wheeling costs, using marginal cost pricing and related computations can be obtained as follows [21].

$$\lambda_W = \frac{\partial F_i}{\partial P_{Gi}} - \frac{\partial F_j}{\partial P_{Gj}} \tag{6.6}$$

where

$\lambda_W$: short-run marginal costs of wheeling.

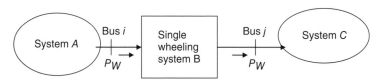

Figure 6.2    Wheeling example.

Equation (6.6) is simply the equation of the spot prices. The total wheeling costs with wheeling power $\Delta P_W$ MW will be

$$\Delta F_W = \lambda_W \Delta P_W = \left[ \frac{\partial F_i}{\partial P_{Gi}} - \frac{\partial F_j}{\partial P_{Gj}} \right] \Delta P_W \tag{6.7}$$

***Embedded Cost Model*** The embedded cost of wheeling methods, used throughout the utility industry, allocates the embedded capital costs and the average annual operation (not production) maintenance costs of existing facilities to a particular wheel; these facilities include transmission, subtransmission, and substation facilities. Happ has given a detailed treatment on all the methods as well as their algorithms. There are four types of embedded methods [22,23]:

(1) Rolled-in-embedded method
    This method assumes that the entire transmission system is used in wheeling, regardless of the actual transmission facilities that carry the wheel. The cost of wheeling as determined by this method is independent of the distance of the wheel, which is the reason that the method is also known as the postage stamp method. The embedded capital costs correspondingly reflect the entire transmission system.

(2) Contract path method
    This method is based upon the assumption that the wheel is confined to flow along a specified electrically continuous path through the wheeling company's transmission system. Changes in flows in facilities that are not along the identified path are ignored. Thus this method is limited to those facilities that lie along the assumed path.

(3) Boundary flow method
    This method incorporates changes in MW boundary flows of the wheeling company due to a wheel, either on a line basis or on a net interchange basis, into the cost of wheeling. Two power flows, executed successively for every year with and without each wheel, yield the changes in either individual boundary line or net interchange MW flows. The load level represented in the power flows can be at peak load or any other appropriate load.

(4) Line-by-line method
    This method considers changes in MW flows due to the wheel in all transmission lines of the wheeling company and the line lengths in miles. Two power flows executed with and without the wheel yield the changes in MW flows in all transmission lines

There are two limitations common to all four embedded cost methods:

(1) The methods consider only the costs of existing transmission facilities.

(2) The methods do not consider changes in production costs as a result of required changes in dispatch and or unit commitment due to the presence of the wheel.

Other cost factors may exist that contribute to the cost of wheeling. In particular, the ATC of the transmission network is not considered.

For example, the economic purchases or sales of power have to be curtailed to accommodate the wheel because of the transmission limits.

***Long-Run Incremental Cost Model***    Long-run incremental transmission costs for wheeling account for

(1) the investment costs for reinforcement to accommodate the wheel or credit for delaying or avoiding reinforcements and

(2) the charge in operating costs and incremental operation and maintenance costs incurred because of the wheel.

There are currently two models for the long-run incremental cost (LRIC) methodologies: standard long-run incremental cost (SLRIC) methodology and long-run fully incremental cost (LRFIC) methodology.

The SLRIC method uses traditional system planning approaches to determine reinforcements that are required, and corresponding investment schedules with and without each wheel, throughout the study period. If more than one wheel is present in the study period, the cost of reinforcement and the change in operating costs have to be accurately allocated to each wheel.

The LRIC method does not allow excess transmission capacity to be used by a wheel but forces reinforcement along the path of the wheel to accommodate it; if more than one wheel is present in the study period, reinforcement is required for each separate wheel [23].

## 6.4    MULTIAREA WHEELING

Multiarea wheeling is a real-world practical concern, because wheeling from a seller to a buyer involves power flow through several intermediate networks. How much power should be wheeled through each path, what wheeling should be applied to each such transaction, and how can these decisions be made optimal?

Consider an interconnected system with multiple intermediate wheeling utilities and multiple seller–buyer couples. An OKA network flow model, which is described in Chapter 5, can be used to represent this energy transaction system [24], where one seller can be treated as one source, and one buyer can be treated as a sink. OKA is able to introduce a super source (seller) and a supper sink (buyer) and make multiple seller–buyer pairs become one simple seller–buyer pair.

Figure 6.3 is a simple system with four intermediate wheeling utilities $W_1$, $W_2$, $W_3$, and $W_4$, and one buyer and seller pair (S-B). There are 10 inter-utility wheeling paths, given by the directed path $b_1$ through $b_{10}$.

Suppose that the energy to be transported through each path is arbitrary, then the computation of wheeling rates for each path can be obtained from the solution of an economic dispatch problem using OKA network flow programming [24]. To decide the optimal power flow on each path, the power flows can be set as variables and the wheeling rates can be used to improve the initial set values. The total operating costs have to be minimized considering the topological structure of multiwheeling

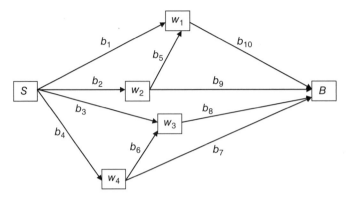

Figure 6.3    Multiarea wheeling topology.

areas and the feasible region of wheeling power flow. The topological relation can be reflected in the following matrix equation.

$$
\begin{bmatrix}
1 & 1 & 1 & 1 & 0 & 0 & 0 & 0 & 0 & 0 \\
1 & 0 & 0 & 0 & 1 & 0 & 0 & 0 & 0 & -1 \\
0 & 1 & 0 & 0 & -1 & 0 & 0 & 0 & -1 & 0 \\
0 & 0 & 1 & 0 & 0 & 1 & 0 & -1 & 0 & 0 \\
0 & 0 & 0 & 1 & 0 & -1 & -1 & 0 & 0 & 0 \\
0 & 0 & 0 & 0 & 0 & 0 & 1 & 1 & 1 & 1
\end{bmatrix}
\begin{bmatrix}
b_1 \\ b_2 \\ b_3 \\ b_4 \\ b_5 \\ b_6 \\ b_7 \\ b_8 \\ b_9 \\ b_{10}
\end{bmatrix}
=
\begin{bmatrix}
1 \\ 0 \\ 0 \\ 0 \\ 0 \\ 1
\end{bmatrix}
$$

The following assumptions are made for the relation [25]:

(1) Power inflow is given a positive sign and power outflow is given a negative sign.

(2) We are only concerned with the sale of unit power from S to B.

Each row–column multiplication represents one power balance equation for a particular utility (there are a total of six utilities in this example).

## 6.5    MAED SOLVED BY NONLINEAR CONVEX NETWORK FLOW PROGRAMMING

### 6.5.1    Introduction

This section proposes a new NLCNFP to solve the problem of security-constrained interconnected MAED. The proposed MAED model considers tie-line security and transfer constraints in each area. In addition, a simple analysis of buying and selling contract in an MAED is also made. The NLCNFP model of security-constrained

MAED is set up and solved by using a combined method of quadratic programming (QP) and network flow programming (NFP). For examining the proposed approach, a network model of four interconnected areas is constructed. Computation results are given in the chapter.

## 6.5.2   NLCNFP Model of MAED

The aim of MAED is to minimize the total production cost of supplying loads to all areas within security constraints. Initially, a basic formulation M-1 is formulated

$$\min F = \sum_{k=1}^{n} \sum_{i=1}^{NG(k)} f_{ik}(P_{Gik}) + h \sum_{k=1}^{n} \sum_{ij \in NT} P_{Lijk} \tag{6.8}$$

such that

$$\sum_{k=1}^{n} \sum_{i=1}^{NG(k)} P_{Gik} - \sum_{k=1}^{n} \sum_{i=1}^{ND(k)} P_{Dik} - P_L = 0 \tag{6.9}$$

$$P_{Gik\,\min} \leq P_{Gik} \leq P_{Gik\,\max} \tag{6.10}$$

$$|\Delta P_{Gik}| \leq \Delta P_{Gik\,GRC} \quad k = 1, \ldots, n; \quad i = 1, \ldots, NG(k) \tag{6.11}$$

$$|P_{ijk}| \leq P_{ijk\,\max} \quad k = 1, \ldots, n; \quad j = 1, \ldots, NL(k) \tag{6.12}$$

$$|P_T| \leq P_{T\,\max} \quad T = 1, \ldots, NT \tag{6.13}$$

where

$f_{ik}$: the generation cost function of $i$th generator in area $k$;

$P_{Gik}$: the active power output of $i$th generator in area $k$;

$P_{Dik}$: the active load at node $i$ in area $k$;

$P_{ijk}$: the active power on the branch $j$ in area $k$;

$P_T$: the active power on the tie −line;

$P_L$: the active power loss of the system;

$P_{Lijk}$: the active power loss of the branch $j$ in the area $k$;

$\Delta P_{Gik\,GRC}$: the limit of the generation rate constraint (GRC);

$NT$: the number of tie lines;

$n$: the number of areas;

$NG(k)$: the number of generators in area $k$;

$ND(k)$: the number of loads in area $k$;

$NL(k)$: the number of transmission lines in area $k$.

Subscripts "min" and "max" stand for the lower and upper bounds of a constraint.

According to Chapter 5, we have the following approximate equations.

$$V \cong 1.0 \ \text{p.u.} \tag{6.14}$$

$$\sin \theta_{ij} \cong \theta_{ij} \tag{6.15}$$

$$\cos \theta_{ij} \cong 1 - \theta_{ij}^2/2 \tag{6.16}$$

Then, the active power loss on the branch $ij$ can be expressed as follows.

$$P_{Lijk} = P_{ijk}^2 Z_{ijk} \tag{6.17a}$$

where

$$Z_{ijk} = \frac{(R_{ijk}^2 + X_{ijk}^2)}{X_{ijk}^2} R_{ijk} \tag{6.18a}$$

$$Pi_{jk} = -b_{ijk} \, \theta_{ijk} \tag{6.19a}$$

$R_{ij}$: the resistance of branch $j$ in area $k$;
$X_{ij}$: the reactance of branch $j$ in area $k$;
$\theta_{ijk}$: the difference of node voltage angles between the sending end and receiving
   end of the branch $j$ in area $k$;
$b_{ijk}$: the susceptance of branch $j$ in area $k$.

The active power loss on tie-lie $T$ can also be expressed as follows.

$$P_{LT} = P_T{}^2 Z_T \tag{6.17b}$$

where

$$Z_T = \frac{(R_T{}^2 + X_T{}^2)}{X_T{}^2} R_T \tag{6.18b}$$

$$P_T = - \, b_T \, \theta_T \tag{6.19b}$$

$R_T$: the resistance of tie-line branch $T$;
$X_T$: the reactance of tie-line branch $T$;
$\theta_T$: the difference in node voltage angles between the sending end and receiving
   end of tie-line branch $T$;
$b_T$: the susceptance of tie-line branch $T$.

Thus the total system power loss can be written as follows.

$$P_L = \sum_{k=1}^{n} \sum_{ij=1}^{Nl(k)} P_{Lijk} + \sum_{T=1}^{NT} P_{LT}$$

$$= \sum_{k=1}^{n} \sum_{ij=1}^{Nl(k)} P_{ijk}^2 Z_{ijk} + \sum_{T=1}^{NT} P_T^2 Z_T \tag{6.20}$$

Similar to Chapter 5, we can get the power flow limit for each branch in area $k$, as well as each tie line.

$$P_{ijk\,max} = \frac{\left[\sqrt{1 + \left(2g_{ijk}P_{ijk}/b_{ijk}^2\right)} - 1\right]}{g_{ijk}} \tag{6.21}$$

$$P_{T\,max} = \frac{\left[\sqrt{1 + \left(2g_T P_T/b_T^2\right)} - 1\right]}{g_T} \tag{6.22}$$

where $g_{ij}$ and $g_T$ are the conductance of branch $j$ in area $k$ and tie line, respectively.

If the KVL is considered in an NFP model of MAED, the voltage equation of the $l$th loop can be written as

$$\sum_{ij}(P_{ijk}Z_{ijk})\mu_{ij,l} = 0 \qquad l = 1, 2, \dots \dots, NM \tag{6.23}$$

where

$NM$: the number of loops in the network;

$\mu_{ij,l}$: the element in the related loop matrix, which takes the value 0 or 1.

Furthermore, assume that the input–output characteristics of the generators in all areas are quadratic functions.

$$f_{ik}(P_{Gik}) = a_{ik}P_{Gik}^2 + b_{ik}P_{Gik} + c_{ik} \tag{6.24}$$

Therefore, we can obtain the following NLCNFP model for the MAED problem (M-2).

$$\min F = \sum_{k=1}^{n}\sum_{i=1}^{NG(k)} (a_{ik}P_{Gik}^2 + b_{ik}P_{Gik} + c_{ik}) + h\sum_{k=1}^{n}\sum_{ij} P_{ijk}^2 Z_{ijk}$$

$$- \lambda_l \sum_{ij}(P_{ijk}Z_{ijk})\mu_{ij,l} \tag{6.25}$$

$$\sum_{k=1}^{n}\sum_{i=1}^{NG(k)} P_{Gik} - \sum_{k=1}^{n}\sum_{i=1}^{ND(k)} P_{Dik} - \left(\sum_{k=1}^{n}\sum_{ij=1}^{Nl(k)} P_{ijk}^2 Z_{ijk} + \sum_{T=1}^{NT} P_T^2 Z_T\right) = 0 \tag{6.26}$$

$$P_{Gik\,min} \le P_{Gik} \le P_{Gik\,max} \tag{6.27}$$

$$|\Delta P_{Gik}| \le \Delta P_{Gik\,GRC} \qquad k = 1, \dots, n; \quad i = 1, \dots, NG(k) \tag{6.28}$$

$$|P_{ijk}| \le \frac{\left[\sqrt{1 + \left(2g_{ijk}P_{ijk}/b_{ijk}^2\right)} - 1\right]}{g_{ijk}} \qquad k = 1, \dots, n; \quad j = 1, \dots, NL(k) \tag{6.29}$$

$$|P_T| \leq \frac{\left[\sqrt{1 + (2g_T P_T / b_T^2)} - 1\right]}{g_T} \qquad T = 1, \dots, NT \qquad (6.30)$$

In the MAED model, equation (6.26) defines the total power balance of multiarea systems. Equation (6.29) is the line security constraint in area $k$. Equation (6.30) is the tie line capacity constraint. Equation (6.27) defines the generator power upper and lower limits. Equation (6.28) is the generation rate constraint and can be written as

$$P_{Gik}^0 - \Delta P_{Gik\,\text{GRC}} \leq P_{Gik} \leq P_{Gik}^0 + \Delta P_{Gik\,\text{GRC}} \qquad (6.31)$$

where $P_{Gik}^0$ is the initial power of $i$th generator in area $k$.

Thus the generation is regulated between two inequality equations (6.27) and (6.31), which can be combined into one expression:

$$\max\{P_{Gik}^0 - \Delta P_{Gik\,\text{GRC}}, P_{Gik\text{min}}\} \leq P_{Gik} \leq \min\{P_{Gik}^0 + \Delta P_{Gik\,\text{GRC}}, P_{Gik\text{max}}\} \quad (6.32)$$

There can be contracts of buying and selling among areas. Suppose area $A$ sells electricity to area $B$, and $P_{AB\,\text{sell}}$ represents the amount of power sold or $P_{BA\,\text{buy}}$ represents the amount of power purchase. The following constraints are introduced into the MAED model.

$$\sum_T P_{TAB} = +P_{AB\,\text{sell}} \qquad (6.33)$$

$$\sum_T P_{TBA} = -P_{BA\,\text{buy}} \qquad (6.34)$$

or

$$(1 - \eta)\% P_{AB\,\text{sell}} \leq \sum_T P_{TAB} \leq (1 + \eta)\% P_{AB\,\text{sell}} \qquad (6.35)$$

$$(1 - \eta)\% P_{BA\,\text{buy}} \leq \left| \sum_T P_{TBA} \right| \leq (1 + \eta)\% P_{BA\,\text{buy}} \qquad (6.36)$$

where

$P_{TAB}$: the tie-line transfer between areas $A$ and $B$, power transfer from the area being considered to be positive if it is an export;

$P_{AB\,\text{sell}}$: the amount of power sold from area $A$ to area $B$;

$P_{BA\,\text{buy}}$: the amount of power purchased;

$\eta$: the trading error that is permitted in interconnected power system operation.

In this way, the MAED model M-2 can be written into the following model M-3 that contains the contract constraints of buying and selling electricity among areas.

$$\min F = \sum_{k=1}^{n} \sum_{i=1}^{NG(k)} (a_{ik}P_{Gik}^2 + b_{ik}P_{Gik} + c_{ik}) + h \sum_{k=1}^{n} \sum_{ij} P_{ijk}^2 Z_{ijk}$$

$$- \lambda_l \sum_{ij} (P_{ijk} Z_{ijk}) \mu_{ij,l}$$

$$+ \beta \left( \sum_{T} P_{TAB} - P_{AB\,sell} \right)^2 + \gamma \left( \left| \sum_{T} P_{TBA} \right| - P_{BA\,buy} \right)^2 \quad (6.37)$$

Subject to

$$\sum_{k=1}^{n} \sum_{i=1}^{NG(k)} P_{Gik} - \sum_{k=1}^{n} \sum_{i=1}^{ND(k)} P_{Dik} - \left( \sum_{k=1}^{n} \sum_{ij=1}^{Nl(k)} P_{ijk}^2 Z_{ijk} + \sum_{T=1}^{NT} P_T^2 Z_T \right) = 0 \quad (6.26)$$

$$\max\{P_{Gik}^0 - \Delta P_{Gik\,GRC}, P_{Gik\,min}\} \le P_{Gik} \le \min\{P_{Gik}^0 + \Delta P_{Gik\,GRC}, P_{Gik\,max}\}$$

$$k = 1, \dots, n; \quad i = 1, \dots, NG(k) \quad (6.32)$$

$$|P_{ijk}| \le \frac{\left[ \sqrt{1 + \left(2g_{ijk}P_{ijk}/b_{ijk}^2\right)} - 1 \right]}{g_{ijk}} \quad k = 1, \dots, n; \quad j = 1, \dots, NL(k)$$
$$(6.29)$$

$$|P_T| \le \frac{\left[ \sqrt{1 + \left(2g_T P_T/b_T^2\right)} - 1 \right]}{g_T} \quad T = 1, \dots, NT \quad (6.30)$$

$$(1 - \eta)\% P_{AB\,sell} \le \sum_{T} P_{TAB} \le (1 + \eta)\% P_{AB\,sell} \quad (6.35)$$

$$(1 - \eta)\% P_{BA\,buy} \le \left| \sum_{T} P_{TBA} \right| \le (1 + \eta)\% P_{BA\,buy} \quad (6.36)$$

where $\beta$ and $\gamma$ are the penalty factors, which are large positive constants.

### 6.5.3  Solution Method

MAED model M-3 is easily changed into a standard model of NLCNFP, that is, model M-4

$$\min C = \sum_{ij} c(f_{ij}) \quad (6.38)$$

such that

$$\sum_{j \in n} (f_{ij} - f_{ji}) = r_i \quad i \in n \quad (6.39)$$

$$L_{ij} \le f_{ij} \le U_{ij} \quad ij \in m \quad (6.40)$$

where

$f_{ij}$: the flow on the arc $ij$ in the network;

$L_{ij}$: the lower bound of the flow on the arc $ij$ in the network;

$U_{ij}$: the upper bound of flow on the arc $ij$ in the network;

$n$: the total number of the nodes in the network;

$m$: the total number of the arcs in the network.

According to Chapter 5 (Section 5.5), the NLCNFP model M-4 can be changed into the following QP model M-5, in which the search direction in the space of the flow variables is to be solved.

$$\min C(D) = \frac{1}{2} D^T G(f) D + g(f)^T D \qquad (6.41)$$

such that

$$AD = 0 \qquad (6.42)$$

$$D_{ij} \geq 0, \quad \text{when } f_{ij} = L_{ij} \qquad (6.43)$$

$$D_{ij} \leq 0, \quad \text{when } f_{ij} = U_{ij} \qquad (6.44)$$

Model M-5 is a special QP model, which has the form of network flow. In order to enhance the calculation speed, we present a new approach, in place of the general QP algorithm, to solve the model M-5. The details of the calculation steps are described in Chapter 5.

### 6.5.4  Test Results

For examining the proposed approach, a network of four interconnected areas is constructed as shown in Figure 6.4. Area $A1$ is an IEEE 30-bus system. It has six generators, 21 loads and 41 transformation branches, in which 1, 2, 5, 8, 11, and 13 are generators. The generators data of IEEE 30-bus system are listed in Table 6.3. The network parameters including network constraints of a 30-bus system are shown in Chapter 5. Parameters of areas $A2$, $A3$, $A4$, and tie lines are given as follows.

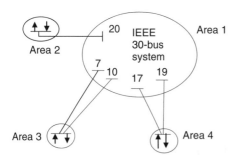

Figure 6.4  The network model of four interconnected power systems.

**TABLE 6.3   Data of Generator Nodes for IEEE 30-Bus System (p.u.)**

| Node | $a_i$ | $b_i$ | $c_i$ | $P_{Gi\min}$ | $P_{Gi\max}$ | $\Delta P_{GiGRC}$ |
|------|-------|-------|-------|--------------|--------------|---------------------|
| 1 | 37.5 | 200 | 0.0 | 0.50 | 2.00 | 0.50 |
| 2 | 175 | 175 | 0.0 | 0.20 | 0.80 | 0.30 |
| 5 | 625 | 100 | 0.0 | 0.15 | 0.50 | 0.15 |
| 8 | 83.4 | 325 | 0.0 | 0.10 | 0.35 | 0.15 |
| 11 | 250 | 300 | 0.0 | 0.10 | 0.30 | 0.15 |
| 13 | 250 | 300 | 0.0 | 0.12 | 0.40 | 0.15 |

Note: The generation cost function is: $f_i = a_i \, P_{Gi}^2 + b_i \, P_{Gi} + c_i$

Fuel cost function and power upper and lower limits are

$$F_{A2} = 80P_{A2}^2 + 175P_{A2} \qquad 0.2 \leq P_{A2} \leq 1.0$$

$$F_{A3} = 90P_{A3}^2 + 150P_{A3} \qquad 0.2 \leq P_{A3} \leq 1.0$$

$$F_{A4} = 600P_{A4}^2 + 300P_{A4} \qquad 0.2 \leq P_{A4} \leq 1.0$$

Loads of areas A2, A3, and A4 are $P_{DA2} + jQ_{DA2} = 0.44 + j0.21$; $P_{DA3} + jQ_{DA3} = 0.312 + j0.14$; and $P_{DA4} + jQ_{DA4} = 0.396 + j0.18$, respectively. Parameters and capacity constraints of the tie line are

$$R_{A2-20} = 0.0340; \quad X_{A2-20} = 0.0680; \quad P_{A2-20T\max} = 0.7$$

$$R_{A3-17} = 0.0192; \quad X_{A3-17} = 0.0575; \quad P_{A3-17T\max} = 0.7$$

$$R_{A3-19} = 0.0192; \quad X_{A3-19} = 0.0575; \quad P_{A3-19T\max} = 0.7$$

$$R_{A4-10} = 0.0267; \quad X_{A4-10} = 0.8200; \quad P_{A4-10T\max} = 0.6$$

$$R_{A4-7} = 0.0267; \quad X_{A4-7} = 0.8200; \quad P_{A4-7T\max} = 0.6$$

The following test cases for MAED are performed in the study, in which the symbol "+" represents the selling contract and "−" represents the purchase contract.

Case 1:   neglecting the buying and selling contract among areas;

Case 2:   considering the buying and selling contract among areas;

$$P_{A3-A1\,\text{sell}} = +0.5; \quad P_{A4-A1\,\text{buy}} = -0.0$$

Case 3:   considering the buying and selling among areas;

$$P_{A3-A1\,\text{sell}} = +0.55; \quad P_{A4-A1\,\text{buy}} = -0.10$$

To evaluate the calculation accuracy, the following performance index (PI) on trading error is proposed, that is,

$$PI_{EAB}\% = \frac{|P_{TAB} - P_{AB\,sell}|}{P_{AB\,sell}} \times \% \tag{6.45}$$

or

$$PI_{EAB}\% = \frac{|P_{TAB} - P_{AB\,buy}|}{P_{AB\,buy}} \times \% \tag{6.46}$$

The calculation results of security-constrained MAED for the above three test cases are listed in Table 6.4. From Table 6.4 we can get
Case 2:

$$P_{TA3-A1} = P_{A3-17} + P_{A3-19} = 0.4086 + 0.0914 = 0.5$$

$$PI_{EA3-A1}\% = 0.0$$

$$P_{TA4-A1} = P_{A4-7} + P_{A4-10} = 0.2088 - 0.2083 = 0.0005$$

$$PI_{EA4-A1}\% = 0.05\%$$

**TABLE 6.4   Test Results of Security-Constrained MAED for Four Interconnected Systems**

| Test Cases | Case 1 (p.u) | Case 2 (p.u.) | Case 3 (p.u.) |
|---|---|---|---|
| Area A1 $P_{G1}$ | 1.1523 | 1.0718 | 1.1146 |
| $P_{G2}$ | 0.3569 | 0.3471 | 0.3539 |
| $P_{G5}$ | 0.1792 | 0.1839 | 0.1833 |
| $P_{G8}$ | 0.1053 | 0.1163 | 0.1124 |
| $P_{G11}$ | 0.1248 | 0.1358 | 0.1319 |
| $P_{G13}$ | 0.1253 | 0.1363 | 0.1324 |
| Area A2 $P_{GA2}$ | 0.8832 | 0.8504 | 0.8684 |
| Area A3 $P_{GA3}$ | 0.9297 | 0.8120 | 0.8620 |
| Area A4 $P_{GA4}$ | 0.2053 | 0.3965 | 0.2964 |
| Total gen. | 04.06176 | 04.04987 | 04.05534 |
| Power losses | 00.07975 | 00.06787 | 00.07333 |
| Total gen. cost ($) | 1041.987 | 1109.621 | 1068.4117 |
| Tie-line power $P_{A2-20}$ | 0.4432 | 0.4104 | 0.4284 |
| $P_{A3-17}$ | 0.4988 | 0.4086 | 0.4487 |
| $P_{A3-19}$ | 0.1189 | 0.0914 | 0.1013 |
| $P_{A4-7}$ | 0.1364 | 0.2088 | 0.1684 |
| $P_{A4-10}$ | -0.3272 | -0.2083 | -0.2680 |
| Line-security | Satisfied | Satisfied | Satisfied |

Case 3:

$$P_{TA3-A1} = P_{A3-17} + P_{A3-19} = 0.4487 + 0.1013 = 0.55$$

$$PI_{EA3-A1}\% = 0.0$$

$$P_{TA4-A1} = P_{A4-7} + P_{A4-10} = 0.1684 - 0.2680 = -0.0996$$

$$PI_{EA4-A1}\% = 0.04\%$$

The maximum trading error is only 0.05%. Therefore, the proposed MAED approach not only satisfies all security constraints, but also has high accuracy.

## 6.6 NONLINEAR OPTIMIZATION NEURAL NETWORK APPROACH

### 6.6.1 Introduction

This section presents a new nonlinear optimization neural network approach to solve the problem of security-constrained interconnected MAED. The optimization neural network (ONN) can be used to solve mathematical programming problems. It has attracted much attention in recent years. In 1986, Tank and Hopfield first proposed an optimization neural network—TH model, which was used to solve linear programming problems. ONN is totally different from traditional optimization methods. It changes the solution of the optimization problem into an equilibrium point (or equilibrium state) of a nonlinear dynamic system and changes optimal criterion into energy functions for a dynamic system. Because of its parallel computational structure and the evolution of dynamics, the ONN approach is superior to traditional optimization methods.

### 6.6.2 The Problem of MAED

According to the previous section, a basic formulation of MAED is formulated as

$$\min F = \sum_{k=1}^{n} \sum_{i=1}^{NG(k)} f_{ik}(P_{Gik}) \tag{6.47}$$

such that

$$\sum_{k=1}^{n} \sum_{i=1}^{NG(k)} P_{Gik} - \sum_{k=1}^{n} \sum_{i=1}^{ND(k)} P_{Dik} - P_L = 0 \tag{6.48}$$

$$P_{Gikmin} \leq P_{Gik} \leq P_{Gikmax} \tag{6.49}$$

$$|\Delta P_{Gik}| \leq \Delta P_{Gik\,GRC} \quad k = 1, \dots, n; \quad i = 1, \dots, NG(k) \tag{6.50}$$

$$|P_{ijk}| \leq P_{ijk\max} \quad k = 1, \ldots, n; \quad ij = 1, \ldots, NL(k) \tag{6.51}$$

$$|P_T| \leq P_{T\max} \quad T = 1, \ldots, NT \tag{6.52}$$

The generation is regulated between two inequality equations (6.49) and (6.50), which can be combined into one expression:

$$\max\{P^0_{Gik} - \Delta P_{Gik\,GRC}, P_{Gik\min}\} \leq P_{Gik} \leq \min\{P^0_{Gik} + \Delta P_{Gik\,GRC}, P_{Gik\max}\} \tag{6.53}$$

There can be contracts of buying and selling among areas. Suppose area $A$ sells electricity to area $B$, and $P_{AB\,sell}$ represents the amount of power sold or $P_{BA\,buy}$ represents the amount of power purchase. The following constraints are introduced into the MAED model, which are the same as in Section 6.5.

$$\sum_T P_{TAB} = +P_{AB\,sell} \tag{6.54}$$

$$\sum_T P_{TBA} = -P_{BA\,buy} \tag{6.55}$$

or

$$(1 - \eta)\%P_{AB\,sell} \leq \sum_T P_{TAB} \leq (1 + \eta)\%P_{AB\,sell} \tag{6.56}$$

$$(1 - \eta)\%P_{BA\,buy} \leq \left|\sum_T P_{TBA}\right| \leq (1 + \eta)\%P_{BA\,buy} \tag{6.57}$$

The above MAED model can be written into the following model M-6, which contains the contract constraints of buying and selling electricity among areas.

$$\min F = \sum_{k=1}^{n} \sum_{i=1}^{NG(k)} f_{ik}(P_{Gik}) + \beta\left(\sum_T P_{TAB} - P_{AB\,sell}\right)^2$$

$$+ \gamma\left(\left|\sum_T P_{TBA}\right| - P_{BA\,buy}\right)^2 \tag{6.58}$$

such that

$$\sum_{k=1}^{n} \sum_{i=1}^{NG(k)} P_{Gik} - \sum_{k=1}^{n} \sum_{i=1}^{ND(k)} P_{Dik} - P_L = 0 \tag{6.59}$$

$$\max\{P^0_{Gik} - \Delta P_{Gik\,GRC}, P_{Gik\min}\} \leq P_{Gik} \leq \min\{P^0_{Gik} - \Delta P_{Gik\,GRC}, P_{Gik\min}\}$$

$$k = 1, \ldots, n; \quad i = 1, \ldots, NG(k) \tag{6.60}$$

$$|P_{bjk}| \leq P_{bjk\max} \quad k = 1, \ldots, n; \quad j = 1, \ldots, NL(k) \tag{6.61}$$

$$|P_T| \leq P_{T\max}    T = 1, \dots, NT \tag{6.62}$$

$$(1 - \eta)\%P_{AB\,\text{sell}} \leq \sum_T P_{TAB} \leq (1 + \eta)\%P_{AB\,\text{sell}} \tag{6.63}$$

$$(1 - \eta)\%P_{BA\,\text{buy}} \leq \left| \sum_T P_{TBA} \right| \leq (1 + \eta)\%P_{BA\,\text{buy}} \tag{6.64}$$

where $\beta$ and $\gamma$ are the penalty factors.

It is noted that there are some different between the above MAED model M-6 and the model M-3 described in the Section 6.5, where some approximations are applied in order to use the NLCNFP algorithm.

### 6.6.3  Nonlinear Optimization Neural Network Algorithm

***Nonlinear Optimization Neural Network Model of MAED***    The above MAED model M-6 can be solved by a new approach of nonlinear optimization neural network (NLONN). The neural network approach is a penalty-minimizing neural network approach with weights based on optimization theory and neural optimization method. It can be used to solve the nonlinear problem with equality and inequality constraints.

The MAED model M-6 can be rewritten into a general form of constrained optimization, that is, model M-7.

$$\min \, f(x) \tag{6.65}$$

such that

$$h_j(x) = 0 \quad j = 1, \dots, m \tag{6.66}$$

$$g_i(x) \geq 0 \quad i = 1, \dots, k \tag{6.67}$$

To change inequality constraints of equation (6.67) into equality constraints, new variables $y_1, \dots \dots, y_m$ (i.e., relaxation variables) are introduced into equation (6.67), In this way, model M-7 can be written as model M-8, that is,

$$\min \, f(x) \tag{6.65}$$

such that

$$h_j(x) = 0 \quad j = 1, \dots, m \tag{6.66}$$

$$g_i(x) - y_i^2 = 0 \quad i = 1, \dots, k \tag{6.68}$$

The optimization neural network is applied to the solution of M-8. The approach is totally different from traditional optimization methods. It changes the solution of optimization problems into an equilibrium point of a nonlinear dynamic system, and changes the optimal criterion into energy functions for a dynamic system. Therefore, the energy function of NLONN needs to be formed at the beginning.

According to optimization theory as described in Reference [26], we can construct the following energy function of neural network for model M-8.

$$E(x, y, \lambda, \mu, S) = f(x) - \mu^T h(x) - \lambda^T [g(x) - y^2]$$

$$+(S/2)\|h(x)\|^2 + (S/2)\|g(x) - y^2\|^2 \tag{6.69}$$

where, $\lambda$, $\mu$ are Lagrange multipliers.

It is possible to construct a different energy function from the above, for example, in an energy function as used in reference [27]. It is noted that a different energy function will produce a different neural network and distinct characteristics. There are two advantages for the proposed NLONN approach. One is that the first three terms in the energy function of equation (6.69) is just an expanded Lagrange function as in conventional nonlinear programming. Methods to guarantee optimal solution of such a function are well understood. Another advantage is due to the quadratic penalties, which are formulated to become part of the energy function (6.69) and equality constraints (6.66)–(6.68). These penalties behave very effectively against any violation of constraint.

Dynamic equations of the neural network can be obtained according to equation (6.69).

$$dx/dt = -\{\nabla_x f(x) + (Sh(x) - \mu)^T \nabla_x h(x) + [S(g(x) - y^2) - \lambda]^T \nabla_x (g(x) - y^2)\} \tag{6.70}$$

$$dy/dt = -\{\nabla_y f(x) + (Sh(x) - \mu)^T \nabla_y h(x) + [S(g(x) - y^2) - \lambda]^T \nabla_y (g(x) - y^2)\} \tag{6.71}$$

$$\partial\mu/\partial t = Sh(x) \tag{6.72}$$

$$\partial\lambda/\partial t = S(g(x) - y^2) \tag{6.73}$$

From equation (6.69), we know that the variables $x$, $y$ are separable. So we can get

$$\min_{x,y} E(x, y, \lambda, \mu, S) = \min_x \min_y E(x, y, \lambda, \mu, S)$$

$$= \min_x E(x, y^*(x, \lambda, \mu, S), \lambda, \mu, S) \tag{6.74}$$

where, $y^*(x, \lambda, \mu, S)$ satisfies the following equation:

$$\min_y E(x, y, \lambda, \mu, S) = E(x, y^*(x, \lambda, \mu, S), \lambda, \mu, S) \tag{6.75}$$

In order to obtain $y^*(x, \lambda, \mu, S)$, we set $dE/dy = 0$. Then, from equation (6.69) we get

$$2y^T[\lambda + Sy^2 - Sg(x)] = 0 \tag{6.76}$$

Obviously, from equation (6.76) we know if $\lambda - Sg(x) \geq 0$, then $y = 0$; if $\lambda - Sg(x) < 0$, then $y = 0$, or $y^2 = (Sg(x) - \lambda)/S$, that is,

$$y^2 = \begin{cases} 0, & \text{if } \lambda - Sg(x) \geq 0 \\ [Sg(x) - \lambda]/S, & \text{if } \lambda - Sg(x) < 0 \end{cases} \tag{6.77}$$

or

$$y^2 - g(x) = \begin{cases} -g(x), & \text{if } -g(x) \geq -\lambda/S \\ -\lambda/S, & \text{if } -g(x) < -\lambda/S \end{cases} \tag{6.78}$$

From equation (6.78), we can get the following expressions.

$$y^2 - g(x) = \max(-g(x), -\lambda/S) \tag{6.79}$$

$$y^2 - g(x) = -\min(g(x), \lambda/S) \tag{6.80}$$

$$g(x) - y^2 = \min(g(x), \lambda/S) \tag{6.81}$$

Substituting equation (6.79) into equation (6.69), we get

$$
\begin{aligned}
E(x, \lambda, \mu, S) &= f(x) - \mu^T h(x) + (S/2)\|h(x)\|^2 - \lambda^T[-\max(-g(x), -\lambda/S)] \\
&\quad + (S/2)\|\max(-g(x), -\lambda/S)\|^2 \\
&= f(x) - \mu^T h(x) + (S/2)\|h(x)\|^2 - (1/2S)[2\lambda^T \max(-Sg(x), -\lambda)] \\
&\quad + (1/2S)\|\max(-Sg(x), -\lambda)\|^2 \\
&= f(x) - \mu^T h(x) + (S/2)\|h(x)\|^2 + (1/2S)\{-\|\lambda\|^2 + \|\lambda\|^2 \\
&\quad + 2\lambda^T \max[-Sg(x), -\lambda] + \|\max[-Sg(x), -\lambda]\|^2\} \\
&= f(x) - \mu^T h(x) + (S/2)\|h(x)\|^2 \\
&\quad + (1/2S)\{\|\lambda + \max[-Sg(x), -\lambda]\|^2 - \|\lambda\|^2\} \\
&= f(x) - \mu^T h(x) + (S/2)\|h(x)\|^2 \\
&\quad + (1/2S)\{\|\max[0, \ \lambda - Sg(x)]\|^2 - \|\lambda\|^2\}
\end{aligned} \tag{6.82}
$$

Substituting equation (6.79) into equation (6.80), we get

$$
\begin{aligned}
dx/dt &= -\{\nabla_x f(x) + [Sh(x) - \mu]^T \nabla_x h(x) \\
&\quad + [S(-\max(-g(x), -\lambda/S) - \lambda]^T \nabla_x g(x)\} \\
&= -\{\nabla_x f(x) + [Sh(x) - \mu]^T \nabla_x h(x) \\
&\quad + [-\max(-Sg(x), -\lambda) - \lambda]^T \nabla_x g(x)\}
\end{aligned}
$$

$$= -\{\nabla_x f(x) + [Sh(x) - \mu]^T \nabla_x h(x)$$

$$- [\max(-g(x), -\lambda) + \lambda]^T \nabla_x g(x)\}$$

$$= -\{\nabla_x f\ (x) + [Sh(x) - \mu]^T \nabla_x h(x)$$

$$-\max[0, \lambda - Sg(x)]^T \nabla_x g(x)\} \tag{6.83}$$

Substituting equation (6.81) into equation (6.73), we get

$$d\lambda/dt = S\ \min(g(x), \lambda/S) = \min[Sg(x), \lambda] \tag{6.84}$$

According to equations (6.82), (6.83), (6.72), and (6.84), we have deduced a new nonlinear optimization neural network model M-9, which can be used to solve the optimization problem with equality and inequality constraints. The NLONN model M-9 can be written as

$$E(x, \lambda, \mu, S) = f(x) - \mu^T h(x) + (S/2)\|h(x)\|^2$$

$$+(1/2S)\{\|\max[0,\ \lambda - Sg(x)]\|^2 - \|\lambda\|^2\} \tag{6.85}$$

$$dx/dt = -\{\nabla_x f(x) + [Sh(x) - \mu]^T \nabla_x h(x)$$

$$-\nabla_x g(x)\max[0, \lambda - Sg(x)]^T\} \tag{6.86}$$

$$d\mu/dt = Sh(x) \tag{6.87}$$

$$d\lambda/dt = \min[Sg(x), \lambda] \tag{6.88}$$

Appendix 6.1 shows that the energy function equation (6.85) in NLONN model M-9 is a Lyapunov function, and the equilibrium point of the neural network corresponds to the optimal solution of the constrained optimization problem M-7.

**Numerical Simulation of NLONN Network**  The first-order Euler method can be used in the numerical analysis of the NLONN network, that is,

$$dZ/dt = [Z(t + \Delta t) - Z(t)]/\Delta t \tag{6.89}$$

$$Z(t + \Delta t) = Z(t) + (dZ/dt)\Delta t \tag{6.90}$$

So dynamic equations (6.86)–(6.88) of the NLONN network can be made equivalent to the following equations:

$$x(t + \Delta t) = x(t) - \Delta t\{\nabla_x f(x(t)) + [Sh(x(t)) - \mu]^T \nabla_x h(x(t))$$

$$-\nabla_x g(x(t))\max[0, \lambda - Sg(x(t))]^T\} \tag{6.91}$$

$$\mu(t + \Delta t) = \mu(t) + \Delta t Sh(x(t)) \tag{6.92}$$

$$\lambda(t + \Delta t) = \lambda(t) + \Delta t\ \min[Sg(x(t)),\ \lambda(t)] \tag{6.93}$$

The calculation steps of the NLONN method are given below.

Step 1:  Select a set of initial values $x(0)$, and parameters $\lambda(0)$, $\mu(0)$, as well as a set of positive ordinal numbers $\{S(k)\}$ $S(k+1) = \rho S(k)$.

Step 2:  Calculate gradients

$$\Phi(x) = \nabla_x E[x(k), \lambda(k), \mu(k), S(k)]$$

$$= \nabla_x f(x(k)) + [S(k)h(x(k)) - \mu(k)]^T \nabla_x h(x(k))]$$

$$- [\max[0, \lambda(k) - S(k)g(x(k))]^T \nabla_x g(x(k)) \tag{6.94}$$

Step 3:  Compute new state

$$x(k+1) = x(k) - \Delta t \ \phi_x(k) \tag{6.95}$$

Step 4:  Perform multiplier iteration

$$\mu(k+1) = \mu(k) + \Delta t \, S(k) \ h(x(k+1)) \tag{6.96}$$

$$\lambda(k+1) = \lambda(k) + \Delta t \min[S(k) \, g(x(k+1)), \ \lambda(k)] \tag{6.97}$$

$$S(k+1) = \rho \, S(k) \tag{6.98}$$

Step 5:  Perform a convergence check, using the criterion

$$\|x(k+1) - x(k)\| \leq \varepsilon_1 \tag{6.99}$$

$$\|\mu(k+1) - \mu(k)\| \leq \varepsilon_2 \tag{6.100}$$

$$\|\lambda(k+1) - \lambda(k)\| \leq \varepsilon_3 \tag{6.101}$$

Stop if equations (6.99)–(6.101) are satisfied. Otherwise let $k = k + 1$, go back to step 2.

## 6.6.4  Test Results

For examining the presented approach, a network of three interconnected areas is constructed as shown in Figure 6.5. Area $A1$ is an IEEE 30-bus system. The generators and loads data of the IEEE 30-bus system are listed in Tables 6.5 and 6.6. The other data and parameters of IEEE 30-bus system are listed in Chapter 5. Parameters of areas $A2$, $A3$, and tie lines are given as follows.

Fuel cost function and power upper and lower limits are

$$F_{31} = 650 \ P_{31}^2 + 325 \ P_{31} \qquad 0.1 \leq P_{31} \leq 0.9$$

$$F_{32} = 30 \ P_{32}^2 + 100 \ P_{32} \qquad 0.1 \leq P_{32} \leq 0.9$$

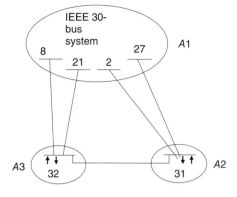

Figure 6.5  The network model of three interconnected power systems.

**TABLE 6.5  Data of Generator Nodes for IEEE 30-Bus System (p.u.)**

| Node | $a_i$ | $b_i$ | $c_i$ | $P_{Gi\,min}$ | $P_{Gi\,max}$ | $\Delta P_{GiGRC}$ |
|------|-------|-------|-------|---------------|---------------|--------------------|
| 1 | 37.5 | 200 | 0.0 | 0.50 | 2.00 | 0.50 |
| 2 | 175 | 175 | 0.0 | 0.20 | 0.80 | 0.30 |
| 5 | 625 | 100 | 0.0 | 0.15 | 0.50 | 0.15 |
| 8 | 83.4 | 325 | 0.0 | 0.10 | 0.35 | 0.15 |
| 11 | 250 | 300 | 0.0 | 0.10 | 0.30 | 0.15 |
| 13 | 250 | 300 | 0.0 | 0.12 | 0.40 | 0.15 |

Note: The generation cost function is: $f_i = a_i P_{Gi}^2 + b_i P_{Gi} + c_i$

**TABLE 6.6  Data of Load Nodes for IEEE 30-Bus System (p.u.)**

| Node No. | Real Power | Reactive Power | Node No. | Real Power | Reactive Power |
|----------|------------|----------------|----------|------------|----------------|
| 1 | 0.000 | 0.000 | 16 | 0.035 | 0.018 |
| 2 | 0.217 | 0.127 | 17 | 0.090 | 0.058 |
| 3 | 0.024 | 0.012 | 18 | 0.032 | 0.009 |
| 4 | 0.076 | 0.016 | 19 | 0.095 | 0.034 |
| 5 | 0.942 | 0.190 | 20 | 0.022 | 0.007 |
| 6 | 0.000 | 0.000 | 21 | 0.175 | 0.112 |
| 7 | 0.228 | 0.109 | 22 | 0.000 | 0.000 |
| 8 | 0.300 | 0.300 | 23 | 0.032 | 0.016 |
| 9 | 0.000 | 0.000 | 24 | 0.087 | 0.067 |
| 10 | 0.058 | 0.020 | 25 | 0.000 | 0.000 |
| 11 | 0.000 | 0.000 | 26 | 0.035 | 0.023 |
| 12 | 0.112 | 0.075 | 27 | 0.000 | 0.000 |
| 13 | 0.000 | 0.000 | 28 | 0.000 | 0.000 |
| 14 | 0.062 | 0.016 | 29 | 0.024 | 0.009 |
| 15 | 0.082 | 0.025 | 30 | 0.106 | 0.019 |

Loads of areas A2 and A3 are $P_{DA2} + jQ_{DA2} = 0.5 + j0.26$ and $P_{DA3} + jQ_{DA3} = 0.4 + j0.21$, respectively. Parameters and capacity constraints of tie line are

$$R_{2-31} = 0.0192; \quad X_{2-31} = 0.0575; \quad P_{2-31T\max} = 0.6$$

$$R_{8-32} = 0.0192; \quad X_{8-32} = 0.0575; \quad P_{8-32T\max} = 0.5$$

$$R_{31-27} = 0.057; \quad X_{31-27} = 0.1737; \quad P_{31-27T\max} = 0.6$$

$$R_{32-21} = 0.057; \quad X_{32-21} = 0.1737; \quad P_{32-21T\max} = 0.5$$

$$R_{31-32} = 0.0192; \quad X_{31-32} = 0.0575; \quad P_{31-32T\max} = 0.5$$

The following test cases of security-constrained MAED are performed in the study.

Case 1:  neglecting the buying and selling among areas;

Case 2:  considering the buying and selling among areas. $P_{A3-A1\,\text{sell}} = 0.4$; $P_{A1-A2\,\text{sell}} = 0.3$;

$$P_{A3-A2\,\text{sell}} = 0.0$$

Case 3:  considering the buying and selling among areas. $P_{A3-A1\,\text{sell}} = 0.32$; $P_{A1-A2\,\text{sell}} = 0.32$;

$$P_{A3-A2\,\text{sell}} = 0.0$$

To evaluate the calculation precision, the following performance index (PI) on trading error is used, that is,

$$PI_{EAB}\% = \frac{|P_{TAB} - P_{AB\,\text{sell}}|}{P_{AB\,\text{sell}}} \times \% \tag{6.102}$$

The calculation results of security-constrained MAED for the above three test cases are listed in Table 6.7. From Table 6.7 we can get

Case 2:

$$P_{TA3-A1} = P_{32-8} + P_{32-21} = 0.172 + 0.228 = 0.4$$

$$PI_{EA3-A1}\% = 0$$

$$P_{TA1-A2} = P_{2-31} + P_{27-31} = 0.4584 - 0.1585 = 0.2999$$

$$PI_{EA1-A2}\% = 0.0333\%$$

$$P_{TA3-A2} = P_{32-31} = 0.0$$

$$PI_{EA3-A2}\% = 0$$

Case 3:

$$P_{TA3-A1} = P_{32-8} + P_{32-21} = 0.1123 + 0.2077 = 0.32$$

TABLE 6.7 Test Results of Security-Constrained MAED for Three Interconnected Systems

| Test Cases | Case 1 (p.u) | Case 2 (p.u.) | Case 3 (p.u.) |
|---|---|---|---|
| Area A1 $P_{G1}$ | 1.5971 | 1.6588 | 1.5951 |
| $P_{G2}$ | 0.4377 | 0.4636 | 0.4304 |
| $P_{G5}$ | 0.2096 | 0.2122 | 0.2133 |
| $P_{G8}$ | 0.2903 | 0.2252 | 0.3324 |
| $P_{G11}$ | 0.1459 | 0.1322 | 0.1748 |
| $P_{G13}$ | 0.1366 | 0.1287 | 0.1699 |
| Area A2 $P_{G31}$ | 0.1000 | 0.2001 | 0.1801 |
| Area A3 $P_{G32}$ | 0.9000 | 0.8000 | 0.7200 |
| Total Gen. | 3.81722 | 3.82081 | 3.81602 |
| Power losses | 0.08322 | 0.08681 | 0.08202 |
| Total gen. cost ($) | 923.0356 | 957.5161 | 974.6212 |
| Tie-line power $P_{32-8}$ | 0.1827 | 0.1720 | 0.1123 |
| $P_{32-21}$ | 0.2364 | 0.2280 | 0.2077 |
| $P_{2-31}$ | 0.4687 | 0.4584 | 0.4624 |
| $P_{27-31}$ | −0.1422 | −0.1585 | −0.1425 |
| $P_{32-31}$ | 0.0808 | 0.0000 | 0.0000 |
| Line security | Satisfied | Satisfied | Satisfied |

$$PI_{EA3-A1}\% = 0$$

$$P_{TA1-A2} = P_{2-31} + P_{27-31} = 0.4624 - 0.1425 = 0.3199$$

$$PI_{EA1-A2}\% = 0.03125\%$$

$$P_{TA3-A2} = P_{32-31} = 0.0$$

$$PI_{EA3-A2}\% = 0$$

The maximum trading error is only 0.0333%. Therefore, the proposed MAED approach not only satisfies all security constraints but also has high precision.

## 6.7 TOTAL TRANSFER CAPABILITY COMPUTATION IN MULTIAREAS

As we analyzed in previous sections, the transfer capability limits affect the wheeling. It is useful to compute the total transfer capability (TTC) of the multiareas.

## 6.7.1   Continuation Power Flow Method

The general method to compute the TTC is the continuation power flow (CPF), or repeated power flow (RPF) method [28–31]. It is sometimes called the perturbation method.

The net active and reactive power injections at the sink and source buses are functions of $\lambda$.

$$P_i = P_{i0} + \lambda L_{Pi} \tag{6.103}$$

$$Q_i = Q_{i0} + \lambda L_{Qi} \tag{6.104}$$

where

$\lambda$: the parameter controlling the amount of injection;
$P_{i0}$: the base case real power injections at the bus;
$Q_{i0}$: the base case reactive power injections at the bus;
$L_{Pi}$: the real power load participation factors;
$L_{Qi}$: the reactive power load participation factors.

The traditional power flow equations augmented by an extra equation for $\lambda$ are expressed as

$$f(\theta, V, \lambda) = 0 \tag{6.105}$$

where

V: the vector of bus voltage magnitudes;
$\theta$: the vector of bus voltage angles.

Once a base case (for $\lambda = 0$) solution is found, the next solution can be predicted by taking an appropriately sized step in a direction tangent to the solution path. The tangent vector is obtained as below.

$$d[f(\theta, V, \lambda)] = f_\theta d\theta + f_V dV + f_\lambda d\lambda \tag{6.106}$$

Since equation (6.106) is rank deficient, an arbitrary value such as 1 can be assigned as one of the elements of the tangent vector $t = [d\theta, dV, d\lambda]^T = \pm 1$, that is $t_k = \pm 1$. Thus,

$$\begin{bmatrix} f_\theta & f_V & f_\lambda \\ & e_k & \end{bmatrix} [t] = \begin{bmatrix} 0 \\ \pm 1 \end{bmatrix} \tag{6.107}$$

where $e_k$ is a row vector with all elements zero, except for the $k$th entry, which is equal to 1.

The new solution after perturbation will then be computed as

$$
\begin{bmatrix} \theta^* \\ V^* \\ \lambda^* \end{bmatrix} = \begin{bmatrix} \theta \\ V \\ \lambda \end{bmatrix} + \varepsilon \begin{bmatrix} d\theta \\ dV \\ d\lambda \end{bmatrix}
\tag{6.108}
$$

Where $\varepsilon$ is a scalar used to adjust the step size.

The new solution obtained in equation (6.108) may violate the limits. Thus it is necessary to correct the continuation parameter. The corrector is a slightly modified Newton power flow algorithm in which the Jacobian matrix is augmented by an equation to account for the continuation parameter.

Let $x = [\theta, V, \lambda]^T$, $x_k = \eta$, then the new set of equations will take the form

$$
\begin{bmatrix} f(x) \\ e_k - \eta \end{bmatrix} = [0]
\tag{6.109}
$$

Therefore, for a specific source/sink transfer case, the steps for computing the TTC are summarized as follows [28]:

**(1)** Input power system data.

**(2)** Select the contingency from the contingency list.

**(3)** Initialize as follows:

(a) Run power flows to ensure that the initial point does not violate any limits.

(b) Set the tolerance for the change of transfer power.

**(4)** Predict the step size of CPF:

(a) Calculate the tangent vector $t = [d\theta, dV, d\lambda)]^T$

(b) Choose the scalar $\varepsilon$ to design the prediction step size.

(c) Make a step of increase of the transfer power to predict the next solution using equation (6.108).

**(5)** Correct the step size of CPF with generator Q limits. Solve equation (6.109).

**(6)** Check for limit violations: Check the solution of the step (5) for violations of operational or physical limits—line flow limit, voltage magnitude limit, and voltage stability limit. If there are violations, reduce the transfer power increment by $\varepsilon = 0.5\varepsilon$; then, go back to step (5) until the change of the transfer power is smaller than the tolerance. The maximum transfer power for the selected contingency is reached. Otherwise, go to the prediction step (4).

**(7)** Check if all contingencies are processed. If they are, compare the maximum transfer powers for all the contingencies and choose the smallest one as the TTC for this specific source/sink transfer case and terminate the procedure. Otherwise, go to step (2).

## 6.7.2 Multiarea TTC Computation

In a multiarea system, it is assumed that each area operates autonomously with its own independent operator. Each area carries out its own CPF calculation and maintains its own detailed system model. Furthermore, each area uses network equivalents to represent the buses in other areas, except for the boundary. One of the equivalent methods is the REI equivalent. The basic idea of the REI equivalent is to aggregate the injections of a group of buses into a single bus. The aggregated injection is distributed to these buses via a radial network called the REI network. After the aggregation, all buses with zero injections are eliminated, yielding the equivalent [32,33]. For example, all PV and PQ buses except for the seller and buyer buses of outer external area are grouped into two different REI equivalent networks, which are assigned the corresponding bus types (PQ or PV) accordingly [28]. In this way, a systemwide TTC can be computed without exchanging the information between each other. However, the admittances of the REI network are functions of the operating point for which the equivalent is constructed. Doing so will also introduce errors in the multiarea TTC result. In light of this, the equivalent has to be properly updated during the TTC computation.

In the case of the multiarea CPF implementation, each area carries out its own CPF, and the continuation parameter for each area may be different at each step. Therefore, a strategy for choosing and updating the continuation parameter that ensures synchronized CPF calculation in different areas is introduced.

Another issue related to updating the equivalents is the generator Q limits. As the power transfer increases at a chosen PQ bus, generator buses will continue to hit their Q limits in succession. As each limit is reached, the generated reactive power will be held at the Q limit, bus type will be switched to PQ, and the bus voltage will become an unknown increasing the dimension of the Jacobian by one. While updating the equivalents, these generator buses that are now of type PQ are grouped with other PQ buses in each area. This will continue until other limits are reached.

A self-adaptive step size control is implemented for the sink area. $\lambda$ is chosen as the continuation parameter when starting from the base case. Then, the continuation parameter is chosen from the voltage increment vector $[dV]^T$. A constant voltage magnitude decrease is used to predict the next solution. Usually, the scalar $\varepsilon$ in equation (6.108) is set as 0.02 [28]. Therefore, a constant decrease in voltage magnitude will result in a large increase in load at the beginning and a small increase in load as the nose point is approached.

After each correction step, the load change at the sink area will be broadcast to all other areas. The continuation parameter remains to be $\lambda$ in all other areas, and the scalar $\varepsilon$ is set as the load change of the sink area at each step. Hence, different areas will have the same load increase at each discrete step of CPF calculation.

If contingencies are considered in the calculation of multiarea TTC, contingencies associated with the tie lines must be co-monitored by all areas. However, contingencies caused by topology changes within individual areas do not have to be modeled directly by others. Instead, when a contingency occurs within one area, only the network model of this area will be changed. As a result, the tie-line power flows and buyer bus voltages calculated from different areas will have very large mismatches during the synchronized computation. After updating the equivalents for the area experiencing the contingency, the updated equivalent buses will reflect the effects of the contingency. This way, other areas can account for the effects of the contingency indirectly.

# APPENDIX A: COMPARISON OF TWO OPTIMIZATION NEURAL NETWORK MODELS

Reference [27] also presented an optimization neural network model, which can be written as M-10.

$$L(S, x) = f(x) + \lambda^T g(x)$$

$$+ \mu^T h(x)(S/2)\|g^+(x)\|^2 + \|h(x)\|^2 \tag{6A.1}$$

$$dx/dt = -\nabla f(x) - \nabla h(x)[S h(x) + \mu]$$

$$- \nabla g(x)^T (S g^+(x) + \lambda) \tag{6A.2}$$

$$d\mu/dt = \varepsilon (S h(x)) \tag{6A.3}$$

$$d\lambda/dt = \varepsilon (S g^+(x)) \tag{6A.4}$$

where, $\varepsilon$ is a very small positive number and

$$g^+(x) = \max[0, g(x)] \tag{6A.5}$$

It is noted that the proposed NLONN model M-9 is different from the traditional optimization neural network model M-10. This can be seen by analyzing the stability and optimization of two neural networks.

## A.1. For Proposed Neural Network M-9

The derivative of the energy function in M-9 with respect to time $t$ can be obtained from the following calculation, that is,

$$\frac{dE}{dt} = \frac{\partial E}{\partial x}\frac{dx}{dt} + \frac{\partial E}{\partial \mu}\frac{d\mu}{dt} + \frac{\partial E}{\partial \lambda}\frac{\partial \lambda}{\partial t}$$

$$= -\left\|\frac{dx}{dt}\right\|^2 - S\|h(x)\|^2 + \frac{1}{S}\{\max[0, \lambda - S g(x)] - \lambda\}^T \min[S g(x), \lambda]$$

$$= -\left\|\frac{dx}{dt}\right\|^2 - S\|h(x)\|^2 + \frac{1}{S}\{\max[-\lambda, S g(x)]\}^T[-\max(-S g(x), -\lambda)]$$

$$= -\left\|\frac{dx}{dt}\right\|^2 - S\|h(x)\|^2 + \frac{1}{S}\|\max[-S g(x), -\lambda]\|^2 \tag{6A.6}$$

Obviously, from equation (6A.6) we can know that $dE/dt \leq 0$. When and only when

$$h(x) = 0; \quad \max[-\lambda, -S g(x)] = 0; \quad dx/dt = 0 \tag{6A.7}$$

then

$$dE/dt = 0 \tag{6A.8}$$

The meaning of $\max[-\lambda, -Sg(x)] = 0$ is that

$$Sg(x) \geq 0 \quad \text{when} \quad \lambda = 0 \tag{6A.9}$$

$$\lambda \geq 0 \quad \text{when} \quad Sg(x) = 0 \tag{6A.10}$$

Equations (6A.9) and (6A.10) are just the Kuhn–Tucker conditions in optimization theory. Thus, $\max[-\lambda, -Sg(x)] = 0$ is tenable.

Certainly, any feasible solutions including the optimal solution satisfy the equation $h(x) = 0$. So from equation (6.86) of M-9 we can get the following expression.

$$dx/dt = -\{\nabla_x f(x) - \mu\nabla_x h(x) - \max[0, \lambda - Sg(x)]\nabla_x g(x)\} \tag{6A.11}$$

According to equations (6A.9) and (6A.10), we can get

$$\max[0, \lambda - Sg(x)]\nabla_x g(x) = \lambda\nabla_x g(x) \tag{6A.12}$$

According to equations (6A.11) and (6A.12), we can get

$$dx/dt = -\{\nabla_x f(x) - \mu\nabla_x h(x) - \lambda\nabla_x g(x)\} \tag{6A.13}$$

If $dx/dt = 0$, when and only when

$$\nabla_x f(x) - \mu\nabla_x h(x) - \lambda\nabla_x g(x) = 0 \tag{6A.14}$$

Equation (6A.14) is just the optimality conditions for the optimization problem M-7. So this condition is tenable. It means that $dx/dt = 0$ is also tenable. Now we have demonstrated that all conditions in equation (6A.7) are satisfied. Therefore, equation (6A.8) is also satisfied. This has proved that the energy function of the proposed NLONN neural network is Lyapunov function. The corresponding neural network is certainly stable and the equilibrium point of neural network corresponds to the optimal solution of the constrained optimization problem M-7.

## A.2. For Neural Network M-10 in Reference [27]

According to equations (6A.1)–(6A.5), the derivative of energy function in M-10 with respect to time $t$ can be obtained from the following calculation, that is,

$$\frac{dL}{dt} = \frac{\partial L}{\partial x}\frac{dx}{dt} + \frac{\partial L}{\partial \mu}\frac{d\mu}{dt} + \frac{\partial L}{\partial \lambda}\frac{\partial \lambda}{\partial t}$$

$$= -\left\|\frac{dx}{dt}\right\|^2 + \varepsilon \cdot S \cdot \|h(x)\|^2 + \varepsilon \cdot S \cdot g^T(x) \cdot g^+(x) \tag{6A.15}$$

as $\varepsilon$ is a very small positive number and $g^+(x) = \max[0, g(x)]$. The last two terms in the right side of equation (6A.15) are not negative. It means that $dL/dt \leq 0$ is untenable all along. Therefore, the stability problem exists in the neural network M-10.

# PROBLEMS AND EXERCISES

1. What is "Wheeling"?

2. What is MAED?

3. State the differences between the optimization neural network and the traditional optimization methods.

4. What is ATC?

5. How is the short-run marginal cost model used to compute the wheeling cost?

6. What is TTC? Is it the same as ATC?

7. There are four utilities with two selling, and two buying. The related data are listed in Tables 6.8 and 6.9.

**TABLE 6.8    Data of Utilities A and B for Exercise 7**

| Utilities Selling Energy | Incremental Cost ($/MWh) | MWh for Sale | Seller's Total Increase in Cost($) |
|---|---|---|---|
| A | 20 | 130 | 2600 |
| B | 26 | 90 | 2340 |

**TABLE 6.9    Data of Utilities C and D for Exercise 7**

| Utilities Buying Energy | Decremental Cost ($/MWh) | MWh for Purchase | Buyer's Total Decrease in Cost($) |
|---|---|---|---|
| C | 32 | 65 | 2080 |
| D | 45 | 155 | 6975 |

Compute the maximum pool savings.

8. For exercise 7, assume that four utilities were scheduled to transact energy by a central dispatching scheme, and 10% of the gross system savings was to be set aside to compensate those systems that provided transmission facilities to the pool. Calculate the maximum pool savings.

9. For exercise 7, assume that four utilities were scheduled to transact energy by a central dispatching scheme, and 15% of the gross system savings was to be set aside to compensate those systems that provided transmission facilities to the pool. Calculate the maximum pool savings.

10. Compare the results of exercises 8 and 9, and analyze the impact of the amount of the gross system savings to the maximum pool savings.

# REFERENCES

1. Zhu JZ. Multiarea power systems economic dispatch using a nonlinear optimization neural network approach. Electr. Pow. Compo. Sys. 2002;31:553–563.
2. Zhu JZ, Momoh JA. Multiarea power systems economic dispatch using nonlinear convex network flow programming. Electr. Pow. Syst. Res. 2001;59(1):13–20.
3. Chang CS, Liew AC, Xu JX, Wang XW, Fan B. Dynamic security constrained multiobjective generation dispatch of longitudinally interconnected power systems using bicriterion global optimization, IEEE/PES Summer Meeting, 1995, SM 578-5.
4. Streiffert D. Multiarea economic dispatch with tie line constraints, IEEE/PES Winter Meeting, 1995, WM 179-2.
5. Wang C, Shahidehpour SM. Power generation scheduling for multiarea hydro-thermal systems with tie line constraints, cascaded reservoirs and uncertain data. IEEE Trans., PAS 1993;8(3):1333–1340.
6. Alsac O, Stott B. Optimal power flow with steady-state security. IEEE Trans., PAS 1974;93:745–751.
7. Zhu JZ, Irving MR. Combined active and reactive dispatch with multiple objectives using an analytic hierarchical process. IEE Proc. C 1996;143(4):344–352.
8. Dommel HW, Tinney WF. Optimal power flow solutions. IEEE trans., PAS 1968;87(10):1866–1876.
9. Sun DI, Ashley B, Hughes A, Tinney WF. Optimal power flow by Newton approach. IEEE Trans., PAS 1984;103(10):2864–2880.
10. Irving MR, Sterling MJH. Economic dispatch of active power with constraint relaxation. IEE Proc. C 1983;130(4).
11. Irving MR, Sterling MJH. Efficient Newton-Raphson algorithm for load flow calculation in transmission and distribution networks. IEE Proc. C 1987;134.
12. Zhu JZ, Xu GY. A new real power economic dispatch method with security. Electr. Pow. Syst. Res. 1992;25(1):9–15.
13. Walters DC, Sheble ZC. Genetic algorithm solution of economic dispatch with valve point loading. IEEE Trans., on Power Syst. 1993;8(3).
14. Salami M, Cain G. Multiple genetic algorithm processor for the economic power dispatch problem, Proceedings of International Conference on genetic algorithm in engineering systems, UK, September, 1995, pp. 188–193.
15. King TD, El-Hawary ME, El-Hawary F. Optimal environmental dispatching of electric power system via an improved Hopfield neural network model. IEEE Trans. on Power Syst. 1995;10(3):1559–1565.
16. Wong KP, Fung CC. Simulated-annealing-based economic dispatch algorithm. IEE Proc. Part C 1993;140(6):509–515.
17. Lee TH, Thorne DH, Hill EF. A transportation method for economic dispatching—application and comparison. IEEE Trans., PAS 1980;99:2372–2385.
18. Hobson E, Fletcher DL, Stadlin WO. Network flow linear programming techniques and their application to fuel scheduling and contingency analysis. IEEE Trans., PAS 1984;103:1684–1691.
19. Wood AJ, Wollenberg BF. *Power Generation, Operation, and Control.* 2nd ed. New York: 1996.
20. Lo KL, Zhu SP. A theory for pricing wheeled power. Electr. Pow. Syst. Res. 1994;28.
21. Caramanis MC, Bohn RE, Schweppe FC. The costs of wheeling and optimal wheeling rates. IEEE Trans. on Power Syst. 1986;PWRS(1).
22. Happ HH. Cost of Wheeling Methodology. IEEE Trans. on Power Syst. 1994;9(1).
23. Purdue ECE Technical Reports, Real time pricing of electric power, 1995, http://docs.lib.purdue.edu/ecetr/120
24. Zhu JZ, Chang CS. Energy transaction scheduling with network constraints and transmission losses, Proceedings of International Conference on advances in power system control, operation and management, APSCOM-97, November 11–14, 1997, Hong Kong.
25. Li YZ, David AK. Optimal multiarea wheeling. IEEE Trans. on Power Syst. 1994;9(1).
26. Bazaraa MS, Aherali HD, Shetty CM. *Nonlinear Programming: Theory and Algorithms.* New York: John Wiley; 1993.
27. Maa CY, Shanblatt MA. A two-phase optimization neural network. IEEE T. Neural Networ 1992;3(6).
28. Min L, Abur A. Total transfer capability computation for multiarea power systems. IEEE Trans. on Power Syst. 2006;21(3):1141–1147.

29. Ajjarapu V, Christy C. The continuation power flow: A tool for steady state voltage stability analysis. IEEE Trans. Power Syst. 1992;7(1):416–423.
30. Canizares CA, Alvarado FL. Point of collapse and continuation methods for large ACDC systems. IEEE Trans. Power Syst. 1993;8(1):1–8.
31. Chiang HD, Flueck AJ, Shah KS, Balu N. CPFLOW: A practical tool for tracing power system steady-state stationary behavior due to load and generation variations. IEEE Trans. Power Syst. 1995;10(2):623–634.
32. Dimo P. *Nodal Analysis of Power System*. Turibrige Wells, Kent, U.K.: Abacus Press; 1975.
33. Tinney WF, Powell WL. The REI approach to power network equivalents, Transl.: 1977 PICA Conference in Toronto, ON, Canada, May 1977, pp. 312–320.

# UNIT COMMITMENT

This chapter first introduces several major techniques for solving the unit commitment (UC) problem, such as the priority method, dynamic programming, and the Lagrange relaxation method. Several new algorithms are then added to tackle UC problems. These are the evolutionary programming-based tabu search method, particle swarm optimization, and the analytic hierarchy process (AHP). A number of numerical examples and analyses are provided in the chapter.

## 7.1 INTRODUCTION

Since generators cannot instantly turn on and produce power, UC must be planned in advance so that enough generation is always available to handle system demand with an adequate reserve margin in the event that generators or transmission lines go out or load demand increases. UC handles the unit generation schedule in a power system for minimizing operating cost and satisfying prevailing constraints such as load demand and system reserve requirements over a set of time periods [1–20]. The classical UC problem is aimed at determining the start-up and shutdown schedules of thermal units to meet the forecast demand over certain time periods (24 h to 1 week) and belongs to a class of combinatorial optimization problems. The methods that have been studied so far fall into roughly three types: heuristic search, mathematical programming, and hybrid methods. Optimization techniques such as the priority list, augmented Lagrangian relaxation, dynamic programming, and the branch-and-bound algorithm have been used to solve the classic UC problem. Genetic algorithms (GAs), simulated annealing (SA), AHP, and particle swarm optimization (PSO) have also been used for the UC problem since the beginning of the last decade.

## 7.2 PRIORITY METHOD

The classic UC problem is to minimize total operational cost and is subject to minimum up- and downtime constraints, crew constraints, unit capability limits, generation constraints, and reserve constraints. Thus the objective function of UC consists of the generation cost function and start-up cost function of the generators.

*Optimization of Power System Operation*, Second Edition. Jizhong Zhu.
© 2015 The Institute of Electrical and Electronics Engineers, Inc. Published 2015 by John Wiley & Sons, Inc.

The former is described in Chapter 4. The latter involves the cost of the energy that brings the unit online.

There are two types of start-up cost models: one brings the unit on-line from a cold start, and the other brings it from bank status, in which the unit is turned off but still close to operating temperature. The start-up cost model when cooling can be expressed as the following exponential function:

$$F_{Sc}(t) = (1 - e^{-t/\alpha}) \times F + C_f \tag{7.1}$$

where

$F_{Sc}$: the cold start cost for the cooling model;

$C_f$: the fixed cost of generator operation including crew expense and maintenance expense;

$F$: the fuel cost;

$t$: time that the unit was cooled;

$\alpha$: thermal time constant for the unit.

The start-up cost model when banking can be expressed as the following linear function:

$$F_{Sb}(t) = F_0 \times t + C_f \tag{7.2}$$

where

$F_{Sb}$: the start-up cost for the banking model;

$F_0$: the cost of maintaining the unit at operating temperature.

The simplest UC solution is to list all combinations of units on and off, as well as the corresponding total cost to create a rank list, and then make the decision according to the rank table. This method is called the priority list. The rank is based on the minimum average production cost of the unit. The average production cost of the unit is defined as

$$\mu = \frac{F(P_G)}{P_G} \tag{7.3}$$

where

$\mu$: the average production cost of the unit;

$F(P_G)$: the generation cost function of the unit;

$P_G$: the generator real power output.

From Chapter 4, the incremental rate of the unit is defined as

$$\lambda = \frac{dF(P_G)}{dP_G} \tag{7.4}$$

When the average production cost of the unit equals the incremental rate of the unit, the corresponding average production cost is called the minimum average production cost $\mu_{min}$. Generally, the power output is close to the rated power when the unit is at the minimum average production cost.

***Example 7.1:***    There are 5 generator units, and the minimum average production cost $\mu_{min}$ is computed as shown in Table 7.1.

The priority order for these units based on the minimum average production cost is shown in Table 7.2.

The steps for using the priority list method are summarized as follows:

Step (1):    Compute the minimum average production cost of all units, and order the units from the smallest value of $\mu_{min}$. Form the priority list.

Step (2):    If the load is increasing during that hour, determine how many units can be started up according to the minimum downtime of the unit. Then, select the top units for turning on from the priority list according to the increase in the load.

Step (3):    If the load is dropping during that hour, determine how many units can be stopped according to the minimum uptime of the unit. Then, select the last units for stopping from the priority list according to the drop in the load.

Step (4):    Repeat the process for the next hour.

There are other priority list methods such as ranking units on the basis of the full-load average production cost of each unit [21] as well as methods based on the incremental cost rate of each unit [22].

**TABLE 7.1    The Minimum Average Production Cost**

| Unit | Minimum Average Production Cost $\mu_{min}$ | Min (MW) | Max (MW) |
|------|----------------|----------|----------|
| G1 | 10.56 | 100 | 400 |
| G2 | 9.76 | 120 | 500 |
| G3 | 11.95 | 100 | 300 |
| G4 | 8.90 | 50 | 600 |
| G5 | 12.32 | 150 | 250 |

**TABLE 7.2    The Priority Order for 5 Units**

| Priority Order | Unit | $\mu_{min}$ | Min (MW) | Max (MW) |
|------|------|-------|----------|----------|
| 1 | G4 | 8.90 | 50 | 600 |
| 2 | G2 | 9.76 | 120 | 500 |
| 3 | G1 | 10.56 | 100 | 400 |
| 4 | G3 | 11.95 | 100 | 300 |
| 5 | G5 | 12.32 | 150 | 250 |

## 7.3 DYNAMIC PROGRAMMING METHOD

Suppose a system has $n$ units. If the enumeration approach is used, there would be $2^n - 1$ combinations. The dynamic programming (DP) method consists in implicitly enumerating feasible schedule alternatives and comparing them in terms of operating costs. Thus DP has many advantages over the enumeration method such as reduction in the dimensionality of the problem.

There are two DP algorithms. They are forward dynamic programming and backward dynamic programming. The forward approach, which runs forward in time from the initial hour to the final hour, is often adopted in UC. The advantages of the forward approach are as follows:

- Generally, the initial state and conditions are known.

- The start-up cost of a unit is a function of the time. Thus the forward approach is more suitable because the previous history of the unit can be computed at each stage.

The recursive algorithm is used to compute the minimum cost in hour $t$ with feasible state $I$, that is,

$$F_{tc}(t, I) = \min_{\{L\}}[F(t, I) + S_c(t - 1, L \Rightarrow t, I) + F_{tc}(t - 1, I)] \tag{7.5}$$

where

$F_{tc}(t, I)$: the total cost from the initial state to hour $t$ state $I$;
$S_c(t - 1, L \Rightarrow t, I)$: the transition cost from state $(t - 1, \ L)$ to state $(t, \ I)$;
$\{L\}$: the set of feasible states at hour $t - 1$;
$F(t, I)$: the production cost for state $(t, \ I)$.

The following constraints should be satisfied for the UC problem solved by dynamic programming.

$$\sum_{i=1}^{n} P_{Gi}^t = P_D^t \tag{7.6}$$

$$x_i^t P_{Gi\,min}^t \leq P_{Gi}^t \leq x_i^t P_{Gi\,max}^t \tag{7.7}$$

where

$P_D^t$: the system load at hour $t$;
$P_{Gi\,min}^t$: the lower limit of the unit power output;
$P_{Gi\,max}^t$: the upper limit of the unit power output:
$x_i^t$: the $0 - 1$ variable.

As we mentioned before, there are $2^n - 1$ combinations or states for $n$ units. The amount of computation is large. We can combine the DP algorithm and priority list method to discard some infeasible states as well as high cost states. In addition, add the unit minimum up- and minimum downtime constraints, which can also reduce the states. For example, before we perform UC using the forward DP algorithm, we first order the units according to the priority list and the unit minimum up/downtime. The first part of the units order is the must-up units, the last part is the must-down units, and middle part is the units ranking based on the minimum average production cost of the rest of units. In this way, the computation amount of DP will be reduced.

***Example 7.2:*** We use the priority list and dynamic programming to solve the UC for a simple four-unit system [21]. The data of the units and the load pattern are listed in Tables 7.3 and 7.4, respectively.

In Table 7.3, the symbol "+" in the initial state means the unit is online, and "−" means the unit is off-line. For example, "8" means the unit has been online for 8 hours, and "−6" means the unit has been off-line for 6 hours.

The number of combinations of the four units is $2^n - 1 = 2^4 - 1 = 15$. If we order the unit combinations or states by the maximum net capacity of each combination, we get Table 7.5.

In the combination of Table 7.5, "1" means committed (unit operating), and "0" means uncommitted (unit shutdown). For example, "0001" for state 1 means the unit 4 is committed, and units 1, 2, 3 are uncommitted. "1001" for state 3 means the units 1 and 4 committed, and units 2 and 3 are uncommitted.

Case 1    Neglecting the constraints of unit minimum up/downtime. Solve the UC problem using the priority list order.

**TABLE 7.3    The Data of Units**

| Unit | Max (MW) | Min (MW) | Cost ($/h) | Ave. Cost | Start-up Cost | Initial State | Min Uptimes (h) | Min Downtimes (h) |
|------|----------|----------|------------|-----------|---------------|---------------|-----------------|-------------------|
| 1 | 80 | 25 | 213.00 | 23.54 | 350 | −5 | 4 | 2 |
| 2 | 250 | 60 | 585.62 | 20.34 | 400 | 8 | 5 | 3 |
| 3 | 300 | 75 | 684.74 | 19.74 | 1100 | 8 | 5 | 4 |
| 4 | 60 | 20 | 252.00 | 28.00 | 0 | −6 | 1 | 1 |

**TABLE 7.4    The Load Pattern**

| Hour | Load (MW) |
|------|-----------|
| 1 | 450 |
| 2 | 530 |
| 3 | 600 |
| 4 | 540 |
| 5 | 400 |
| 6 | 280 |
| 7 | 290 |
| 8 | 500 |

**TABLE 7.5   The Ordering of the Unit Combinations**

| State | Unit Combination | Max Net Capacity (MW) |
|---|---|---|
| 15 | 1 1 1 1 | 690 |
| 14 | 1 1 1 0 | 630 |
| 13 | 0 1 1 1 | 610 |
| 12 | 0 1 1 0 | 550 |
| 11 | 1 0 1 1 | 440 |
| 10 | 1 1 0 1 | 390 |
| 9 | 1 0 1 0 | 380 |
| 8 | 0 0 1 1 | 360 |
| 7 | 1 1 0 0 | 330 |
| 6 | 0 1 0 1 | 310 |
| 5 | 0 0 1 0 | 300 |
| 4 | 0 1 0 0 | 250 |
| 3 | 1 0 0 1 | 140 |
| 2 | 1 0 0 0 | 80 |
| 1 | 0 0 0 1 | 60 |
| 0 | 0 0 0 0 | 0 |
| (Unit) | 1 2 3 4 | |

In Case 1, units are committed in order until the load is satisfied. The total cost for the interval is the sum of the eight dispatch costs plus the transitional costs for starting any units. It can be known from the average production cost in Table 7.3 that the priority order for the four units are unit 3, unit 2, unit 1, unit 4. All possible commitments start from state 12 as the load at first hour is 450 MW, and maximum net capacity from state 1 to state 11 is only 440 MW. In addition, state 13 is discarded as it does not satisfy the order of priority list. The UC results for the priority ordered method are listed in Table 7.6.

Case 2   Neglecting the constraints of unit minimum up/downtime, Solve the UC problem using dynamic programming.

Case 2, first select the feasible states using the priority list order. For first 4 h, the feasible states have only 12, 14, and 15 in Table 7.5. For the last 4 hours, the feasible states have 5, 12, 14, and 15. Thus, the total feasible states are: {5, 12, 14, 15}, and the initial state is 12. According to the recursive algorithm of dynamic programming, we can compute the minimum total cost.

$$F_{tc}(t, I) = \min_{\{L\}}[F(t, I) + S_c(t - 1, L \Rightarrow t, I) + F_{tc}(t - 1, I)]$$

For $t = 1$ :   {L} = {12} and {I} = {12, 14, 15}

$$F_{tc}(1, 12) = F(1, 12) + S_c(0, 12 \Rightarrow 1, 12) + F_{tc}(0, 12)$$

**TABLE 7.6   UC Results by Priority List**

| Hour | Load (MW) | Units On-Line | Generation Cost |
|------|-----------|---------------|-----------------|
| 1 | 450 | Units 3 and 2 | 9208 |
| 2 | 530 | Units 3 and 2 | 10648.36 |
| 3 | 600 | Units 3, 2, and 1 | 12265.36 |
| 4 | 540 | Units 3 and 2 | 10828.36 |
| 5 | 400 | Units 3 and 2 | 8308.36 |
| 6 | 280 | Unit 3 | 5573.54 |
| 7 | 290 | Unit 3 | 5748.14 |
| 8 | 500 | Units 3 and 2 | 10108.36 |

$$= F(1, 12) + S_c(0, 12 \Rightarrow 1, 12) + 0 = 9208 + 0 = 9208$$

$$F_{tc}(1, 14) = F(1, 14) + S_c(0, 14 \Rightarrow 1, 14) + F_{tc}(0, 14) = 9493 + 350 = 9843$$

$$F_{tc}(1, 15) = F(1, 15) + S_c(0, 15 \Rightarrow 1, 15) + F_{tc}(0, 15) = 9861 + 350 = 10211$$

For $t = 2$ :   $\{L\} = \{12, 14\}$ and $\{I\} = \{12, 14, 15\}$

$$F_{tc}(2, 15) = \min_{\{12,14\}} [F(2, 15) + S_c(1, L \Rightarrow 2, 15) + F_{tc}(1, L)]$$

$$= 11301 + \min \begin{bmatrix} (350 + 9208) \\ (0 + 9843) \end{bmatrix} = 20859$$

and so on.

The UC results are the same as those in case 1.

## 7.4   LAGRANGE RELAXATION METHOD

Since the enumeration approach is involved in UC solved by the dynamic programming method, the computation burden is huge for large power systems with many generators. The priority list is very simple and has fast calculation speed, but it may discard the optimum scheme. The Lagrange relaxation method can overcome the aforementioned disadvantages.

The mathematical problem of the UC can be expressed as follows.

**1.** Objective function

$$\min \sum_{t=1}^{T} \sum_{i=1}^{n} [F_i(P_{Gi}^t)x_i^t + F_{si}(t)x_i^t] = F(P_{Gi}^t, x_i^t) \qquad (7.8)$$

2. Constraints

(a) Load balance equation

$$\sum_{i=1}^{n} P_{Gi}^t x_i^t = P_D^t, \quad t = 1, 2, \ldots, T \tag{7.9}$$

(b) Generator power output limits

$$x_i^t P_{Gi\,min}^t \leq P_{Gi}^t \leq x_i^t P_{Gi\,max}^t, \quad t = 1, 2, \ldots, T \tag{7.10}$$

(c) Power reserve constraint

$$\sum_{i=1}^{n} P_{Gi\,max} x_i^t \geq P_D^t + P_R^t, \quad t = 1, 2, \ldots, T \tag{7.11}$$

(d) Minimum up/downtime

$$(U_{t-1,i}^{up} - T_i^{up})(x_i^{t-1} - x_i^t) \geq 0, \quad t = 1, 2, \ldots, T, \; i = 1, 2, \ldots, n \tag{7.12}$$

$$(U_{t-1,i}^{down} - T_i^{down})(x_i^t - x_i^{t-1}) \geq 0, \quad t = 1, 2, \ldots, T, \; i = 1, 2, \ldots, n \tag{7.13}$$

where

$F_{Si}$: the start-up cost of unit $i$ at time period $t$;

$P_R^t$: the power reserve at time period $t$;

$T_i^{up}$: the minimum up time for unit $i$ in hours;

$T_i^{down}$: the minimum downtime for unit $i$ in hours;

$U_{t-1,i}^{up}$: the number of consecutive uptime periods until time period $t$, measured in hours;

$U_{t-1,i}^{down}$: the number of consecutive downtime periods until time period $t$, measured in hours.

The UCP has two kinds of constraints: separable and coupling constraints. Separable constraints such as capacity and minimum up- and downtime constraints are related with one single unit. On the other hand, coupling constraints involve all units. A change in one unit affects the other units. The power balance and power reserve constraints are examples of coupling constraints. The Lagrange relaxation framework relaxes the coupling constraints and incorporates them into the objective function by a dual optimization procedure. Thus the objective function can be separated into independent functions for each unit, subject to unit capacity and minimum up- and downtime constraints. The resulting Lagrange function of the UCP is as follows:

$$L(P, x, \lambda, \beta) = F(P_{Gi}^t, x_i^t) + \sum_{t=1}^{T} \lambda_t \left( P_D^t - \sum_{i=1}^{n} P_{Gi}^t x_i^t \right)$$

$$+ \sum_{t=1}^{T} \beta_t \left( P_D^t + P_R^t - \sum_{i=1}^{n} P_{Gi\,max} x_i^t \right) \tag{7.14}$$

The UC problem becomes the minimization of the Lagrange function (7.14), subject to constraints (7.10), (7.12), and (7.13). For the sake of simplicity, we have used the symbol $P$, without the subscripts $Gi$ and $t$, to denote any appropriate vector of elements $P^t_{Gi}$. The symbols $x$, $\lambda$, and $\beta$ are handled in the same way. The LR approach requires minimizing the Lagrange function given as

$$q(\lambda, \beta) = \min_{P,x} L(P, x, \lambda, \beta) \qquad (7.15)$$

Since $q(\lambda, \beta)$ provides a lower bound for the objective function of the original problem, the LR method requires to maximize the objective function over the Lagrange multipliers:

$$q^*(\lambda, \beta) = \max_{\lambda, \beta} q(\lambda, \beta) \qquad (7.16)$$

After eliminating constant terms such as $\lambda_t P^t_D$ and $\beta_t(P^t_D + P^t_R)$ in equation (7.14), equation (7.15) can be written as

$$q(\lambda, \beta) = \min_{P,x} \sum_{i=1}^{n} \sum_{t=1}^{T} \{[F_i(P^t_{Gi}) + F_{Si}(t)]x^t_i - \lambda_t P^t_{Gi} x^t_i - \beta_t P_{Gi\,max} x^t_i\} \qquad (7.17)$$

subject to

$$x^t_i P^t_{Gi\,min} \leq P^t_{Gi} \leq x^t_i P^t_{Gi\,max}, \quad t = 1, 2, \ldots, T$$

$$(U^{up}_{t-1,i} - T^{up}_i)(x^{t-1}_i - x^t_i) \geq 0, \quad t = 1, 2, \ldots, T, \ i = 1, 2, \ldots, n$$

$$(U^{down}_{t-1,i} - T^{down}_i)(x^t_i - x^{t-1}_i) \geq 0, \quad t = 1, 2, \ldots, T, \ i = 1, 2, \ldots, n$$

There are two basic steps for the Lagrange procedure to solve the UC problem. They are

1. Initializing the Lagrange multipliers with values that try to make $q(\lambda, \beta)$ larger.
2. Assuming the values of the Lagrange multipliers in step (1) are fixed and the Lagrange function (L) is minimized by adjusting $P^t_{Gi}$ and $x^t_i$.

This minimization is done separately for each unit, and different techniques such as LP and dynamic programming can be used. The solutions for the $N$ independent subproblems are used in the master problem to find a new set of Lagrange multipliers. This involves dual optimization. As we know, for dual optimization, if the function to be optimized is convex, and the variables are continuous, then the maximization of the dual function gives a result that is identical to the one obtained by minimizing the primal function. However, for the UC problem, the variables $0 - 1$ that indicate the status of the units are integer variables, which are neither continuous nor non-convex. Thus the dual theory is not exactly satisfied in the UC problem. The application of the dual optimization method to the UC problem has been given the name *Lagrange relaxation*. A gap exists between the results of the maximization of the dual function and minimization of the primal function. The aim of the Lagrange relaxation method is to reduce the duality gap by iterations. If a criterion is prespecified, this iterative procedure continues until the duality gap criterion is met. The duality gap is also used as a measure of convergence. If the relative duality gap between the primal and the dual solutions is less than a specific tolerance, it is considered that the optimum has been reached. The process then ends with finding a feasible UC schedule.

Actually, the multipliers can be updated by using a subgradient method with a scaling factor and tuning constants which are determined heuristically. This method is as follows:

A vector $g$ is called a subgradient of $L(\cdot)$ at $\lambda^*$ if

$$L(\lambda) \leq L(\lambda^*) + (\lambda - \lambda^*)^T g \tag{7.18}$$

If the subgradient is unique at a point $\lambda$, then it is the gradient at that point. The set of all subgradients at $\lambda$ is called the subdifferential, $\partial L(\lambda)$, and is a closed convex set. A necessary and sufficient condition for optimality in subgradient optimization is $0 \in \partial L(\lambda)$. The value of $\lambda$ can be adjusted by the subgradient optimization algorithm as follows.

$$\lambda_t^{k+1} = \lambda_t^k + \alpha g^k \tag{7.19}$$

where, $g^k$ is any subgradient of $L(\cdot)$ at $\lambda_t^k$. The step size, $\alpha$, has to be chosen carefully to achieve good performance by the algorithm. Here $g^k$ is calculated as follows

$$g^k = \frac{\partial L(\lambda_t^k)}{\partial L\lambda_t^k} = P_D^t - \sum_{i=1}^n x_i^k P_{Gi}^t \tag{7.20}$$

***Example 7.3:*** The data of the three units, four hours, UC problem are as follows; the problem is solved using the Lagrange relaxation technique [21].

1. Units data

$$F_1(P_{G1}) = 0.002P_{G1}^2 + 10P_{G1} + 500$$

$$F_2(P_{G2}) = 0.0025P_{G2}^2 + 8P_{G2} + 300$$

$$F_3(P_{G3}) = 0.005P_{G3}^2 + 6P_{G3} + 100$$

$$100 \leq P_{G1} \leq 600$$

$$100 \leq P_{G2} \leq 400$$

$$50 \leq P_{G3} \leq 200$$

2. Hourly load data are shown in Table 7.7
   For simplification, there are no start-up costs and minimum up- or downtime constraints. The results of several iterations are shown in Tables 7.8–7.13,

**TABLE 7.7    Hourly Load Data**

| Hour ($t$) | Load $P_D^t$ (MW) |
|---|---|
| 1 | 170 |
| 2 | 520 |
| 3 | 1100 |
| 4 | 330 |

starting from an initial condition where all $\lambda^t$ values are set to zero. An economic dispatch is performed for each hour, provided there is sufficient generation committed that hour. The primal value $J^*$ represents the total generation cost summed over all hours as calculated by economic dispatch. $q(\lambda)$ stands for the dual value. The duality gap will be $J^* - q^*$, or the relative duality gap will be $\frac{J^*-q^*}{q^*}$.

For iteration 1, $q(\lambda) = 0$, $j^* = 40,000$, and $\frac{J^*-q^*}{q^*} = undefined$. In the next iteration, the $\lambda^t$ values have been increased as 1.7, 5.2, 11.0, and 3.3. The results as well as the relative duality gap for the several iterations are shown in the Tables 7.9–7.13.

For iteration 2 (Table 7.9), $q(\lambda) = 14,982$, $j^* = 40,000$, and $\frac{J^*-q^*}{q^*} = 1.67$.

For iteration 3 (Table 7.10), $q(\lambda) = 18,344$, $j^* = 36,024$, and $\frac{J^*-q^*}{q^*} = 0.965$.

For iteration 4 (Table 7.11), $q(\lambda) = 19,214$, $j^* = 28,906$, and $\frac{J^*-q^*}{q^*} = 0.502$.

For iteration 5 (Table 7.12), $q(\lambda) = 19,532$, $j^* = 36,024$, and $\frac{J^*-q^*}{q^*} = 0.844$.

For iteration 6 (Table 7.13), $q(\lambda) = 19,442$, $j^* = 20,170$, and $\frac{J^*-q^*}{q^*} = 0.037$.

After 10 iterations, $q(\lambda) = 19,485$, $j^* = 20,017$, and $\frac{J^*-q^*}{q^*} = 0.027$. The relative duality gap is still not zero. The solution will not converge to a final value. Therefore, a tolerance for the relative duality gap should be introduced if the Lagrange relaxation algorithm is used. It means that when $\frac{J^*-q^*}{q^*} \leq \epsilon$ the Lagrange relaxation algorithm will be stopped.

# 7.5    EVOLUTIONARY PROGRAMMING-BASED TABU SEARCH METHOD

## 7.5.1    Introduction

Tabu search (TS) is a powerful optimization procedure that has been successfully applied to a number of combinatorial optimization problems. It has the ability to avoid entrapment in local minima. The TS method uses a flexible memory system (in contrast to "memoryless" systems, such as simulated annealing and genetic algorithm, and rigid memory system such as in branch-and-bound). Specific attention is given

**TABLE 7.8 Iteration 1**

| Hour | $\lambda$ | $u_1$ | $u_2$ | $u_3$ | $P_{G1}$ | $P_{G2}$ | $P_{G3}$ | $\Delta P$ | $P_{G1}^{ed}$ | $P_{G2}^{ed}$ | $P_{G3}^{ed}$ |
|------|-----------|-------|-------|-------|----------|----------|----------|------------|---------------|---------------|---------------|
| 1 | 0 | 0 | 0 | 0 | 0 | 0 | 0 | 170 | 0 | 0 | 0 |
| 2 | 0 | 0 | 0 | 0 | 0 | 0 | 0 | 520 | 0 | 0 | 0 |
| 3 | 0 | 0 | 0 | 0 | 0 | 0 | 0 | 1100 | 0 | 0 | 0 |
| 4 | 0 | 0 | 0 | 0 | 0 | 0 | 0 | 330 | 0 | 0 | 0 |

Where, $\Delta P = P_D^t - \sum_{i=1}^{n} P_{Gi}^t x_i^t$

**TABLE 7.9 Iteration 2**

| Hour | $\lambda$ | $u_1$ | $u_2$ | $u_3$ | $P_{G1}$ | $P_{G2}$ | $P_{G3}$ | $\Delta P$ | $P_{G1}^{ed}$ | $P_{G2}^{ed}$ | $P_{G3}^{ed}$ |
|------|-----------|-------|-------|-------|----------|----------|----------|------------|---------------|---------------|---------------|
| 1 | 1.7 | 0 | 0 | 0 | 0 | 0 | 0 | 170 | 0 | 0 | 0 |
| 2 | 5.2 | 0 | 0 | 0 | 0 | 0 | 0 | 520 | 0 | 0 | 0 |
| 3 | 11.0 | 0 | 1 | 1 | 0 | 400 | 200 | 500 | 0 | 0 | 0 |
| 4 | 3.3 | 0 | 0 | 0 | 0 | 0 | 0 | 330 | 0 | 0 | 0 |

**TABLE 7.10 Iteration 3**

| Hour | $\lambda$ | $u_1$ | $u_2$ | $u_3$ | $P_{G1}$ | $P_{G2}$ | $P_{G3}$ | $\Delta P$ | $P_{G1}^{ed}$ | $P_{G2}^{ed}$ | $P_{G3}^{ed}$ |
|------|-----------|-------|-------|-------|----------|----------|----------|------------|---------------|---------------|---------------|
| 1 | 3.4 | 0 | 0 | 0 | 0 | 0 | 0 | 170 | 0 | 0 | 0 |
| 2 | 10.4 | 0 | 1 | 1 | 0 | 400 | 200 | −80 | 0 | 320 | 200 |
| 3 | 16.0 | 1 | 1 | 1 | 600 | 400 | 200 | −100 | 500 | 400 | 200 |
| 4 | 6.6 | 0 | 0 | 0 | 0 | 0 | 0 | 330 | 0 | 0 | 0 |

**TABLE 7.11 Iteration 4**

| Hour | $\lambda$ | $u_1$ | $u_2$ | $u_3$ | $P_{G1}$ | $P_{G2}$ | $P_{G3}$ | $\Delta P$ | $P_{G1}^{ed}$ | $P_{G2}^{ed}$ | $P_{G3}^{ed}$ |
|------|-----------|-------|-------|-------|----------|----------|----------|------------|---------------|---------------|---------------|
| 1 | 5.1 | 0 | 0 | 0 | 0 | 0 | 0 | 170 | 0 | 0 | 0 |
| 2 | 10.24 | 0 | 1 | 1 | 0 | 400 | 200 | −80 | 0 | 320 | 200 |
| 3 | 15.8 | 1 | 1 | 1 | 600 | 400 | 200 | −100 | 500 | 400 | 200 |
| 4 | 9.9 | 0 | 1 | 1 | 0 | 380 | 200 | −250 | 0 | 130 | 200 |

**TABLE 7.12 Iteration 5**

| Hour | $\lambda$ | $u_1$ | $u_2$ | $u_3$ | $P_{G1}$ | $P_{G2}$ | $P_{G3}$ | $\Delta P$ | $P_{G1}^{ed}$ | $P_{G2}^{ed}$ | $P_{G3}^{ed}$ |
|------|-----------|-------|-------|-------|----------|----------|----------|------------|---------------|---------------|---------------|
| 1 | 6.8 | 0 | 0 | 0 | 0 | 0 | 0 | 170 | 0 | 0 | 0 |
| 2 | 10.08 | 0 | 1 | 1 | 0 | 400 | 200 | −80 | 0 | 320 | 200 |
| 3 | 15.6 | 1 | 1 | 1 | 600 | 400 | 200 | −100 | 500 | 400 | 200 |
| 4 | 9.4 | 0 | 0 | 1 | 0 | 0 | 200 | 130 | 0 | 0 | 200 |

**TABLE 7.13    Iteration 6**

| Hour | $\lambda$ | $u_1$ | $u_2$ | $u_3$ | $P_{G1}$ | $P_{G2}$ | $P_{G3}$ | $\Delta P$ | $P_{G1}^{\text{ed}}$ | $P_{G2}^{\text{ed}}$ | $P_{G3}^{\text{ed}}$ |
|---|---|---|---|---|---|---|---|---|---|---|---|
| 1 | 8.5 | 0 | 0 | 1 | 0 | 0 | 200 | −30 | 0 | 0 | 170 |
| 2 | 9.92 | 0 | 1 | 1 | 0 | 384 | 200 | −64 | 0 | 320 | 200 |
| 3 | 15.4 | 1 | 1 | 1 | 600 | 400 | 200 | −100 | 500 | 400 | 200 |
| 4 | 10.7 | 0 | 1 | 1 | 0 | 400 | 200 | −270 | 0 | 130 | 200 |

to the short-term memory component of TS, which has provided solutions superior to the best obtained with other methods for a variety of problems.

Research endeavors, therefore, have been focused on efficient, near-optimal UC algorithms, which can be applied to large-scale power systems and have reasonable storage and computation time requirements. The major limitations of the numerical techniques are the problem dimensions, large computational time, and complexity in programming.

The LR approach introduced in the previous section to solve the short-term UC problems was found to provide a faster solution but will fail to obtain solution feasibility and solution quality problems and becomes complex if the number of units increases.

Evolutionary programming (EP) is capable of determining the global or near-global solution. EP is based on the basic genetic operation of human chromosomes. It operates with stochastic mechanics, which combine offspring creation based on the performance of current trial solutions and competition and selection based on the successive generations, from a considerably robust scheme for large-scale real-valued combinational optimization. This section will introduce the EP-based TS method to solve the UC problem.

## 7.5.2    Tabu Search Method

The same mathematical model of the UC problem in Section 7.4 is adopted.

The UC problem is a combinatorial problem with integer and continuous variables. It can then be decomposed into two subproblems: a combinatorial problem in integer variables and a nonlinear optimization problem in output power variables. The tabu search (TS) method is used to solve the combinatorial optimization, whereas the nonlinear optimization is solved via a quadratic programming method [14]. The steps of the TS are as follows.

Step 1:    Assume that the fuel costs are fixed for each hour and all the generators share the loads equally.

Step 2:    By optimum allocation, find the initial feasible solution on unit status.

Step 3:    Take the demand as the control parameter.

Step 4:    Generate the trial solution.

Step 5: Calculate the total operating cost as the sum of the running cost and start-up–shutdown cost.

Step 6: Tabulate the fuel cost for each unit for every hour.

Neighbors should be randomly generated about the trial solution. Because of the constraints in the UCP, this is not a simple matter. The most difficult constraints to satisfy are the minimum up/downtimes. The TS algorithm requires a starting feasible schedule, which satisfies all constraints of the system and the units. This schedule is randomly generated.

Once a trial solution is obtained, the corresponding total operating cost is determined. Since the production cost is a quadratic function, a quadratic programming method can be used to solve the subproblem. The start-up cost is then calculated for the given schedule. The calculation is stopped if the following conditions are satisfied.

- The load balance constraints are satisfied.
- The spinning reserve constraints are satisfied.

The tabu list (TL) is controlled by the trial solutions in the order in which they are made. Each time a new element is added to the "bottom" of a list, the oldest element on the list is dropped from the "top." Empirically, TL sizes, which provide good results, often grow with the size of the problem and stronger restrictions are generally coupled with smaller sizes [14]. Best sizes of TL lie in an intermediate range between these extremes. In some applications, a simple choice of TL size in a range centered on 7 seems to be quite effective.

Another important criterion of TS arises when the move under consideration has been found to be tabu. Associated with each entry in the TL, there is a certain value for the evaluation function called the "aspiration level." Normally, the aspiration level criteria are designed to override tabu status if a move is "good enough" [14].

## 7.5.3 Evolutionary Programming

Evolutionary programming (EP) is a mutation-based evolutionary algorithm applied to discrete search spaces. Real-parameter EP is similar in principle to evolution strategy (ES), in which normally distributed mutations are performed in both algorithms. Both algorithms encode mutation strength (or variance of the normal distribution) for each decision variable and a self-adapting rule is used to update the mutation strengths. For the case of evolutionary strategies, Fogel remarks "evolution can be categorized by several levels of hierarchy: the gene, the chromosome, the individual, the species, and the ecosystem" [24–26]. Thus, while genetic algorithms stress models of genetic operators, ES emphasizes mutational transformation that maintains behavioral linkage between each parent and its offspring at the level of the individual.

The general EP algorithm is shown below [15,24–26].

**1.** The initial population is determined by setting

$$s_i = S_i \sim U(a_k, b_k)^k, \quad i = 1, \ldots, m \tag{7.21}$$

where

$S_i$: a random vector;

$s_i$: the outcome of the random vector;

$U(a_k, b_k)^k$: a uniform distribution ranging over $[a_k, b_k]$ in each of $k$ dimensions;

$m$: the number of parents.

2. Each $s_i$ is assigned a fitness score

$$\varphi(s_i) = G(F(s_i), v_i), \quad i = 1, \ldots, m \tag{7.22}$$

where $F$ maps $s_i \to R$ and denotes the true fitness of $s_i$. $v_i$ represents random alteration in the instantiation of $s_i$. $G(F(s_i), v_i)$ describes the fitness score to be assigned. In general, the functions $F$ and $G$ can be as complex as required. For example, $F$ may be a function not only of a particular $s_i$ but also of other members of the population, conditioned on a particular $s_i$.

3. Each $s_i$ is altered and assigned to $s_{i+m}$ such that

$$s_{i+m} = s_{i,j} + N(0, \beta_j \varphi(s_i) + z_j), \quad j = 1, \ldots, k \tag{7.23}$$

where $N(0, \beta_j \varphi(s_i) + z_j)$ represents a Gaussian random variable, $\beta_j$ is a constant of proportionality of scale $\varphi(s_i)$, and $z_j$ represents an offset to guarantee a minimum amount of variance.

4. Each $s_{i+m}$ is assigned a fitness score

$$\varphi(s_{i+m}) = G(F(s_{i+m}), v_{i+m}), \quad i = 1, \ldots, m \tag{7.24}$$

5. For each $s_i, i = 1, \ldots, 2m$, a value $w_i$ is assigned according to

$$w_i = \sum_{t=1}^{c} w_t^* \tag{7.25}$$

$$w_t^* = \begin{cases} 1, & \text{if } \varphi\left(s_i^*\right) \le \varphi(s_i) \\ 0, & \text{otherwise} \end{cases} \tag{7.26}$$

where $c$ is the number of competitions.

6. The solutions $s_i, i = 1, \ldots, 2m$ are ranked in descending order of their corresponding values $w_i$. The first $m$ solutions are transcribed along with their corresponding values $\varphi(s_i)$ to be the basis of the next generation.

7. The process proceeds to step (3) unless the available execution time is exhausted or an acceptable solution has been discovered.

Applying the aforementioned evolutionary programming to UC problem, the calculation steps are shown below.

1. Initialize the parent vector $p = [p_1, p_2, \ldots, p_n]$, $i = 1, 2, \ldots, N_p$ such that each element in the vector is determined by $p_j \sim \text{random}(p_{j\min}, p_{j\max})$, $j = 1, 2, \ldots, N$ with one generator as dependent generator.

2. Calculate the overall objective function of the UC problem using the trail vector $p_i$ and find the minimum of the objective function $F_{Ti}$.

3. Create the offspring trail solution $p_i'$ as follows.

   (a) Compute the standard deviation

$$\sigma_j = \beta \left( \frac{F_{Tij}}{\min \left( F_{Ti} \right)} \right) (P_{j\max} - P_{j\min}) \qquad (7.27)$$

   (b) Add a Gaussian random variable $N(0, \sigma_j^2)$ to all of the state variables of $p_i$, to get $p_i'$.

4. Select the first $N_p$ individuals from the total $2N_p$ individuals of both $p_i$ and $p_i'$ through evaluating each trail vector by $W_{pi} = \text{sum}(W_x)$, where $x = 1, 2, \ldots, N_p$, $i = 1, 2, \ldots, 2N_p$ such that

$$W_x = \begin{cases} 1, & \text{if } \dfrac{F_{Tij}}{F_{Tij} + F_{Tir}} < \text{random} (0, 1) \\ 0, & \text{otherwise} \end{cases} \qquad (7.28)$$

5. Sort the $W_{pi}$ in descending order and the first $N_p$ individuals will survive and be transcribed along with their elements to form the basis of the next generation.

6. Go back to step 2 until a maximum number of generations $N_m$ is reached.

### 7.5.4  Evolutionary Programming-Based Tabu-Search for Unit Commitment

In the TS technique for solving the UC problem, the initial operating schedule status in terms of maximum real power generation of each unit is given as input. As we know that TS is used to improve any given status by avoiding entrapment in local minima, the offspring obtained from the EP algorithm is given as input to TS, and the refined status is obtained. Considering the features of EP and TS algorithms, the EP-based TS method is used for solving UC problems.

1. Get the demand for 24 hours and number of iterations to be carried out.

2. Generate a population of parents ($N$) by adjusting the existing solution to the given demand to the form of state variables.

3. Unit downtime makes a random recommitment.

4. Check for constraint in the new schedule by TS. If the constraints are not met, then repair the schedule. A repair mechanism to restore the feasibility of the constraints is applied; this is described as follows.

   ○ Pick at random one of the OFF units at one of the violated hours.

- ○ Apply the rules in Section 7.5.2 to switch the selected unit from OFF to ON keeping the feasibility of the downtime constraints.
- ○ Check for the reserve constraints at this hour. Otherwise, repeat the process at the same hour for another unit.

5. Solve the master problem of UC and calculate the total production cost for each parent.

6. Add the Gaussian random variable to each state variable and, hence, create an offspring. This will further undergo some repair operations. After these, the new schedules are checked in order to verify that all constraints are met.

7. Improve the status of the evolved offspring, and verify the constraints by TS.

8. Formulate the rank for the entire population.

9. Select the best $N$ number of population for the next iteration.

10. Has the iteration count been reached? If yes, go to step 11, otherwise, go to step 2.

11. Select the best population (s) by evolutionary strategy.

12. Print the optimum schedule.

# 7.6 PARTICLE SWARM OPTIMIZATION FOR UNIT COMMITMENT

## 7.6.1 Algorithm

Particle swarm optimization (PSO) was introduced by Kennedy and Eberhart in 1995 [23] as an alternative to GAs. The PSO technique has since then turned out to be a competitor in the field of numerical optimization. Similar to GA, a PSO consists of a population refining its knowledge of the given search space. PSO is inspired by particles moving around in the search space. The individuals in a PSO thus have their own positions and velocities. These individuals are denoted as particles. Traditionally, PSO has no crossover between individuals, has no mutation, and particles are never substituted by other individuals during the run. Instead, the PSO refines its search by attracting the particles to positions with good solutions. Each particle remembers its own best position found so far in the exploration. This position is called the *personal best* and is denoted by $P_{bi}^t$ in equation (7.29). Additionally, among these $P_{bi}^t$, there is only one particle that has the best fitness, called the *global best* and is denoted by $P_{gbi}^t$ in equation (7.29). The velocity and position update equations of PSO are given by

$$V_i^t = wV_i^{t-1} + C_1 \times r_1 \times (P_{bi}^{t-1} - X_i^{t-1}) + C_2 \times r_2 \times (P_{gbi}^{t-1} - X_i^{t-1}) \qquad (7.29)$$

$$X_i^t = X_i^{t-1} + V_i^t \quad i = 1, \dots, N_D \qquad (7.30)$$

where

w: the inertia weight;

$C_1$, $C_2$: the acceleration coefficients;

$N_D$: the dimension of the optimization problem (number of decision variables);

$r1$, $r2$: two separately generated uniformly distributed random numbers between 0 and 1;

$X$: the position of the particle;

$V_i$: the velocity of the $i$th dimension.

PSO has the following key features compared with the conventional optimization algorithms.

- It only requires a fitness function to measure the "quality" of a solution instead of complex mathematical operations, such as the gradient, Hessian, or matrix inversion. This reduces the computational complexity and relieves some of the restrictions that are usually imposed on the objective function, such as differentiability, continuity, or convexity.
- It is less sensitive to a good initial solution because it is a population-based method.
- It can be easily incorporated with other optimization tools to form hybrid ones.
- It has the ability to escape local minima because it follows probabilistic transition rules.

More interesting PSO advantages can be emphasized when compared to other members of evolutionary algorithms, such as the following.

- It can be easily programmed and modified with basic mathematical and logic operations.
- It is inexpensive in terms of computation time and memory.
- It requires less parameter tuning.
- It works with direct real-valued numbers, which eliminates the need to do binary conversion of a classical canonical genetic algorithm.

The simplest version of PSO lets every individual move from a given point to a new one that is a weighted combination of the individual's best position ever found and of the individual's best position $P_{bi}^t$. The choice of the PSO algorithm's parameters (such as the inertia weight) seems to be of utmost importance for the speed and efficiency of the algorithm.

If economic power dispatch (EPD) is also considered in the UC, a hybrid PSO (HPSO) can be used [20]. The blending real-valued PSO (solving EPD) with binary-valued PSO (solving UC) are operated independently and simultaneously. The binary PSO (BPSO) is made possible with a simple modification to the particle swarm algorithm. This BPSO solves binary problems in a manner similar to the traditional method. In the binary particle swarm, $X_i$ and $P_{bi}^t$ can take on values of 0 or 1 only. The velocity $V_i$ will determine a probability threshold. If $V_i$ is higher, the individual is more likely to choose 1, and lower values favor 0. Such a threshold needs to stay in the range [0.0, 1.0]. One straightforward function for accomplishing this is common in neural networks. The function is called the *sigmoid function* and is defined as follows:

$$s(V_i) = \frac{1}{1 + \exp(-V_i)} \tag{7.31}$$

The function squashes its input into the requisite range and has properties that make it agreeable to be used as a probability threshold. A random number (drawn from a uniform distribution between 0.0 and 1.0) is then generated, whereby $X_i$ is set to 1 if the random number is less than the value from the sigmoid function, that is

$$X_i = \begin{cases} 1, & \text{if } r < s\left(V_i\right) \\ 0, & \text{otherwise} \end{cases} \tag{7.32}$$

In the UC problem, $X_i$ represents the on or off state of generator $i$. To ensure that there is always some chance of a bit flipping (on and off of generators), a constant $V_{\max}$ is selected the start of a trial to limit the range of $V_i$. A large $V_{\max}$ results in a low frequency of the changing state of the generator, whereas a small value increases the frequency of on/off of a generator.

## 7.6.2 Implementation

The mathematical model of the UC problem, which is described in Section 7.4, can be expressed as the general form.

$$\min f(x) \tag{7.33}$$

such that

$$h_j(x) = 0 \quad j = 1, \ldots, m \tag{7.34}$$

$$g_i(x) \geq 0 \quad i = 1, \ldots, k \tag{7.35}$$

To handle the infeasible solutions, the cost function is used to evaluate a feasible solution, that is,

$$\Phi_f(x) = f(x) \tag{7.36}$$

The constraint violation measure $\Phi_u(x)$ for the $r + m$ constraints are usually defined as

$$\Phi_u(x) = \sum_{i=1}^{r} g_i^+(x) + \sum_{j=1}^{m} |h_j^+(x)| \tag{7.37}$$

or

$$\Phi_u(x) = \frac{1}{2}\left[ \sum_{i=1}^{r} \left(g_i^+(x)\right)^2 + \sum_{j=1}^{m} (h_j^+(x))^2 \right] \tag{7.38}$$

where

$g_i^+(x)$: the magnitude of the violation of the $i$th inequality constraint;
$h_j^+(x)$: the magnitude of the violation of the $j$th equality constraint;
    r: the number of inequality constraints;
    m: the number of equality constraints.

Then the total evaluation of an individual $x$, which can be interpreted as the error (for a minimization problem) of an individual $x$, is obtained as

$$\Phi(x) = \Phi_f(x) + \gamma \ \Phi_u(x) \tag{7.39}$$

where $\gamma$ is a penalty parameter of a positive (or negative) constant for the minimization (or maximization) problem, respectively. By associating a penalty with all constraint violations, a constrained problem is transformed to an unconstrained problem such that we can deal with candidates that violate the constraints to generate potential solutions without considering the constraints.

According to equation (7.39), we formulate the objective of the UC problem as a combination of total production cost as the main objective with power balance and spinning reserve as inequality constraints, then we get

$$\Phi(x) = F\left(P_{Gi}^t, x_i^t\right) + \frac{\gamma}{2} \sum_{t=1}^{T}\left[C_1\left(P_D^t - \sum_{i=1}^{n} P_{Gi}^t x_i^t\right)^2 + C_2\left(P_D^t + P_R^t - \sum_{i=1}^{n} P_{Gi\max}^t x_i^t\right)^2\right] \tag{7.40}$$

The penalty factor $\gamma$ is computed at the $k$th generation defined by

$$\gamma = \gamma_0 + \log(k+1) \tag{7.41}$$

The choice of $\gamma$ determines the accuracy and speed of convergence. From the experiment, a greater value of $\gamma$ increases its speed and convergence rate. For this reason, a value of 100 is selected for $\gamma_0$. The pressure on the infeasible solution can be increased with the number of generations, as discussed in the Kuhn–Tucker optimality theorem, and the penalty function theorem provides guidelines to choose the penalty term. In equation (7.40), $C_1$ is set to 1 if a violation to constraint (7.9) occurs and $C_1 = 0$ whenever equation (7.9) is not violated. Similarly, $C_2$ is also set to 1 whenever a violation of equation (7.11) is detected, and it remains 0 otherwise.

Substituting equation (7.8) in equation (7.40), we get

$$\Phi(x) = \sum_{t=1}^{T} \sum_{i=1}^{n} \left[F_i\left(P_{Gi}^t\right) x_i^t + F_{si}(t) x_i^t\right]$$

$$+ \frac{\gamma}{2} \sum_{t=1}^{T}\left[C_1\left(P_D^t - \sum_{i=1}^{n} P_{Gi}^t x_i^t\right)^2 + C_2\left(P_D^t + P_R^t - \sum_{i=1}^{n} P_{Gi\max}^t x_i^t\right)^2\right]$$

$$= \sum_{t=1}^{T} \left\{ \begin{array}{l} \displaystyle\sum_{i=1}^{n} \left[ F_i \left( P_{Gi}^t \right) + F_{si}(t) \right] x_i^t \\ \displaystyle + \frac{\gamma}{2} \left[ C_1 \left( P_D^t - \sum_{i=1}^{n} P_{Gi}^t x_i^t \right)^2 + C_2 \left( P_D^t + P_R^t - \sum_{i=1}^{n} P_{Gi\,\mathrm{max}}^t x_i^t \right)^2 \right] \end{array} \right\}$$

(7.42)

Equation (7.42) is the fitness function for evaluating every particle in the population of PSO for time period $T$. The initial values of power are generated randomly within power limits of a generator. As particles explore the searching space, starting from initial values which are generated randomly within the power limit as shown in equation (7.10), they do encounter cases whereby the power generated exceeds the boundary (minimum or maximum capacity) and therefore violate the constraint in equation (7.10). To avoid the boundary violation, we reinitialize the value whenever it is greater than the maximum capacity or smaller than the minimum capacity of a generator. Again, the re-initialization is done within the power limits of a generator.

The minimum up- and downtimes can be easily handled. As the solution is based upon the best particle ($P_{gbi}^t$) in the history of the entire population, constraints are taken care of by forcing the binary value in $P_{gbi}^t$ to change its state whenever either the minimum up or minimum down constraint is violated. However, this may change the current fitness, which is evaluated using equation (7.42). It implies that the current $P_{gbi}^t$ might no longer be the best among all the other particles. To avoid this situation, $P_{gbi}^t$ will be revaluated using the same equation. Ramping can be incorporated by adding the ramping cost into the total production cost in equation (7.8).

## 7.7 ANALYTIC HIERARCHY PROCESS

The classical UC problem is aimed at determining the start-up and shutdown schedules of thermal units to meet the forecast demand over certain time periods (24 h to 1 week) and belongs to a class of combinatorial optimization problems. The previous sections introduced several methods.

Although these techniques are effective for the problem posed, they do not handle network constraints and bidding issues. The section addresses future UC requirements in a deregulated environment where network constraints, reliability, value of generation, and variational changes in demands and other costs may be factors.

The classical UC Lagrange method cannot solve the problem because of combinatorial explosion. Accordingly, as an initial approach to solve this complex problem, we attempt to find a method for solving UC considering network limitation and generation bids as a daily operational planning problem. This approach supports the decision making effectively of ranking units in terms of their values by using the AHP and the analytic network process (ANP) techniques. The scheduled generation over time is studied as input into the optimal power flow (OPF) problem for optimal dispatch within the network and generation constraint. The OPF problem will be discussed in depth in Chapter 8.

## 7.7.1 Explanation of Proposed Scheme

The basic concept of the proposed optimal generation scheduling is as follows:

First, it is assumed that the ranking of generating units, and their priority as well as demand are known. As a result, the preferred generators for competitive scheduling and pricing will be known. Therefore, the number of generators whose fuel consumption constraints must be considered can be reduced considerably. This reduces the difficulties of UC and optimal power flow. The proposed scheme addresses adequate ranking and prioritizing of units before optimizing the pricing of generation units to meet a given demand. By incorporating the interaction of factors, such as load demand, generating cost curve, bid/sale price, unit up/down cost, and the relative importance of different generation units, the scheme can be implemented to address the technical and nontechnical constraints in the UC problem. This information is easily augmented with the optimization scheme for effective optimal decisions. The scheme consists of the three following stages:

- ranking of units in terms of their values by AHP/ANP;
- checking the constraints by the rule–based method;
- solving the optimization problem by interior point optimal power flow.

Next, for all generators committed, the network availability for power transfer, the constraints on start-up and shutdown, and generated output and reserve are determined for daily operational planning. In the daily UC calculation, a Lagrange method is used without network constraints. Since the majority of connected generators include network constraints and other equipment limitation to ensure feasibility, an OPF technique based on the modified quadratic interior point (MQIP) method [27] is adopted for solving the resulting optimal generation scheduling problem. This gives the proposed scheme a significant advantage over classical heuristic or Lagrange methods. Further work to evaluate this technique is ongoing for multi-utility areas where reliability and stability constraints on the networks are requirements.

According to the above discussion, the scheme for optimal generation scheduling can be represented as illustrated Figure 7.1.

## 7.7.2 Formulation of Optimal Generation Scheduling

***Objective Functions*** In general, in UC problems, the objective function to be minimized is the sum of the operation and start-up costs. First, the fuel cost of the generation is a function of its output $P_i$.

For simplicity, we assume the generation production cost is a quadratic function. Thus the total generation cost can be expressed as

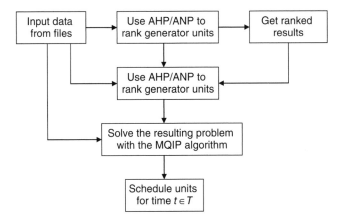

Figure 7.1   Scheme for optimal generation scheduling.

$$F_g(P_{gi}(t)) = \sum_{i=1}^{NG}(a_iP_{gi}(t)^2 + b_iP_{gi}(t) + c_i)$$     (7.43)

where $P_{gi}(t)$ is the real power output of the $i$th generator in period $t$.

$P_{gi}(t)$ is assumed to be within the maintenance schedule, that is, considered to be at an acceptable efficiency to meet the prescribed load. It should be noted that machines being committed are not operating at 100% efficiency owing to imperfect operating conditions and aging.

The start-up cost, on the other hand, increases with shutdown time of generator. We assume that the boiler and turbine cool down after shutdown and the cost of preheating increases with shutdown time and is embedded in $F_{Si}(t)$ (start-up cost of generator $i$ at time $t$).

Therefore, if the number of generators is $NG$, and the duration of the period under consideration is $T$, then the objective function is

$$\min F = \sum_{i=1}^{NG}\sum_{t=1}^{T}\left[\left(a_iP_{gi}(t)^2 + b_iP_{gi}(t) + c_i\right) + F_{Si}(t)\right]x_i(t)$$     (7.44)

**Constraints**   The constraints can be classified as coupling constraints and local constraints. The coupling constraints are related to all generators (in service) under consideration, regardless of age or efficiency, and the following are considered.

*Demand–Supply Balance Constraint*   The sum of the generator outputs must be equal to the demand $P_D(t)$

$$\sum_{i=1}^{NG}(x_i(t)P_{gi}(t)) = P_D(t)$$     (7.45)

Again $x_i(t)$ is a $0 - 1$ variable expressing the state, that is, 0: shutdown and 1: start-up of the $i$th generator in period $t$.

*Reserve Power Constraint* In order to deal with unpredictable disturbances (interruption of generation and transmission lines or unexpected increase in demand), the output of generators in operation must increase, and hence, the instantaneous reserve power shown in the equation below must be required

$$\sum_{i=1}^{NG}(X_{si}(t)r_{si}(t)) \geq R_S(t) \tag{7.46}$$

where $r_{si}(t)$ is the contribution of unit $i$ to spinning reserve at hour $t$, and $R_S(t)$ is the operational reserve requirement at period $t$.

*Generator Output Constraint* When the generator is in the midst of start-up, its output must be between the upper limit $P_{gi\max}$ and lower limit $P_{gi\min}$.

$$x_i(t)P_{gi\min} \leq P_{gi}(t) \leq x_i(t)P_{gi\max} \tag{7.47}$$

For unit ramp rate conditions,

$$P_{gi}(t) - P_{gi}(t-1) \leq UP_{gi}; \quad \text{for unit ramp up of unit } i \tag{7.48}$$

$$P_{gi}(t-1) - P_{gi}(t) \leq DR_{gi}; \quad \text{for unit ramp down of unit } i \tag{7.49}$$

For each selected generator for the bid, the constraint on bid price for unit $i$ at period $t$ is

$$B_{gi}(t) > BP_{gi\min}(t); \quad i \in NG \tag{7.50}$$

where $B_{gi}(t)$ is bid price of unit $i$ at time $t$.

**Network Limitation** To account for network limitation during UC dispatch, the network and operation constraints are specified as additional constraints:

*Power Flow Equation:* The power flow equation at bus $i$ with losses are given as

$$P_{gi}(t) - P_{di}(t) = F_{pi}(V, \theta, t) \tag{7.51}$$

$$Q_{gi}(t) - Q_{di}(t) = F_{qi}(V, \theta, t) \tag{7.52}$$

where

$$F_{pi}(t) = V_i(t)\sum_{j=1}^{NG}(V_j(t)Y_{ij}\cos(\theta_i - \theta_j - \delta_{ij})) \tag{7.53}$$

$$F_{qi}(t) = V_i(t) \sum_{j=1}^{NG} (V_j(t)Y_{ij} \sin(\theta_i - \theta_j - \delta_{ij})) \tag{7.54}$$

The transformer taps in the circuit should be within limits to minimum loss or voltage deviation

$$T_{i\min} \le T_i(t) \le T_{i\max} \tag{7.55}$$

where

$T_{i\min}$: the minimum tap ratio of the transformer;
$T_{i\max}$: the maximum tap ratio of the transformer.

The minimum operation time and minimum shutdown due to fatigue limit of the generator are

$$t_{\text{up min}} \le t_i \le t_{\text{up max}} \tag{7.56}$$

$$t_{\text{down min}} \le t_i \le t_{\text{down max}} \tag{7.57}$$

The limits on flow are defined as

$$\frac{V_i^2 + V_j^2 - 2V_i V_j \cos(\theta_i - \theta_j)}{Z_L^2} \le I_{L\max}^2 \tag{7.58}$$

where

$Z_L$: the impedance of transmission line;
$I_{L\max}$: the maximum current limit of the transmission line.

In addition, each generator is also required to maintain one of the following generator limits for reactive power constraints:

$$x_i(t)Q_{gi\min} \le Q_{gi}(t) \le x_i(t)Q_{gi\max} \tag{7.59}$$

$$V_{gi\min}(t) \le V_{gi}(t) \le V_{gi\max}(t) \tag{7.60}$$

Further, for load buses, we have the following constraint:

$$V_{di\min}(t) \le V_{di}(t) \le V_{di\max}(t) \tag{7.61}$$

The problem posed can be solved by many optimization methods such as Lagrange relaxation methods, heuristic rules, and optimal power flow with decomposition techniques. The Lagrange method utilizes the following primal problem:

Given

$$\min F(x_i(t), P_{gi}(t), F_{si}(t)) \tag{7.62}$$

such that

1. local coupling constraints (7.45) to (7.49);
2. power flow constraints (7.51) and (7.54), given as

$$g_i(x_i(t), P_{gi}(t)) \le 0, \quad i = 1, \dots, NG$$

The function $F$ expresses the sum of fuel consumption and start-up cost. Using the Lagrange multiplier, we determine $\lambda$ and $\mu$, which are introduced in the Lagrange function as follows:

$$L[x_i(t), P_{gi}(t), \lambda(t), \mu(t)] = F[x_i(t), P_{gi}(t), F_{Si}(t)]$$

$$- \lambda(t) \sum_{i=1}^{NG} (x_i(t)P_{gi}(t) - P_D(t))$$

$$+ \mu(t) \sum_{i=1}^{NG} (x_i(t)P_{gimax}(t) - R_s(t)) \tag{7.63}$$

This is usually converted to a dual problem where

$$\max\{\min L[x_i(t), P_{gi}(t), \lambda(t), \mu(t)]\} \tag{7.64}$$

such that.

$$g_i(x_i(t), P_{gi}(t)) \le 0 \tag{7.65}$$

To include the network constraints and bidding of generators, a new UC–based OPF/AHP is proposed [7], namely, we solve for the UC problem over time using OPF to account for the network voltage, transformer, and flow constraints. Application of the MQIP optimization method solves for the optimal operating point at each time period. The second phase of the algorithm uses AHP/ANP to determine the value and merit of each generation bid to be submitted for commitment.

### 7.7.3 Application of AHP to Unit Commitment

*AHP Algorithm* The AHP is a decision-making approach [28–30]. It presents alternatives and criteria, evaluates trade-off, and performs a synthesis to arrive at a final decision. AHP is especially appropriate for cases that involve both qualitative and quantitative analyses. The ANP is an extension of AHP. It makes decisions when alternatives depend on criteria with multiple interactions.

The steps of the AHP algorithm may be written as follows:

Step 1: Set up a hierarchy model.

Step 2: Form a judgment matrix.
The value of elements in the judgment matrix reflects the user's knowledge about the relative importance between every pair of factors.

Step 3: Calculate the maximal eigenvalue and the corresponding eigenvector of the judgment matrix.

Step 4: Hierarchy ranking and consistency check of results.

We can perform the hierarchy ranking according to the value of elements in the eigenvector, which represents the relative importance of the corresponding factor. The consistency index of a hierarchy ranking $CI$ is defined as

$$CI = \frac{\lambda_{max} - n}{n - 1} \tag{7.66}$$

where $\lambda_{max}$ is the maximal eigenvalue of the judgment matrix, $n$ is the dimension of the judgment matrix.

The stochastic consistency ratio is defined as

$$CR = \frac{CI}{RI} \tag{7.67}$$

where $RI$ is a set of given average stochastic consistency indices and $CR$ is the stochastic consistency ratio.

For matrices with dimensions ranging from one to nine, respectively, the values of RI will be as follows:

| $n$ : | 1 | 2 | 3 | 4 | 5 | 6 | 7 | 8 | 9 |
|---|---|---|---|---|---|---|---|---|---|
| $RI$ | 0.00 | 0.00 | 0.58 | 0.90 | 1.12 | 1.24 | 1.32 | 1.41 | 1.45 |

It is obvious that for a matrix with dimension of one or two, it is not necessary to check the stochastic consistency ratio. Generally, the judgment matrix is satisfied if the stochastic consistency ratio, $CR < 0.10$.

It is possible to precisely calculate the eigenvalue and the corresponding eigenvector of a matrix. But this would be time consuming. Moreover, it is not necessary to precisely compute the eigenvalue and the corresponding eigenvector of the judgment matrix. The reason is that the judgment matrix itself, which is formed by the subjective judgment of the user, has some range of error. Therefore, the following two approximate approaches are adopted to compute the maximal eigenvalue and the corresponding eigenvector.

**(A)** Root Method

(1) Multiply all elements of each row in the judgment matrix

$$M_i = \Pi_i X_{ij}, i = 1, \ldots, n; j = 1, \ldots, n \tag{7.68}$$

where

   n: the dimension of the judgment matrix A;
   $X_{ij}$: an element in the judgment matrix A.

(2) Calculate the nth root of $M_i$

$$W_i^* = \sqrt[n]{M_i}, i = 1, \dots, n \tag{7.69}$$

We can obtain the vector

$$W^* = [W_1^*, W_2^*, \dots, W_n^*]^T \tag{7.70}$$

(3) Normalize the vector $W^*$

$$W_i = \frac{W_i^*}{\displaystyle\sum_{j=1}^{n} W_j^*} \quad i = 1, \dots, n \tag{7.71}$$

In this way, we obtain the eigenvector of the judgment matrix A, that is,

$$W = [W_1, W_2, \dots, W_n]^T \tag{7.72}$$

(4) Calculate the maximal eigenvalue $\lambda_{max}$ of the judgment matrix

$$\lambda_{max} = \sum_{i=1}^{n} \frac{(AW)_j}{nW_i} \quad j = 1, \dots, n \tag{7.73}$$

where $(AW)_i$ represents the $i$th element in vector $AW$.

***Example 7.4:***   Compute the maximal eigenvalue $\lambda_{max}$ and the corresponding eigenvector for the following judgment matrix.

$$A = \begin{bmatrix} 1 & 1/5 & 1/3 \\ 5 & 1 & 3 \\ 3 & 1/3 & 1 \end{bmatrix}$$

The calculation steps of the root method are as follows.

1. Multiply all elements of each row in the judgment matrix

$$M_1 = 1 \times \frac{1}{5} \times \frac{1}{3} = \frac{1}{15} = 0.067$$

$$M_2 = 5 \times 1 \times 3 = 15$$

$$M_3 = 3 \times \frac{1}{3} \times 1 = 1$$

**2.** Calculate the $n$th root of $M_i$

$$W_1^* = \sqrt[3]{M_1} = \sqrt[3]{0.067} = 0.405$$

$$W_2^* = \sqrt[3]{M_2} = \sqrt[3]{15} = 2.466$$

$$W_3^* = \sqrt[3]{M_3} = \sqrt[3]{1} = 1$$

We can obtain the vector

$$W^* = [W_1^*, W_2^*, W_3^*]^T = [0.405,\ 2.466,\ 1]^T$$

**3.** Normalize the vector $W^*$

$$\sum_{j=1}^{3} W_j^* = 0.405 + 2.466 + 1 = 3.871$$

$$W_1 = \frac{W_1^*}{\sum\limits_{j=1}^{3} W_j^*} = \frac{0.405}{3.871} = 0.105$$

$$W_2 = \frac{W_2^*}{\sum\limits_{j=1}^{3} W_j^*} = \frac{2.466}{3.871} = 0.637$$

$$W_3 = \frac{W_3^*}{\sum\limits_{j=1}^{3} W_j^*} = \frac{1}{3.871} = 0.258$$

The eigenvector of the judgment matrix $A$ is obtained, that is

$$W = [W_1, W_2, W_3]^T = [0.105,\ 0.637,\ 0.258]^T$$

**4.** Calculate the maximal eigenvalue $\lambda_{\max}$ of the judgment matrix

$$AW = \begin{bmatrix} 1 & 1/5 & 1/3 \\ 5 & 1 & 3 \\ 3 & 1/3 & 1 \end{bmatrix} \begin{bmatrix} 0.105 \\ 0.637 \\ 0.258 \end{bmatrix}$$

$$AW_1 = 1 \times 0.105 + \frac{1}{5} \times 0.637 + \frac{1}{3} \times 0.258 = 0.318$$

$$AW_2 = 5 \times 0.105 + 1 \times 0.637 + 3 \times 0.258 = 1.936$$

$$AW_3 = 3 \times 0.105 + \frac{1}{3} \times 0.637 + 1 \times 0.258 = 0.785$$

$$\lambda_{\max} = \sum_{i=1}^{n} \frac{(AW)_j}{nW_i} = \frac{(AW)_1}{3W_1} + \frac{(AW)_2}{3W_2} + \frac{(AW)_3}{3W_3}$$

$$= \frac{0.318}{3 \times 0.105} + \frac{1.936}{3 \times 0.637} + \frac{0.785}{3 \times 0.258}$$

$$= 3.037$$

**(B)** Sum Method

(1) Normalize every column in the judgment matrix

$$X_{ij}^* = \frac{X_{ij}}{\sum\limits_{k=1}^{n} X_{kj}} \quad i,j = 1,\dots,n \tag{7.74}$$

Now the judgment matrix $A$ is changed into a new matrix $A^*$, in which each column has been normalized.

(2) Add the all elements of each row in matrix $A^*$

$$W_i^* = \sum_{j=1}^{n} X_{ij}, i = 1,\dots,n \tag{7.75}$$

(3) Normalizing the vector $W^*$, we have

$$W_i = \frac{W_i^*}{\sum\limits_{j=1}^{n} W_j^*} i = 1,\dots,n \tag{7.76}$$

Hence, we obtain the eigenvector of the judgment matrix $A$,

$$W = [W_1, W_2, \dots, W_n]^T \tag{7.77}$$

(4) Calculate the maximal eigenvalue $\lambda_{\max}$ of the judgment matrix

$$\lambda_{\max} = \sum_{i=1}^{n} \frac{(AW)_j}{nW_i} \, j = 1, \ldots, n \tag{7.78}$$

where $(AW)_i$ represents the $i$th element in vector $AW$.

***Example 7.5:*** The judgment matrix $A$ is the same as in Example 7.4. Compute the maximal eigenvalue $\lambda_{\max}$ and the corresponding eigenvector using the sum method. The calculation steps are as follows.

**1.** Normalize every column in the judgment matrix

$$\sum_{k=1}^{3} X_{k1} = 1 + 5 + 3 = 9$$

$$X_{11}^* = \frac{X_{11}}{\sum\limits_{k=1}^{3} X_{k1}} = \frac{1}{9} = 0.111$$

$$X_{21}^* = \frac{X_{21}}{\sum\limits_{k=1}^{3} X_{k1}} = \frac{5}{9} = 0.556$$

$$X_{31}^* = \frac{X_{31}}{\sum\limits_{k=1}^{3} X_{k1}} = \frac{3}{9} = 0.333$$

$$\sum_{k=1}^{3} X_{k2} = \frac{1}{5} + 1 + \frac{1}{3} = 1.533$$

$$X_{12}^* = \frac{X_{12}}{\sum\limits_{k=1}^{3} X_{k2}} = \frac{0.2}{1.533} = 0.130$$

$$X_{22}^* = \frac{X_{22}}{\sum\limits_{k=1}^{3} X_{k2}} = \frac{0.2}{1.533} = 0.652$$

$$X_{32}^* = \frac{X_{32}}{\sum\limits_{k=1}^{3} X_{k2}} = \frac{0.333}{1.533} = 0.217$$

$$\sum_{k=1}^{3} X_{k3} = \frac{1}{3} + 3 + 1 = 4.333$$

$$X_{13}^* = \frac{X_{13}}{\sum\limits_{k=1}^{3} X_{k3}} = \frac{0.333}{4.333} = 0.077$$

$$X_{23}^* = \frac{X_{23}}{\sum\limits_{k=1}^{3} X_{k3}} = \frac{3}{4.333} = 0.692$$

$$X_{33}^* = \frac{X_{33}}{\sum\limits_{k=1}^{3} X_{k3}} = \frac{1}{4.333} = 0.231$$

Now the judgment matrix A is changed into a new matrix $A^*$, in which each column has been normalized.

$$A^* = \begin{bmatrix} 0.111 & 0.130 & 0.077 \\ 0.556 & 0.652 & 0.692 \\ 0.333 & 0.217 & 0.231 \end{bmatrix}$$

2. Add the all elements of each row in matrix $A^*$

$$W_1^* = \sum_{j=1}^{3} X_{1J}^* = 0.111 + 0.130 + 0.077 = 0.317$$

$$W_2^* = \sum_{j=1}^{3} X_{2j}^* = 0.556 + 0.652 + 0.692 = 1.900$$

$$W_3^* = \sum_{j=1}^{3} X_{3j}^* = 0.333 + 0.217 + 0.231 = 0.781$$

3. Normalizing the vector $W^*$, we have

$$\sum_{j=1}^{3} W_j^* = 0.317 + 1.900 + 0.781 = 2.998$$

$$W_1 = \frac{W_1^*}{\sum\limits_{j=1}^{3} W_j^*} = \frac{0.317}{2.998} = 0.106$$

$$W_2 = \frac{W_2^*}{\displaystyle\sum_{j=1}^{3} W_j^*} = \frac{1.900}{2.998} = 0.634$$

$$W_3 = \frac{W_3^*}{\displaystyle\sum_{j=1}^{3} W_j^*} = \frac{0.781}{2.998} = 0.261$$

The eigenvector of the judgment matrix A is obtained as follows:

$$W = [W_1, W_2, W_3]^T = [0.106,\ 0.634,\ 0.261]^T$$

**4.** Calculate the maximal eigenvalue $\lambda_{\max}$ of the judgment matrix

$$AW = \begin{bmatrix} 1 & 1/5 & 1/3 \\ 5 & 1 & 3 \\ 3 & 1/3 & 1 \end{bmatrix} \begin{bmatrix} 0.106 \\ 0.634 \\ 0.261 \end{bmatrix}$$

$$AW_1 = 1 \times 0.106 + \frac{1}{5} \times 0.634 + \frac{1}{3} \times 0.261 = 0.320$$

$$AW_2 = 5 \times 0.106 + 1 \times 0.634 + 3 \times 0.261 = 1.941$$

$$AW_3 = 3 \times 0.106 + \frac{1}{3} \times 0.634 + 1 \times 0.261 = 0.785$$

$$\lambda_{\max} = \sum_{i=1}^{n} \frac{(AW)_j}{nW_i} = \frac{(AW)_1}{3W_1} + \frac{(AW)_2}{3W_2} + \frac{(AW)_3}{3W_3}$$

$$= \frac{0.320}{3 \times 0.106} + \frac{1.941}{3 \times 0.634} + \frac{0.785}{3 \times 0.261}$$

$$= 3.036$$

It is noted from examples 7.4 and 7.5 that both the root method and the sum method can achieve similar results.

***AHP-Based Unit Commitment***   According to the theory of AHP/ANP, the following AHP/ANP model in Figure 7.2 is devised to handle ranking of the generator units.

The hierarchical network model of ranking of units consists of three sections:

**1.** the unified ranking of units;

**2.** the ranking criteria or performance indices, in which the $\text{PI}_C$ reflects the relative importance of units;

**3.** the generating units $G_1, \ldots, G_m$.

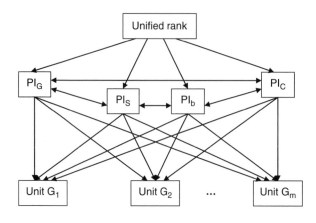

Figure 7.2 Hierarchical network model of units rank.

The performance indices $PI_G, PI_S$, and $PI_b$ are defined as

$$PI_G = \frac{1}{F_{gi}(P_{gi}(t))} \tag{7.79}$$

$$PI_S = \frac{1}{F_{Si}(t)} \tag{7.80}$$

$$PI_b = \frac{1}{BP_{gi}(t)} \tag{7.81}$$

The four ranking criteria $PI_G$, $PI_S$, $PI_b$, and $PI_C$ are interacted. The basic principle of AHP/ANP is to calculate the eigenvector of the alternatives for each criterion. For qualitative factors such as the relative importance of units and criteria, the corresponding eigenvectors can be obtained by computing the judgment matrix. The judgment matrix can be formed on the basis of some scaling method such as the 9-scaling method. For two performance indices $A$ and $B$, their relationship can be expressed as follows if the 9-scaling method is used.

If both performance indices $A$ and $B$ are equally important, then the scaling factor will be "1."

If performance index $A$ is slightly more important compared with performance index $B$, then the scaling factor of $A$ to $B$ will be "3."

If performance index $A$ is more important than performance index $B$, then the scaling factor of $A$ to $B$ will be "5."

If performance index $A$ is far more important than performance index $B$, then the scaling factor of $A$ to $B$ will be "7".

If performance index $A$ is extremely important compared with performance index $B$, then the scaling factor of $A$ to $B$ will be "9".

Naturally, "2," "4," "6," "8" are the medians of two neighboring judgments, respectively.

Using the above 9-scaling, the judgment matrix for representing the relative importance of the four criteria is given in Table 7.14.

The ranking results of units for each time stage will be obtained from AHP/ANP calculation. The list of unit ranking shows the priority of units to be committed at each time stage. However, it has not considered constraints such as system real power balance and system spinning reserve requirement. This chapter adopts the rule-based method to solve this problem.

AHP/ANP is used to decide total ranking of all units for each time stage, and the rule-based system decides the commitment state of units according to the system power balance and system spinning reserve requirement. So the final UC results are obtained through the communication between AHP/ANP ranking and the rule-based constraint checking.

As mentioned above, the priority ranking of all units for each time stage can be obtained by AHP/ANP. This priority rank considers the nontechnical constraints and nonquantitative factors, but it does not involve the constraints of power balance and reserve requirements in the UC. Therefore, the rule-based method is used to coordinate this problem. The implementation steps of the rule-based UC are as follows.

Step (1) Select the number 1 unit from the priority rank of units at hour $t$.

Step (2) Check the constraints of the ramp up/down of the unit.
If the constraints are satisfied, go to step 4.

Step (3) If the constraints of the ramp up/down of the unit are not satisfied, discard this unit at hour $t$. Select the next unit from the priority rank of units, and go to step 2.

Step (4) Check the power balance. If system power can be balanced, go to step (5). Otherwise, add one more unit according to the priority of units, and go to step (2).

Step (5) Check the spinning reserve at hour $t$. If the system has enough spinning reserve, go to the next step. Otherwise, add one more unit according to the priority rank of units, and go to step 2.

Step (6) Stop. All units that were not selected as well as those that have been discarded in the selection will not be committed at hour $t$. The other units will be committed at hour $t$.

**TABLE 7.14   Judgment Matrix A–PI**

| A | $PI_G$ | $PI_S$ | $PI_b$ | $PI_C$ |
|---|---|---|---|---|
| $PI_G$ | 1 | 3 | 1 | 3 |
| $PI_S$ | 1/3 | 1 | 1/2 | 1/2 |
| $PI_b$ | 1 | 1/2 | 1 | 2 |
| $PI_C$ | 1/3 | 2 | 1/2 | 1 |

***Mathematical Demonstration of AHP*** It is noted that the AHP method highly relies on the judgment matrix, which is formed according to the experiences of the users using some scaling method. It is possible that consistency is not obtained. The higher the order of the judgment matrix, the more serious this problem becomes. In this case, a series of problems that must be addressed:

1. Does a single maximal eigenvalue of the judgment exist?
2. Are all the components of the eigenvector of the judgment matrix corresponding to the maximal eigenvalue positive?
3. Is it necessary to check the consistency of the judgment matrix?

*Maximal Eigenvalue and Corresponding Eigenvector of Judgment Matrix* To answer these questions, let us calculate the maximal eigenvalue and corresponding eigenvector of the judgment matrix.

Generally, the judgment matrix $A$ has the following characteristics:

$$a_{ij} > 0$$

$$a_{ji} = \frac{1}{a_{ij}}, \quad i \neq j$$

$$a_{ii} = 1$$

$$i, j = 1, 2, \ldots, n \qquad (7.82)$$

where

$a_{ij}$: the element of the judgment matrix $A$.
n: the dimension of the judgment matrix.

Obviously, the judgment matrix $A$ is positive. Naturally, it is also a nonnegative and irreducible matrix [31,32].

According to Reference [33], we can prove that the judgment matrix is primitive [31]. Therefore, the judgment matrix $A$ has a largest positive eigenvalue $\lambda_{max}$, which is unique and the eigenvector $W$ of matrix $A$ corresponding to the maximal eigenvalue $\lambda_{max}$ has positive components and is essentially unique by the theorem of Perron–Frobenius and the properties of the judgment matrix [31].

*Consistency of judgment matrix* We first give the definition of the consistency matrix.

***Definition*** We say matrix $A = [a_{ij}]$ is consistent if there exist $a_{ij} = \frac{a_{ik}}{a_{jk}}$, *for all* $i, j$, and $k$.

If a positive matrix $A$ is consistent, it has the following properties:

**(a)**

$$a_{ij} = \frac{1}{a_{ji}}$$

$$a_{ii} = 1$$

$$i, j = 1, 2, \ldots, n \tag{7.83}$$

**(b)** The transposition of $A$ is also consistent.

**(c)** Each row in $A$ can be obtained by multiplying any row by a positive number.

**(d)** The maximal eigenvalue of A is $\lambda_{max} = n$. The other eigenvalues of $A$ are all zero.

**(e)** If the eigenvector of $A$ corresponding to the largest eigenvalue $\lambda_{max}$ is $X = [X_1, X_2, \ldots, X_n]^T$,

$$a_{ij} = \frac{X_i}{X_j}; \quad i, j = 1, 2, \ldots, n \tag{7.84}$$

Now, we discuss the case that the elements of the positive consistent matrix are perturbed but still satisfy the property (a). Obviously, the judgment matrix, which we presented in this section, is such a case.

Suppose the eigenvector of the judgment matrix $A$ corresponding to the maximal eigenvalue $\lambda_{max}$ is $W = [W_1, W_2, \ldots, W_n]^T$. Let

$$a_{ij} = \left( \frac{W_i}{W_j} \right) \times \epsilon_{ij}; \quad i, j = 1, 2, \ldots, n \tag{7.85}$$

where

$$\epsilon_{ii} = 1,$$

$$\epsilon_{ij} = \frac{1}{\epsilon_{ji}} \tag{7.86}$$

When $\epsilon_{ij} = 1$ for all $i$ and $j$, equation (7.85) is converted into equation (7.84). In this case, the judgment matrix is consistent. When $\epsilon_{ij} \neq 1$ $(i \neq j, \ i, j = 1, 2, \ldots, n)$, the judgment matrix $A$ is regarded as a perturbed matrix based on the consistency.

According to the property (d) of the consistent positive matrix and $n$ eigenvalues of the judgment matrix, $\lambda_1 (= \lambda_{max}), \lambda_2, \ldots, \lambda_n$, we can obtain

$$\sum_i \lambda_i = n, \quad i = 1, 2, \ldots, n \tag{7.87}$$

We define the following equation as a matrix which reflects that the judgment matrix deviations from the consistent matrix:

$$\mu = -\left(\frac{1}{n-1}\right) \sum_i \lambda_i, \quad i = 1, 2, \ldots, n \tag{7.88}$$

From equation (7.87), we get

$$\mu = \frac{\lambda_{\max} - n}{n - 1} \tag{7.89}$$

In fact, we can obtain the following theorem.

***Theorem 1***   If the positive eigenvector of the judgment matrix $A$ corresponding to the largest eigenvalue $W = [W_1, W_2, \ldots, W_n]^T$, $a_{ij} = \left(\frac{W_i}{W_j}\right) \times \epsilon_{ij}$, $\epsilon_{ij} > 0$, we have

$$\mu = -1 + \left(\frac{1}{n(n-1)}\right) \sum_{1 \leq i \leq j \leq n} \left[\epsilon_{ij} + \frac{1}{\epsilon_{ij}}\right] \tag{7.90}$$

***Proof.***   According to Perron–Frobenius' theorem, we obtain

$$\lambda_{\max} = \sum_j a_{ij} \left(\frac{W_j}{W_i}\right), \quad i, j = 1, 2, \ldots, n \tag{7.91}$$

$$\lambda_{\max} - 1 = \sum_{j \neq i} a_{ij} \left(\frac{W_j}{W_i}\right), \quad i, j = 1, 2, \ldots, n \tag{7.92}$$

then

$$n\lambda_{\max} - n = \sum_{1 \leq i \leq j \leq n} \left[a_{ij} \left(\frac{W_j}{W_i}\right) + a_{ji} \left(\frac{W_i}{W_j}\right)\right] \tag{7.93}$$

Consequently, we get

$$\mu = \frac{\lambda_{\max} - n}{n - 1} = -1 + \frac{1}{n(n-1)} \sum_{1 \leq i \leq j \leq n} \left[a_{ij} \left(\frac{W_j}{W_i}\right) + a_{ji} \left(\frac{W_i}{W_j}\right)\right] \tag{7.94}$$

Substitute $a_{ij} = \left(\frac{W_i}{W_j}\right) \times \epsilon_{ij}$ into equation (7.94), completing the proof of Theorem 1.

We know from Theorem 1 that the smallest extremum of $\mu$ is zero under the condition of $\epsilon_{ij} = 1$ for all $i$ and $j$.

**Theorem 2**  Let $\lambda_{max}$ be the maximal eigenvalue of the judgment matrix $A$. Then

$$\lambda_{max} \geq n \tag{7.95}$$

Let

$$\epsilon_{ij} = 1 + \delta_{ij}$$

$$\delta_{ij} > -1 \tag{7.96}$$

Then

$$a_{ij} = \frac{W_j}{W_i} + \left(\frac{W_i}{W_j}\right)\delta_{ij} \tag{7.97}$$

$\delta_{ij}$ can thus be regarded as the relative change in the disturbed consistency matrix.
From equation (7.94), we have

$$\mu = \left(\frac{1}{n(n-1)}\right)\sum_{1 \leq i \leq j \leq n}\left[\frac{\delta_{ij}^2}{1 + \delta_{ij}}\right] \tag{7.98}$$

According to equations (7.89) and (7.98), we can obtain Theorem 2.
When $\delta = \max_{ij}\delta_{ij}$,

$$\lambda_{max} - n < \frac{1}{n}\sum_{1 \leq i \leq j \leq n}\delta_{ij}^2 \leq \frac{(n-1)\delta^2}{2} \tag{7.99}$$

From equations (7.95) and (7.99), we have

$$n \leq \lambda_{max} \leq n + \frac{(n-1)\delta^2}{2} \tag{7.100}$$

Therefore, in order to make the judgment matrix nearly consistent, we always hope that $\mu$ is close to zero, or $\lambda_{max}$ is close to $n$. Generally, the smaller $\delta_{ij}$ is, the closer $\lambda_{max}$ is to $n$. This is why we check the consistency of the judgment matrix when we apply the AHP to power system problems.

**Example 7.6:**    The proposed approach is examined with the IEEE 39-bus test system, which is taken from Reference [7]. The test system has 10 generators, that is, G30, G31, G32, G33, G34, G35, G36, G37, G38, and G39. The daily load demands are given as in Table 7.15. The generating unit data are given as in Table 7.16. Table 7.17 shows the bid price of generation power over a set of time periods.
The calculation results of UC are listed in Tables 7.18 and 7.19. Table 7.18 is the UC schedule obtained from AHP/ANP and rule-based method. It has not considered

**TABLE 7.15 Daily Load Demands in MW**

| Hour | $P_D$ | $R_S$ | Hour | $P_D$ | $R_S$ | Hour | $P_D$ | $R_S$ |
|------|-------|-------|------|-------|-------|------|-------|-------|
| 1 | 4878 | 244 | 9 | 6341 | 317 | 17 | 6524 | 326 |
| 2 | 5061 | 253 | 10 | 6585 | 329 | 18 | 6585 | 329 |
| 3 | 5183 | 259 | 11 | 6707 | 335 | 19 | 6402 | 320 |
| 4 | 5486 | 274 | 12 | 6768 | 338 | 20 | 6219 | 311 |
| 5 | 5610 | 281 | 13 | 6707 | 335 | 21 | 5792 | 290 |
| 6 | 5792 | 290 | 14 | 6646 | 332 | 22 | 5486 | 274 |
| 7 | 5853 | 293 | 15 | 6585 | 329 | 23 | 5183 | 259 |
| 8 | 6079 | 503 | 16 | 6463 | 323 | 24 | 4939 | 247 |

**TABLE 7.16 Generating Unit Data**

| Unit no. | $a_i$ | $b_i$ | $c_i$ | $P_{i\max}$ | $P_{i\min}$ | $F_{Si}(t)$ |
|----------|-------|-------|-------|-------------|-------------|-------------|
| 30 | 0.834 | 2.50 | 0.00 | 500.0 | 0.00 | 800 |
| 31 | 0.650 | 0.00 | 0.00 | 999.0 | 0.00 | 900 |
| 32 | 0.834 | 0.00 | 0.00 | 700.0 | 0.00 | 850 |
| 33 | 0.824 | 0.00 | 0.00 | 700.0 | 0.00 | 850 |
| 34 | 0.814 | 0.00 | 0.00 | 700.0 | 0.00 | 850 |
| 35 | 0.804 | 0.00 | 0.00 | 700.0 | 0.00 | 850 |
| 36 | 0.830 | 0.00 | 0.00 | 700.0 | 0.00 | 850 |
| 37 | 0.800 | 0.00 | 0.00 | 700.0 | 0.00 | 850 |
| 38 | 0.650 | 0.00 | 0.00 | 900.0 | 0.00 | 870 |
| 39 | 0.600 | 0.00 | 0.00 | 1200.0 | 0.00 | 920 |

**TABLE 7.17 Bid Price of Power Generation Over a Set of Time Period in Dollars Per MW Per Hour**

| Unit | 0–3 | 4–6 | 7–9 | 10–12 | 13–15 | 16–18 | 19–21 | 22–24 |
|------|-----|-----|-----|-------|-------|-------|-------|-------|
| 30 | 40 | 42 | 38 | 45 | 42 | 36 | 38 | 44 |
| 31 | 26 | 29 | 32 | 28 | 26 | 30 | 32 | 28 |
| 32 | 30 | 32 | 33 | 30 | 34 | 36 | 33 | 36 |
| 33 | 32 | 34 | 32 | 36 | 34 | 32 | 36 | 38 |
| 34 | 42 | 38 | 37 | 34 | 36 | 38 | 40 | 45 |
| 35 | 31 | 33 | 35 | 32 | 34 | 36 | 35 | 37 |
| 36 | 29 | 31 | 34 | 37 | 35 | 39 | 41 | 43 |
| 37 | 35 | 37 | 39 | 35 | 37 | 40 | 37 | 39 |
| 38 | 33 | 35 | 37 | 39 | 41 | 37 | 42 | 45 |
| 39 | 24 | 26 | 28 | 28 | 30 | 32 | 30 | 28 |

TABLE 7.18  Unit Commitment Without Transmission Security
and Voltage Constraints

| Unit no. | Hour (0–24) |
|----------|-------------|
| 30 | 0 0 0 0 0 0 0 0 0 0 0 0 0 0 0 0 0 0 0 0 0 0 0 0 |
| 31 | 1 1 1 1 1 1 1 1 1 1 1 1 1 1 1 1 1 1 1 1 1 1 1 1 |
| 32 | 0 0 1 1 1 1 1 1 1 1 1 1 1 1 1 1 1 1 1 1 1 1 1 1 |
| 33 | 1 1 1 1 1 1 1 1 1 1 1 1 1 1 1 1 1 1 1 1 1 1 1 1 |
| 34 | 0 0 0 0 0 0 1 1 1 1 1 1 1 1 1 1 1 1 1 0 0 0 |
| 35 | 1 1 1 1 1 1 1 1 1 1 1 1 1 1 1 1 1 1 1 1 1 1 1 1 |
| 36 | 1 1 1 1 1 1 1 1 1 1 1 1 1 1 1 1 1 0 0 0 0 0 |
| 37 | 0 0 0 0 0 0 1 0 1 1 1 1 1 1 1 1 1 1 1 1 1 0 |
| 38 | 1 1 1 1 1 1 1 1 1 1 1 1 1 1 1 1 1 1 1 1 1 1 1 1 |
| 39 | 1 1 1 1 1 1 1 1 1 1 1 1 1 1 1 1 1 1 1 1 1 1 1 1 |

TABLE 7.19  Unit Commitment with Transmission Security and
Voltage Constraints

| Unit no. | Hour (0–24) |
|----------|-------------|
| 30 | 0 0 0 0 0 0 0 0 0 0 0 0 0 0 0 0 0 0 0 0 0 0 0 0 |
| 31 | 1 1 1 1 1 1 1 1 1 1 1 1 1 1 1 1 1 1 1 1 1 1 1 1 |
| 32 | 1 1 1 1 1 1 1 1 1 1 1 1 1 1 1 1 1 1 1 1 1 1 1 1 |
| 33 | 1 1 1 1 1 1 1 1 1 1 1 1 1 1 1 1 1 1 1 1 1 1 1 1 |
| 34 | 0 0 0 0 0 0 0 1 1 1 1 1 1 1 1 1 1 1 1 0 0 0 |
| 35 | 1 1 1 1 1 1 1 1 1 1 1 1 1 1 1 1 1 1 1 1 1 1 1 1 |
| 36 | 1 1 1 1 1 1 1 1 1 1 1 1 1 1 1 1 1 0 0 1 0 0 |
| 37 | 0 0 0 0 1 1 1 0 1 1 1 1 1 1 1 1 1 1 1 1 1 1 |
| 38 | 1 1 1 1 1 1 1 1 1 1 1 1 1 1 1 1 1 1 1 1 1 1 1 1 |
| 39 | 1 1 1 1 1 1 1 1 1 1 1 1 1 1 1 1 1 1 1 1 1 1 1 1 |

the voltage security and transmission security constraints. The corresponding power
flow solution also violates voltage limits and transmission security limits.

From Table 7.18, we find that power flows at hours 1, 2, 4, 5, 8, 22, and 24 are
infeasible. Table 7.19 is the final UC schedule with OPF corrections. It satisfies the
voltage security and transmission security constraints. The total generation cost for
UC schedule in Table 19 is \$11 391.00. If the commitment states of units are taken
as the input of OPF, the total optimal generation cost will be reduced to \$11 159.60.

## PROBLEMS AND EXERCISES

**1.** What is UC?

**2.** What is the different between UC and ED?

3. What is the minimum average production cost of a unit?

4. How is the priority list method used to solve the UC problem?

5. State the features of dynamic programming–based UC.

6. What are the key features of PSO compared with the conventional optimization algorithms?

7. What is the duality gap in the Lagrange relaxation method?

8. Suppose the production cost functions of five generating units are as follows

$$F_1 = 0.0005P_{G1}^2 + 0.6P_{G1} + 9 \text{ Btu/h}$$

$$F_2 = 0.0013P_{G2}^2 + 0.5P_{G2} + 6 \text{ Btu/h}$$

$$F_3 = 0.0008P_{G3}^2 + 0.7P_{G3} + 5 \text{ Btu/h}$$

$$F_4 = 0.0010P_{G4}^2 + 0.6P_{G4} + 7 \text{ Btu/h}$$

$$F_5 = 0.0007P_{G5}^2 + 0.8P_{G5} + 4 \text{ Btu/h}$$

The power output limits of the five units are

$$100 \le P_{G1} \le 500 \text{ MW}$$

$$150 \le P_{G2} \le 300 \text{ MW}$$

$$150 \le P_{G3} \le 400 \text{ MW}$$

$$100 \le P_{G4} \le 350 \text{ MW}$$

$$100 \le P_{G5} \le 450 \text{ MW}$$

Compute the average production cost for each unit.
Write the priority order list for the five units.

9. For a simple four-unit system, the data of the units and the load pattern are listed in Tables 7.20 and 7.21, respectively. Solve the unit commitment problem.

**TABLE 7.20  The Data of Units for Exercise 9**

| Unit | Max (MW) | Min (MW) | Cost ($/h) | Ave. Cost | Start-up Cost | Initial State | Min Uptimes (h) | Min Downtimes (h) |
|------|----------|----------|------------|-----------|---------------|---------------|-----------------|-------------------|
| 1 | 100 | 30 | 213.00 | 23.54 | 350 | −5 | 4 | 2 |
| 2 | 200 | 50 | 585.62 | 20.34 | 400 | 8 | 5 | 3 |
| 3 | 250 | 70 | 684.74 | 19.74 | 1100 | 8 | 5 | 4 |
| 4 | 50 | 20 | 252.00 | 28.00 | 0 | −6 | 1 | 1 |

**TABLE 7.21    The Load Pattern for Exercise 9**

| Hour | Load (MW) |
|------|-----------|
| 1 | 450 |
| 2 | 500 |
| 3 | 650 |
| 4 | 550 |
| 5 | 400 |
| 6 | 260 |

10. Compute the maximal eigenvalue $\lambda_{max}$ and the corresponding eigenvector for the following judgment matrix.

$$A = \begin{bmatrix} 1 & 1/7 & 1/4 \\ 7 & 1 & 3 \\ 4 & 1/3 & 1 \end{bmatrix}$$

# REFERENCES

1. Cohen AI, Yoshimura M. A branch and bound algorithm for unit commitment. IEEE Trans. Power Syst. 1982;PAS–101:444–451.
2. Cohen AI, Wan SH. A method for solving the fuel constrained unit commitment. IEEE Trans. Power Syst. 1987;1:608–614.
3. Snyder WL, Powell HD, Rayburn C. Dynamic programming approach to unit commitment. IEEE Trans. Power Syst. 1987;PWRS-2:339–350.
4. Vemuri S, Lemonidis L. Fuel constrained unit commitment. IEEE Trans. Power Syst. 1992;7:410–415.
5. Ruzic S, Rajakovic N. A new approach for solving extended unit commitment problem. IEEE Trans. Power Syst. 1991;6:269–277.
6. Allen EH, Ilic MD, Stochastic unit commitment in a deregulated utility industry, in Proceedings on 29th North America Power Symposium, Laramie, Wyoming, October 1997, pp. 105–112.
7. Momoh JA, Zhu JZ. Optimal generation scheduling based on AHP/ANP. IEEE Trans. on Systems, Man, and Cybernetics—Part B 2003;33(3):531–535.
8. Lauer GS, Bertsekas DP, Sandell NR Jr, Posbergh TA. Solution of large-scale optimal unit commitment problems. IEEE Trans. Automat. Contr. 1982;AC-28:1–11.
9. Merlin A, Sandrin P. A new method for commitment at electricitéde France. IEEE Trans. Power Syst. 1983;PAS-102:1218–1255.
10. Ouyang Z, Shahiderpour SM. Short term unit commitment expert system. Int. J. Elect. Power Syst. Res. 1990;20:1–13.
11. Su CC, Hsu YY. Fuzzy dynamic programming: an application to unit commitment. IEEE Trans. Power Syst. 1991;6:1231–1237.
12. Sasaki H, Watanabe M, Kubokawa J, Yorino N. A solution method of unit commitment by artificial neural networks. IEEE Trans. Power Syst. 1992;7:974–981.
13. Padhy NP. Unit commitment using hybrid models: a comparative study for dynamic programming, expert systems, fuzzy system and genetic algorithms. Int. J. Elect. Power Energy Syst. 2000;23(1):827–836.

14. Mantawy AH, Youssef YL, Abdel-Magid L, Shokri SZ, Selim Z. A unit commitment by Tabu search. Proc. Inst. Elect. Eng. Gen. Transm. Dist. 1998;145(1):56–64.

15. Juste KA, Kita H, Tanaka E, Hasegawa J. An evolutionary programming solution to the unit commitment problem. IEEE Trans. Power Syst. 1999;14:1452–1459.

16. Yang HT, Yang PC, Huang CL. Evolutionary programming based economic dispatch for units with nonsmooth fuel cost functions. IEEE Trans. Power Syst. 1996;11:112–117.

17. Mantawy AH, Abdel-Magid YL, Selim SZ. Integrating genetic algorithm, Tabu search and simulated annealing for the unit commitment problem. IEEE Trans. Power Syst. 1999;14:829–836.

18. Rajan CCA, Mohan MR. An evolutionary programming-based tabu search method for solving the unit commitment problem. IEEE Trans. Power Syst. 2004;19(1):577–585.

19. Balci HH, Valenzuela JF. Scheduling electric power generators using particle swarm optimization combined with the lagrangian relaxation method. Int. J. Appl. Math. Comput. Sci. 2004;14(3):411–421.

20. Ting TO, Rao MVC, Loo CK. A novel approach for unit commitment problem via an effective hybrid particle swarm optimization. IEEE Trans. Power Syst. 2006;21(1):411–418.

21. Wood AJ, Wollenberg B. *Power Generation Operation and Control*. 2nd ed. New York: Wiley; 1996.

22. Li WY. *Power Systems Security Economic Operation*. Chongqing: Chongqing University Press; 1989.

23. Kennedy J, Eberhart R, Particle swarm optimization. Presented at Proceedings of IEEE International Conference on neural networks. [Online]. Available: http://www.engr.iupui.edu/~shi/Conference/psopap4.html.

24. Fogel DB. *Evolutionary Computation, Toward a New Philosophy of Machine Intelligence*. Piscataway, NJ: IEEE Press; 1995.

25. Back T. *Evolutionary Algorithms in Theory and Practice*. New York: Oxford University Press; 1996.

26. Fogel LJ, Owens AJ, Walsh MJ. *Artificial Intelligence Through Simulated Evolution*. New York: Wiley; 1996.

27. Momoh JA, Zhu JZ. Improved interior point method for OPF problems. IEEE Trans. on Power Syst. 1999;14(3):1114–1120.

28. Zhu JZ, Irving MR. Combined active and reactive dispatch with multiple objectives using an analytic hierarchical process. IEE Proc. -C 1996;143(4):344–352.

29. Zhu JZ, Momoh JA. Optimal VAr pricing and VAr placement using analytic hierarchy process. Elec. Power Syst. Res. 1998;48:11–17.

30. Satty TL. *The Analytic Hierarchy Process*. Canada: McGraw Hill, Inc; 1980.

31. Zhu JZ, Irving MR, The development of Combined active and reactive dispatch, Part I: Mathematical model and algorithm, Research Report No.1, Brunel Institute of Power Systems, Brunel University, March, 1995.

32. Ledermann W. *Handbook of Applicable Mathematics*. Algebra. Vol. I. New Jersey: Wiley-Interscience; 1980.

33. Wilf HS. *Mathematics for the Physical Sciences*. New Jersey: Wiley; 1962.

# CHAPTER 8

# *OPTIMAL POWER FLOW*

This chapter selects several classic optimal power flow (OPF) algorithms and describes their implementation details. These algorithms include traditional methods such as Newton method, gradient method, linear programming, as well as the latest methods such as modified interior point (IP) method, analytic hierarchy process (AHP), and particle swarm optimization (PSO) method.

## 8.1 INTRODUCTION

The OPF was first introduced by Carpentier in 1962 [1]. The goal of OPF is to find the optimal settings of a given power system network that optimizes the system objective functions such as total generation cost, system loss, bus voltage deviation, emission of generating units, number of control actions, and load shedding while satisfying its power flow equations, system security, and equipment operating limits. Different control variables, some of which are the generators' real power outputs and voltages, transformer tap changing settings, phase shifters, switched capacitors, and reactors, are manipulated to achieve an optimal network setting based on the problem formulation.

According to the selected objective functions, and constraints, there are different mathematical formulations for the OPF problem. They can be broadly classified as follows [1–65].

1. Linear problem in which objectives and constraints are given in linear forms with continuous control variables.

2. Nonlinear problem where either objectives or constraints or both combined are nonlinear with continuous control variables.

3. Mixed-integer linear problems with both discrete and continuous control variables.

Various techniques were developed to solve the OPF problem. The algorithms may be classified into three groups: (1) conventional optimization methods, (2) intelligence search methods, and (3) nonquantitative approach to address uncertainties in objectives and constraints.

*Optimization of Power System Operation*, Second Edition. Jizhong Zhu.
© 2015 The Institute of Electrical and Electronics Engineers, Inc. Published 2015 by John Wiley & Sons, Inc.

## 8.2   NEWTON METHOD

### 8.2.1   Neglecting Line Security Constraints

If the line security constraints are neglected, the OPF problem with real and reactive power variables can be represented as follows:

$$\min \ F = \sum_{i=1}^{NG} f_i(P_{Gi}) \tag{8.1}$$

such that

$$P_i(V, \theta) = P_{Gi} - P_{Di} \tag{8.2}$$

$$Q_i(V, \theta) = Q_{Gi} - Q_{Di} \tag{8.3}$$

$$P_{Gi\,\min} \le P_{Gi}(V, \theta) \le P_{Gi\,\max} \tag{8.4}$$

$$Q_{Gi\,\min} \le Q_{Gi}(V, \theta) \le Q_{Gi\,\max} \tag{8.5}$$

$$V_{i\,\min} \le V_i \le V_{i\,\max} \tag{8.6}$$

where

$P_{Gi}$: the real power output of the generator connecting to bus $i$
$Q_{Gi}$: the reactive power output of the generator connecting to bus $i$
$P_{Di}$: the real power load connecting to bus $i$
$Q_{Di}$: the reactive power load connecting to bus $i$
$P_i$: the real power injection at bus $i$
$Q_i$: the reactive power injection at bus $i$
$V_i$: the voltage magnitude at bus $i$
$f_i$: the generator fuel cost function.

The subscripts "min" and "max" in the equations represent the lower and upper limits of the constraint, respectively.

Equations (8.2) and (8.3) are power flow equations, and can be written as follows.

$$P_i(V, \theta) = V_i \sum_{j=1}^{N} V_j(G_{ij} \cos \theta_{ij} + B_{ij} \sin \theta_{ij}) \tag{8.7}$$

$$Q_i(V, \theta) = V_i \sum_{j=1}^{N} V_j(G_{ij} \sin \theta_{ij} - B_{ij} \cos \theta_{ij}) \tag{8.8}$$

Substituting equations (8.7) and (8.8) in equations (8.2)–(8.6), we get

$$\min\ F(V,\theta) \tag{8.9}$$

s.t.

$$W_{Pi} = V_i \sum_{j=1}^{N} V_j(G_{ij}\cos\theta_{ij} + B_{ij}\sin\theta_{ij}) - P_{Gi} + P_{Di} = 0 \tag{8.10}$$

$$W_{Qi} = V_i \sum_{j=1}^{N} V_j(G_{ij}\sin\theta_{ij} - B_{ij}\cos\theta_{ij}) - Q_{Gi} + Q_{Di} = 0 \tag{8.11}$$

$$W_{PMi} = V_i \sum_{j=1}^{N} V_j(G_{ij}\cos\theta_{ij} + B_{ij}\sin\theta_{ij}) - P_{Gimax} \le 0 \tag{8.12}$$

$$W_{PNi} = V_i \sum_{j=1}^{N} V_j(G_{ij}\cos\theta_{ij} + B_{ij}\sin\theta_{ij}) - P_{Gimin} \ge 0 \tag{8.13}$$

$$W_{QMi} = V_i \sum_{j=1}^{N} V_j(G_{ij}\sin\theta_{ij} - B_{ij}\cos\theta_{ij}) - Q_{Gimax} \le 0 \tag{8.14}$$

$$W_{QNi} = V_i \sum_{j=1}^{N} V_j(G_{ij}\sin\theta_{ij} - B_{ij}\cos\theta_{ij}) - Q_{Gimin} \ge 0 \tag{8.15}$$

$$W_{VMi} = V_i - V_{imax} \le 0 \tag{8.16}$$

$$W_{VNi} = V_i - V_{imin} \ge 0 \tag{8.17}$$

We construct the new augmented objective function by introducing the constraints (8.10)–(8.17) into the original objective function (8.9) with penalty factors.

$$L(X) = F(X) + \sum_{i=1}^{N} r_{Pi} W_{Pi}^2(X) + \sum_{i=1}^{N} r_{Qi} W_{Qi}^2(X) + \sum_{i=1}^{N} r_{Vi} W_{Vi}^2(X) \tag{8.18}$$

where

$X$: the vector that consists of $V$ and $\theta$

$W_{Pi}$: includes all constraints related to real power variables such as equations (8.10), (8.12) and (8.13)

$W_{Qi}$: includes all constraints related to reactive power variables such as equations (8.11), (8.14) and (8.15)

$W_{Vi}$: includes all constraints related to voltage variables such as equations (8.16) and (8.17)

$r_{Pi}$: the penalty factor for violated constraints related to real power variable; for no constraint violation, $r_{Pi} = 0$

$r_{Qi}$: the penalty factor for violated constraints related to reactive power variable; for no constraint violation, $r_{Qi} = 0$

$r_{Vi}$: the penalty factor for violated constraints related to voltage variable; for no constraint violation, $r_{Vi} = 0$

$N$: the total number of buses.

In this way, the OPF problem represented in equations (8.1)–(8.6) becomes an unconstrained optimization problem (8.18). It is noted that only violated constraints are introduced in equation (8.18) as the penalty factor will be zero if the constraint is not violated. The unconstrained optimization problem can be solved by the Newton method or the Hessian matrix method (see Appendix in Chapter 4).

***Calculation of Hessian Matrix and Gradient***   From equation (8.18) as well as equations (8.10)–(8.17), we can get the gradient and Hessian matrix of the augmented objective function as follows.

*Gradient*

$$\frac{\partial L}{\partial V_j} = \frac{\partial F}{\partial V_j} + 2\left[\sum_{i=1}^{N} r_{Pi}W_{Pi}\frac{\partial P_i}{\partial V_j} + \sum_{i=1}^{N} r_{Qi}W_{Qi}\frac{\partial Q_i}{\partial V_j} + r_{Vj}W_{Vj}\right] \tag{8.19}$$

$$\frac{\partial L}{\partial \theta_j} = \frac{\partial F}{\partial \theta_j} + 2\left[\sum_{i=1}^{N} r_{Pi}W_{Pi}\frac{\partial P_i}{\partial \theta_j} + \sum_{i=1}^{N} r_{Qi}W_{Qi}\frac{\partial Q_i}{\partial \theta_j}\right] \tag{8.20}$$

*Hessian Matrix*

$$\frac{\partial^2 L}{\partial V_i^2} = \frac{\partial^2 F}{\partial V_i^2} + 2\sum_{i=1}^{N} r_{Pi}\left[W_{Pi}\frac{\partial^2 P_i}{\partial V_i^2} + \left(\frac{\partial P_i}{\partial V_j}\right)^2\right]$$
$$+ 2\sum_{i=1}^{N} r_{Qi}\left[W_{Qi}\frac{\partial^2 Q_i}{\partial V_i^2} + \left(\frac{\partial Q_i}{\partial V_j}\right)^2\right] + 2r_{Vj} \tag{8.21}$$

$$\frac{\partial^2 L}{\partial V_j\partial V_k} = \frac{\partial^2 F}{\partial V_j\partial V_k} + 2\sum_{i=1}^{N} r_{Pi}\left[W_{Pi}\frac{\partial^2 P_i}{\partial V_j\partial V_k} + \frac{\partial P_i}{\partial V_j}\frac{\partial P_i}{\partial V_k}\right]$$
$$+ 2\sum_{i=1}^{N} r_{Qi}\left[W_{Qi}\frac{\partial^2 Q_i}{\partial V_j\partial V_k} + \frac{\partial Q_i}{\partial V_j}\frac{\partial Q_i}{\partial V_k}\right] \quad j \neq k \tag{8.22}$$

$$\frac{\partial^2 L}{\partial V_j \partial \theta_k} = \frac{\partial^2 F}{\partial V_j \partial \theta_k} + 2\sum_{i=1}^{N} r_{Pi} \left[ W_{Pi} \frac{\partial^2 P_i}{\partial V_j \partial \theta_k} + \frac{\partial P_i}{\partial V_j} \frac{\partial P_i}{\partial \theta_k} \right]$$

$$+ 2\sum_{i=1}^{N} r_{Qi} \left[ W_{Qi} \frac{\partial^2 Q_i}{\partial V_j \partial \theta_k} + \frac{\partial Q_i}{\partial V_j} \frac{\partial Q_i}{\partial \theta_k} \right] \quad j \neq k \qquad (8.23)$$

$$\frac{\partial^2 L}{\partial V_j \partial \theta_j} = \frac{\partial^2 F}{\partial V_j \partial \theta_j} + 2\sum_{i=1}^{N} r_{Pi} \left[ W_{Pi} \frac{\partial^2 P_i}{\partial V_j \partial \theta_j} + \frac{\partial P_i}{\partial V_j} \frac{\partial P_i}{\partial \theta_j} \right]$$

$$+ 2\sum_{i=1}^{N} r_{Qi} \left[ W_{Qi} \frac{\partial^2 Q_i}{\partial V_j \partial \theta_j} + \frac{\partial Q_i}{\partial V_j} \frac{\partial Q_i}{\partial \theta_j} \right] \qquad (8.24)$$

$$\frac{\partial^2 L}{\partial \theta_i^2} = \frac{\partial^2 F}{\partial \theta_i^2} + 2\sum_{i=1}^{N} r_{Pi} \left[ W_{Pi} \frac{\partial^2 P_i}{\partial \theta_i^2} + \left( \frac{\partial P_i}{\partial \theta_i} \right)^2 \right]$$

$$+ 2\sum_{i=1}^{N} r_{Qi} \left[ W_{Qi} \frac{\partial^2 Q_i}{\partial \theta_i^2} + \left( \frac{\partial Q_i}{\partial \theta_i} \right)^2 \right] \qquad (8.25)$$

$$\frac{\partial^2 L}{\partial \theta_j \partial \theta_k} = \frac{\partial^2 F}{\partial \theta_j \partial \theta_k} + 2\sum_{i=1}^{N} r_{Pi} \left[ W_{Pi} \frac{\partial^2 P_i}{\partial \theta_j \partial \theta_k} + \frac{\partial P_i}{\partial \theta_j} \frac{\partial P_i}{\partial \theta_k} \right]$$

$$+ 2\sum_{i=1}^{N} r_{Qi} \left[ W_{Qi} \frac{\partial^2 Q_i}{\partial \theta_j \partial \theta_k} + \frac{\partial Q_i}{\partial \theta_j} \frac{\partial Q_i}{\partial \theta_k} \right] \qquad (8.26)$$

where the derivatives of the bus power injection with respect to variables $V$ and $\theta$ can be obtained from the power flow equations, that is,

$$V_j \frac{\partial P_i}{\partial V_j} = \begin{cases} V_i V_j \left( G_{ij} \cos \theta_{ij} + B_{ij} \sin \theta_{ij} \right) & i \neq j \\ V_i^2 G_{ii} + P_i & i = j \end{cases} \qquad (8.27)$$

$$\frac{\partial P_i}{\partial \theta_j} = \begin{cases} V_i V_j \left( G_{ij} \sin \theta_{ij} - B_{ij} \cos \theta_{ij} \right) & i \neq j \\ -V_i^2 B_{ii} - Q_i & i = j \end{cases} \qquad (8.28)$$

$$V_j \frac{\partial Q_i}{\partial V_j} = \begin{cases} V_i V_j \left( G_{ij} \sin \theta_{ij} - B_{ij} \cos \theta_{ij} \right) & i \neq j \\ -V_i^2 B_{ii} - Q_i & i = j \end{cases} \qquad (8.29)$$

$$\frac{\partial Q_i}{\partial \theta_j} = \begin{cases} -V_i V_j \left( G_{ij} \cos \theta_{ij} + B_{ij} \sin \theta_{ij} \right) & i \neq j \\ -V_i^2 G_{ii} + P_i & i = j \end{cases} \qquad (8.30)$$

$$\frac{\partial^2 P_i}{\partial V_i^2} = \begin{cases} 0 & i \neq j \\ 2G_{ii} & i = j \end{cases} \qquad (8.31)$$

$$\frac{\partial^2 P_i}{\partial V_j \partial V_k}_{j \neq k} = \begin{cases} 0 & i \neq j, i \neq k \\ G_{ij} \cos \theta_{ij} + B_{ij} \sin \theta_{ij} & i = k \\ G_{ik} \cos \theta_{ik} + B_{ik} \sin \theta_{ik} & i = j \end{cases} \tag{8.32}$$

$$V_j \frac{\partial^2 P_i}{\partial V_j \partial \theta_j} = \begin{cases} V_i V_j \left( G_{ij} \sin \theta_{ij} - B_{ij} \cos \theta_{ij} \right) & i \neq j \\ -V_i^2 B_{ii} - Q_i & i = j \end{cases} \tag{8.33}$$

$$\frac{\partial^2 P_i}{\partial V_j \partial \theta_k}_{j \neq k} = \begin{cases} 0 & i \neq j, i \neq k \\ V_i \left( -G_{ij} \sin \theta_{ij} + B_{ij} \cos \theta_{ij} \right) & i = k \\ V_k (G_{ik} \sin \theta_{ik} - B_{ik} \cos \theta_{ik}) & i = j \end{cases} \tag{8.34}$$

$$\frac{\partial^2 P_i}{\partial \theta_i^2} = \begin{cases} V_i V_j \left( -G_{ij} \cos \theta_{ij} - B_{ij} \sin \theta_{ij} \right) & i \neq j \\ V_i^2 G_{ii} - P_i & i = j \end{cases} \tag{8.35}$$

$$\frac{\partial^2 P_i}{\partial \theta_j \partial \theta_k}_{j \neq k} = \begin{cases} 0 & i \neq j, i \neq k \\ V_i V_j \left( G_{ij} \cos \theta_{ij} + B_{ij} \sin \theta_{ij} \right) & i = k \\ V_i V_k (G_{ik} \cos \theta_{ik} + B_{ik} \sin \theta_{ik}) & i = j \end{cases} \tag{8.36}$$

$$\frac{\partial^2 Q_i}{\partial V_i^2} = \begin{cases} 0 & i \neq j \\ -2B_{ii} & i = j \end{cases} \tag{8.37}$$

$$\frac{\partial^2 Q_i}{\partial V_j \partial V_k}_{j \neq k} = \begin{cases} 0 & i \neq j, i \neq k \\ G_{ij} \sin \theta_{ij} - B_{ij} \cos \theta_{ij} & i = k \\ G_{ik} \sin \theta_{ik} - B_{ik} \cos \theta_{ik} & i = j \end{cases} \tag{8.38}$$

$$V_j \frac{\partial^2 Q_i}{\partial V_j \partial \theta_j} = \begin{cases} V_i V_j \left( -G_{ij} \cos \theta_{ij} - B_{ij} \sin \theta_{ij} \right) & i \neq j \\ -V_i^2 G_{ii} + P_i & i = j \end{cases} \tag{8.39}$$

$$\frac{\partial^2 Q_i}{\partial V_j \partial \theta_k}_{j \neq k} = \begin{cases} 0 & i \neq j, i \neq k \\ V_i \left( G_{ij} \cos \theta_{ij} + B_{ij} \sin \theta_{ij} \right) & i = k \\ -V_k (G_{ik} \cos \theta_{ik} + B_{ik} \sin \theta_{ik}) & i = j \end{cases} \tag{8.40}$$

$$\frac{\partial^2 Q_i}{\partial \theta_i^2} = \begin{cases} -V_i V_j \left( G_{ij} \sin \theta_{ij} - B_{ij} \cos \theta_{ij} \right) & i \neq j \\ -V_i^2 B_{ii} - Q_i & i = j \end{cases} \tag{8.41}$$

$$\frac{\partial^2 Q_i}{\partial \theta_j \partial \theta_k}_{j \neq k} = \begin{cases} 0 & i \neq j, i \neq k \\ V_i V_j \left( G_{ij} \sin \theta_{ij} - B_{ij} \cos \theta_{ij} \right) & i = k \\ V_i V_k (G_{ik} \sin \theta_{ik} - B_{ik} \cos \theta_{ik}) & i = j \end{cases} \tag{8.42}$$

***Computation of Search Direction*** The formula for the search direction in the Newton method or the Hessian matrix method is

$$S^k = -[H(X^k)]^{-1}g(X^k) \tag{8.43}$$

where

g: the gradient of the augmented function
H: the Hessian matrix of the augmented function
S: the search direction.

The advantage of the Hessian matrix method is fast convergence. The disadvantage is that it is required to compute the inverse of the Hessian matrix, which leads to an expensive memory and calculation burden. Thus we rewrite equation (8.43) as follows.

$$H(X^k)S^k = -g(X^k) \tag{8.44}$$

For a given gradient and Hessian matrix of the objective function at $X^k$, the search direction $S^k$ can be obtained by solving equation (8.44) by the Gauss elimination method. Since the Hessian matrix of the augmented function is a sparse matrix in the OPF problem, the sparsity programming technique can be used.

The iteration calculation based on the search direction is as follows.

$$X^{k+1} = X^k + \beta^k S^k \tag{8.45}$$

where $\beta$ is a scalar step length.

The iteration calculation will be stopped if the following convergence condition is satisfied.

$$\|X^{k+1} - X^k\| \leq \varepsilon_1 \tag{8.46}$$

or

$$\frac{|L(X^{k+1}) - L(X^k)|}{|L(X^k)|} \leq \varepsilon_2 \tag{8.47}$$

where $\varepsilon_1, \varepsilon_2$ are the permitted tolerances.

***Steps of the Newton Method*** The calculation steps of the Newton method are summarized as follows.

(1) The initial values for the penalty factors are given.

(2) The permitted calculation tolerances are given.

(3) Solve: the initial power flow to get the values of the state variables $X^0$ and set the iteration number $k = 0$.

(4) Compute: the augmented objective function $L(X^k)$, its gradient $g^k$, and Hessian matrix $H^k$.

(5) Compute the search direction $S^k$ according to equation (8.43).

**(6)** Compute the step length $\beta$ using quadratic interpolation.

**(7)** Compute the new state variable $X^{k+1}$ according to equation (8.45).

**(8)** Compute the augmented objective function $L(X^{k+1})$, its gradient $g^{k+1}$, and Hessian matrix $H^{k+1}$, and check the convergence conditions. If either equation (8.46) or (8.47) is met, go to next step. Otherwise, set $k = k + 1$ and go back to step 5).

**(9)** Check whether all constraints are met. If yes, stop the calculation. Otherwise, double the penalty factor for the violated constraint, and reset $k = 0$. Go back to step 4).

## 8.2.2 Consider Line Security Constraints

The line power constraints can be expressed as

$$P_{l\min} \leq P_l \leq P_{l\max} \tag{8.48}$$

where $P_l$ is the power flow at the line $l$ from bus $j$ to bus $k$.

Similarly, the above constraint can be written as

$$W_{PMl} = P_l - P_{l\max} \leq 0 \tag{8.49}$$

$$W_{PNl} = P_l - P_{l\min} \geq 0 \tag{8.50}$$

We use $W_{Pl}$ to express the above line power constraints and introduce it into the augmented objective function (8.18). The new objective function will be

$$L^*(X) = L(X) + \sum_{l=1}^{Nl} r_{Pl} W_{Pl}^2(X) \tag{8.51}$$

where

$r_{Pl}$: the penalty factor for violated line security constraints. If there is no line power flow constraint violation, $r_{Pl} = 0$.

$Nl$: the total number of lines.

Since the augmented objective function includes a new penalty term on line power flow violation, the gradient and Hessian matrix equations (8.19)–(8.26) will be updated to add the corresponding term, that is,

$$\frac{\partial L^*}{\partial V_j} = \frac{\partial L}{\partial V_j} + 2\sum_{l=1}^{Nl} r_{Pl} W_{Pl} \frac{\partial P_l}{\partial V_j} \tag{8.52}$$

$$\frac{\partial L^*}{\partial \theta_j} = \frac{\partial L}{\partial \theta_j} + 2\sum_{l=1}^{Nl} r_{Pl} W_{Pl} \frac{\partial P_l}{\partial \theta_j} \tag{8.53}$$

$$\frac{\partial^2 L^*}{\partial V_i^2} = \frac{\partial^2 L}{\partial V_i^2} + 2\sum_{l=1}^{Nl} r_{Pl} \left[ W_{Pl} \frac{\partial^2 P_l}{\partial V_i^2} + \left( \frac{\partial P_l}{\partial V_j} \right)^2 \right] \tag{8.54}$$

$$\frac{\partial^2 L^*}{\partial V_j \partial V_k} = \frac{\partial^2 L}{\partial V_j \partial V_k} + 2\sum_{l=1}^{Nl} r_{Pl} \left[ W_{Pl} \frac{\partial^2 P_l}{\partial V_j \partial V_k} + \frac{\partial P_l}{\partial V_j} \frac{\partial P_l}{\partial V_k} \right] \quad j \neq k \tag{8.55}$$

$$\frac{\partial^2 L^*}{\partial V_j \partial \theta_k} = \frac{\partial^2 L}{\partial V_j \partial \theta_k} + 2\sum_{l=1}^{Nl} r_{Pl} \left[ W_{Pl} \frac{\partial^2 P_l}{\partial V_j \partial \theta_k} + \frac{\partial P_l}{\partial V_j} \frac{\partial P_l}{\partial \theta_k} \right] \quad j \neq k \tag{8.56}$$

$$\frac{\partial^2 L^*}{\partial V_j \partial \theta_j} = \frac{\partial^2 L}{\partial V_j \partial \theta_j} + 2\sum_{l=1}^{Nl} r_{Pl} \left[ W_{Pl} \frac{\partial^2 P_l}{\partial V_j \partial \theta_j} + \frac{\partial P_l}{\partial V_j} \frac{\partial P_l}{\partial \theta_j} \right] \tag{8.57}$$

$$\frac{\partial^2 L^*}{\partial \theta_i^2} = \frac{\partial^2 L}{\partial \theta_i^2} + 2\sum_{l=1}^{Nl} r_{Pl} \left[ W_{Pl} \frac{\partial^2 P_l}{\partial \theta_i^2} + \left( \frac{\partial P_l}{\partial \theta_i} \right)^2 \right] \tag{8.58}$$

$$\frac{\partial^2 L^*}{\partial \theta_j \partial \theta_k} = \frac{\partial^2 L}{\partial \theta_j \partial \theta_k} + 2\sum_{l=1}^{Nl} r_{Pl} \left[ W_{Pl} \frac{\partial^2 P_l}{\partial \theta_j \partial \theta_k} + \frac{\partial P_l}{\partial \theta_j} \frac{\partial P_l}{\partial \theta_k} \right] \quad j \neq k \tag{8.59}$$

Let the branch admittance of the line $l$ be $g_{jk} + jb_{jk}$; neglecting the line charging, the line power flow can be expressed as

$$P_l = P_{jk} = V_j^2 g_{jk} - V_j V_k (g_{jk} \cos \theta_{jk} + b_{jk} \sin \theta_{jk}) \tag{8.60}$$

The derivatives of the line power with respect to variables $V$ and $\theta$ in equations (8.52)–(8.59) can be obtained from equation (8.60).

$$\frac{\partial P_l}{\partial V_j} = g_{jk}(2V_j - V_k \cos \theta_{jk}) - b_{jk} V_k \sin \theta_{jk} \tag{8.61}$$

$$\frac{\partial P_l}{\partial V_k} = -g_{jk} V_j \cos \theta_{jk} - b_{jk} V_j \sin \theta_{jk} \tag{8.62}$$

$$\frac{\partial P_l}{\partial \theta_j} = g_{jk} V_j V_k \sin \theta_{jk} - b_{jk} V_j V_k \cos \theta_{jk} \tag{8.63}$$

$$\frac{\partial P_l}{\partial \theta_k} = -g_{jk} V_j V_k \sin \theta_{jk} + b_{jk} V_j V_k \cos \theta_{jk} \tag{8.64}$$

$$\frac{\partial^2 P_l}{\partial V_j^2} = 2g_{jk} \tag{8.65}$$

$$\frac{\partial^2 P_l}{\partial V_k^2} = 0 \tag{8.66}$$

$$\frac{\partial^2 P_l}{\partial V_j \partial V_k} = -g_{jk} \cos \theta_{jk} - b_{jk} \sin \theta_{jk} \tag{8.67}$$

$$\frac{\partial^2 P_l}{\partial V_j \partial \theta_j} = g_{jk} V_k \sin \theta_{jk} - b_{jk} V_k \cos \theta_{jk} \tag{8.68}$$

$$\frac{\partial^2 P_l}{\partial V_k \partial \theta_j} = g_{jk} V_j \sin \theta_{jk} - b_{jk} V_j \cos \theta_{jk} \tag{8.69}$$

$$\frac{\partial^2 P_l}{\partial V_j \partial \theta_k} = -g_{jk} V_k \sin \theta_{jk} - b_{jk} V_k \cos \theta_{jk} \tag{8.70}$$

$$\frac{\partial^2 P_l}{\partial V_k \partial \theta_k} = -g_{jk} V_j \sin \theta_{jk} + b_{jk} V_j \cos \theta_{jk} \tag{8.71}$$

$$\frac{\partial^2 P_l}{\partial \theta_j^2} = \frac{\partial^2 P_l}{\partial \theta_k^2} = g_{jk} V_j V_k \cos \theta_{jk} + b_{jk} V_j V_k \sin \theta_{jk} \tag{8.72}$$

$$\frac{\partial^2 P_l}{\partial \theta_j \partial \theta_k} = -g_{jk} V_j V_k \cos \theta_{jk} - b_{jk} V_j V_k \sin \theta_{jk} \tag{8.73}$$

The same calculation steps given in the precious section can be used when line power flow constraints are considered.

***Example 8.1:***   The test example is a 5-bus system, which is taken from reference [17]. The data of generators are shown in Table 8.1. The generator fuel cost is a quadratic function, that is, $f_i = a_i P_{Gi}^2 + b_i P_{Gi} + c_i$. The other data and parameters are shown in Figure 8.1, where the p.u. is used. Table 8.2 is the initial power flow results with the initial system cost of \$4518.04. The OPF results solved by Newton method are shown in Table 8.3. The system minimum cost is \$4236.5.

**TABLE 8.1   Data of Generators for 5-Bus System**

| Unit No. | $a_i$ | $b_i$ | $c_i$ | $P_{Gimin}$ | $P_{Gimax}$ | $Q_{Gimax}$ | $Q_{Gimax}$ |
|---|---|---|---|---|---|---|---|
| 1 | 44.4 | 351 | 50 | 2.0 | 3.5 | 1.5 | 2.5 |
| 2 | 40.0 | 389 | 50 | 4.0 | 5.5 | 1.0 | 2.0 |

**TABLE 8.2   Initial Power Flow Results for 5-Bus System**

| Bus No. | $P_i$ | $Q_i$ | $V_i$ | $\theta_i$ |
|---|---|---|---|---|
| 1 | 2.5794 | 2.2993 | 1.05 | 0 |
| 2 | 5.0 | 1.8130 | 1.05 | 21.84 |
| 3 | −1.6 | −0.8 | 0.8621 | −4.38 |
| 4 | −2.0 | −1.0 | 1.0779 | 17.85 |
| 5 | −3.7 | −1.3 | 1.0364 | −4.28 |

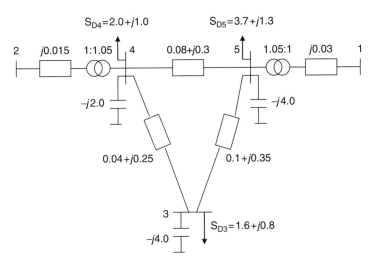

Figure 8.1   A 5-bus system.

**TABLE 8.3   OPF Results by Newton Method for 5-Bus System**

| Bus No. | $P_i$ | $Q_i$ | $V_i$ | $\theta_i$ | $V_{imax}$ | $V_{imin}$ |
|---------|--------|--------|--------|---------|---------|---------|
| 1 | 3.4351 | 2.0707 | 1.0999 | 0 | 1.1 | 0.9 |
| 2 | 3.9997 | 1.2000 | 1.0634 | 8.67 | 1.1 | 0.9 |
| 3 | −1.6 | −0.8 | 0.9324 | −10.96 | 1.1 | 0.9 |
| 4 | −2.0 | −1.0 | 1.1003 | 5.59 | 1.1 | 0.9 |
| 5 | −3.7 | −1.3 | 1.1000 | −5.13 | 1.1 | 0.9 |

## 8.3   GRADIENT METHOD

### 8.3.1   OPF Problem without Inequality Constraints

The OPF problem without inequality constraints can be represented as follows.

$$\min F = \sum_{i=1}^{NG} f_i(P_{Gi})$$

such that

$$P_i(V, \theta) = P_{Gi} - P_{Di}$$

$$Q_i(V, \theta) = Q_{Gi} - Q_{Di}$$

Before we solve the above OPF problem, we first define the state variables $X$ as

$$X = \begin{bmatrix} \left.\begin{matrix} \theta \\ V \end{matrix}\right\} & \text{on each } PQ \text{ bus} \\ \theta & \text{on each } PV \text{ bus} \end{bmatrix} \qquad (8.74)$$

And all specified variables $Y$ as

$$Y = \begin{bmatrix} \left.\begin{matrix} \theta_{\text{ref}} \\ V_{\text{ref}} \end{matrix}\right\} & \text{on reference bus} \\ \left.\begin{matrix} P_D \\ Q_D \end{matrix}\right\} & \text{on each } PQ \text{ bus} \\ \left.\begin{matrix} P_G \\ V_G \end{matrix}\right\} & \text{on each } PV \text{ bus} \end{bmatrix} \qquad (8.75)$$

Some of the parameters of the $Y$ vector are adjustable, such as generator power output and generator bus voltage, and some of them are fixed, such as $P$ and $Q$ at each load bus. Thus, vector $Y$ can be partitioned into a vector $U$ of control parameters and a vector $W$ of fixed parameters,

$$Y = \begin{bmatrix} U \\ W \end{bmatrix} \qquad (8.76)$$

Then the power flow equations can be expressed as

$$g(X,Y) = \begin{bmatrix} \left.\begin{matrix} P_i(V,\theta) - (P_{Gi} - P_{Di}) \\ Q_i(V,\theta) - (Q_{Gi} - Q_{Di}) \end{matrix}\right\} & \text{on each bus} \\ P_k(V,\theta) - (P_{Gk} - P_{Dk}) & \begin{matrix} \text{on each } PV \text{ bus } k, \text{ not} \\ \text{including the reference bus} \end{matrix} \end{bmatrix} \qquad (8.77)$$

Thus the OPF problem without inequality constraints can be expressed as

$$\min \ f(X,U) \qquad (8.78)$$

such that

$$g(X,U,W) = 0 \qquad (8.79)$$

The unconstrained Lagrange function for the OPF problem is obtained.

$$L(X,U,W) = f(X,U) + \lambda^T g(X,U,W) \qquad (8.80)$$

or

$$L(X, U, W) = \sum_{\substack{i=1 \\ i \neq \text{ref}}}^{NG} f_i(P_{Gi}) + f_{\text{ref}}[P_{\text{ref}}(V, \theta)] + [\lambda_1, \lambda_1, \ldots, \lambda_m] \begin{bmatrix} P_i(V, \theta) - P_{\text{inet}} \\ Q_i(V, \theta) - Q_{\text{inet}} \\ P_k(V, \theta) - P_{k\text{net}} \\ \vdots \end{bmatrix}$$

(8.81)

where

$$P_{\text{inet}} = P_{Gi} - P_{Di}$$

$$Q_{\text{inet}} = Q_{Gi} - Q_{Di}$$

The number of Lagrange multipliers is $m$ as there are $m$ power flow equations. According to the necessary conditions for a minimum, we get

$$\nabla L_X = \frac{\partial L}{\partial X} = \frac{\partial f}{\partial X} + \left[\frac{\partial g}{\partial X}\right]^T \lambda = 0$$

(8.82)

$$\nabla L_U = \frac{\partial L}{\partial U} = \frac{\partial f}{\partial U} + \left[\frac{\partial g}{\partial U}\right]^T \lambda = 0$$

(8.83)

$$\nabla L_\lambda = \frac{\partial L}{\partial \lambda} = g(X, U, W) = 0$$

(8.84)

Since the objective function itself is not a function of the state variable except for the reference bus, the derivatives of the objective function with respect to the state variables become

$$\frac{\partial f}{\partial X} = \begin{bmatrix} \dfrac{\partial f_{\text{ref}}(P_{\text{ref}})}{\partial P_{\text{ref}}} \dfrac{\partial P_{\text{ref}}}{\partial \theta_1} \\ \dfrac{\partial f_{\text{ref}}(P_{\text{ref}})}{\partial P_{\text{ref}}} \dfrac{\partial P_{\text{ref}}}{\partial V_1} \\ \vdots \end{bmatrix}$$

(8.85)

The $\frac{\partial g}{\partial X}$ in equation (8.82) is the Jacobian matrix for the Newton power flow, which was discussed in Chapter 2, that is,

$$\frac{\partial g}{\partial X} = \begin{bmatrix} \dfrac{\partial P_1}{\partial \theta_1} & \dfrac{\partial P_1}{\partial V_1} & \dfrac{\partial P_1}{\partial \theta_2} & \dfrac{\partial P_1}{\partial V_2} & \cdots \\ \dfrac{\partial Q_1}{\partial \theta_1} & \dfrac{\partial Q_1}{\partial V_1} & \dfrac{\partial Q_1}{\partial \theta_2} & \dfrac{\partial Q_1}{\partial V_2} & \cdots \\ \dfrac{\partial P_2}{\partial \theta_1} & \dfrac{\partial P_2}{\partial V_1} & & & \cdots \\ \dfrac{\partial Q_2}{\partial \theta_1} & \dfrac{\partial Q_2}{\partial V_1} & & & \cdots \\ \vdots & & & & \vdots \end{bmatrix}$$

(8.86)

Equation (8.83) is the gradient of the Lagrange function with respect to the control variables, in which the vector $\frac{\partial f}{\partial U}$ is a vector of derivatives of the objective function with respect to the control variables.

$$
\frac{\partial f}{\partial U} = \begin{bmatrix} \dfrac{\partial f_1\,(P_1)}{\partial P_1} \\[2ex] \dfrac{\partial f_2(P_2)}{\partial P_2} \\[1ex] \vdots \end{bmatrix}
\tag{8.87}
$$

The other term in equation (8.83), $\frac{\partial g}{\partial U}$, consists of a matrix of all zeros with some $-1$ terms on the diagonals, which correspond to equations in $g(X, U, W)$ where a control variable is present.

The solution steps of the gradient method of OPF are as follows [2,13]:

1. A set of fixed parameters $W$ is given. Assume a starting set of control variables $U$.

2. Solve a power flow. This makes sure that equation (8.84) is satisfied.

3. Solve equation (8.82) for lambda:

$$
\lambda = - \left[ \left( \frac{\partial g}{\partial X} \right)^T \right]^{-1} \frac{\partial f}{\partial X}
\tag{8.88}
$$

4. Substitute $\lambda$ in equation (8.83) and compute the gradient of the Lagrange function with respect to the control variables.

$$
\nabla L_U = \frac{\partial f}{\partial U} + \left[ \frac{\partial g}{\partial U} \right]^T \lambda = \frac{\partial f}{\partial U} + \left[ \frac{\partial g}{\partial U} \right]^T \left\{ - \left[ \left( \frac{\partial g}{\partial X} \right)^T \right]^{-1} \frac{\partial f}{\partial X} \right\}
$$

$$
= \frac{\partial f}{\partial U} - \left[ \frac{\partial g}{\partial U} \right]^T \left[ \left( \frac{\partial g}{\partial X} \right)^T \right]^{-1} \frac{\partial f}{\partial X}
\tag{8.89}
$$

The gradient will give the direction of maximum increase in the cost function as a function of the adjustments in each of the control variables. Since the objective is minimization of the cost function, it needs to move in the negative direction of the gradient.

5. If $|\nabla L_U|$ is sufficiently small, the minimum has been reached. Otherwise, go to the next step.

**6.** Find a new set of control parameters from

$$U^{k+1} = U^k + \Delta U = U^k - \beta |\nabla L_U| \qquad (8.89)$$

where $\beta$ is the step length. Go back to step 2 and use the new values of the control variables.

## 8.3.2   Consider Inequality Constraints

***Inequality Constraints on Control Parameters***   The inequality constraints on control parameters such as generator bus voltage limits can be expressed as follows.

$$U_{min} \leq U \leq U_{max} \qquad (8.90)$$

These constraints can be easily handled during the calculation of the new control parameters in equation (8.89). If the control variable $i$ exceeds one of its limits, it will be set to the corresponding limit, that is,

$$U_i^{k+1} = \begin{cases} U_{imax}, & \text{if } U_i^k + \Delta U_i > U_{imax} \\ U_{imin}, & \text{if } U_i^k + \Delta U_i < U_{imin} \\ U_i^k + \Delta U_i, & \text{otherwise} \end{cases} \qquad (8.91)$$

At the minimum the components $\frac{\partial f}{\partial U}$ of $\nabla L_U$ will be

$$\frac{\partial f}{\partial U_i} = 0, \text{ if } U_{imin} < U_i < U_{imax}$$

$$\frac{\partial f}{\partial U_i} \leq 0, \text{ if } U_i \geq U_{imax} \qquad (8.92)$$

$$\frac{\partial f}{\partial U_i} \geq 0, \text{ if } U_i \leq U_{imin}$$

The Kuhn–Tucker theorem proves that the conditions of equation (8.92) are necessary for a minimum, provided the functions involved are convex.

***Functional Inequality Constraints***   The upper and lower limits on the state variables such as bus voltages on $PQ$ buses can also be functional inequality constraints, which can be expressed as

$$h(X, U) \leq 0 \qquad (8.93)$$

Compared with the inequality constraints on control variables, the functional inequality constraints are difficult to handle and the method can become very time consuming or practically impossible in some situations. Basically, a new direction that is different from the negative gradient must be found when confronting a functional inequality

**TABLE 8.4    OPF Results by Gradient Method for 5-Bus System**

| Bus No. | $P_i$ | $Q_i$ | $V_i$ | $\theta_i$ | $V_{imax}$ | $V_{imin}$ |
|---|---|---|---|---|---|---|
| 1 | 3.4351 | 2.0359 | 1.0938 | 0 | 1.1 | 0.9 |
| 2 | 3.9987 | 1.2487 | 1.0650 | 8.53 | 1.1 | 0.9 |
| 3 | −1.6 | −0.8 | 0.9300 | −11.10 | 1.1 | 0.9 |
| 4 | −2.0 | −1.0 | 1.1014 | 5.45 | 1.1 | 0.9 |
| 5 | −3.7 | −1.3 | 1.0944 | −5.18 | 1.1 | 0.9 |

constraint. The often used method is the penalty method, in which the objective function is augmented by penalties for functional constraint violations. This forces the solution back sufficiently close to the constraint. The reasons for the penalty method being selected are as follows.

1. Generally, functional constraints are seldom rigid limits in the strict mathematical sense but are, rather, soft limits. For example, $V \le 1.0$ on a $PQ$ bus means $V$ should not exceed 1.0 by too much, and $V = 1.01$ may still be permissible. The penalty method produces just such soft limits.

2. The penalty method adds very little to the algorithm, as it simply amounts to adding terms to $\frac{\partial f}{\partial X}$, and also to $\frac{\partial f}{\partial U}$ if the functional constraint is also a function of $U$.

3. It produces feasible power flow solutions, with the penalties signaling the trouble spots, where poorly chosen rigid limits would exclude solutions.

***Example 8.2:***    The test example is a 5-bus system, which was shown in Figure 8.1 in Example 8.1. The data and parameters of the system are the same as in Example 8.1. The OPF results solved by the gradient method are shown in Table 8.4. The system minimum cost is $4235.7

## 8.4    LINEAR PROGRAMMING OPF

The early LP-based OPF method was limited to network-constrained economic power dispatch, which we introduced in Chapter 5. The earliest versions used the fixed constraint approximations, based on the purely DC power flow. Later on, incremental formulations were introduced, whereby constraint linearization is iterated with AC power flow, to model and enforce the constraints exactly [18]. The advantages of the LP-based OPF are

1. reliability of the optimization;

2. ability to recognize problem infeasibility quickly, so that appropriate strategies can be put into effect;

3. the range of operating limits can be easily accommodated and handled, including contingency constraints;

**4.** convergence to engineering accuracy is rapid, and also accepted when the changes in the controls have become very small.

Large-scale application of LP-based methods has traditionally been limited to network-constrained real and reactive dispatch calculations whose objectives are separable, comprising the sum of convex cost curves. The accuracy of calculation may be lost if the oversimplified approximation is adopted in LP-based OPF. The piecewise linear segmentation of the generator fuel cost curve should be good for avoiding this problem. The piecewise approach can fit an arbitrary curve convexly to any desired accuracy with a sufficient number of segments. Originally, a separable LP variable had to be used for each segment, and the resulting large problems with multisegments cost curve modeling were prohibitively time and storage consuming. The difficulty was alleviated considerably by a separable programming procedure that uses a single variable per cost curve, regardless of the number of the segments. However, the number of segments still affects the solution speed and precision. If the segment sizes are large, the following issues may be appeared.

**1.** Even a very small change in an OPF problem can cause some optimized controls to jump to adjacent segment breakpoints.
**2.** Discrete jumps between segment breakpoints occasionally produce solution oscillations when iterating with AC power flow.

The technique of successive segment refinement can be used to overcome the above problems. The idea is that the nonlinear cost curves are approximated with relatively large segments at the beginning. Then, in each subsequent iteration, each cost curve is modeled with a smaller segment size, until the final degree of refinement has been reached.

For LP-based OPF, in addition to the linearization of the objective function, the constraints also need to be linearized. Generally, the linearized power flow equations are used in LP-based OPF, either based on a linear sensitivity matrix or on the fast decoupled power flow model. The latter can be written as

$$[B']\Delta\theta = \Delta P \qquad (8.94)$$

$$[B'']\Delta V = \Delta Q \qquad (8.95)$$

These provide accurate enforcement of the network constraints in the real or reactive subproblems through the iterative process. The real power subproblem in OPF based on equation (8.94) is restricted to the "real power" constraints that are strong functions of angle "$\theta$," and the reactive power subproblem in OPF based on equation (8.95) is restricted to the "reactive power" constraints that are strong functions of the magnitude of voltage $V$. Tests on a large power system have demonstrated that successive constrained P- and Q-subproblems for OPF are effective in achieving practical overall optimization. If only a real power subproblem is considered in OPF, it becomes the security-constrained economic power dispatch, which was introduced in Chapter 5.

For inequality constraints in LP-based OPF, the sensitivity approach is used to express each selected constraint in terms of the control variables. Let $U$, $X$, and $P$ be the control, state variables, and bus power injections, respectively. $Y$ is the constraint whose sensitivities are to be computed. The incremental relationships between these variables are

$$\Delta Y = C\Delta X + D\Delta U \tag{8.96}$$

$$\Delta P = [B]\Delta U \tag{8.97}$$

$$\Delta X = [A]^{-1}\Delta P \tag{8.98}$$

From the above equations, we get the following sensitivity vector.

$$\frac{\Delta Y}{\Delta U} = C[A]^{-1}[B]\Delta U + D \tag{8.99}$$

The row vectors $C$ and $D$ are usually extremely sparse, and are specific to the particular constraint $Y$. The power flow Jacobian matrix $[A]$ and matrix $[B]$ are constant throughout the OPF iteration. The main work in calculating the sensitivity vector from equation (8.99) is the repeat solution $C[A]^{-1}$ using fast-forward substitution.

After the above handlings on OPF objective function and constraints, the linear OPF model can be constructed and, consequently, solved by linear programming algorithm.

## 8.5   MODIFIED INTERIOR POINT OPF

### 8.5.1   Introduction

OPF calculations determine optimal control variables and system quantities for efficient power system planning and operation. OPF has now become a useful tool in power system operation as well as in planning. Over the years, different objective functions have emerged, and the constraints and size of systems to be solved have increased. An efficient OPF tool is required to solve both the operations problem and the planning problem. The operational OPF problem, considering t time duration from one-half hour to a day, consists of many objective functions such as economic dispatch and loss minimization. For volt-ampere reactive (VAR) planning, the time duration can be up to 5 years. VAR planning can also consider the operational cost of losses, thus forming a hybrid planning/operation problem.

An OPF package must handle large, interconnected power systems. In some instances, the area to be optimized needs to be identified and the type of optimization needs to be established before optimization. Generally, the available OPF packages do not determine the type of problem, nor do they recommend the type of objective or identify the area to be optimized. Also, in most OPF packages, the model is predetermined and cannot be modified by the user without access to the source code [27]. An OPF package that allows the user to pick certain constraints from a specified list is useful for adapting the package to the user's needs.

To implement the above requirements, a more versatile OPF package is necessary. Obviously, the conventional OPF algorithms are limited and too slow for this purpose. The increasing burden being imposed on optimization is handled by rapidly advancing computer technology as well as through development of more efficient algorithms exploiting the sparse nature of the power system structure. The IP method is one of the most efficient algorithms as evident from the list of references [27–45]. The IP method classification is a relatively new optimization approach that was applied to solve power system optimization problems in the late 1980s and early 1990s. This method is essentially a linear programming method; and as expected, linear programming problems dominate IP classification. When compared with other well-known linear programming techniques, IP methods maintain their accuracy while achieving great advantages in speed of convergence of as much as $12:1$ in some cases. However, IP methods, in general, suffer from bad initial, termination, and optimality criteria and, in most cases, are unable to solve nonlinear and quadratic objective functions. The extended quadratic interior point (EQIP) method described here can handle quadratic objective functions subject to linear and nonlinear constraints.

The optimization technique used in this section is an improved quadratic interior point (IQIP) method. The IQIP method features a general starting point (rather than a good point as in the former EQIP as well as general IP methods) that is even faster than the EQIP optimization scheme. Consequently, the OPF approach described in this section offers great improvements in speed, accuracy, and convergence in solving multi-objective and multi-constraint optimization problems. It is also capable of solving global optimization of an interconnected system and a partitioned system for local optimization. The scheduled generation, transformer taps, bus voltages, and reactors are used to achieve a feasible and optimized power flow solution.

## 8.5.2 OPF Formulation

***Objective Functions*** Three objective functions are considered. They are fuel cost minimization, VAR planning, and loss minimization.

**(1)** Fuel cost minimization

$$\min F_g = \sum_{i=1}^{NG}(a_i P_{gi}^2 + b_i P_{gi} + c_i) \tag{8.100}$$

**(2)** VAR planning

$$\min F_q = \sum_{i=1}^{Nc} S_{ci}(q_{ci}^{\text{tot}} - q_{ci}^{\text{exist}}) - \sum_{i=1}^{Nr} S_{ri}(q_{ri}^{\text{tot}} - q_{ri}^{\text{exist}}) + S_\omega P_L \tag{8.101}$$

**(3)** Loss minimization

$$\min P_L = F(P_{g\,\text{slack}}) \tag{8.102}$$

where

$P_{gi}$: the real power generation at generator $i$

$P_L$: system real power loss

$P_{g\,\text{slack}}$: the real power of slack generator

$S_c$: the cost of unit capacitive VAR

$S_r$: the cost of unit inductive VAR

$q_c$: the capacitive VAR support

$q_r$: the inductive VAR support

$l$: the contingency case, $l = 0$, means the intact case or base case

$S_w$: the coupling coefficient between the VAR and loss portions in the VAR planning objective function.

**Constraints** The linear and nonlinear constraints that include voltage, flows, real generation, reactive sources, and transformer taps are considered as follows.

$$P_{gi,l}^{\min} \leq P_{gi,l} \leq P_{gi,l}^{\max}, \quad i \in NG \tag{8.103}$$

$$\sum_{i=1}^{NG} P_{gi} = \sum_{k=1}^{ND} P_{dk} + P_L \tag{8.104}$$

$$P_{gi} - P_{di} - F_i(V, \theta, T) = 0$$
$$i = 1, 2, \ldots, N_{\text{bus}}, \quad i \neq \text{Slack} \tag{8.105}$$

$$Q_{gi} - Q_{di} - G_i(V, \theta, T) = 0$$
$$i = 1, 2, \ldots, N_{\text{bus}}, \quad i \neq \text{Slack} \tag{8.106}$$

$$\frac{V_i^2 + V_j^2 - 2V_i V_j \cos(\theta_i - \theta_j)}{Z_L(l)^2} - I_{L\max}{}^2(l) \leq 0 \tag{8.107}$$

$$l = 0, 1, 2, \ldots, Nl$$

$$Q_{gi\min} \leq Q_{gi} \leq Q_{gi\max}, \quad i \in NG \tag{8.108}$$

$$0 \leq q_{ci}^{\text{exist}} \leq q_{ci\max}^{\text{exist}}, \quad i \in VAR \text{ sites} \tag{8.109}$$

$$0 \leq q_{ri}^{\text{exist}} \leq q_{ri\max}^{\text{exist}}, \quad i \in VAR \text{ sites} \tag{8.110}$$

$$q_{ci}^{\text{tot}} - q_{ci}^{\text{exist}} \geq 0, \quad i \in VAR \text{ sites} \tag{8.111}$$

$$q_{ri}^{\text{tot}} - q_{ri}^{\text{exist}} \geq 0, \quad i \in VAR \text{ sites} \tag{8.112}$$

$$V_{gi\min} \leq V_{gi} \leq V_{gi\max}, \quad i \in NG \tag{8.113}$$

$$V_{di\,\min} \leq V_{di} \leq V_{di\,\max}, \quad i \in ND \tag{8.114}$$

$$T_{i\min} \leq T_i \leq T_{i\max}, \quad i \in NT \tag{8.115}$$

$$P_{\text{slack}} = F_{\text{slack}}(V, \theta, T) \tag{8.116}$$

where

$P_{dk}$: real power load at load bus $k$
$Q_{di}$: reactive power load at load bus $i$
$V_{gi}$: the voltage magnitude at generator bus $i$
$V_{di}$: voltage magnitude at load bus $i$
$Q_{gi}$: VAR generation of generator $i$
$Z_L$: the impedance of transmission line $L$
$I_{L\max}$: the maximal current limit through transmission line $L$
$T$: the transformer tap position
$\theta$: the bus voltage angle
$P_L$: the system real power loss
$NG$: number of generation buses
$NT$: number of transformer branches
$ND$: number of load buses
$N_{\text{bus}}$: number of total network buses
$\phi_i$: the angle of phase shifter transformer $i$
$NM\phi$: adjustment numbers of phase shifter
$Nl$: the set of the outage line ($l = 0$ means no line outage).

The subscripts "min" and "max" stand for the lower and upper bounds of a constraint, respectively.

We can pick certain constraints from equations (8.103)–(8.116) according to the particular needs of the practical system. Generally, the constraints in equations (8.103)–(8.108) and (8.113)–(8.115) are considered for economic dispatch. The constraints in equations (8.104)–(8.116) are considered for VAR planning. For loss minimization, the constraints in equations (8.104)–(8.108) and (8.113)–(8.116) are considered.

### 8.5.3   IP OPF Algorithms

***General Interior Point Algorithm***   The OPF problem can be expressed in general form as follows:

$$\min \ f(x) \tag{8.117}$$

such that

$$d(x) \geq 0$$

$$x \geq 0 \tag{8.118}$$

There are several primal–dual IP methods. Here we introduce the logarithmic barrier function-based IP method. For the above problem, the logarithmic barrier function is given by

$$b(x, \mu) = f(x, \mu) - \mu \sum_{j=1}^{m} \ln d_j(x) - \mu \sum_{i=1}^{n} \ln x_i \tag{8.119}$$

where

$\mu$: a positive parameter
$m$: the number of constraints
$n$: the number of variables.

The barrier gradient and Hessian are

$$\nabla b(x, \mu) = g - \mu B^T D^{-1} I - (\mu X^{-1} I) \tag{8.120}$$

$$\nabla^2 b(x, \mu) = \nabla^2 f - \sum_{j=1}^{m} \frac{\mu}{d_j} \nabla^2 d_j + \mu B^T D^{-2} B + \mu X^{-2} \tag{8.121}$$

where

$I$: a column vector of ones
$D$: diagonal matrix diag$\{d(x)\}$
$X$: diagonal matrix diag$\{x\}$.

The solution to the above problem can be obtained via a sequence of solutions to the unconstrained subproblem.

$$\text{Minimize } b(x, \mu) \tag{8.122}$$

According to Kuhn–Tucker conditions, we have

$$\nabla b(x, \mu) = 0 \tag{8.123}$$

$$\nabla^2 b(x, \mu) = 0 \text{ is positive definite} \tag{8.124}$$

$$\lim_{\mu \to 0} (x_\mu) = x^*$$

$$\lim_{\mu \to 0} \frac{\mu}{x_{j\mu}} = s_j^* \tag{8.125}$$

$$\lim_{\mu \to 0} \frac{\mu}{d_j(x_\mu)} = z_j^*$$

where $s_j^*$ and $z_j^*$ are the Lagrange multipliers. The points $(x_\mu)$ define a barrier trajectory, or a local central path for equation (8.125). If we introduce the slack variable

$$v_\mu = d(x_\mu), \quad v_\mu \geq 0 \tag{8.126}$$

and define

$$z_\mu = \mu D(x_\mu)^{-1}I, \quad z_\mu \geq 0 \tag{8.127}$$

$$s_\mu = \mu X_\mu^{-1}I, \quad s_\mu \geq 0 \tag{8.128}$$

then the central path is equivalent to

$$g_\mu - B_\mu^T z_\mu - s_\mu = 0 \tag{8.129}$$

$$d_\mu - v_\mu = 0 \tag{8.130}$$

$$\nabla^2 f_\mu - \sum_{j=1}^{m} z_{j\mu} \nabla^2 d_{j\mu} + B_\mu^T V_\mu^{-1} Z_\mu B_\mu + X_\mu^{-1} S_\mu = 0 \tag{8.131}$$

$$V_\mu z_\mu = \mu I, \quad v_\mu, z_\mu \geq 0 \tag{8.132}$$

$$X_\mu s_\mu = \mu I, \quad x_\mu, s_\mu \geq 0 \tag{8.133}$$

The above nonlinear equations can be expressed as follows, which hold at $(x_\mu, v_\mu, z_\mu, s_\mu)$

$$\begin{bmatrix} -g + B^T z - s \\ d - v \\ Vz - \mu I \\ Sx - \mu I \end{bmatrix} = 0 \tag{8.134}$$

Applying Newton's method to the above, we obtain

$$\begin{bmatrix} -W & B^T \\ B & 0 \end{bmatrix} \begin{bmatrix} \Delta x \\ \Delta z \end{bmatrix} + \begin{bmatrix} \Delta s \\ -\Delta v \end{bmatrix} = \begin{bmatrix} g - B^T z - s \\ v - d \end{bmatrix} \tag{8.135}$$

and

$$V\Delta z + Z\Delta v = \mu I - Zv \tag{8.136}$$

$$S\Delta x + X\Delta s = \mu I - Xs \tag{8.137}$$

The solution of the above linear systems can be obtained as follows. First, compute the $\Delta s$ and $\Delta v$.

$$\Delta v = -v - Z^{-1}V\Delta z + \mu Z^{-1}I \tag{8.138}$$

$$\Delta s = -s - X^{-1}S\Delta x + \mu X^{-1}I \tag{8.139}$$

Then substitute the above two equations in equation (8.135) to get the augmented system

$$\begin{bmatrix} -D_x & B^T \\ B & Z^{-1}V \end{bmatrix} \begin{bmatrix} \Delta x \\ \Delta z \end{bmatrix} = \begin{bmatrix} g - B^T z - \mu X^{-1}I \\ \mu Z^{-1}I - d \end{bmatrix} \tag{8.140}$$

where

$$D_x = W + X^{-1}S \tag{8.141}$$

Solving the above equation, we get $\Delta z$ as bellow.

$$\Delta z = -V^{-1}ZB\Delta x + V^{-1}(\mu I - Zd) \tag{8.142}$$

The solution $\Delta x$ can be obtained by solving the following normal system.

$$-K\Delta x = h \tag{8.143}$$

where

$$K = D_x + B^T V^{-1}ZB \tag{8.144}$$

$$h = g - B^T z + B^T V^{-1}(Zd - \mu I) - \mu X^{-1}I \tag{8.145}$$

*Calculation of the Step Length*   It should be noticed that if started far from a solution (or the start point is not good), the primal–dual IP methods may fail to converge to a solution [31–39]. For this reason, primal–dual methods usually use a merit function in order to induce convergence. There are, however, problems associated with the merit function, particularly with the choice of the penalty parameter [66]. The filter technique [42] may be used to handle the convergence issue.

There are two competing aims in the primal–dual solution of equation (8.117). The first aim is to minimize the objective, and the second is the satisfaction of the constraints. These two conflicting aims can be written as

$$\min \ f(x) \tag{8.146}$$

s.t.

$$\min \ \delta = (d - v)^2 \tag{8.147}$$

A merit function usually combines equations (8.146) and (8.147) into a single objective. Instead, we see equations (8.146) and (8.147) as two separate objectives, similar to multi-objective optimization. However, the situation here is different as it is essential to find a point where $d = v$ if possible. In this sense, the second objective has priority. Nevertheless, we will make use of the principle of domination from multi-objective programming in order to introduce the concept of the filter.

***Definition 1 [66]*** A pair $(f^k, \delta^k)$ is said to dominate another pair $(f^j, \delta^j)$ if and only if $f^k \leq f^j$, and $\delta^k \leq \delta^j$

In the context of the primal–dual method, this implies that the $k$th iterate is at least as good as the $j$th iterate with respect to equations (8.146) and (8.147). Next, we define the filter which will be used in the line search to accept or reject a step.

***Definition 2 [66]*** A filter is a list of pairs $(f^j, \delta^j)$ such that no pair dominates any other. A point $(f^k, \delta^k)$ is said to be accepted for inclusion in the filter if it is not dominated by any point in the filter.

The filter therefore accepts any point that either improves optimality or infeasibility.

In most primal–dual methods, separate step lengths are used for the primal and dual variables [67]. A standard ratio test is used to ensure that nonnegative variables remain nonnegative

$$\alpha_P = \min\{\alpha_x, \alpha_v\} \tag{8.148}$$

$$\alpha_D = \min\{\alpha_z, \alpha_s\} \tag{8.149}$$

where

$$\alpha_j = \min\left\{1, \quad 0.9995 \times \min\left\{\frac{\omega_j}{-\Delta\omega_j}, \quad \text{if } \Delta\omega_j < 0\right\}\right\}$$

$$\omega = x, v, z, s \tag{8.150}$$

The step lengths in the above are successively halved until the following iteration becomes acceptable to the filter.

$$x' = x + \alpha_P \Delta x \tag{8.151}$$

$$v' = v + \alpha_P \Delta v \tag{8.152}$$

$$z' = z + \alpha_D \Delta z \tag{8.153}$$

$$s' = s + \alpha_D \Delta s \tag{8.154}$$

*Selection of the Barrier Parameter* Another important issue in the primal–dual method is the choice of the barrier parameter. Many methods are based on approximate complementarity where the centering parameter is fixed *a priori* [68]. Mehrotra [69] suggested a scheme for linear programming in which the barrier parameter is estimated dynamically during iteration. The heuristic originally proposed in [69] may be used. First, the Newton equations system is solved with the barrier $\mu$ set to zero. The direction obtained in this case, $(\Delta x^\alpha, \Delta v^\alpha, \Delta z^\alpha, \Delta s^\alpha)$, is called the *affine-scaling direction*. The barrier parameter is estimated dynamically from the estimated reduction in the complementarity gap along the affine-scaling direction.

$$\mu = \left(\frac{g^\alpha}{z^T v + s^T x}\right)^2 \left(\frac{z^T v + s^T x}{m + n}\right) \tag{8.155}$$

where

$$g^\alpha = (z + \alpha_D^\alpha \Delta z^\alpha)^T (v + \alpha_P^\alpha \Delta v^\alpha) + (s + \alpha_D^\alpha \Delta s^\alpha)^T (x + \alpha_P^\alpha \Delta x^\alpha) \tag{8.156}$$

The step lengths in the affine-scaling direction are obtained using equations (8.155) and (8.156). To avoid numerical instability, the above equation is used to compute $\mu$ when the absolute complementarity gap $z^T v + s^T x \geq 1$. But if $z^T v + s^T x \leq 1$, we use following equation to compute $\mu$, that is,

$$\mu = \left(\frac{1}{m+n}\right)^2 \left(\frac{z^T v + s^T x}{m+n}\right) \tag{8.157}$$

**The Improved Quadratic Interior Point Method**  The OPF model discussed in this section is a nonlinear mathematical programming problem. It can be reduced by an elimination procedure. The reduction of the OPF model is based on the linearized load flow around the base load flow solution for a small perturbation. The reduced OPF model can be expressed as

$$\min F = \frac{1}{2}X^T Q X + G^T X + C \tag{8.158}$$

such that

$$AX = B$$

$$X \geq 0 \tag{8.159}$$

Equation (8.158) is a scalar objective function which corresponds to the objective functions of OPF. Equation (8.159) corresponds to constraints (8.103)–(8.116) with linearization handling. $X$ in (8.158) and (8.159) is a vector of controllable variables, which is defined as $X = [V_g^T, T^T, P_g^T]^T$ in economic dispatch, $X = [V_g^T, T^T, q_c^T, q_r^T, P_L^T]^T$ in VAR planning, or $X = [V_g^T, T^T, P_L^T]^T$ in loss minimization.

The model (8.158)–(8.159) has a quadratic objective function subject to the linear constraints that satisfy the basic requirements of the quadratic interior point (QIP) scheme. The barrierlike IP methods discussed in previous section and the enhanced projection method used in QIP have the enough speed and accuracy to solve OPF problems such as economic dispatch, loss minimization and VAR optimization. However, the effectiveness of these IP methods depends on a good starting point [27]. The IQIP) is presented in this section. It features a general starting point (rather than a good point) and faster convergence. The calculation steps of IQIP are as follows.

S1: Given a starting point $X_1$
S2: $X_1 := A X_1$
S3: $\Delta := B - A X_1$
S4: $\Delta\text{max} := \max |\Delta i|$
S5: If $\Delta \max < \varepsilon_0$, go to S10. Otherwise, go to the next step.
S6: $U := [A_1(A_1 A_1^T)^{-1}]\Delta$
S7: $R := \min \{Ui\}$

S8: If $R + 1 \geq 0$,   $X_1 := X_1 \times (1 + U)$, go to S3. Otherwise, go to the next step.

S9: $QB := -1/R$, $X_1 := X_1^*(1 + QB^*U)$, go to S3.

S10: $D_k := \text{diag}[x_1, x_2, \ldots, x_n]$

S11: $B_k := AD_k$

S12: $dp^k := [B_k^T(B_kB_k^T)^{-1}B_k - 1]D_k[QX^k + G]$

S13: $\beta_1 := -\frac{1}{\gamma}$, $\gamma < 0$; $\beta_1 := 10^6$, $\gamma \geq 0$ where $\gamma = \min[dp_j^k]$

S14:
$\beta_2 := \frac{(dp^k)^T(dp^k)}{W}$, if $W > 0$; $\beta_2 := 10^6$, if $W \leq 0$

where $W = (D_kdp^k)^TQ(D_kdp^k)$

S15:
$X^{k+1} := X^k + \alpha(\beta D_kdp^k)$,

where $\beta = \min[\beta_1, \beta_2]$; $\alpha(< 0)$ is a variable step.

Set $k := k + 1$, and go to S11. End when $dp^k < m$, where $k$ is the iteration counter.

The partitioning scheme and optimization modules are adopted here. The partitioning scheme provides the objective function and the optimizable area. The optimization module selects the default constraints for the selected objective unless other specified. The user can add or remove constraints from the default constraint set (equations (8.103)–(8.116)). The optimization is carried out using the IQIP method described earlier. The nonlinear constraints are handled via successive linearization in conjunction with an area power flow.

IQIP handles the initial value of the state variables before optimization so that it can solve the bad initial conditions encountered in other IP methods. Consequently, IQIP has a faster convergence speed than other IP methods. IQIP achieves an optimum in the linearized space, while the power flow adjusts for the approximation caused by the linearization. The check of the power flow mismatch should be performed in the optimization area first. In this way, the optimization calculation accuracy will be increased. It ensures local optimization with all violations removed. Then the check of the power flow mismatch will be performed in the whole system including the external areas, which adjusts the changes in the boundary injections caused by the local optimization. The overall scheme ensures a local optimum, with no violation in the optimized area, while satisfying a global power flow. The local optimum will be the global optimum if there is only one area in the system.

If the region formed by the constraints is very narrow, the solution may be declared infeasible. Three options are available for infeasibility handling. They are

(1) The bounds option, which allows the program to widen the bounds on violating soft constraints. The new limits or a percentage increase/decrease from the current limits can be prespecified by the user for all objective functions.

(2) The VAR option I, which allows the program to add new VAR sites at buses with big contributions to improving system performance (only for VAR optimization).

(3) The VAR option II, which allows the program to add new VAR sites at buses with severe voltage violations (only for VAR optimization).

For economic dispatch or loss minimization, if infeasibility is detected, the bounds option is selected. The bounds on violating constraints are widened accordingly. For VAR optimization, or planning, if infeasibility is detected, the VAR option I is first selected, and the new VAR sites are added at buses with big contributions to improve system performance such as reducing system loss or voltage violations. If further infeasibilities occur, the VAR option II is selected, and other new VAR sites are added at buses with severe voltage violations.

***Simulation Calculations***   The simulation examples are taken from reference [27]. The two IP–based OPF methods are tested on an IEEE 14-bus system, and a modified IEEE 30-bus systems. One is the EQIP and the other, the IQIP. For comparison, the solution method of MINOS is also used to solve the OPF problem with the same data and same conditions. MINOS is a Fortran-based optimization package developed by Stanford University, which is designed to solve large-scale optimization problems. The solution method in the MINOS program is a reduced gradient algorithm or a projected augmented Lagrange algorithm.

The data and parameters of the 14-bus system were shown in Chapter 3. The optimization data used for simulating the IEEE 14-bus system using the three objective functions are given in Tables 8.5–8.7.

**TABLE 8.5   Generator Data for 14-Bus System (p.u.)**

| Unit No. | $a$ | $b$ | $c$ | $P_{gimin}$ | $P_{gimax}$ |
|---|---|---|---|---|---|
| 1 | 0.0784 | 0.1350 | 0.0000 | 0.0000 | 3.0000 |
| 2 | 0.0834 | 0.2250 | 0.0000 | 0.0000 | 1.3000 |
| 6 | 0.0875 | 0.1850 | 0.0000 | 0.2000 | 2.0000 |

**TABLE 8.6   Capacitive VAR Data for 14-Bus System (p.u.)**

| VAR Site Bus | Fixed Unit Cost | Variable Unit Cost | Max. Capacitive VAR | Max. Inductive VAR |
|---|---|---|---|---|
| 5 | 2.3500 | 0.1500 | 0.8000 | 0.0000 |
| 9 | 3.4500 | 0.2000 | 0.8000 | 0.0000 |
| 13 | 3.4500 | 0.2000 | 0.8000 | 0.0000 |

**TABLE 8.7   Inductive VAR Data for 14-Bus System (p.u.)**

| VAR Site Bus | Fixed Unit Cost | Variable Unit Cost | Max. Capacitive Var | Max. Inductive VAR |
|---|---|---|---|---|
| 5 | 6.0000 | 0.2500 | 0.4000 | 0.0000 |
| 9 | 6.0000 | 0.2500 | 0.4000 | 0.0000 |
| 13 | 6.0000 | 0.2500 | 0.4000 | 0.0000 |

Table 8.5 represents the generator data used for the IEEE 14-bus system. Tables 8.6 and 8.7 represent the capacitor and inductor VAR allocation data of the IEEE 14-bus system, respectively.

In the following calculations, optimization iteration will be stopped when the difference in the objective value $\Delta F$ is less than $\varepsilon$ ($\varepsilon = 10^{-6}$).

*Sample Set of Results Using IQIP/EQIP/MINOS Options (Minimization of the Total Generation Cost as Objective Function)* Three test cases are given here for the 14-bus system for OPF with minimization of generation cost as the objective function (i.e., objective 1 in the OPF model in Section 8.5.2). The initial values of real power for three cases are different as shown in Table 8.8. The comparisons of results for the three test cases using IQIP/EQIP/MINOS methods are listed in Tables 8.9–8.11.

**TABLE 8.8  Three Test Cases for OPF Objective 1**

| Initial Value | Case 1 | Case 2 | Case 3 |
|---|---|---|---|
| $P_{G1}$ | 0.0000 | 0.0000 | 0.0000 |
| $P_{G2}$ | 0.4000 | 0.3500 | 0.0000 |
| $P_{G6}$ | 0.7000 | 0.7000 | 0.7000 |
| $V_{G1}$ | 1.0500 | 1.0500 | 1.0500 |
| $V_{G2}$ | 1.0450 | 1.0450 | 1.0450 |
| $V_{G6}$ | 1.0500 | 1.0500 | 1.0500 |

**TABLE 8.9  Optimization Results and Comparison for Case 1 (p.u.)**

| Control Option | IQIP | EQIP | MINOS |
|---|---|---|---|
| $P_{G1}$ | 1.53414 | 2.18319 | – |
| $P_{G2}$ | 0.93357 | 0.34326 | – |
| $P_{G6}$ | 0.38141 | 0.35392 | – |
| $V_{G1}$ | 1.05000 | 1.05000 | – |
| $V_{G2}$ | 1.04997 | 1.04683 | – |
| $V_{G6}$ | 1.05000 | 1.05000 | – |
| $T_{4-7}$ | 0.98454 | 0.97513 | – |
| $T_{4-9}$ | 1.01278 | 0.98307 | – |
| $T_{5-6}$ | 0.98454 | 0.94992 | – |
| Total $P_G$ | 2.84912 | 2.88037 | – |
| Power loss | 0.10912 | 0.14037 | – |
| Total $P_G$ cost | 0.757856 | 0.827207 | – |
| Objective value | 0.757856 | 0.827207 | – |
| PF mismatch | 0.1402E-6 | 0.4370E-4 | – |
| Iteration no. | 12 | 26 | – |
| CPU time (s) | 30.0 | 252.9 | No convergence |

**TABLE 8.10 Optimization Results and Comparison for Case 2 (p.u.)**

| Control Option | IQIP | EQIP | MINOS |
|---|---|---|---|
| $P_{G1}$ | 1.65313 | 2.21476 | – |
| $P_{G2}$ | 0.84114 | 0.31538 | – |
| $P_{G6}$ | 0.35920 | 0.35192 | – |
| $V_{G1}$ | 1.05000 | 1.05000 | – |
| $V_{G2}$ | 1.04997 | 1.04588 | – |
| $V_{G6}$ | 1.04996 | 1.05000 | – |
| $T_{4-7}$ | 0.98208 | 0.97525 | – |
| $T_{4-9}$ | 1.01269 | 0.98293 | – |
| $T_{5-6}$ | 0.98853 | 0.94962 | – |
| Total $P_G$ | 2.85347 | 2.88206 | – |
| Power loss | 0.11347 | 0.14206 | – |
| Total $P_G$ cost | 0.7632329 | 0.8340057 | – |
| Objective value | 0.7632329 | 0.8340057 | – |
| PF mismatch | 0.1866E-4 | 0.4357E-4 | – |
| Iteration no. | 12 | 26 | – |
| CPU time (s) | 30.2 | 253.8 | No convergence |

**TABLE 8.11 Optimization Results and Comparison for Case 3 (p.u.)**

| Control Option | IQIP | EQIP | MINOS |
|---|---|---|---|
| $P_{G1}$ | 1.55607 | 1.58973 | – |
| $P_{G2}$ | 0.93372 | 0.88235 | – |
| $P_{G6}$ | 0.36034 | 0.37895 | – |
| $V_{G1}$ | 1.05000 | 1.05000 | – |
| $V_{G2}$ | 1.04993 | 1.05000 | – |
| $V_{G6}$ | 1.04956 | 1.04987 | – |
| $T_{4-7}$ | 1.00047 | 0.99398 | – |
| $T_{4-9}$ | 1.00715 | 1.01298 | – |
| $T_{5-6}$ | 0.99392 | 0.97887 | – |
| Total $P_G$ | 2.85319 | 2.85100 | – |
| Power loss | 0.11319 | 0.11100 | – |
| Total $P_G$ cost | 0.760950 | 0.758355 | – |
| Objective value | 0.760950 | 0.758355 | – |
| PF mismatch | 0.9630E-6 | 0.1622E-4 | – |
| Iteration no. | 3 | 11 | – |
| CPU time (s) | 21.3 | 35.9 | No convergence |

It can be observed from Tables 8.9–8.11 that the MINOS method cannot converge for these test cases, while the other two methods evaluated the optimization solutions. The improved IQIP method has high accuracy, fewer iteration numbers, and fast calculation speed compared with OPF based on the EQIP method. The maximum speed ratio between IQIP and EQIP can reach 1:8 (See Table 8.9 and Table 8.10). If the initial starting point is good (as in case 3), the OPF based on the EQIP method has the fastest convergence speed but the convergence speed is still slower than that of IQIP-based OPF. Meanwhile, for the same iteration number, the objective value obtained by IQIP is less than that by EQIP. Therefore, the improved IQIP method is superior to the EQIP method. It features a general starting point and fast convergence.

Since the MINOS program cannot converge under specific operating conditions and constraints, the other test case, the 30-bus system, is used to further demonstrate the effectiveness of the IQIP method. The data and parameters of the 30-bus system are taken from reference [3]. The optimization results and comparison for IQIP/EQIP/MINOS methods are listed in Table 8.12. It can be observed that the

**TABLE 8.12    Optimization Results and Comparison for IEEE 30-Bus System (p.u.)**

| Control Option | IQIP | EQIP | MINOS |
|---|---|---|---|
| $P_{G1}$ | 0.73357 | 0.73921 | 0.75985 |
| $P_{G2}$ | 0.59838 | 0.59999 | 0.38772 |
| $P_{G5}$ | 0.61117 | 0.61412 | 0.66590 |
| $P_{G11}$ | 0.58787 | 0.57562 | 0.60000 |
| $P_{G13}$ | 0.34092 | 0.34321 | 0.40355 |
| $V_{G1}$ | 1.05000 | 1.05000 | 1.05000 |
| $V_{G2}$ | 1.04999 | 1.05000 | 1.03984 |
| $V_{G5}$ | 1.04998 | 1.05000 | 1.01709 |
| $V_{G11}$ | 1.04867 | 1.04915 | 1.05000 |
| $V_{G13}$ | 1.05000 | 1.05000 | 1.05000 |
| $T_{6-9}$ | 1.05160 | 1.08149 | 1.05461 |
| $T_{6-10}$ | 1.07615 | 1.01465 | 0.92151 |
| $T_{4-12}$ | 1.06768 | 1.09528 | 1.03377 |
| $T_{28-27}$ | 0.97443 | 0.94345 | 0.97217 |
| Total $P_G$ | 2.87190 | 2.87215 | 2.87120 |
| Power loss | 0.03790 | 0.03815 | 0.03720 |
| Total $P_G$ cost | 0.657582 | 0.658195 | 0.657258 |
| Objective value | 0.657582 | 0.658195 | 0.657258 |
| PF mismatch | 0.9447E-6 | 0.3988E-4 | 0.5734E-7 |
| Iteration no. | 7 | 12 | 9 |
| CPU time (s) | 147.0 | 267.4 | 567.9 |

**TABLE 8.13   Initial Voltages on Load Bus for 14-Bus System (p.u.)**

| Bus No. | Initial $V$ | $V_{min}$ | $V_{max}$ |
|---------|-------------|-----------|-----------|
| 3 | 0.94410 | 0.95000 | 1.05000 |
| 5 | 0.99220 | 0.95000 | 1.05000 |
| 7 | 0.94250 | 0.95000 | 1.05000 |
| 8 | 0.93270 | 0.95000 | 1.05000 |
| 9 | 0.93330 | 0.95000 | 1.05000 |
| 10 | 0.93910 | 0.95000 | 1.05000 |
| 13 | 0.98720 | 0.95000 | 1.05000 |
| 14 | 0.93530 | 0.95000 | 1.05000 |

proposed IQIP method has the fastest convergence speed, followed by the EQIP method. The MINOS method has the slowest convergence speed.

*Sample Set of Results Using IQIP/EQIP/MINOS Options (VAR Optimal Placement as Objective Function)*   The test case given here is for the 14-bus system for OPF with VAR optimal placement as the objective function (i.e., objective 2 in the OPF model in Section 8.5.2). The initial voltages on load buses are shown in Table 8.13. The optimization results and comparisons for the IQIP/EQIP/MINOS methods are listed in Table 8.14.

It is observed from Table 8.14 that both IQIP and EQIP have almost the same optimization results, which are better than those obtained from the MINOS method. The comparison of the results shows that the three methods alleviate the voltage violations satisfactorily. The convergence speed of the IQIP method ranks first, followed by EQIP method. The MINOS method ranks last.

*Sample Set of Results Using IQIP/EQIP/MINOS Options (Loss Minimization as Objective Function)*   The test case given here is for the 14-bus system for OPF with loss minimization as the objective function (i.e., objective 3 in the OPF model in Section 8.5.2). The optimization results and comparison for loss minimization using IQIP/EQIP/MINOS methods are listed in Table 8.15.

From Table 8.15, it can be seen that IQIP and EQIP have almost the same optimization results for the loss minimization objective. In view of loss reduction, load voltage modification, and convergence speed, both IQIP and EQIP methods appear superior to the MINOS method. Similarly, the IQIP method has the fastest convergence speed for loss minimization.

## 8.6   OPF WITH PHASE SHIFTER

The problem of power system security has received considerable attention in the deregulated power industry. To meet the load demand in a power system and satisfy

TABLE 8.14    Optimization Results and Comparison for Objective 2 (p.u.)

| Control Option | IQIP | EQIP | MINOS |
|---|---|---|---|
| $V_{G1}$ | 1.05000 | 1.05000 | 1.05000 |
| $V_{G2}$ | 1.05000 | 1.05000 | 1.04248 |
| $V_{G6}$ | 1.05000 | 1.05000 | 1.04430 |
| $T_{4-7}$ | 0.97001 | 0.97000 | 0.97000 |
| $T_{4-9}$ | 0.96001 | 0.96001 | 0.96000 |
| $T_{5-6}$ | 1.03000 | 1.03000 | 0.93000 |
| $V_{D3}$ | 0.98340 | 0.98340 | 0.97610 |
| VD5 | 1.02600 | 1.02600 | 1.02030 |
| VD7 | 1.00200 | 1.00200 | 0.99530 |
| VD8 | 0.99270 | 0.99280 | 0.98600 |
| VD9 | 0.98970 | 0.98970 | 0.98300 |
| VD10 | 0.99130 | 0.99130 | 0.98470 |
| VD13 | 1.02180 | 1.02180 | 1.01580 |
| VD14 | 0.98320 | 0.98320 | 0.97670 |
| Power loss | 0.110866 | 0.110868 | 0.110459 |
| Objective value | 0.110866 | 0.110868 | 0.110459 |
| PF mismatch | 0.1596E-6 | 0.4634E-8 | 0.4225E-6 |
| Iteration no. | 4 | 4 | 8 |
| CPU time (s) | 115.9 | 150.4 | 184.4 |

the stability and reliability criteria, either the existing transmission lines must be utilized more efficiently, or new line(s) should be added to the system. The latter is often impractical. The reason is that building a new power transmission line is in many countries a very time-consuming process and sometimes an impossible task, because of environmental problems. Therefore, the first alternative provides an economically and technically attractive solution to the power system security problem by use of some efficient controls, such as controllable series capacitors, phase shifters, load shedding, and so on. This chapter introduces power system security enhancement through OPF with a phase shifter. The objective functions of OPF include minimum line overloads and minimum adjustment of the number of phase shifters. It is noted that general OPF calculations are hourly based and the control variables of OPF are continuous. However, the calculations of phase shifters are daily based. The control variables associated with phase shifter transformers are discrete. To solve this problem, a rule-based OPF with a phase shifter scheme can be adopted for practical system operations [25].

**TABLE 8.15  Optimization Results and Comparison for Loss Minimization (p.u.)**

| Control Option | IQIP | EQIP | MINOS |
|---|---|---|---|
| VG1 | 1.05000 | 1.05000 | 1.05000 |
| VG2 | 1.05000 | 1.05000 | 1.02837 |
| VG6 | 1.05000 | 1.05000 | 1.03330 |
| $T_{4-7}$ | 0.97001 | 0.97001 | 0.97000 |
| $T_{4-9}$ | 0.96001 | 0.96001 | 0.96000 |
| $T_{5-6}$ | 1.03000 | 1.02999 | 1.03000 |
| VD5 | 1.02600 | 1.02600 | 1.00930 |
| VD9 | 0.98970 | 0.98970 | 0.97040 |
| VD13 | 1.02180 | 1.02180 | 1.00430 |
| Initial loss | 0.1164598 | 0.1164598 | 0.1164598 |
| Final loss | 0.1108663 | 0.1108664 | 0.1118670 |
| Objective value | 0.1108663 | 0.1108664 | 0.1118670 |
| PF mismatch | 0.4132E−6 | 0.4634E−8 | 0.4339E−6 |
| Iteration no. | 3 | 3 | 8 |
| CPU time (s) | 22.2 | 27.0 | 70.7 |

### 8.6.1  Phase Shifter Model

A phase shifter model can be represented by an equivalent circuit, which is shown in Figure 8.2(a). It consists of an admittance in series with an ideal transformer having a complex turn ratio $k\angle\phi$.

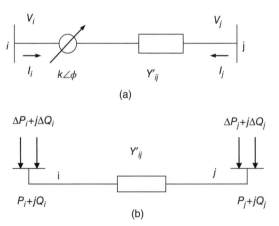

Figure 8.2  Phase shifter model.

The mathematical model of the phase shifter can be derived from Figure 8.2(a), that is,

$$\begin{bmatrix} \widetilde{I}_i \\ \widetilde{I}_j \end{bmatrix} = \begin{bmatrix} Y'_{ij} + Y_i & -Y'_{ij} \\ -Y'_{ij} & Y'_{ij} + Y_j \end{bmatrix} \begin{bmatrix} \widetilde{V}_i \\ \widetilde{V}_j \end{bmatrix} \tag{8.160}$$

where

$$Y_i = Y'_{ij} \left[ \frac{1}{k^2} - 1 + \left( 1 - \frac{1}{k\angle(-\phi)} \right) \frac{V_j}{V_i} \right] \tag{8.161}$$

$$Y_j = Y'_{ij} \left[ \left( 1 - \frac{1}{k\angle\phi} \right) \frac{V_i}{V_j} \right] \tag{8.162}$$

It is seen from equation (8.160) that the mathematical model of the phase shifter makes the Y bus unsymmetrical. To make the Y bus symmetrical, the phase shifter can be simulated by installing additional injections at the buses. The additional injections can be simplified as follows.

$$\Delta P_i = |V_i||V_j|B'_{ij} \ \cos(\theta_i - \theta_j)\sin\phi_{ij}$$

$$\Delta P_j = -|V_i||V_j|B'_{ij} \cos(\theta_i - \theta_j)\sin\phi_{ij}$$

$$\Delta Q_i = |V_i||V_j|B_{ij} \sin(\theta_i - \theta_j)\sin\phi_{ij}$$

$$\Delta Q_j = |V_i||V_j|B'_{ij} \sin(\theta_i - \theta_j)\sin\phi_{ij}$$

where

$I_i, P_i$: current and real power flow at bus $i$
$I_j, P_j$: current and real power flow at bus $j$
$Q_i$: reactive power at bus $i$
$Q_j$: reactive power at bus $j$
$V_i\angle\theta_i$: complex voltage at bus $i$
$V_j\angle\theta_j$: complex voltage at bus $j$
$k\angle\phi$: complex turn ratio of the phase shifter
$Y'_{ij} = G'_{ij} + jB'_{ij}$: series admittance of the line $ij$.

Therefore, the phase shifter model can be simulated by increasing the injections at the terminal buses as shown in Figure 8.2(b).

## 8.6.2 Rule-Based OPF with Phase Shifter Scheme

### OPF Formulation with Phase Shifter

*Objective Functions*   As a result of installation of the phase shifter, the system will have lots of benefits such as overload release, system loss reduction, generation cost reduction, generation adjustment reduction, and so on. All these benefits may be selected as objective functions for OPF with a phase shifter. However, the primary purpose of installing a phase shifter is to remove the line overload. Thus the minimal line overload is selected as the primary objective function. In addition, as the adjustment numbers of phase shifters are limited in practical systems, the minimal adjustment number of phase shifters is also selected as the objective function. Two objective functions are given as follows.

**(1)** Minimal line overloads

$$\min \ F_o = \sum_{ij=1}^{NB} (P_{ij}(t) - P_{ij\max})^2 \tag{8.163}$$

where

$F_o$: the overload objective function

$P_{ij}(t)$: the overload flow on transmission line $ij$ at time stage $t$;

$P_{ij\max}$: transmission limit of line $ij$;

$NB$: set of overload lines.

**(2)** Minimal adjustment number of phase shifters

$$\min \ F_\phi = \sum_{i=1}^{NS} W_i \phi_i \tag{8.164}$$

where

$F_\phi$: phase shifter adjustment objective function

$\phi_i$: the angle of the phase shifter transformer

$W_i$: priority coefficient of the phase shifter transformers

$NS$: set of phase shifter transformers.

$NG$: set of generators.

*Constraints*   In addition to the general linear/nonlinear constraints, the constraints relating to phase shifter variables such as phase shifter angle and maximal adjustment numbers should be included in the OPF formulation with phase shifter. The candidate constraints are as follows:

Constraint 1:   Real power flow equation

Constraint 2:   Reactive power flow equation

Constraint 3:    Upper and lower limits of real power output of the generators

Constraint 4:    Upper and lower limits of reactive power output of the generators

Constraint 5:    Upper and lower limits of node voltages

Constraint 6:    Available transfer capacity of the transmission lines

Constraint 7:    Upper and lower limits of transformer taps

Constraint 8:    Upper and lower limits of phase shifter taps

Constraint 9:    Maximal adjustment times of phase shifters per day.

It is noted that constraints 8 and 9 are the phase shifter constraints that were used in the rule-based search technique, and the limits of all control and state variables are determined for the specific system under study.

The above-mentioned OPF model with phase shifter is a nonlinear mathematical programming problem. It can be reduced by an elimination procedure and solved by the IQIP method, which was introduced in the previous section.

**Rule-Based Scheme**    To determine the best location for installing the phase shifter, sensitivity analysis is adopted. The formulation of sensitivity analysis of the objective function with respect to the phase shifter variable can be expressed as follows.

$$S_{F-\phi} = \frac{\partial F_o}{\partial \phi_i} = \frac{F_o(0) - F_o(\phi_i)}{|\Delta\phi_i|} \tag{8.165}$$

where

$F_o(0)$: the total line overload before phase shifter $i$ is installed

$F_o(\phi_i)$: the total line overload after phase shifter $i$ is installed.

In equation (8.165), the value of sensitivity $S_{F-\phi}$ will be greater than zero if power violation is reduced by use of the phase shifter, that is, $F_o(\phi_i) < F_o(0)$. Obviously, if phase shifter $i$ is not helpful in alleviating line overload, $F_o(\phi_i) \geq F_o(0)$. In this case, we define the value of the sensitivity $S_{F-\phi} = 0$.

In the rule-based system, the following rules are defined.

Rule 1:    If the system operates in the normal state without load change, then none of the existing phase shifters will change tap.

Rule 2:    If the system load increases or the system operates in contingency state, then judge:

If no line overload appeared, then none of the existing phase shifters will change tap.

If line overload occurred in system, then go to rule 3 to adjust the tap of some phase shifters.

Rule 3:    If the phase shifter leads to maximal overload reduce at time stage $t$, then this phase shifter will be recommended at this time.

Rule 4: If phase shifters $i$ and $j$ lead to same overload reduce at time stage $t$, then check the other benefits:

If phase shifter $i$ make less generation cost benefit than phase shifter $j$, then phase shifter $j$ will be recommended at this time.

If phase shifter $i$ make less system loss benefit than phase shifter $j$, then phase shifter $j$ will be recommended at this time.

Rule 5: Phase shifter $i$ is recommended and the line overloads are still exist, then the next priority phase shifter in the rank will be joined to remove the violations until no phase shifters are available.

Rule 6: If OPF suggests a solution, and RBS confirms that phase shifter constraints are met, then the problem is solved at this time stage.

Rule 7: If RBS checks the OPF solution and the OPF solution violates the phase shifter constraints, then freeze the corresponding tap of the phase shifter.

Rule 8: If RBS checks the state of the phase shifters and phase shifter $k$ has a frozen tap, then phase shifter $k$ will be out of service in the subsequent time stages.

A phase shifter tap will be frozen when the tap number of the phase shifter at time reaches its maximum. The IQIP algorithm then uses the fixed or scheduled tap value for the phase shifter that is determined by the rule based engine. The solution steps of the integrated algorithm for OPF with phase shifter are as follows.

Step 1: Assume several contingencies.

Step 2: OPF calculation without phase shifter for each given contingency from time stage $t$ ($t = 1$, first time stage).

Step 3: Judge whether the OPF is solvable. If the answer is "Yes," there is no need to use a phase shifter. If "No," go to step 4.

Step 4: Contingency analysis through power flow calculation. Check the overload state of lines.

Step 5: Conduct a sensitivity analysis for obtaining a list of phase shifter ranking according to the amount of releasing the line overload for each phase shifter. Then decide the corresponding weighting factor.

Step 6: OPF calculation with the available phase shifter.

Step 7: Use the rule based method to check the operation limitation of the phase shifter. Calculate the operation times, $NM\phi_i = NM\phi_i + 1$, if the phase shifter $i$ is operated in this time stage.

Step 8: If $NM\alpha_i(t) = NM\phi_{imax}$, freeze the corresponding taps of the phase shifter. That is, this phase shifter will be out of service in subsequent time.

Step 9: Check the time stages. If $t = t_{max}$ (e.g., 24 h), stop. Otherwise, $t = t + 1$ and go to step 2.

Finally, in the search technique, the phase shifters are adjusted sequentially and their direction of adjustments are governed by the impact on the primary objective function of minimal line overload. The engineering rules are such that the least number of phase shifters are adjusted at a time, provided that they have the greatest impact in reducing the line flow overloads. The phase shifter constraints, which are handled by the rule-based search technique, are adjusted to produce discrete settings and in turn pass on to the IQIP module of the algorithm.

***Example 8.3:***    The integrated scheme of OPF with phase shifter is tested on the IEEE 30-bus system. The data and parameters of the 30-bus system are the same as in the previous section, and the limits of the installed phase shifters were taken as 10° [25].

The total system load of the IEEE 30-bus system is 283.4 MW. The corresponding load-scaling factor (LSF) is 1.0. The daily load demands of the IEEE 30-bus system are shown in Table 8.16. To determine the degree of line violations at the line $L_{i-j}$, the following performance index is defined [25].

$$PI_{ij} = \frac{P_{ij} - P_{ijmax}}{P_{ijmax}}, \quad ij \in \text{NOL} \tag{8.166}$$

where

$PI_{ij}$: the performance index of line overloads
$P_{ij}$: the overload flow on transmission line
NOL: the set of overloaded lines.

Through power flow analysis for each time stage, line overloads only appeared at hours 8, 15, 16, 17, 18, and 19, which are peak load periods. The violation amounts of line flow for each time stage are summarized in Table 8.17.

The line overloads will become more serious if system contingency scenarios are considered. Therefore, OPF with phase shifter adjustment should be employed for enhancing power system security.

**TABLE 8.16    Daily Load Curve for IEEE 30-Bus System**

| Time (h) | Load (LSF) | Time Stage | Load (LSF) | Time Stage | Load (LSF) |
|---|---|---|---|---|---|
| 1 | 0.90 | 9 | 1.30 | 17 | 1.50 |
| 2 | 0.96 | 10 | 1.15 | 18 | 1.55 |
| 3 | 1.00 | 11 | 1.10 | 19 | 1.40 |
| 4 | 1.05 | 12 | 1.05 | 20 | 1.20 |
| 5 | 1.10 | 13 | 1.16 | 21 | 1.12 |
| 6 | 1.15 | 14 | 1.30 | 22 | 1.03 |
| 7 | 1.30 | 15 | 1.40 | 23 | 0.96 |
| 8 | 1.40 | 16 | 1.45 | 24 | 0.90 |

**TABLE 8.17 Total Power Flow Violation Without Contingency**

| Time (h) | Overload (MW) | Time (h) | Overload (MW) | Time (h) | Overload (MW) |
|---|---|---|---|---|---|
| 1–7 | 0.00 | 15 | 5.12 | 18 | 13.08 |
| 8 | 5.12 | 16 | 6.78 | 19 | 5.12 |
| 9–14 | 0.00 | 17 | 9.62 | 20–24 | 0.00 |

**TABLE 8.18 Summary of Contingency Analysis**

| Outage line | $L_{12-14}$ | $L_{10-21}$ | $L_{22-25}$ | $L_{24-27}$ | $L_{29-30}$ |
|---|---|---|---|---|---|
| Overloaded lines | $L_{1-2}$ | $L_{1-2}$ | $L_{1-2}$ | $L_{1-2}$ | $L_{1-2}$ |
| | $L_{6-8}$ | $L_{6-8}$ | $L_{6-8}$ | $L_{6-8}$ | $L_{6-8}$ |
| | $L_{9-10}$ | $L_{9-11}$ | $L_{9-11}$ | $L_{9-10}$ | $L_{9-10}$ |
| | $L_{9-11}$ | $L_{10-20}$ | $L_{10-20}$ | $L_{9-11}$ | $L_{9-11}$ |
| | | | | $L_{10-21}$ | $L_{27-30}$ |
| Overloaded time stage | T8 | T7–9 | T8 | T8 | T8 |
| | T15–T19 | T14–T20 | T15–T19 | T15–T19 | T15–T19 |
| Total line MW violation | 50.68 | 102.76 | 52.73 | 57.18 | 50.53 |

For the purpose of simulation, the following line contingency scenarios are given, that is, $L_{12-14}$, $L_{10-21}$, $L_{22-25}$, $L_{24-27}$, and $L_{29-30}$.

Table 8.18 provides the summary of contingency analysis and shows the total power violations for all time stages. It can be observed from Table 8.18 that the line $L_{10-21}$ outage is the most serious contingency case, where the total line violation is 107.26 MW.

Table 8.19 gives the details of contingency calculation under the peak load (at hour 18). The calculation results show that although the contingency ranks for different time stages are not totally the same, the selected worst contingency case is the same, that is, the line $L_{10-21}$ outage. The worst scenario for this example is that the line $L_{10-21}$ outage happens under peak load (at hour 18).

To determine the priority of the phase shifters, the sensitivity analysis of the phase shifters is conducted under the peak load and the worst contingency cases. Simulation results show that system security will be greatly enhanced if the phase shifters are installed at locations $L_{1-3}$, $L_{2-4}$, $L_{2-6}$, $L_{6-8}$, $L_{10-22}$, $L_{15-18}$, $L_{24-25}$, respectively.

For the specified worst contingency, it can be seen from Table 8.20 that the best three locations for installing phase shifters are $L_{10-22}$, $L_{15-18}$, $L_{24-25}$.

Table 8.21 lists the results of phase shifter adjustments during the operation period (24 h) based on OPF. Simulation results show that all the line overloads are removed because of the use of the phase shifters.

**TABLE 8.19    Contingency Analysis Results at Peak Load Time Stage 18**

| Outage Line | Overload Line (MW) | Line Flow Limit (MW) | Overload Index (PI) | Power Violation | Contingency Ranking |
|---|---|---|---|---|---|
| $L_{12-14}$ | $L_{1-2}$ | 130 | 0.144 | 33.63 | 4 |
| | $L_{6-8}$ | 55 | 0.167 | | |
| | $L_{9-10}$ | 65 | 0.042 | | |
| | $L_{9-11}$ | 65 | 0.046 | | |
| $L_{10-21}$ | $L_{1-2}$ | 130 | 0.144 | 43.38 | 1 |
| | $L_{6-8}$ | 55 | 0.176 | | |
| | $L_{9-11}$ | 65 | 0.034 | | |
| | $L_{10-20}$ | 32 | 0.390 | | |
| $L_{22-25}$ | $L_{1-2}$ | 130 | 0.144 | 31.665 | 5 |
| | $L_{6-8}$ | 55 | 0.187 | | |
| | $L_{9-11}$ | 65 | 0.021 | | |
| | $L_{10-20}$ | 16 | 0.096 | | |
| $L_{24-27}$ | $L_{1-2}$ | 130 | 0.139 | 38.53 | 2 |
| | $L_{6-8}$ | 55 | 0.135 | | |
| | $L_{9-10}$ | 65 | 0.045 | | |
| | $L_{9-11}$ | 65 | 0.063 | | |
| | $L_{10-21}$ | 32 | 0.188 | | |
| $L_{29-30}$ | $L_{1-2}$ | 130 | 0.144 | 33.86 | 3 |
| | $L_{6-8}$ | 55 | 0.167 | | |
| | $L_{9-10}$ | 65 | 0.037 | | |
| | $L_{9-11}$ | 65 | 0.027 | | |
| | $L_{27-30}$ | 19 | 0.108 | | |

## 8.7    MULTIPLE OBJECTIVES OPF

The OPF problem may have many objectives, creating complications in the implementation because these objectives do not have a consistent goal to pursue in order to reach the optimum solution. This section introduces the OPF problem, which is a fully coupled active and reactive dispatch or combined active and reactive dispatch (CARD). The purpose of the OPF is to achieve the overall objective of minimum generation cost and to improve the distribution of reactive power and voltage, subject to constraints that ensure system security. Security is defined as the maintenance of individual circuit flows, generator real and reactive power output, and system voltages within limits under normal system conditions and contingency cases. Five different objective functions are considered [11]. They are minimization of generator fuel cost, maximization of reactive power reserve margins, voltage maximization, avoidance

**TABLE 8.20 Ranking of Phase Shifter Locations Based on Sensitivity Analysis (LSF = 1.55, Outage line $L_{10-21}$)**

| Phase Shifter Location $(L_{ij})$ | Phase Shifter Angle (deg.) | Over-Loaded Lines $(L_{ij})$ | Line Flow Limit (MW) | Performance Indices $(PI_{ij})$ | Sensitivity values $S_{ij}$ (MW/degree) | Phase Shifter Ranking $(Rk_{ij})$ |
|---|---|---|---|---|---|---|
| $L_{1-3}$ | +5 | $L_{6-8}$ | 55 | 0.172 | 1.87 | 7 |
| | | $L_{9-11}$ | 65 | 0.026 | | |
| | | $L_{10-22}$ | 32 | 0.382 | | |
| $L_{6-8}$ | +1 | $L_{1-2}$ | 130 | 0.145 | 2.30 | 5 |
| | | $L_{9-11}$ | 65 | 0.033 | | |
| | | $L_{10-22}$ | 32 | 0.383 | | |
| $L_{15-18}$ | −3 | $L_{6-8}$ | 55 | 0.147 | 4.45 | 3 |
| | | $L_{9-11}$ | 65 | 0.007 | | |
| | | $L_{10-22}$ | 32 | 0.257 | | |
| $L_{2-4}$ | +1 | $L_{1-2}$ | 130 | 0.125 | 1.99 | 6 |
| | | $L_{6-8}$ | 55 | 0.178 | | |
| | | $L_{9-11}$ | 65 | 0.039 | | |
| | | $L_{10-22}$ | 32 | 0.393 | | |
| $L_{10-22}$ | +1 | $L_{6-8}$ | 55 | 0.160 | 15.5 | 1 |
| | | $L_{9-11}$ | 65 | 0.009 | | |
| | | $L_{10-20}$ | 16 | 0.055 | | |
| | | $L_{10-21}$ | 32 | 0.094 | | |
| $L_{2-6}$ | +3 | $L_{6-8}$ | 55 | 0.169 | 3.15 | 4 |
| | | $L_{9-11}$ | 65 | 0.019 | | |
| | | $L_{10-22}$ | 32 | 0.383 | | |
| $L_{24-25}$ | +3 | $L_{9-11}$ | 65 | 0.003 | 7.87 | 2 |
| | | $L_{24-27}$ | 32 | 0.040 | | |

of voltage collapse, and improvement in the ability of the system to maintain higher system load level. The analytic hierarchical process (AHP) is pursued to handle these objectives during the implementation of CARD.

## 8.7.1 Formulation of Combined Active and Reactive Dispatch

***Objective Functions*** Five objective functions that are used in CARD are as follows [11,12].

**TABLE 8.21    Results of Phase Shifter Adjustments**

| Time (h) | Phase Shifter Site (Located at Line $L_{ij}$) | Phase Shifter Angle (degree) | Overload (MW) |
|---|---|---|---|
| 1–6 | None | – | – |
| 7 | $L_{10-22}$ | +1 | 0.00 |
| 8 | $L_{10-22}$ | +1 | 0.00 |
| 9 | $L_{10-22}$ | +1 | 0.00 |
| 10–13 | None | – | – |
| 14 | $L_{10-22}$ | +1 | 0.00 |
| 15 | $L_{10-22}$ | +1 | 0.00 |
| 16 | $L_{10-22}$ | +1 | 0.00 |
| 17 | $L_{24-25}$ | +1 | 0.00 |
| 18 | $L_{10-22}$ | +1 | 0.00 |
|  | $L_{24-25}$ | +1 | 0.00 |
|  | $L_{15-18}$ | −2 | 0.00 |
| 19 | $L_{10-22}$ | +1 | 0.00 |
| 20 | $L_{10-22}$ | +1 | 0.00 |
| 21–24 | None | None | – |

*Minimization of Generation Fuel Costs*   Generally, the generation fuel cost can be expressed as a quadratic function:

$$F_1 = \sum_{j \in \text{NSTEP}} \sum_{i \in NG} (a_i P_{ij}^2 + b_i P_{ij} + c_i)\tau_j \tag{8.167}$$

where

NG: the number of generators
NSTEP: the number of time steps
$\tau_j$: the approximate integration coefficients.

Linearizing equation (8.167), we get

$$\Delta F_1 = \sum_{j \in \text{NSTEP}} \sum_{i \in NG} (2a_i P_{ij} + b_i)\Delta P_{ij}\tau_j \tag{8.168}$$

If the generation fuel costs are modeled by linear functions relating monetary units to energy supplied, the following expression can be used.

$$\Delta F_1 = \sum_{j \in \text{NSTEP}} \sum_{i \in NG} (c_i \Delta P_{ij})\, \tau_j \tag{8.169}$$

where

$$\tau_1 = 0.5T_1$$

$$\tau_2 = 0.5\,(T_1 + T_2)$$

$$\cdots$$

$$\tau_{\text{NSTEP}} = 0.5\,T_{\text{NSTEP}}, \quad \text{and}$$

$$T_j = \quad \text{duration of time stage } j$$

The time factors $\tau_j$ correspond to the integration of fuel costs over the operation period by means of the trapezoidal rule.

*Maximization of Reactive Power Reserve Margins*   This objective aims to maximize the reactive power reserve margins and seeks to distribute the reserve among the generators and static VAR compensators (SVCs) in proportion to ratings. It can be expressed as

$$F_2 = \sum_{j \in \text{NSTEP}} \sum_{i \in NG} \left( \frac{Q_{ij}^2}{Q_{i\max}} \right) \tag{8.170}$$

Linearizing the above equation, we get

$$\Delta F_2 = 2 \sum_{j \in \text{NSTEP}} \sum_{i \in NG} \left( \frac{Q_{ij}^0 \Delta Q_{ij}}{Q_{i\max}} \right) \tag{8.171}$$

*Maximization of Load Voltage*   This objective aims to optimize the voltage profile by maximizing the sum of the load voltage.

$$\Delta F_3 = \sum_{j \in \text{NSTEP}} \sum_{i \in ND} \Delta V_{ij} \tag{8.172}$$

where ND is the number of loads.

*Avoidance of Voltage Collapse*   This objective aims to optimize the voltage profile by maximizing the voltage collapse proximity indicator for the whole system. It can be expressed as

$$\Delta F_4 = \sum_{j \in \text{NSTEP}} \sum_{k \in \text{NCTG}} \Delta \lambda_{kj} \tag{8.173}$$

where $\lambda_{kj}$ is a scalar (to be maximized) less than any bus voltage collapse proximity indicator at time stage $j$, contingency $k$ ($k = 0$, refer to the base case).

*Ability to Maintain Higher System Load Level*  This objective aims to allow the generators to respond efficiently to system load changes by optimizing the ability of the system to maintain higher system load level, meanwhile constraining generators within their reactive limits. It can be expressed as

$$\Delta F_5 = \sum_{j \in \text{NSTEP}} \sum_{k \in \text{NCTG}} \Delta \alpha_{kj} \tag{8.174}$$

where $\alpha_{kj}$ is a system load increment (to be maximized) at time stage $j$, contingency $k$.

The objective function of CARD can be written as

$$\Delta F = w_1 \Delta F_1 + w_2 \Delta F_2 + w_3 \Delta F_3 + w_4 \Delta F_4 + w_5 \Delta F_5 \tag{8.175}$$

where $w_i$ is the weighting coefficient of the $i$th objective function. The calculation of $w_i$ will be discussed later.

**Constraints**  At each time step, the following constraints are taken into account:

1. *Active power constraints:*
   - The active power balance equation
   - The generator active power upper and lower limits
   - The generator active power reserve upper and lower limits group import and export constraints
   - The active power-reserve relationship constraints
   - The system active power reserve constraint
   - The upper and lower limits of line active power flow.
2. *Reactive power constraints:*
   - The reactive power balance equation
   - The generator reactive power upper and lower limits
   - Network voltage limits
   - The transformer tap changer ranges
   - $Q - VZ$ characteristics of SVCs
   - The additional constraints aimed at avoiding voltage collapse
   - The additional constraints aimed at improving the ability of the system to maintain higher system load.
3. *Constraints that are a combined function of active and reactive power:*
   - The generator capability chart limits (other than simple MW or MVAr limits)
   - The branch current flow limits, modeled at the midpoint of the branch.
   - The additional constraints aimed at improving the ability of the system to maintain higher system load taking into account generator capability chart limits.

Some of the constraints are straightforward constraints (constraints regarding system variables) and others are functional constraints that are stated as follows.

*Group Limits* Station limits and approximate network security limits may be expressed by a number of group import and export constraints:

$$\left( \sum_i P_{ij} \right) - P_{Dj\,\text{local}} \leq P_{\text{exp}} \tag{8.176}$$

$$\left( \sum_i P_{ij} \right) - P_{Dj\,\text{local}} \geq P_{\text{imp}} \tag{8.177}$$

Writing the above equations as incremental form, we have

$$\sum_i \Delta P_{ij} \leq P_{\text{exp}} - \sum_i P_{ij0} + P_{Dj\,\text{local}} \tag{8.178}$$

$$\sum_i \Delta P_{ij} \geq P_{\text{imp}} - \sum_i P_{ij0} + P_{Dj\,\text{local}} \tag{8.179}$$

where $P_{Dj\,\text{local}}$ is the local load demand within the group at time stage $j$.

*Spinning Reserve Constraints* The reserve available from a generator may be modeled as a trapezoidal function of generation [11,12]. The allocation of the corresponding independent variable $\Delta R_{ij}$ is then subject to

$$R_{i\text{min}} - R_{ij0} \leq \Delta R_{ij} \leq R_{i\text{max}} - R_{ij0} \tag{8.180}$$

$$\Delta R_{ij} + \Delta P_{ij} \leq P_{i\text{max}} - P_{ij0} - R_{ij0} \tag{8.181}$$

$$\sum_{\text{gen}} \Delta R_{ij} \geq S_{\text{total}} - \sum_{\text{gen}} R_{ij0} \tag{8.182}$$

*Operating Chart Limits for Generators* The ability of generators to absorb reactive power is generally limited by the machine minimum excitation limit. A further limit is determined so as to provide an adequate margin of safety for the machine thermal limit. A simplified generator capability chart can be defined in which the leading and lagging limits of machine reactive output are expressed as a function of the real power output. Using a trapezoidal approximation, this can be represented as

$$P_{ij} + \left( \frac{\beta_{i1}}{\alpha_{i1}} \right) Q_{ij} - \beta_{i1} \leq 0 \tag{8.183}$$

$$P_{ij} + \left( \frac{\beta_{i2}}{\alpha_{i2}} \right) Q_{ij} - \beta_{i2} \leq 0 \tag{8.184}$$

Linearizing the above equations around the current operating point, we obtain

$$\Delta P_{ij} + \left(\frac{\beta_{i1}}{\alpha_{i1}}\right) \Delta Q_{ij} + P_{ij0} + \left(\frac{\beta_{i1}}{\alpha_{i1}}\right) Q_{ij0} - \beta_{i1} \leq 0 \tag{8.185a}$$

$$\Delta P_{ij} + \left(\frac{\beta_{i2}}{\alpha_{i2}}\right) \Delta Q_{ij} + P_{ij0} + \left(\frac{\beta_{i2}}{\alpha_{i2}}\right) Q_{ij0} - \beta_{i2} \leq 0 \tag{8.185b}$$

where $\alpha_{i1}$, $\alpha_{i2}$ are the intersections with the $Q$-axis, and $\beta_{i1}, \beta_{i2}$ are the intersection with the $P$-axis.

*Maintaining Higher System Load Constraints* Every generator $i$ should contribute its share of reactive power output to meet a prospective increase in system demand in such a way that the generator output does not exceed its reactive limits:

$$Q_{ij} + \left(\frac{\delta Q_{ij}}{\delta \alpha_j}\right) \alpha_j \leq Q_{ijmax} \tag{8.186}$$

When considering generators with active power control, the operating chart limits for the generators are taken into account.

Linearizing equation (8.186) around the current operating point, we obtain

$$\Delta Q_{ij} + \left(\frac{\delta Q_{ij}}{\delta \alpha_j}\right) \Delta \alpha_j \leq Q_{ijmax} - Q_{ij0} - \left(\frac{\delta Q_{ij}}{\delta \alpha_j}\right) \alpha_{j0} \tag{8.187}$$

where $\frac{\delta Q_{ij}}{\delta \alpha_j}$ represents the change in the reactive power output of generator $i$ as a fraction of the change in load demand at time stage $j$ evaluated using a load flow algorithm. $\alpha_j$ represents the increase in system demand.

*Avoidance of Voltage Collapse Constraints* For a network with $n$ buses, Thevenin's equivalent impedance looking into the port between bus $i$ and ground is $Z_{ii} \angle \theta_i$, which equals the $i$th diagonal element of $[Z] = [Y]^{-1}$. Therefore, for permissible power transfer to the load at bus $i$ we must have $Z_i/Z_{ii} > 1$, where $Z_i \angle \gamma_i$ is the impedance for load $i$ ($Z_i = V_i^2 \cos \gamma_i / P_i$).

The idea is to constrain the voltage collapse proximity indicators at the load nodes in order to maintain an acceptable system voltage profile. This has been done by finding a parameter greater than 1 at each time interval, such that the voltage collapse proximity indicators at the load nodes specified by the user are greater than this parameter. These parameters $\lambda_j$ form part of the objective function. The corresponding constraints can be written as

$$Z_i/Z_{ii} \geq \lambda^2 \tag{8.188}$$

$$V_i \geq \lambda \sqrt{Z_{ii} P_i / \cos \gamma_i} \tag{8.189}$$

Linearizing equation (8.189) around the current operating point, we obtain

$$\Delta V_{ij} - \Delta \lambda_j \sqrt{Z_{ii} P_i / \cos \gamma_i} \geq -V_{ij0} + \Delta \lambda_{j0} \sqrt{Z_{ii} P_i / \cos \gamma_i} \tag{8.190}$$

*Static VAR Compensators*   SVCs are high-speed variable reactive power sources and sinks connected to the system. Their electrical characteristic is such that MVAR output (or absorption) is related to voltage in a linear manner; normally, for a small change in voltage, the compensator will go from zero to full output. This is known as the slope. Thus the constraint of SVCs can be modeled as

$$V_{ij\text{min}} \leq V_{ij} - a_i Q_{ij} \leq V_{ij\text{max}} \tag{8.191}$$

$$Q_{ij\text{min}} \leq Q_{ij} \leq Q_{ij\text{max}} \tag{8.192}$$

The linearized incremental model is

$$V_{ij\text{min}} - V_{ij0} + a_i Q_{ij0} \leq \Delta V_{ij} - a_i \Delta Q_{ij} \leq V_{ij\text{max}} - V_{ij0} + a_i Q_{ij0} \tag{8.193}$$

where $a_i$ is the slope.

*Dynamic Constraints*   In the dynamic dispatch case, additional generation rate limit constraints can be considered:

$$-P_{rdi}T_j \leq P_{ij} - P_{i(j-1)} \leq P_{rui}T_j \tag{8.194}$$

The linearized incremental form of the above equation is

$$-P_{rdi}T_j \leq P_{ij0} + \Delta P_{ij} - P_{i(j-1)0} + \Delta P_{i(j-1)} \leq P_{rui}T_j \tag{8.195}$$

where, $P_{rdi}$, $P_{rui}$ are the vector limits for decreasing and increasing output, respectively, and $T_j$ is the length of the time step.

For every contingency at every time step, the constraints regarding the slack bus will be included in addition to the constraints for the normal case.

## 8.7.2   Solution Algorithm

**AHP Model of CARD**   Obviously, the mathematical model of CARD mentioned in Section 8.7.1 is a linear model based on a multi-objective function. It is not appropriate to use an equal weighting coefficient for the various kinds of objectives in (8.175) because the importance of these objectives is different in a practical power system. Therefore, the weighting coefficients of the various objective functions in the CARD model must be determined before CARD can be executed. However, it is very difficult to decide precisely the weighting coefficient of each objective in the CARD model unless only one or two objectives are considered. There are two reasons for this: one is that the objectives are interrelated and interact with each other. Another reason is that the relative importance of these objectives is not the same, not only for different power systems but also within the same power system in different circumstances. An analytic hierarchical process was recommended to solve this challenging problem [11].

The principle and method of the AHP were introduced in Chapter 7. AHP transforms the complex problem into rank calculation within the hierarchy structure. In the ranking computation, the ranking in each hierarchy can also be converted into the judgment and comparison of a series of pairs of factors. The judgment matrix can be formed according to the quantified judgment of pairs of factors using some ratio scale method. Consequently, the value of the weighting coefficients of all factors can be obtained through calculating the maximal eigenvalue and the corresponding eigenvector of the judgment matrix. The judgment matrix A of the CARD hierarchy model can be written as follows:

$$
A = \begin{bmatrix}
1 & \dfrac{W_1}{W_2} & \dfrac{W_1}{W_3} & \dfrac{W_1}{W_4} & \dfrac{W_1}{W_5} \\[2ex]
\dfrac{W_2}{W_1} & 1 & \dfrac{W_2}{W_3} & \dfrac{W_2}{W_4} & \dfrac{W_2}{W_5} \\[2ex]
\dfrac{W_3}{W_1} & \dfrac{W_3}{W_2} & 1 & \dfrac{W_3}{W_4} & \dfrac{W_3}{W_5} \\[2ex]
\dfrac{W_4}{W_1} & \dfrac{W_4}{W_2} & \dfrac{W_4}{W_3} & 1 & \dfrac{W_4}{W_5} \\[2ex]
\dfrac{W_5}{W_1} & \dfrac{W_5}{W_2} & \dfrac{W_5}{W_3} & \dfrac{W_5}{W_4} & 1
\end{bmatrix}
\tag{8.196}
$$

where $W_i$ is the weighting coefficient of the $i$th sub-objective in the hierarchy model of CARD.

The AHP algorithm and the selection of the judgment matrix can be found in Chapter 7.

***Solution Algorithm***   The solution algorithm adopted for the AHP-based CARD may be described as follows:

1. Either, perform a merit order dispatch, or use an existing active power generation pattern provided by the user to satisfy active power demand. The same active generation pattern applies for contingency cases.

2. Perform a Newton–Raphson power flow for normal and defined contingency cases at every time step. If power flow analysis only is required, then stop; otherwise proceed to step (3).

3. For every contingency case, at every time step, include a new set of variables and constraints relating that case to the variables and constraints of the intact case.

4. Set up a hierarchy model for CARD.

5. Form a judgment matrix according to the experiences and needs of the user.

6. Perform the AHP calculation to obtain the optimum weighting coefficients of the various objective functions.

7. Linearize the objective function and constraints around the operating point.

**8.** Execute the LP algorithm (sparse dual revised simplex method with relaxation) to obtain the optimum state of the linearized system.

**9.** Apply constraint limit squeezing automatically, or as necessary, depending on the option to be selected.

**10.** Iterate between LP and power flow until the system converges.

The AHP-based CARD algorithm is designed to satisfy the following convergence criteria simultaneously:

- The consistency of the weighting coefficients is satisfactory.
- No violation of constraint limit occurs.
- Changes in control variables over two consecutive iterations are within specified tolerances.
- Changes in objective function value over two consecutive iterations are within specified tolerance.

## 8.8 PARTICLE SWARM OPTIMIZATION FOR OPF

As already discussed, various traditional optimization techniques were developed to solve the OPF problem. Some of these techniques have excellent convergence characteristics and some are widely used in the industry. It is noted that each technique may be tailored to suit a specific OPF optimization problem on the basis of the mathematical nature of the objectives and/or constraints. In addition, some of these techniques might converge to local solutions instead of global ones if the initial guess happens to be in the neighborhood of a local solution. This occurs as a result of using Kuhn–Tucker conditions as termination criteria to detect stationary points. This practice is commonly used in most commercial nonlinear optimization programs [70].

In recent years, a new optimization method—PSO is applied to solve OPF problem [59–64]. This section introduces several major PSO methods that are used in OPF.

### 8.8.1 Mathematical Model

Generally, the following OPF model is used in various PSO approaches. The objective function may be one of following.

**(1)** Fuel cost minimization

$$\text{min } F_g = \sum_{i=1}^{NG} (a_i P_{gi}^2 + b_i P_{gi} + c_i) \tag{8.197}$$

**(2)** Fuel emission minimization

$$\text{min } E_g = \sum_{i=1}^{NG} (\alpha_i P_{gi}^2 + \beta_i P_{gi} + \gamma_i) \tag{8.198}$$

**(3)** Loss minimization

$$\min P_L = \sum_{l=1}^{NL} P_l \qquad (8.199)$$

**(4)** Voltage deviation minimization at load buses

$$\min \text{VD} = \sum_{i=1}^{ND} (V_i - V_i^{\text{sp}})^2 \qquad (8.200)$$

where

$V_i^{\text{sp}}$: the prespecified reference value at load bus $i$
$P_{gi}$: the real power generation at generator $i$
$P_L$: the system real power loss
$P_l$: the real power loss on line $l$
$P_{g\,\text{slack}}$: the real power of the slack generator
$a_i, b_i, c_i$: the coefficients of generator fuel cost
$\alpha_i, \beta_i, \gamma_i$: the coefficients of generator emission function
VD: the total voltage deviation at load buses
NG: the number of generating units
ND: the number of load buses
NL: the number of lines

The constraints are as follows.

$$P_{gi} - P_{di} - f_{Pi}(V, \theta, T) = 0 \qquad (8.201)$$

$$Q_{gi} - Q_{di} - f_{Qi}(V, \theta, T) = 0 \qquad (8.202)$$

$$P_{gi\min} \le P_{gi} \le P_{gi\max}, \quad i \in NG \qquad (8.203)$$

$$Q_{gi\min} \le Q_{gi} \le Q_{gi\max}, \quad i \in NG \qquad (8.204)$$

$$Q_{ci\min} \le Q_{ci} \le Q_{ci\max}, \quad i \in NC \qquad (8.205)$$

$$V_{gi\min} \le V_{gi} \le V_{gi\max}, \quad i \in NG \qquad (8.206)$$

$$V_{di\,\min} \le V_{di} \le V_{di\,\max}, \quad i \in ND \qquad (8.207)$$

$$T_{i\min} \le T_i \le T_{i\max}, \quad i \in NT \qquad (8.208)$$

$$S_{Lj} \le S_{Lj\max}, \quad j \in NL \qquad (8.209)$$

where

$S_{Lj}$: the transmission line loadings
$S_{ljmax}$: the limit of transmission line loadings
$Q_{di}$: Switchable VAR compensations at bus $i$
$NC$: the number of switchable VAR sources
$V_{gi}$: the voltage magnitude at generator bus $i$.

The subscripts "min" and "max" stand for the lower and upper bounds of a constraint, respectively.

Several PSO methods can be used to solve the above mentioned OPF problem, which are introduced in the next section.

## 8.8.2   PSO Methods [59,71–75]

The PSO introduced in Chapter 7 has been used to solve the unit commitment. Here, we focus on applying PSO methods to solve the OPF problem.

***Conventional Particle Swarm Optimization***   In PSO algorithms, each particle moves with an adaptable velocity within the regions of decision space and retains a memory of the best position it ever encountered. The best position ever attained by each particle of the swarm is communicated to all other particles. The conventional PSO assumes an $n$-dimensional search space $S \subset R^n$, where $n$ is the number of decision variables in the optimization problem, and a swarm consisting of $N$ particles.

In PSO, a number of particles form a swarm that evolve or fly throughout the problem hyperspace to search for optimal or near optimal solution. The coordinates of each particle represent a possible solution with two vectors associated with it, the position $X$ and velocity $V$ vectors. During their search, particles interact with each other in a certain way to optimize their search experience. There are different variants of the particle swarm paradigms but the most general one is the $P_{gb}$ model where the whole population is considered as a single neighborhood throughout the optimization process. In each iteration, the particle with the best solution shares its position coordinates ($P_{gb}$) information with the rest of the swarm.

Thus, the variables are defined as follows.

The position of the $i$th particle at time $t$ is an $n$-dimensional vector denoted by

$$X_i(t) = (x_{i,1}, x_{i,2}, \ldots, x_{i,n}) \in S \qquad (8.210)$$

The velocity of this particle at time $t$ is also an $n$-dimensional vector denoted by

$$V_i(t) = (v_{i,1}, v_{i,2}, \ldots, v_{i,n}) \in S \qquad (8.211)$$

The best previous position of the $i$th particle at time $t$ is a point in $S$, which is denoted by

$$P_i = (p_{i,1}, p_{i,2}, \cdots, p_{i,n}) \in S \tag{8.212}$$

The global best position ever attained among all particles is a point in $S$, which is denoted by

$$P_{gb} = (p_{gb,1}, p_{gb,2}, \cdots, p_{gb,n}) \in S \tag{8.213}$$

Then, each particle updates its coordinates on the basis of its own best search experience ($P_i$) and $P_{gb}$ according to the following velocity and position update equations.

$$V_i^{t+1} = wV_i^t + C_1 \times r_1 \times (P_i - X_i^t) + C_2 \times r_2 \times (P_{gb} - X_i^t) \tag{8.214}$$

$$X_i^{t+1} = X_i^t + V_i^{t+1} \tag{8.215}$$

where

    $w$: inertia weight

$C_1, C_2$: acceleration coefficients

  $r_1, r_2$: two separately generated uniformly distributed random numbers in the range [0,1] added in the model to introduce stochastic nature.

The inertia weighting factor for the velocity of a particle is defined by the inertial weight approach

$$w^t = w_{max} - \frac{w_{max} - w_{min}}{t_{max}} \times t \tag{8.216}$$

where, $t_{max}$ the maximum number of iterations, and $t$ is the current number of iterations. $w_{max}$ and $w_{min}$ are the upper and lower limits of the inertia weighting factor, respectively.

Moreover, in order to guarantee the convergence of the PSO algorithm, the constriction factor $k$ is defined as

$$k = \frac{2}{|2 - \varphi - \sqrt{\varphi^2 - 4\varphi}|} \tag{8.217}$$

where $\varphi = C_1 + C_2$, $\varphi \geq 4$.

In this constriction factor approach (CFA), the basic system equations of the PSO (8.214), (8.215) can be considered as difference equations. Therefore, the system dynamics, namely, the search procedure, can be analyzed by the eigenvalue analysis and can be controlled so that the system behavior has the following features:

**(1)** The system converges.

**(2)** The system can search different regions efficiently.

In the CFA, $\varphi$ must be greater than 4.0 to guarantee stability. However, as $\varphi$ increases, the factor $k$ decreases and diversification is reduced, yielding slower response. Therefore, we choose 4.1 as the smallest $\varphi$ that guarantees stability but yields the fastest response. It has been observed that $4.1 \leq \varphi \leq 4.2$ leads to good solutions [59].

***Passive Congregation–Based PSO*** According to the local-neighborhood variant of the PSO algorithm (L-PSO) [75], each particle moves toward its best previous position and toward the best particle in its restricted neighborhood. As the local-neighborhood leader of a particle, its nearest particle (in terms of distance in the decision space) with the better evaluation is considered. Since the CFA generates higher-quality solutions in the basic PSO, some enhancements are presented. Specifically, Parrish and Hammer [76] have proposed mathematical models to show how these forces organize the swarms. These can be classified in two categories: the *aggregation* and the *congregation* forces.

Aggregation refers to the swarming of particles by nonsocial, external physical forces. There are two types of aggregation: passive aggregation and active aggregation. Passive aggregation is a swarming by physical forces, such as the water currents in the open sea group, the plankton [76].

Congregation, on the other hand, is a swarming by social forces, which is the source of attraction of a particle to others and is classified in two types: social and passive. Social congregation usually happens when the swarm's fidelity is high, such as genetic relation. Social congregation necessitates active information transfer, for example, ants that have high genetic relation use antennal contacts to transfer information about location of resources.

According to references [59,75,76], passive congregation is an attraction of a particle to other swarm members, where there is no display of social behavior because particles need to monitor both environment and their immediate surroundings such as the position and the speed of neighbors. Such information transfer can be employed in the passive congregation. A hybrid L-PSO with a passive congregation operator (PAC) is called an *LPAC PSO* [59]. Moreover, the global variant–based passive congregation PSO (GPAC) can also be enhanced with the CFA.

The swarms of the enhanced GPAC and LPAC are manipulated by the following velocity update.

$$V_i^{t+1} = k \ [w^t V_i^t + C_1 \times r_1 \times (P_i - X_i^t) + C_2 \times r_2 \times (P_k - X_i^t) + C_3 \times r_3 \times (P_r - X_i^t)]$$

$$i = 1, 2, \ldots, N \tag{8.218}$$

where

$C_1, C_2, C_3$: the cognitive, social, and passive congregation parameters, respectively

$P_i$: the best previous position of the $i$th particle

$P_k$: either the global best position ever attained among all particles in the case of enhanced GPAC or the local best position of particle $i$, namely, the position of its nearest particle $k$ with better evaluation in the case of LPAC

$P_i$: the position of passive congregator (position of a randomly chosen particle $r$).

The positions are updated using the same equation (8.215). The positions of the $i$th particle in the $n$-dimensional decision space are limited by the minimum and maximum positions expressed by vectors

$$X_{imin} \leq X_i \leq X_{imax} \tag{8.219}$$

The velocities of the $i$th particle in the $n$-dimensional decision space are limited by

$$V_{imax} \leq V_i \leq V_{imax} \tag{8.220}$$

where the maximum velocity in the $m$th dimension of the search space is computed as

$$V_{imax}^m = \frac{s_{imax}^m - s_{imin}^m}{Nr}, \quad m = 1, 2, \dots, n \tag{8.221}$$

Where, $s_{imax}^m$, and $s_{imin}^m$ are the limits in the $m$-dimension of the search space. The maximum velocities are constricted in small intervals in the search space for better balance between exploration and exploitation. $Nr$ is a chosen number of search intervals for the particles. It is an important parameter in the enhanced GPAC and LPAC PSO algorithms. A small $Nr$ facilitates global exploration (searching new areas), while a large one tends to facilitate local exploration. A suitable value for the $Nr$ usually provides balance between global and local exploration abilities and consequently results in a reduction of the number of iterations required to locate the optimum solution. The basic steps of the enhanced GPAC and LPAC are listed in the following [59].

Step (1) Generate a swarm of $N$ particles with uniform probability distribution, initial positions $X_i(0)$, and velocities $V_i(0)$, $(i = 1, 2, \dots, N)$, and initialize the random parameters. Evaluate each particle $i$ using objective function $f$ (e.g., to be minimized).

Step (2) For each particle $i$, calculate the distance $d_{ij}$ between its position and the positions of all other particles:

$$d_{ij} = \|X_i - X_j\| (i = 1, 2, \dots, N, i \neq j)$$

where $X_i$ and $X_j$ are the position vectors of particle $i$ and particle $j$, respectively.

Step (3)    For each particle $i$, determine the nearest particle, particle $k$, with better evaluation than its own, that is, $d_{ik} = \min_j(d_{ij}), f_k \leq f_j$ and set it as the leader of particle $i$.

In the case of enhanced GPAC, particle $k$ is considered as the global best.

Step (4)    For each particle $i$, randomly select a particle $r$ and set it as passive congregator of particle $i$.

Step (5)    Update the velocities and positions of particles using (8.218) and (8.215), respectively.

Step (6)    Check if the limits of positions in equation (8.219) and velocities in equations (8.220) and (8.221) are enforced. If the limits are violated, then they are replaced by the respective limits.

Step (7)    Evaluate each particle using the objective function $f$. The objective function $f$ is calculated by running a power flow. In the case where for a particle no power flow solution exists, an error is returned and the particle retains its previous achievement.

Step (8)    If the stopping criteria are not satisfied, go to Step (2).

The enhanced GPAC and LPAC PSO algorithms will be terminated if one of the following criteria is satisfied: (i) no improvement of the global best in the last 30 generations is observed, or (ii) the maximum number of allowed iterations is achieved.

Finally, we can indicate that the last term of equation (8.218), added in the conventional PSO velocity update equation (8.214), displays the information transferred via passive congregation of particle with a randomly selected particle $r$. This passive congregation operator can be regarded as a stochastic variable that introduces perturbations to the search process. For each particle $i$, the perturbation is proportional to the distance between itself and a randomly selected particle $r$ rather than an external random number, namely, the turbulence factor introduced in [77]. The CFA helps the convergence of algorithm more than the turbulence factor because (i) in the early stages of the process, where distance between particles is large, the turbulence factor should be large, avoiding premature convergence; and (ii) in the last stages of the process, as the distance between particles becomes smaller, the turbulence factor should be smaller too, enabling the swarm to converge in the global optimum [77] Therefore, LPAC is more capable of probing the decision space, avoiding suboptimums and improving information propagation in the swarm than other conventional PSO algorithms.

***Coordinated Aggregation Based PSO***   The coordinated aggregation is a completely new operator introduced in the swarm, where each particle moves considering only the positions of particles with better achievements than its own, with the exception of the best particle, which moves randomly. The coordinated aggregation can be considered as a type of active aggregation where particles are attracted only by places with the most food.

Let $X_i(t)$ and $X_j(t)$ be the positions of particle $i$ and particle $j$ at iterative cycle $t$, respectively. The differences between the positions of particles $i$ and $j$, $X_i(t) - X_j(t)$ are defined as *coordinators* of particle velocity. The ratios of differences between the achievement of particle $i$, $A(X_i)$ and the better achievements by particles $j$, $A(X_j)$ to the sum of all these differences are called the achievement's *weighting factors* $\omega_{ij}^t$

$$\omega_{ij} = \frac{A(X_j) - A(X_i)}{\sum_l A(X_l) - A(X_i)}, \quad j, l \in \Omega_i \tag{8.222}$$

where $\Omega_i$ represents the set of particles $j$ with better achievement than particle $i$.

The velocity of particle is adapted by means of coordinators multiplied by weighting factors.

The steps of the coordinated aggregation–based PSO (CAPSO) algorithm are listed below [59].

Step (1)   *Initialization:* Generate $N$ particles. For each particle $i$, choose the initial position $X_i(0)$ randomly. Calculate its initial achievement $A(X_i(0))$ using the objective function $f$ and find the maximum $A_g(0) = \max_i A(X_i(0))$ called the *global best achievement*. Then, particles update their positions in accordance with the following steps.

Step (2)   *Swarm's manipulation:* The particles, except the best of them, regulate their velocities in accordance with the equation

$$V_i^{t+1} = w^t V_i^t + \sum_j r_j \omega_{ij}^t (X_j^t - X_i^t) \quad j \in \Omega_i, i = 1, 2, \dots, N \tag{8.223}$$

Where, $\omega_{ij}^t$ are the achievement's weighting factors; and the inertia weighting factor $w^t$ is defined by equation (8.216). The role of the inertia weighting factor is considered critical for the CAPSO convergence behavior. It is employed to control the influence of the previous history of the velocities on the current one. Accordingly, the inertia weighting function regulates the trade-off between the global and local exploration abilities of the swarms.

Step (3)   *Best particle's manipulation (craziness):* The best particle in the swarm updates its velocity using a *random coordinator* calculated between its position and the position of a randomly chosen particle in the swarm. The manipulation of the best particle seems like the crazy agents or the turbulence factor introduced in [77] and helps the swarm escape from the local minima.

Step (4)   Check if the limits of velocities in equations (8.220) and (8.221) are enforced. If the limits are violated, then they are replaced by the respective limits.

Step (5)   *Position update:* The positions of particles are updated using equation (8.215). Check if the limits of positions in equation (8.219) are enforced.

Step (6)   *Evaluation:* Calculate the achievement $A(X_i(t))$ of each particle using the objective function $f$. The achievement is calculated by running a power flow. In the case where, for a particle, no power flow solution exists, an error is returned and the particle retains its previous achievement.

Step (7)  If the stopping criteria are not satisfied, go to Step (2). The CAPSO algorithm will be terminated if no more improvement of the global best achievement in the last 30 generations is observed or the maximum number of allowed iterations is achieved.

Step (8)  *Global optimal solution:* Choose the optimal solution as the global best achievement.

### 8.8.3  OPF Considering Valve Loading Effects

Generally, the generator fuel cost function in the OPF model ignores the valve point loading that introduces rippling effects to the actual input–output curve. The overall fuel cost function for a number of thermal generating units are modeled by a quadratic function, which is shown in equation (8.197). The valve effects can be expressed as a sine function [49] and added to equation (8.197), that is,

$$\min\ F_g = \sum_{i=1}^{NG} [a_i P_{gi}^2 + b_i P_{gi} + c_i + |e_i \sin(f_i(P_{gimin} - P_{gi}))|] \tag{8.224}$$

This more accurate modeling adds more challenges to most derivative-based optimization algorithms in finding the global solution because the objective is no longer convex nor differentiable everywhere.

A hybrid particle swarm optimization (HPSO) approach can be used to solve this problem [64]. This approach combines the PSO technique with the Newton–Raphson based power flow program in which the former technique is used as a global optimizer to find the best combinations of the mixed-type control variables, while the latter serves as a minimizer to reduce the nonlinear power flow equations mismatch. The Newton–Raphson method used in this implementation is the one with the full Jacobian evaluated and updated at each iteration. The HPSO utilizes a population of particles or possible solutions to explore the feasible solution hyperspace in its search for an optimal solution. Each particle's position is used as a feasible initial guess for the power flow subroutine. This mechanism of multiple initial solutions can provide better probability of detecting an optimal solution to the power flow equations that would globally minimize a given objective function. The importance of such hybridization is signified by realizing the fact that in a transmission system, the solution to the power flow equation is not unique, that is, multiple solutions within the stability margins may exist and only one can globally optimize a certain objective.

The same OPF constraints as in equations (8.201)–(8.209) are used here. Within the context of PSO applications to the OPF, inequality constraints that represent the permissible operating range of each optimization variable are typically handled in the following two ways [59–64]:

**(1)** Set to limit approach (SLA): If any optimization variable exceeds its upper or lower bound, the value of the variable is set to the violated limit. This resembles the idea found in operating all generating units at equal incremental principle to reach economic dispatch, which was described in Chapter 4. It is important to note that PSO has some randomness in the update equation that might cause several variables to exceed their limits during the optimization process. Thus, this approach may fix multiple optimization variables to their operating limits for which a global solution may not be reached. Also this approach fails to utilize the memory element that each particle has once it exceeds its boundaries.

(2) Penalty factor: The other approach is to use penalty factors to incorporate the inequality constraints with the objective, which we used in Section 8.2 in this chapter. The main problem with this approach is introducing new parameters that need to be properly selected in order to reach acceptable PSO performance. Values of the penalty factors are problem dependent; thus, this approach requires proper adjustments of the penalty factors in addition to tuning the PSO parameters.

Another approach is combining these two methods to handle the inequality constraints [64]. It combines the ideas of preserving feasible solution and infeasible solution rejection methods to retain only feasible solutions throughout the optimization process without the need to introduce penalty factors in the objective function. In most of the evolutionary computation optimization methods that employ the infeasible solution rejection method to handle constraints, any solution candidate among the population is randomly re-initialized once it crosses the boundaries of the feasible region. The majority of methods do not have memory elements associated with each candidate in the population. However, in the case of HPSO, each particle has a memory element $(P_i)$ that recalls the best-visited location through its own flying experience to search for the optimal solution and may use this information once it violates the problem boundaries. Thus, this hybridization makes use of the memory element that each particle has to maintain its feasibility status. This restoration operation keeps the infeasible particle *alive* as a possible candidate that could locate the optimal solution instead of a complete rejection that eliminates its potential in the swarm.

For the control variables in equations (8.201)–(8.209), there are two types: continuous and discrete. The continuous variables are initialized with uniformly distributed pseudorandom numbers that take the range of these variables, for example, $P_{gi} = \text{random}[P_{gimin}, \; P_{gimax}]$ and $V_i = \text{random}[V_{imin}, \; V_{imax}]$.

However, in the case of the discrete variables, an additional operator is needed to account for the distinct nature of these variables. A rounding operator is included to ensure that each discrete variable is rounded to its nearest decimal integer value that represents the physical operating constraint of a given variable. Each transformer tap setting is rounded to its nearest decimal integer value of 0.01 by utilizing the rounding operator as: $\text{round}(\text{random}[T_{imin}, T_{imax}], 0.01)$. The same principle applies to the discrete reactive injection as a result of capacitor banks with the difference being the step size, that is, $\text{round}(\text{random}[Q_{cimin}, Q_{cimax}], 1)$. This ensures that the fitness of each solution is measured only when all elements of the solution vector are properly represented to reflect the real world nature of each variable. Since the

**TABLE 8.22 Data of the Generator for IEEE 30-Bus System**

| Unit | 1 | 2 | 3 | 4 | 5 | 6 |
|---|---|---|---|---|---|---|
| Bus no. | 1 | 2 | 22 | 27 | 23 | 13 |
| A | 0.02 | 0.0175 | 0.0625 | 0.00834 | 0.025 | 0.025 |
| B | 2 | 1.75 | 1 | 3.25 | 3 | 3 |
| C | 0 | 0 | 0 | 0 | 0 | 0 |
| E | 300 | 200 | 150 | 100 | 200 | 200 |
| F | 0.2 | 0.22 | 0.42 | 0.3 | 0.35 | 0.35 |
| $P_{min}$ (MW) | 0 | 0 | 0 | 0 | 0 | 0 |
| $P_{max}$ (MW) | 80 | 80 | 50 | 55 | 30 | 40 |
| $Q_{min}$ (Mvar) | −20 | −20 | −15 | −15 | −10 | −15 |
| $Q_{max}$ (Mvar) | 150 | 60 | 62.5 | 48.7 | 40 | 44.7 |

particle update equations have some uniformly distributed random operators built into them and because of the addition of two different types of vectors, the rounding operator is called again after each update to act only on the discrete variables as round($T_i$, 0.01) and round($Q_{ci}$, 1). Once the rounding process is over, all solution elements go through a feasibility check. This simple rounding method guarantees that power flow calculations and fitness measurements are obtained only when all problem variables are properly addressed and their nature types are accounted for.

***Example 8.4:*** The example is extracted from [64]. The test system is the IEEE 30-bus system with modified unit data and bus data, which are shown in Tables 8.22 and 8.23. The line data are the same as in Table 5.6 in Chapter 5. There are two capacitors banks installed at bus 5 and bus 24 with ratings of 19 and 4 MVAR, respectively. A series of experiments were conducted to properly tune the HPSO parameters to suit the targeted OPF problem. The most noticeable observation from this groundwork is that the optimal settings for $C_1$ and $C_2$ are found to be 1.0. These values are relatively small as most of the values reported in the previously related work are in the range 1.4–2 [59–63]. The best settings for the number of particles and particle's maximum velocity ($V_{max}$) are 20 and 0.1 respectively. The inertia weight is kept fixed throughout the simulation process between the upper and lower bounds of 0.9 and 0.4, respectively.

The following three cases are considered.

Case 1: Considering only the continuous control variables. The objective is to minimize the generator fuel costs, which are the quadratic fuel cost functions. The OPF results solved by HPSO are listed in Table 8.24. For comparison, the OPF results solved by sequential quadratic programming (SQP) are also listed in Table 8.24. The comparison of the results shows that HPSO achieved better solution when only continuous optimization variables are used.

**TABLE 8.23  Bus Data for IEEE 30-Bus System (p.u.)**

| Bus No. | $P_D$ | $Q_D$ | $V_{min}$ | $V_{max}$ | Bus no. | $P_D$ | $Q_D$ | $V_{min}$ | $V_{max}$ |
|---|---|---|---|---|---|---|---|---|---|
| 1 | 0.000 | 0.000 | 0.95 | 1.1 | 16 | 0.035 | 0.016 | 0.90 | 1.05 |
| 2 | 0.217 | 0.127 | 0.95 | 1.1 | 17 | 0.090 | 0.058 | 0.90 | 1.05 |
| 3 | 0.024 | 0.012 | 0.90 | 1.05 | 18 | 0.032 | 0.009 | 0.90 | 1.05 |
| 4 | 0.076 | 0.016 | 0.90 | 1.05 | 19 | 0.095 | 0.034 | 0.90 | 1.05 |
| 5 | 0.942 | 0.190 | 0.90 | 1.05 | 20 | 0.022 | 0.007 | 0.90 | 1.05 |
| 6 | 0.000 | 0.000 | 0.90 | 1.05 | 21 | 0.175 | 0.112 | 0.90 | 1.05 |
| 7 | 0.228 | 0.109 | 0.90 | 1.05 | 22 | 0.000 | 0.000 | 0.95 | 1.1 |
| 8 | 0.300 | 0.300 | 0.90 | 1.05 | 23 | 0.032 | 0.016 | 0.95 | 1.1 |
| 9 | 0.000 | 0.000 | 0.90 | 1.05 | 24 | 0.087 | 0.067 | 0.90 | 1.05 |
| 10 | 0.058 | 0.020 | 0.90 | 1.05 | 25 | 0.000 | 0.000 | 0.90 | 1.05 |
| 11 | 0.000 | 0.000 | 0.90 | 1.05 | 26 | 0.035 | 0.023 | 0.90 | 1.05 |
| 12 | 0.112 | 0.075 | 0.90 | 1.05 | 27 | 0.000 | 0.000 | 0.95 | 1.1 |
| 13 | 0.000 | 0.000 | 0.95 | 1.1 | 28 | 0.000 | 0.000 | 0.90 | 1.05 |
| 14 | 0.062 | 0.016 | 0.90 | 1.05 | 29 | 0.024 | 0.009 | 0.90 | 1.05 |
| 15 | 0.082 | 0.025 | 0.90 | 1.05 | 30 | 0.106 | 0.019 | 0.90 | 1.05 |

**TABLE 8.24  OPF Results of IEEE 30-Bus System for Case 1 and 2**

| Case | Case 1 | Case 1 | Case 2 |
|---|---|---|---|
| Method | SQP | PSO | PSO |
| $P_{g1}$ | 41.51 | 43.611 | 42.180 |
| $P_{g2}$ | 55.4 | 58.060 | 57.013 |
| $P_{g13}$ | 16.2 | 17.555 | 17.305 |
| $P_{g22}$ | 22.74 | 22.998 | 22.025 |
| $P_{g23}$ | 16.27 | 17.056 | 17.872 |
| $P_{g27}$ | 39.91 | 32.567 | 35.060 |
| $V_{g1}$ | 0.982 | 1.000 | 1.000 |
| $V_{g2}$ | 0.979 | 1.000 | 0.999 |
| $V_{g13}$ | 1.064 | 1.059 | 1.061 |
| $V_{g22}$ | 1.016 | 1.012 | 1.071 |
| $V_{g23}$ | 1.026 | 1.021 | 1.076 |
| $V_{g27}$ | 1.069 | 1.037 | 1.10 |
| $Q_{c5}$ | – | – | 4.000 |
| $Q_{c24}$ | – | – | 8.000 |
| $T_{6-9}$ | – | – | 0.900 |
| $T_{6-10}$ | – | – | 0.950 |
| $T_{4-12}$ | – | – | 0.930 |
| $T_{27-28}$ | – | – | 0.950 |
| Total cost ($/h) | 576.892 | 575.411 | 574.143 |
| Total losses (MW) | 2.860 | 2.647 | 2.255 |

Case 2:  Considering both the continuous and discrete control variables. The test system is modified by introducing four tap-changing transformers between buses 6–9, 6–10, 4–12, and 27–28. The operating range of all transformers is set between 0.9–1.05 with a discrete step size of 0.01. The capacitor banks at buses 5 and 24 are also considered as new discrete control variables with a range of 0–40 MVAR and a step size of 1. With this modification, the problem now has both continuous and discrete control variables that can be troublesome to most conventional optimization methods. The results are shown in the last column in Table 8.24.

Case 3:  Considering the valve loading effects. The fuel cost function is augmented with an additional sine term as in equation (8.224). HPSO is applied to solve this kind of optimization problems. Table 8.25 lists the results obtained using different swarm sizes. Increasing the swarm's size improved the HPSO performance in achieving better results at the expense of computational time.

**TABLE 8.25  OPF Results of IEEE 30-Bus System for Case 3**

| Swarm Size | 20 | 30 | 100 |
|---|---|---|---|
| Method | PSO | PSO | PSO |
| $P_{g1}$ | 47.068 | 47.059 | 47.126 |
| $P_{g2}$ | 42.911 | 42.359 | 71.366 |
| $P_{g13}$ | 8.790 | 35.902 | 8.972 |
| $P_{g22}$ | 44.728 | 37.359 | 37.391 |
| $P_{g23}$ | 8.983 | 8.826 | 8.993 |
| $P_{g27}$ | 42.044 | 20.959 | 20.777 |
| $V_{g1}$ | 1.000 | 1.000 | 1.000 |
| $V_{g2}$ | 1.099 | 1.009 | 1.097 |
| $V_{g13}$ | 1.091 | 1.017 | 1.037 |
| $V_{g22}$ | 1.087 | 1.082 | 0.982 |
| $V_{g23}$ | 1.048 | 1.057 | 1.048 |
| $V_{g27}$ | 1.029 | 1.080 | 1.088 |
| $Q_{c5}$ | 33.000 | 16.000 | 29.000 |
| $Q_{c24}$ | 35.000 | 15.000 | 12.000 |
| $T_{6-9}$ | 1.040 | 1.010 | 1.020 |
| $T_{6-10}$ | 1.010 | 1.000 | 0.950 |
| $T_{4-12}$ | 1.040 | 0.990 | 1.020 |
| $T_{27-28}$ | 0.990 | 1.030 | 1.040 |
| Total cost ($/h) | 658.416 | 645.333 | 615.250 |

# PROBLEMS AND EXERCISES

**1.** What is OPF?

**2.** State several major constraints used in OPF calculation.

**3.** What is CARD?

**4.** What is the difference between OPF and SCED?

**5.** Compare QIP-based OPF with general IP-based OPF

**6.** What is the role of the phase shifter in OPF calculation?

**7.** State the differences of several OPF methods: Newton method, gradient method, linear programming, IP, and PSO.

**8.** State "True" or "False"

8.1 OPF generally includes both real power optimization and reactive power optimization.

8.2 OPF is nonlinear model, and it cannot be solved by LP method.

8.3 OPF is an economic dispatch method.

8.4 Reactive power optimization is a simplified OPF.

8.5 Both OPF and ED must consider reactive power and voltage constraints.

8.6 All IP methods can only solve linearized OPF.

**9.** A 5-bus system is shown in Figure 8.1. The data of generators are shown in Table 8.26. The generator fuel cost is a quadratic function, that is, $f_i = a_i P_{Gi}^2 + b_i P_{Gi} + c_i$.

**TABLE 8.26    Data of Generators**

| Unit No. | $a_i$ | $b_i$ | $c_i$ | $P_{Gimin}$ | $P_{Gimax}$ | $Q_{Gimax}$ | $Q_{Gimax}$ |
|----------|-------|-------|-------|-------------|-------------|-------------|-------------|
| 1 | 46.2 | 360 | 60 | 2.0 | 3.5 | 1.5 | 2.5 |
| 2 | 39.0 | 380 | 60 | 4.0 | 6.0 | 1.0 | 2.0 |

The other data and parameters are shown in Figure 8.1, except for the load data, which are

$$S_{D3} = 3.5 + j1.1, S_{D4} = 2.1 + j1.0, S_{D5} = 1.5 + j0.6$$

Use the Newton method to solve the OPF.

**10.** The system and the corresponding data are the same as above. Use the gradient method to solve the OPF.

# REFERENCES

1. Carpentier J. Contribution e létude do dispatching economique. Bull. Soc. Franc. Elect. 1962;431–447.
2. Dommel HW, Tinney WF. Optimal power flow solutions. IEEE Trans. on Power Syst. 1968;87(10):1866–1876.

3. Alsac O, Stott B. Optimal power flow with steady-state security. IEEE Trans. on Power Syst. 1974;93:745–75.

4. Sun DI, Ashley B, Hughes A, Tinney WF. Optimal power flow by Newton approach. IEEE Trans. Power Syst. 1984;103:2864–2880.

5. Irving MR, Sterling MJH. Economic dispatch of active power with constraint relaxation. IEE Proc., Part C 1983;130(4).

6. Dias LG, El-Hawary ME. Security-constrained OPF: influence of fixed tap transformer fed loads. IEEE Trans. Power Syst. 1991;6(4):1366–1372.

7. Zhu JZ, Xu GY. Network flow model of multi-generation plan for on-line economic dispatch with security. Modeling, Simulation & Control, A 1991;32(1):49–55.

8. Zhu JZ, Xu GY. Comprehensive investigation of real power economic dispatch with N and N-1 security, Proc. of 1991 Intern. Conf. on Power Systems Technology, Beijing, 1991.

9. Zhu JZ, Xu GY. A new real power economic dispatch method with security. Electr. Power Syst. Res. 1992;25(1):9–15.

10. Papalexopoulos AD, Imparato CF, Wu FF. Large scale optimal power flow: effects of initialization decoupling and discretization. IEEE Trans. on Power Syst. 1989;4:748–759.

11. Zhu JZ, Irving MR. Combined active and reactive dispatch with multiple objectives using an analytic hierarchical process. IEE Proc., Part C 1996;143(4):344–352.

12. Chebbo AM, Irving MR. Combined active and reactive dispatch, Part I: problem formulation and solution. IEE Proc., Part C 1995;142(4):393–400.

13. Wood AJ, Wollenberg B. *Power Generation Operation and Control.* 2nd ed. New York: Wiley; 1996.

14. Zhu JZ, Chang CS. Security-constrained multiarea economic load dispatch using nonlinear optimization neural network approach, 1997 Intern. Conf. on Intelligent System Applications to Power Systems, Seoul Korea, July, 1997.

15. Lee TH, Thorne DH, Hill EF. A transportation method for economic dispatching—application and comparison. IEEE Trans. on Power Syst. 1980;99:2372–2385.

16. Hobson E, Fletcher DL, Stadlin WO. Network flow linear programming techniques and their application to fuel scheduling and contingency analysis. IEEE Trans. on Power Syst. 1984;103: 1684–1691.

17. Li WY. *Secure Economic Operation of Power Systems.* Chongqing University Press; 1989.

18. Alsac O, Bright J, Prais M, Stott B. Further developments in LP-based optimal power flow. IEEE Trans. on Power Syst. 1990;5:697–711.

19. Chen YL, Liu CC. Optimal Multi-Objective VAR Planning using an Interactive Satisfying Method IEEE PES, 1994 Summer Meeting, July, 1994, pp. 24–28.

20. Mamandur K, Chenoweth R. Optimal control of reactive power flow for improvements in voltage profiles and for real power loss minimization. IEEE Trans. on Power Syst. 1981;100:3185–3194.

21. Mansour MO, Abdel-Rahman TM. Non-linear VAR optimization using decomposition and coordination. IEEE Trans. on Power Syst. 1984;103:246–255.

22. Acha E, Perez HA, Esquivel CR. Advanced transformer control modeling in an optimal power flow using Newton's method. IEEE Trans. on Power Syst. 2000;15(1):290–298.

23. Zhu JZ, Irving MR. A new approach to secure economic power dispatch. Int. J. Electr. Power Energy. Syst. 1998;20(8):533–538.

24. Zhu JZ, Momoh JA. Multi-area power systems economic dispatch using nonlinear convex network flow programming. Electr. Power Syst. Res. 2001;59(1):13–20.

25. Momoh JA, Zhu JZ, Boswell GD, Hoffman S. Power system security enhancement by OPF with phase shifter. IEEE Transactions on Power Syst. 2001;16(2):287–293.

26. Momoh JA, Zhu JZ, Dolce JL. Optimal allocation with network limitation for autonomous space power system. AIAA Journal—J. Propul. Power 2000;16(6):1112–1117.

27. Momoh JA, Zhu JZ. Improved interior point method for OPF problems. IEEE Transactions on Power Syst. 1999;14(3):1114–1120.

28. Momoh JA, Adapa R, El-Hawary ME. A review of selected optimal power flow literature to 1993—I: nonlinear and quadratic programming approaches. IEEE Trans. Power Syst. 1999;14(1): 96–104.

29. Momoh JA, El-Hawary ME, Adapa R. A review of selected optimal power flow literature to 1993—II: Newton, linear programming and interior point methods. IEEE Trans. Power Syst. 1999;14(1):105–111.

30. Clements KA, Davis PW, Frey KD. An Interior Point Algorithm for Weighted Least Absolute value Power System State Estimation, IEEE PES Winter Meeting, 1991.

31. Ponnambalam K, Quintana VH, Vannelli A. A Fast Algorithm for Power System Optimization Problems Using an Interior Point Method, IEEE PES Winter Meeting, 1991.

32. Vargas LS, Quintana VH, Vannelli A. A Tutorial Description of an Interior Point Method and its Application to Security-Constrained Economic Dispatch, IEEWPES 1992 Summer Meeting.

33. Momoh JA, Guo SX, Ogbuobiri CE, Adapa R. The quadratic interior point method for solving power system optimization problems. IEEE Trans. on Power Syst. 1994;9.

34. Lu NC, Unum MR. Network constrained security control using an interior point algorithm. IEEE Trans. on Power Syst. 1993;8.

35. Granville S. Optimal reactive dispatch through interior-point methods. IEEE Trans. Power Syst. 1994;9:136–146.

36. Momoh JA, Brown GF, Adapa RA. Evaluation of Interior Point Methods and Their Application to Power Systems Economic Dispatch, Proc. of North American Power Symposium, Washington, D.C., October 1993, pp. 116–123.

37. Wei H, Sasaki H, Yokoyama R. An application of interior point quadratic programming algorithm to power system optimization problems. IEEE Trans. on Power Syst. 1996;11:260–266.

38. Granville S, Mello JCO, Melo ACG. Application of interior point methods to power flow un-solvability. IEEE Trans. on Power Syst. 1996;11:1096–1103.

39. Wu YC, Debs AS, Marsten RE. A direct nonlinear predictor-corrector primal–dual interior point algorithm for optimal power flows. IEEE Trans. Power Syst. 1994;9:876–883.

40. Torres GL, Quintana VH. An interior-point method for nonlinear optimal power flow using voltage rectangular coordinates. IEEE Trans. Power Syst. 1998;13:1211–1218.

41. Castronuovo ED, Campagnolo JM, Salgado R. Optimal power flow solutions via interior point method with high-performance computation techniques, in *Proc. 13th PSCC in Trondheim*, June 28–July 2, 1999, pp. 1207–1213.

42. Jabr RA, Coonick AH, Cory BJ. A prime-dual interior point method for optimal power flow dispatching. IEEE Trans. Power Syst. 2002;17(3):654–662.

43. Xie K, Song YH, Stonham J, Yu E, Liu G. Decomposition model and interior point methods for optimal spot pricing of electricity in deregulation environments. IEEE Trans. Power Syst. 2000;15(1): 39–50.

44. Zhu JZ, Yan W, Chang CS, Xu GY. Reactive power optimization using an analytic hierarchical process and a nonlinear optimization neural network approach. IEE Proc. Gener., Transm. Distrib. 1998;145(1):89–96.

45. Zhu JZ, Momoh JA. Optimal VAR pricing and VAR placement using analytic hierarchy process. Electr. Power Syst. Res. 1998;48(1):11–17.

46. Zhu JZ, Xiong XF. Optimal reactive power control using modified interior point method. Electr. Power Syst. Res. 2003;66:187–192.

47. Zhu JZ. Multi-area power systems economic dispatch using a nonlinear optimization neural network approach. Electr. Pow. Compo. Sys. 2002;31:553–563.

48. Kulworawanichpong T, Sujitjorn S. Optimal power flow using tabu search. IEEE Power Eng. Rev. 2002;22(6):37–55.

49. Walters DC, Sheble GB. Genetic algorithm solution of economic dispatch with valve point loading. IEEE Trans. Power Syst. 1993;8(3):1325–1332.

50. Wong KP, Li A, Law TMY. Advanced constrained genetic algorithm load flow method. IEE Proc. C 1999;146(6):609–618.

51. Abdul-Rahman KH, Shahidehpour SM. Application of fuzzy sets to optimal reactive power planning with security constraints. IEEE Trans. Power Syst. 1994;9(2):589–597.

52. Lai LL, Ma JT. Application of evolutionary programming to reactive power planning-comparison with nonlinear programming approach. IEEE Trans. Power Syst. 1997;12(1):198–206.

53. Jwo WS, Liu CW, Liu CC, Hsiao YT. Hybrid expert system and simulated annealing approach to optimal reactive power planning. IEE Proc. Gener., Transm. Distrib. 1995;142(4):381–385.

54. Zhang WJ, Li FX, Tolbert LM. Review of reactive power planning: objectives, constraints, and algorithms. IEEE Trans. Power Syst. 2007;22(4):2177–2186.

55. Yokoyama R, Bae SH, Morita T, Sasaki H. Multiobjective optimal generation dispatch based on probability security criteria. IEEE Trans. Power Syst. 1988;3(1):317–324.

56. Lin W-M, Cheng F-S, Tsay M-T. An improved Tabu search for economic dispatch with multiple minima. IEEE Trans. Power Syst. 2002;17:108–112.

57. Deeb N. Simulated annealing in power systems, in *Proc. IEEE Int. Conf. Man and Cybernetics*, October 1992, vol. 2, pp. 1086–1089.

58. Korsak AJ. On the question of uniqueness of stable load-flow solutions. IEEE Trans. Power App. Syst. 1972;PAS-91(3):1093–1100.

59. Vlachogiannis JG, Lee KY. A comparative study on particle swarm optimization for optimal steady-state performance of power systems. IEEE Trans. Power Syst. 2006;21(4):1718–1728.

60. Park JB, Lee KS, Shin JR, Lee KY. A particle swarm optimization for economic dispatch with nonsmooth cost functions. IEEE Trans. Power Syst. 2005;20(1):34–42.

61. Gaing ZL. *Constrained optimal power flow by mixed-integer particle swarm optimization*, in *Proc. IEEE Power Eng. Soc. General Meeting*, San Francisco, CA, 2005, pp. 243–250.

62. Zhao B, Guo CX, Cao YJ. *Improved particle swarm optimization algorithm for OPF problems*, in *Proc. IEEE/PES Power Systems Conf. Exposition*, New York, 2004, pp. 233–238.

63. Abido MA. Optimal power flow using particle swarm optimization. Int. J. Elect. Power Energy Syst. 2002;24(7):563–571.

64. AlRashidi MR, El-Hawary ME. Hybrid particle swarm optimization approach for solving the discrete OPF problem considering the valve loading effects. IEEE Trans. Power Syst. 2007;22(4):2030–2038.

65. Kodsi SKM, Cañizares CA. Application of a stability-constrained optimal power flow to tuning of oscillation controls in competitive electricity markets. IEEE Trans. Power Syst. 2007;22(4):1944–1954.

66. Fletcher R, Leyffer S. Nonlinear programming without a penalty function, Univ. of Dundee, Dundee, U.K., Numeric. Anal. Rep. NA/171, September 22, 1997.

67. Wright SJ. *Primal Dual Interior Point Methods*. Philadelphia, PA: SIAM; 1997.

68. Gay DM, Overton ML, Wright MH. A primal–dual interior method for non-convex nonlinear programming, Computing Sciences Research Center, Bell Laboratories, Murray Hill, NJ, Tech. Rep. 97-4-08, July 1997.

69. Mehrotra S. On the implementation of a primal–dual interior point method. SIAM J. Optim. 1992;2:575–601.

70. Avriel M, Golany B. *Mathematical Programming for Industrial Engineers*. New York: Marcel Dekker; 1996.

71. Eberhart R, Kennedy J. A new optimizer using particle swarm theory, in *Proc. 6th Int. Symp. Micro Machine and Human Science*, Nagoya, Japan, 1995, pp. 39–43.

72. Kennedy J, Eberhart R. Particle swarm optimization, in *Proc. IEEE Int. Conf. Neural Networks*, Perth, Australia, 1995, vol. 4, pp. 1942–1948.

73. Hu X, Shi Y, Eberhart R. Recent advances in particle swarm, in *Proc. Congr. Evolutionary Computation*, Portland, OR, 2004, vol. 1, pp. 90–97.

74. Eberhart RC, Shi Y. Guest editorial special issue on particle swarm optimization. IEEE Trans. Evol. Comput. 2004;8(3):201–203.

75. Kennedy J, Eberhart RC. *Swarm Intelligence*. San Francisco, CA: Morgan Kaufmann; 2001.
76. Parrish JK, Hammer WM. *Animal Groups in Three Dimensions*. Cambridge, U.K.: Cambridge University Press; 1997.
77. He S, Wu QH, Wen JY, Saunders JR, Patton PC. A particle swarm optimizer with passive congregation. Biosyst. 2004;78:135–147.

CHAPTER *9*

# STEADY-STATE SECURITY REGIONS

Steady-state security region analysis is important in power system operation. This chapter presents the concept and definition of the security region, and introduces several major methods used in steady-state security region analysis: the security corridor, the traditional expansion method, the enhanced expansion method, linear programming, and the fuzzy set theory.

## 9.1 INTRODUCTION

In the steady state, a power system is designed by the so-called power flow equations or the steady-state network relationships. Given a set of power injections (generators, loads), the power flow equations may be solved to obtain the operation point (voltages, angles). Therefore, a lot of power flow calculations are needed in the traditional steady-state security analysis, and the corresponding amount of computations is very huge. A method of steady-state security analysis—"steady-state security regions" has been attracting a lot of attention over the last decades [1–14]. The main idea of security regions is to obtain a set of security injections explicitly so that for security assessment one need only check whether a given injection vector lies within the security region. By doing so, the solution of power flow equations can be avoided.

The approach for steady-state security regions of power systems was first proposed by Hnyilicza et al. in 1975 [1]. Fischl et al. developed methods to identify steady-state security regions [2,3]. The idea of steady-state security regions was expanded by Banakar and Galiana, who suggest a method to construct the so-called "security corridors" for security assessment [5]. The previous security region, which was formed by using the active constraints, was implicit and there was difficulty in using it in power system security analysis and security operation. Wu and Kumagai deduced a hyperbox to approximately express the steady-state security regions, so that the disadvantages of the former methods for security regions can be overcome [6]. However, such steady-state security regions were very conservative. To avoid being conservative, Liu proposed an expanding method to obtain the hyperbox, which tended to achieve maximal security regions [7]. The expanding speed, however, was very slow because of the adoption of fixed expanding steps.

*Optimization of Power System Operation*, Second Edition. Jizhong Zhu.
© 2015 The Institute of Electrical and Electronics Engineers, Inc. Published 2015 by John Wiley & Sons, Inc.

**365**

Moreover, the fuzzy branch power constraints and $N - 1$ security constraints had not been considered in these investigations of steady-state security regions.

Zhu proposed a new expanding method of the steady-state security regions of power systems based on the fast decoupled load flow model [8,9]. For the first time, the fuzzy branch power constraints and the $N - 1$ security constraints are introduced into the study of the steady-state security regions [10–12]. Recently, Zhu also applied the optimization method to compute the steady-state security regions [13,14].

## 9.2 SECURITY CORRIDORS

### 9.2.1 Concept of Security Corridor [4,5]

In terms of $x$, the rectangular coordinate components of the complex bus voltages, the load flow equations can be expressed by

$$z = [L(x)]\, x \tag{9.1}$$

where $L(x)$ is a real matrix equal to half the Jacobian of the load flow equations and $z$ is the vector of specified nodal injections. Without loss of generality, one can assume that there is no mixed (hybrid) bus in the system, which implies that

$$z = \begin{bmatrix} u \\ -d \end{bmatrix} \tag{9.2}$$

where $u$ is the vector of control variables (voltage levels at the generation buses and real power generations at the PV buses), $d$ is the demand vector (real and reactive loads at PQ buses).

In terms of $x$, a load flow–dependent variable can be expressed in the general form

$$y = x^T[Y]x \tag{9.3}$$

where $Y$ represents the functional dependence of $y$ on the network parameters, which is the sparse, constant, symmetric matrix. In the conventional load flow formulation, the line power flows, reactive power generations, the square of voltage levels at the load buses, and the real power injection at the slack bus are among the dependent variables.

Considering the network constraints, equation (9.3) will be restricted as follows:

$$y_{j\min} \le y_j \le y_{j\max}, \quad j = 1, \dots, N_{dp} \tag{9.4}$$

where $y_{j\min}, y_{j\max}$ are the lower and upper bounds of the constraint, respectively. $N_{dp}$ is the total number of such dependent variables in the system.

Since each point in the $x$-space can be mapped into the $z$-space through equation (9.1), one can define the set of all injections $z$, which satisfy a specific operating constraint. For instance, the set $z^j$ defined by

$$z^j = \{z | z = [L(x)]x;\ x \in x^j\} \tag{9.5}$$

represents the map of the following set:

$$x^j = \{x | y_{j\min} \le x^T [y_j] x \le y_{j\max}\} \tag{9.6}$$

This is into the $z$-space. Let $S_z$ be the set of all the injections satisfying the various operating constraints on the intact system. It can be defined as follows:

$$S_z = H_z \cap \left( \bigcap_{j=1}^{2N_{dp}} z^j \right) \tag{9.7}$$

The hyperbox $H$ is defined by the known limitations on the control variables and conservative bounds on the load variables, namely,

$$H_z = \{z | z_{\min} \le z \le z_{\max}\} \tag{9.8}$$

If we select an expansion point, the constraints (9.4) can be explicitly expressed through a Taylor series expansion of $y$.

A more demanding security set is the invulnerability set. This set contains all the injection vectors that do not violate any of the system's operating limits, while it is intact or subjected to a list of probable outages.

Since the variations of the loads $d(t)$ can be predicted using a bus-load forecast, and a control vector $u(t)$ can be computed which satisfies the security requirements, a predicted trajectory of the injection vector $z(t)$ can be established. Therefore, one can introduce the concept of a security corridor. Such a corridor can be thought of as a "tube of varying width" in $z$-space surrounding the predicted trajectory and lying entirely within the security region $S_z$. The security corridor $E_S^c$ has two important properties:

**(1)** It is characterized by a very small number of inequalities compared to $S_z$.

**(2)** Since the security corridor $E_S^c$ is a subset of $S_z$ with some "width" in all directions, the actual trajectory can deviate from the predicted one while still remaining inside $E_S^c$ and hence in $S_z$.

The security corridor then permits the monitoring of security by the very simple task of verifying that the actual injection vector $z$ belongs to $E_S^c$. If $z$ is inside the corridor, it becomes unnecessary to test all other security inequalities or to run repeated load flows. In the infrequent cases when the actual trajectory deviates beyond the limits of the corridor, a conventional security analysis based on load flow computations would have to be carried out. The advantage gained is that most of the time quite wide excursions in the trajectory are needed to go outside the security corridor. The typical periodic and stochastic load behavior will normally not violate the security corridor limits. Finally, the security corridor greatly facilitates the computation as well as verification of the effectiveness of control actions such as emergency or preventive rescheduling.

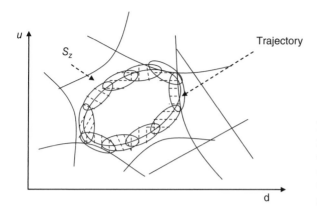

Figure 9.1   A pictorial representation of a nominal daily trajectory and its associated security corridor (shaded) inside the security set.

The corridor can be characterized via a small number of overlapping ellipsoids whose centers lie on the predicted trajectory [5]. A pictorial illustration of such an arrangement is given in Figure 9.1.

Since the ellipsoids are expressible by simple, explicit functions and they can be oriented to lie along the trajectory, they seem to be the logical choice for this purpose. The $N$ ellipsoids forming the corridor are defined by

$$E^i = \{z | (z - z_i)^T [A_i](z - z_i) \le c_i\} \qquad i = 1, \dots, N \qquad (9.9)$$

where $A$ is a constant, symmetric, positive definite matrix representing the orientation of $E^i$. The vector $z_i$ represents the center of $E^i$, while the constant $c_i$ controls its size. The union of the ellipsoids, denoted by $E^c$, forms the corridor, that is,

$$E^c = \bigcup_{i=1}^{N} E^i \qquad (9.10)$$

The secure part of $E^c$ is then referred to as the "security corridor," and is defined by

$$E_s^c = S_z \cap E^c = \bigcup_{i=1}^{N} E_s^i \qquad (9.11)$$

where $E_s^i$ is the secure part of $E^i$.

## 9.2.2   Construction of Security Corridor [5]

It can be seen from equations (9.9) and (9.10) that the key to constructing a security corridor is to select $z_i$ and $A$. In order to have the predicted trajectory surrounded by the corridor, the centers of the ellipsoids $z_i$, $i = 1, \dots, N$, must be on the trajectory. These points should be selected inside $S_z$ so that $E$ is not empty.

The number of ellipsoids $N$ needed to cover a trajectory is small when the ellipsoids are oriented properly along the trajectory. Let the unit tangent to the trajectory at $z_i$ be represented by $a_i$. The ellipsoid $E^i$ is laid along the trajectory by making sure that its major axis lies along $a_i$. This can be accomplished by defining $A_i$ as follows.

$$[A_i] = \lambda_{imax}[I] - (\lambda_{imax} - \lambda_{imin})[a_i a_i^T], \quad \lambda_{imax} > \lambda_{imin} > 0 \tag{9.12}$$

One can easily show that the eigenvalues of $A_i$ are all $\lambda_{imax}$ except one which is $\lambda_{imin}$, and that the engenvector corresponding to $\lambda_{imin}$ is $a_i$. In addition, the storage requirements of $A_i$ being very low, its inverse can be analytically obtained as follows:

$$[A_i]^{-1} = \frac{[I] + \left(\frac{\lambda_{imax}}{\lambda_{imin}} - 1\right)[a_i a_i^T]}{\lambda_{imax}} \tag{9.13}$$

According to the expression of the security corridor in equation (9.11) and the expression of the security sets in equation (9.7), we get

$$E_s^i = S_z \cap E^i = H_z \cap \left(\bigcap_{j=1}^{2N_{dp}} z^j\right) \cap E^i \tag{9.14}$$

or

$$E_s^i = H_z \cap \left\{\bigcap_{j=1}^{2N_{dp}} z^j \cap E^i\right\} \tag{9.15}$$

For a relatively small $E^i$, (i.e., small $c_i$) the majority of the sets $z^j$ contain the entire set $E^i$. For such sets one can write

$$z^j \supset E^i \Rightarrow z^j \bigcap E^i = E^i \tag{9.16}$$

Those few that intersect $E^i$ must be identified and characterized explicitly. This can be accomplished by solving the following optimization problem:

$$\min c_{ij} = (z - z_i)^T [A_i](z - z_i), \quad z \in \text{Ext}(z^j) \text{ for } j = 1, \dots, 2N_{dp} \tag{9.17}$$

Since $z_i \in S_z$, the intersection $z^j \cap E^i$ is always nonempty. In terms of $x$, the above problem can be written as

$$\min c_{ij} = \{[L(x)]x - z_i\}^T [A_i]\{[L(x)]x - z_i\} \tag{9.18}$$

such that

$$y_{jmin} \le x^T [y_j]x \le y_{jmax} \tag{9.19}$$

To simplify the above optimization problem, $z^j$ can be approximated as follows.

$$\hat{z}^j = \{z|D_j^T(x_j)z \le y_{j\,limit}\} \tag{9.20}$$

The solution to equation (9.17) with $z \in \text{Ext}(\hat{z}^j)$ is simply

$$\hat{c}_{ij}^* = [y_{j\,limit} - D_j^T(x_j)z_i]^2/\delta_{ij} \tag{9.21}$$

where $x_i$ is the load flow solution to $z_i$ and

$$\delta_{ij} = D_j^T(x_i)[A_i]^{-1}D_j(x_i) \tag{9.22}$$

Thus the corresponding approximated security corridor $E_s^i$ is expressed explicitly as follows.

$$\widehat{E}_s^i = H_z \bigcap E^j \bigcap \{z|D_j^T(x_i)z \le y_{j\,\text{limit}}, \ \forall j \in I^i\} \tag{9.23}$$

where $I^i$ is an integer set, and its elements are defined as follows.

$$j \in I^i \quad \text{if } \widehat{c}_{ij}^* < c_i$$

It is noted that this approximation requires that the solution point is relatively close to $z_i$.

A relatively simple but sufficiently indicative measure of the size of $E^i$ is $\Delta P_{d\max}\%$, the maximum percentage change that the total real demand $P_d$ can have inside $E^i$ with respect to $P_{di}$, the total demand at $z_i$. To compute this quantity, we need to solve the following problem.

$$\max P_d = -\alpha^T z \tag{9.24}$$

such that

$$(z - z_i)^T[A_i](z - z_i) = c_i \tag{9.25}$$

The entries of the vector $\alpha$ are either zero or 1, with ones appearing at locations which correspond to real power demands $z$.

In summary, the steps of constructing a security corridor are as follows.

1. Choose $z_i$ from the trajectory and run a load flow to make sure that $z_i \in S_z$.
2. Compute $a_i$ and define the matrix $A_i$.
3. Compute values of $\widehat{c}_{ij}^*$, $j = 1, 2, \ldots, N_{dp}$ using equation (9.21) and tabulate them in ascending order.
4. Decide on $N_{i\max}$, the maximum number of elements that $I^i$ can have.
5. Assign to $c_i$ the first $N_{i\max} + 1$ values of $\widehat{c}_{ij}^*$ in the list, one at a time. For each value, compute and tabulate $\Delta P_{d\max}\%$, as well as the times when the trajectory enters and leaves the resulting $E^i$.
6. Compare the results to establish what value of $c_i$ chosen from those examined, could offer a reasonable $\Delta P_{d\max}\%$ and sufficient overlapping with $E^{i-1}$, while the number of elements in $I^i$ is small ($\le N_{i\max}$). If such a $c_i$ cannot be found, then either change the eigenvalues of $A_i$ or choose $z_i$ closer to $z_{i-1}$ and repeat the relevant steps.

Note that in the last step, it is assumed that the value of $c_{i-1}$ is already fixed, and the time when the trajectory enters and leaves $E^{i-1}$ as well as its associated $\Delta P_{d\max}\%$ are known. Sufficient overlapping is achieved between $E^i$ and $E^{i-1}$ when a significant portion (normally 25%) of the time spent by the trajectory inside $E^{i-1}$ is also part of the time that it spends inside $E^i$. Since the trajectory is usually available in a piecewise linear form, the computation of the trajectory's "arrival" and "departure" times for a given ellipsoid is quite simple to calculate.

The number of elements in $I^i$ is limited here by $N_{imax}$ mainly because of the non-sparsity of the vectors $D_j(x_i)$, $j \in I^i$ $f : ..; (\sim i)$, $i = 1, \ldots, N$, which have to be computed and stored. The vectors $D_j(x_i)$ can be obtained by performing a single constant Jacobian Newton power flow iteration, that is,

$$[L(x_0)]^T D_j(x_0) = [Y_j]x_0 \tag{9.26}$$

## 9.3    TRADITIONAL EXPANSION METHOD

### 9.3.1    Power Flow Model

Given a power system, suppose the total number of branches is $m$; the total number of buses is $n$. Bus $n$ is the slack bus, buses 1 to $n_d$ are load buses and buses $n_d + 1$ to $n - 1$ are PV buses (the number of PV buses is $NG$). According to fast decoupled power flow, the active power flow equations can be written as follows.

$$[P] = [B'][\theta] \tag{9.27}$$

$$[\theta_L] = [A]^T[\theta] \tag{9.28}$$

where $P$ is the vector of active power injections, $\theta$ is the vector of node voltage angle, $\theta_L$ is the vector of node voltage angle differences across lines, and $A$ is the relation matrix between the nodes and branches.

From equations (9.27) and (9.28) we can obtain

$$[\theta_L] = [A]^T[B']^{-1}[P] \tag{9.29}$$

where

$$B_{ij}' = -1/X_{ij} \tag{9.30}$$

$$B_{ii}' = \sum_{\substack{j=1 \\ j \neq i}}^{n} \left( \frac{1}{X_{ij}} \right) \tag{9.31}$$

$X_{ij}$ and $B_{ij}$ are the reactance and susceptance of branch $ij$, respectively.

If we use reactive injection current to replace the reactive injection power, the reactive power flow equations can be written as follows.

$$[I] = [B''][V] \tag{9.32}$$

$$[V] = [B'']^{-1}[I] \tag{9.33}$$

where

$$I_i \approx \frac{Q_i}{V_i} \tag{9.34}$$

$$B''_{ij} = -\frac{X_{ij}}{R_{ij}^2 + X_{ij}^2} \tag{9.35}$$

$$B''_{ii} = \sum_{\substack{j=1 \\ j \neq i}}^{n} (-B_{ij}) \tag{9.36}$$

### 9.3.2   Security Constraints

The following security constraints will be considered in the study of steady-state security regions:

**(1)** Generator power output constraints

$$P_{Gi\min} \leq P_{Gi} \leq P_{Gi\max} \tag{9.37}$$

$$\frac{Q_{Gi\min}}{V_i} \leq I_{Gi} \leq \frac{Q_{Gi\max}}{V_i} \tag{9.38}$$

For the slack bus unit, the power output constraints are

$$P_{Gn\min} \leq -\sum_{i=1}^{n-1} P_i \leq P_{Gn\max} \tag{9.39}$$

$$\frac{Q_{Gn\min}}{V_n} \leq -\sum_{i=1}^{n-1} I_i \leq \frac{Q_{Gn\max}}{V_n} \tag{9.40}$$

where subscripts "min" and "max" represent the lower and upper bounds of the constraints, respectively.

**(2)** Branch power flow constraints

$$-\theta_{ij\max} \leq \theta_{ij} \leq \theta_{ij\max} \tag{9.41}$$

In the normal operation status of power systems, the branch reactive power constraints can be neglected.

### 9.3.3   Definition of Steady-State Security Regions

The aim of steady-state security analysis is to analyze and check whether all elements in the system would operate within constraints as defined by a given set of input data and information. Therefore, the steady-state security regions can be represented by a set of power injections which satisfy the power flow equations and security constraints.

$$R_P = \{P/\exists \theta \in R, \text{ and } (f_P(\theta) = P) \in R\} \tag{9.42}$$

$$R_Q = \{I/\exists V \in R, \text{ and } (f_Q(V) = I) \in R\} \tag{9.43}$$

where $R_P$ and $R_Q$ are the active and reactive power steady-state security regions, $R$ is the set of security constraints, and $f$ is the set of load flows.

On one hand, the calculation methods for active and reactive power steady-state security regions are the same. On the other hand, the active power security is relatively more important because the reactive power problem is generally a local issue. Thus we focus on active security regions in this chapter.

In practical terms, it is desired to obtain each security region to cover as many operating points as possible. Hence, the idea of maximal security region was proposed. $\Omega_P^* \in R_P$ is said to be a *maximal security region* if there exists no hyperbox $\Omega_P$ in $R_P$, such that $\Omega_P$ strictly contains $\Omega_P^*$, that is, $\Omega_P^* \not\subset \Omega_P$. In other words, a hyperbox $\Omega_P^*$ is maximal if it is impossible to extend it in any dimension with $R_P$.

### 9.3.4  Illustration of the Calculation of Steady-State Security Region

Generally, the expanding method is used to compute the maximal security region. The idea is to select the initial operation point first, and then expand the initial point by adding the fixed step until we reach the limit of any of the constraints.

For example, there is a simple system with two generators. The feasible region can be shown in Figure 9.2. The steady-state security region obtained by the expanding method is shown in Figure 9.3.

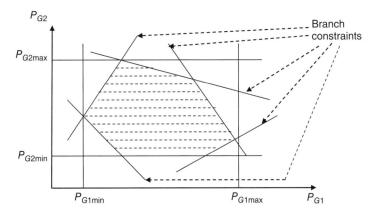

Figure 9.2   Feasible region of illustrating system.

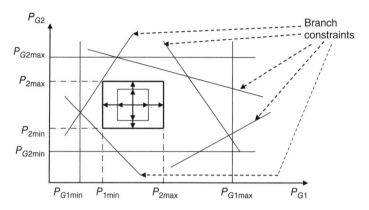

Figure 9.3   Security region obtained by the expanding method.

**TABLE 9.1   The Security Region Results for IEEE 6-Bus System (p.u.)**

| Regions | $P_{G4}$ | $P_{G5}$ | $P_{G6}$ |
|---|---|---|---|
| $P_{imax}$ | 3.7500 | 2.6490 | 2.5510 |
| $P_{imin}$ | 2.4490 | 1.4000 | 0.0000 |

**TABLE 9.2   The Security Region Results for IEEE 30-Bus System (p.u.)**

| Regions | $P_{G2}$ | $P_{G5}$ | $P_{G8}$ | $P_{G11}$ | $P_{G13}$ |
|---|---|---|---|---|---|
| $P_{imax}$ | 0.7120 | 0.4020 | 0.3500 | 0.3000 | 0.4000 |
| $P_{imin}$ | 0.4280 | 0.1500 | 0.1480 | 0.1000 | 0.1770 |

### 9.3.5   Numerical Examples

The expanding method for computing power steady-state security region is further illustrated by IEEE 6-bus and 30-bus systems. The parameters of systems are taken from the references [6,8,11]. The obtained security regions for two systems are shown in Tables 9.1 and 9.2, respectively.

## 9.4   ENHANCED EXPANSION METHOD

### 9.4.1   Introduction

Since computing speed is very slow in the previous expanding methods, a new expanding method is presented in this section. In this expanding method, security constraints are divided into two groups and the expanding calculations are first carried out in the first group of constraints with small constraint margins. In additional, the failure probability of branch temporary overload and the capability of tapping the potentialities for branch power capacity are considered on the basis of fuzzy sets. Furthermore, an idea of "$N - 1$ constraint zone" is also adopted to calculate the all $N - 1$ security constraints so as to reduce the computation burden.

## 9.4.2 Extended Steady-State Security Region

***Security Constraints***  The same power flow model as in Section 9.3 is used here. The following security constraints will be taken in the study of steady-state security regions:

**(1)** Generator active power output constraints

$$P_{Gimin} \leq P_{Gi} \leq P_{Gimax} \tag{9.44}$$

$$P_{Gnmin} \leq -\sum_{i=1}^{n-1} P_i \leq P_{Gnmax} \tag{9.45}$$

**(2)** Fuzzy branch load flow constraints

$$-\theta_{ijmax} \leq \theta_{ij} \leq \theta_{ijmax} \tag{9.46}$$

or

$$-P_{ijmax}/b_{ij} \leq \theta_{ij} \leq P_{ijmax}/b_{ij} \tag{9.47}$$

When the limits of branch power flow are not determined beforehand, equations (9.46) and (9.47) cannot be directly adopted. During the stage of planning and system design, values of branch power flow limits are given to allow for some margin of security and reliability. In fact, it is possible to tap extra potentialities of branch power flow capacity in some cases, so as to allow some margins to be expanded. However, over-tapping of potentialities for branch power flow capacity will lead to some problems such as high power losses and unreliability. Hence, it is conceptually sound to replace equation (9.46) or (9.47) by fuzzy constraints. By changing each bilateral inequality constraint into two single inequality constraints, the branch active power constraints can be expressed as follows.

$$\mu_{\theta_{ij}}(\theta_{ij}) = \begin{cases} 1, & \text{if } \theta_{ij} \leq \theta_{ijmax} \\ L\left(\theta_{ijmax}, \theta'_{ijmax}; \theta_{ij}\right), & \text{if } \theta_{ijmax} \leq \theta_{ij} \leq \theta'_{ijmax} \\ 0, & \text{if } \theta_{ij} \geq \theta'_{ijmax} \end{cases} \tag{9.48}$$

where $L$ is a droop function in which $\theta_{ijmax}$, $\theta'_{ijmax}$ are its parameters, and $L = 1$ when $\theta_{ij} = \theta_{ijmax}$, $L = 0$ when $\theta_{ij} = \theta'_{ijmax}$. The fuzzy branch power constraint is as shown in Figure 9.4, in which $\theta'_{ijmax}$ represents the tapping limit of potentialities for the branch power capacity.

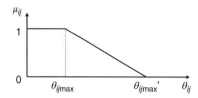

Figure 9.4    Fuzzy branch power flow constraint.

Substituting equation (9.29) in equations (9.46) and (9.47), and similarly changing each bilateral inequality constraint into two single inequality constraints, the fuzzy branch active power constraints can be expressed as follows.

$$[A_1][P] \leq [\theta] \tag{9.49}$$

where

$$[A_1] = [A]^T [B']^{-1} \tag{9.50}$$

Dividing the matrix $A_1$ into the generator node submatrix and the load node submatrix, that is, $A_G$ and $A_d$, equation (9.49) can be written as follows.

$$[A_G][P_G] \leq [\theta_G] \tag{9.51}$$

where

$$[\theta_G] = [\theta] - [A_d][P_d] \tag{9.52}$$

According to Figure 9.4, equation (9.52) can be implemented with fuzzy operation under the $\lambda$-cut of fuzzy set (see the following).

**Definition of Steady-State Security Regions**    As defined in Section 9.3, the active power steady-state security regions can be represented by a set of active power injections, which satisfy the load flow equations and security constraints.

$$R_P = \{P/\exists \theta \in R, \text{ and } (f_P(\theta) = P) \in R\} \tag{9.53}$$

where $R$ includes the set of fuzzy security constraints.

In addition, from the economic point of view, the operating regions expressed in terms of power injections may still be conservative. This is because load variations are allowed in constructing the regions, which can be known from the definition of power injection $P_i = P_{Gi} - P_{Di}$. The bigger the range of the load positive variations (i.e., increase), the smaller the obtained region is (i.e., more conservative). If the load demands are fixed at, say, the base values, a security region in terms of the generators, which is equivalent to the region of power injections under the load determination, can be considered.

### 9.4.3   Steady-State Security Regions with N−1 Security

$N - 1$ security means that the line flows will not exceed the settings of protective devices for the intact lines when any branch has an outage. Many works have been done pertaining to $N - 1$ security in the study of power system economic dispatch [15–19], but less in the study of steady-state security regions.

The $N - 1$ steady-state security region is defined as a set of node power injections that satisfies not only the load flow equations and $N$ security constraints but also $N - 1$ security constraints.

$$R_{PN} = \{P/\exists \theta \in R_N, \text{ and } (f_P(\theta) = P) \in R_N\} \tag{9.54}$$

where $R_{PN}$ is the active power steady-state security region with $N - 1$ security. $R_N$ is the set of $N$ and $N - 1$ security constraints.

Obviously, the crux of the $N - 1$ steady-state security regions is to perform the $N - 1$ security analysis (i.e., the calculation of $N - 1$ security constraints). The "$N - 1$ constrained zone," which is discussed in Chapter 5, will be coordinated with the steady-state security regions.

### 9.4.4   Consideration of the Failure Probability of Branch Temporary Overload

In Section 9.4.2, we discussed the problem of tapping the potentialities of branch power flow capacity. In fact, it corresponds to the problem of whether the branch may temporary overload in a practical power system under some case. Therefore, the value of $\theta'_{ij\max}$, which is the limit of the capability of tapping the potentialities for branch power capacity, will be determined according to the particular case of a practical power system.

Suppose the average overloading time of a branch is AOT. The average overloading ratio of the branch can be written as follows.

$$\eta_{ij} = \frac{1}{(\text{AOT})_{ij}} \quad ij \in NL \tag{9.55}$$

It is assumed that the failure probability of branch temporary overloading is Poisson distributed. It can be expressed as

$$p_{ij} = 1 - e^{-\eta_{ij}T} \tag{9.56}$$

where, $p_{ij}$ is the failure probability of branch $ij$ overload; $T$ is system operation time.

Obviously, the average overloading time AOT of the branch is random and uncertain. It is dealt with a fuzzy number in this case. If AOT is a trapezoidal fuzzy number, as shown in Figure 9.5, the fuzzy number $\eta_{ij}$ from equation (9.55) is also trapezoidal. Moreover, the fuzzy failure probability of branch temporary overload $p_{ij}$ computed from equation (9.56) is also dealt with a trapezoidal fuzzy number.

The all branches' failure probability $p_{ij}$ ($ij = 1, 2, \ldots \ldots NL$) under any $\lambda$-cut of fuzzy set $\mu$ can be obtained from equation (9.56). A ranking list, which reflects the

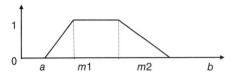

Figure 9.5 Trapezoidal fuzzy number.

relative capability of tapping potentialities for different branches, is acquired according to the value of $p_{ij}$. So the limit of the capability of tapping the potentialities for branch power capacity $\theta'_{ijmax}$ can be determined easily according to the ranking list.

To acquire higher security and reliability when fuzzy steady-state security regions are used in the practical operation of power systems, the branches with big failure probability will not be allowed to be temporarily overloaded. It means that the branch power capacity of these branches cannot be tapped by the potentialities, that is, $\theta'_{ijmax} = \theta_{ijmax}$ in this case. Therefore, we define a performance index $PI$. If the failure probability of branch overload under the $\lambda$-cut of fuzzy set $\mu$ is bigger than $PI$, that is,

$$\mu_{pij} > PI \tag{9.57}$$

then the corresponding branches will not be allowed to be temporarily overloaded.

### 9.4.5 Implementation

In the enhanced expanding method, security constraints are divided into two groups and the expanding calculations are first carried out in the first group of constraints with small constraint margins. If the maximal region is not obtained after the calculation is finished in the first group, the expanding computation will be continued in the second group until the security regions cannot be further expanded.

***Method of Step-Size Calculation*** Assume that there are $m$ inequalities in equation (9.47), in which the $i$th inequality constraint (under $\lambda$-cut of the fuzzy set) is as follows.

$$\sum_j a_{Gij} P_{Gi} < \mu_{\theta ij}(\theta_{ij}) \quad i = 1, ....., NG \tag{9.58}$$

Suppose $\Omega$ is a hyperbox, in which the generator power outputs are control variables. If not all summits of $\Omega$ have reached the boundary of $R$, $\Omega$ can still be expanded by solving the following $m$ equations.

$$\sum_i a_{Gij} P^*_{Gi} = \mu_{\theta ij}(\theta_{ij}) \quad j = 1, ... ... , m \tag{9.59}$$

where

$$P^*_{Gi} = \begin{cases} P_{imax} + \varepsilon, & \text{if } a_{Gij} > 0 \\ P_{imin} - \varepsilon, & \text{if } a_{Gij} < 0 \\ 0, & \text{if } a_{Gij} = 0 \end{cases} \tag{9.60}$$

$P^*_{Gi}$ and $\varepsilon$ can be obtained from equation (9.60). Let $\varepsilon_{min} = \min\{\varepsilon_j, j = 1, 2, \ldots \ldots, m\}$, which is taken as the calculation step, the new expanding security region can be obtained as follows.

$$\Omega = \{P_G/P^*_{imin} \leq P_{Gi} \leq P^*_{imax}\} \qquad i = 1, 2, \ldots NG \qquad (9.61)$$

$$P^*_{imin} = P_{imin} - \varepsilon_{min} \qquad (9.62)$$

$$P^*_{imax} = P_{imax} + \varepsilon_{min} \qquad (9.63)$$

***Steps of New Expanding Method***   The calculation steps of the new expanding method are given as follows [8].

Step 1:   Select the generators' operating point $P_{Gi}{}^0$ as the initial expanding point. Then the initial security regions can be expressed as

$$\Omega^0 = \{P_G/P^0_{imin} \leq P_{Gi} \leq P^0_{imax}, \; i = 1, \ldots, \; NG\} \qquad (9.64)$$

$$P^0_{imin} = P^0_{imax} = P_{Gi}{}^0 \qquad (9.65)$$

Let iteration number $K = 0$, and mark the variables (or indices) $V_i^M = V_i^m = 1$, $i = 1, \ldots, NG$.

Step 2:   Obtain $\varepsilon_j$ $(j = 1, \ldots \ldots, m)$ according to the method of step calculation from equation (9.60). Then $\varepsilon_{min}$ can be found. A threshold value is defined as follows.

$$\varepsilon_h = \frac{\varepsilon_{min}}{\beta} \qquad (9.66)$$

where $\beta$ is a constant.

Then, the $m$ constraints will be divided into two groups based on the threshold value $\varepsilon_h$. Suppose there are $m_1$ constraints with $\varepsilon \leq \varepsilon_h$ (called group one), and there are $m_2$ constraints with $\varepsilon > \varepsilon_h$ (called group two).

Step 3:   Calculate $\varepsilon_{min}$ in the $m_1$ constraints of group one, that is,

$$P^*_{Gi} = \begin{cases} P_{imax} + \varepsilon, & \text{if } a_{Gij} > 0, \text{ and } V_i^M \neq 0 \\ P_{imax} & \text{if } a_{Gij} > 0, \text{ and } V_i^M = 0 \\ P_{imin} - \varepsilon, & \text{if } a_{Gij} < 0, \text{ and } V_i^m \neq 0 \\ P_{imin}, & \text{if } a_{Gij} < 0, \text{ and } V_i^m = 0 \\ 0, & \text{if } a_{Gij} = 0 \end{cases} \qquad (9.67)$$

then $\varepsilon_{min} = \min\{\varepsilon_j\}$ $(j = 1, 2, \ldots, m_1)$,

Step 4: Let $K = k$, then the security regions can be obtained, that is,

$$\Omega^k = \{P_G/P_{i\min}{}^k \le P_{Gi} \le P_{i\max}{}^k, \quad i = 1, \ldots\ldots, NG\} \qquad (9.68)$$

$$P_{i\max}{}^k = P_{i\max}{}^{k-1} + \varepsilon_{\min} \; V_i^M \qquad (9.69)$$

$$P_{i\min}{}^k = P_{i\min}{}^{k-1} - \varepsilon_{\min} \; V_i^m \qquad (9.70)$$

Step 5: Find the inequality constraint with $\varepsilon_j = \varepsilon_{\min}$ and let the corresponding $V_i^M = V_i^m = 0$.

Step 6: Stop if $V_i^M = V_i^m = 0$ for all $i = 1, \; 2, \; \ldots\ldots, NG$. Otherwise, let $k = k + 1$, go back to step 3.

Step 7: If $k = m_1$ but some $V_i^M$ and $V_i^m$ are still not zero, step 3–6 will be repeated in the second group, which contains $m_2$ constraints, that is, until $V_i^M = V_i^m = 0$ for all $i = 1, \ldots\ldots, NG$.

In this way, the maximal security regions are obtained as follows.

$$\Omega = \{P_G/P_{i\min} \le P_{Gi} \le P_{i\mathrm{Max}}, \quad i = 1, \ldots\ldots, NG\} \qquad (9.71)$$

### 9.4.6 Test Results and Analysis

The enhanced steady-state security region technique including the model and its algorithm are tested with the IEEE 6-bus and 30-bus systems. Suppose the system operation time is 150 h. The performance index of branch failure probability $PI$ is 0.085.

To enhance the calculation speed, the following two measures are adopted in the new expanding method. The first is the adoption of calculation step (not fixed step), and the second is that the constraints are divided into two groups based on the threshold value shown in the equation (9.54). Obviously, the value of $\beta$ will produce some effect in expanding speed. We found from a great number of numerical examples and calculations that satisfactory results can be obtained when $\beta$ is selected as a gold separation constant, that is, $\beta = 0.618$.

The IEEE 6-bus system contains eight branches. The average overloading time ACTs of the branches are assumed as in Table 9.3. The failure probability of branch temporary overloading for the IEEE 6-bus system can be computed and shown in Table 9.4. It can be known from Table 9.4 that the values of failure probability for all branches are less than $PI$. It means that the power capacity for all branches in the IEEE 6-bus system can be tapped by the potentialities. The fuzzy line power capacities are given as: $P_{ij\max} = 1.0, 3.0, 3.0, 1.6, 1.6, 0.95, 3.0, 0.25$, respectively; and $P'_{ij\max} = 1.08, 3.5, 3.3, 2.0, 1.8, 1.3, 3.5, 0.28$, respectively.

The IEEE 30-bus system contains 41 branches. The corresponding average overloading times AOTs of the branches are assumed as in Table 9.5. The failure probability of branch temporary overloading for the IEEE 30-bus system can be computed and are shown in Table 9.6.

**TABLE 9.3    The Average Overloading Time for IEEE 6-Bus System**

| Branch No. | AOT (h) | | | |
|---|---|---|---|---|
| | $a$ | $m1$ | $m2$ | $b$ |
| 1 | 1834 | 1868 | 1898 | 1922 |
| 2 | 1888 | 1922 | 1959 | 2027 |
| 3 | 1845 | 1882 | 1907 | 1949 |
| 4 | 2081 | 2127 | 2150 | 2190 |
| 5 | 1992 | 2048 | 2081 | 2123 |
| 6 | 2108 | 2152 | 2196 | 2240 |
| 7 | 1888 | 1922 | 1959 | 2027 |
| 8 | 1854 | 1896 | 1919 | 1961 |

**TABLE 9.4    The Branch Failure Probability for IEEE 6-Bus System**

| Branch No. | $p_{ij}$ | | | |
|---|---|---|---|---|
| | $a$ | $m1$ | $m2$ | $b$ |
| 1 | 0.075 | 0.076 | 0.077 | 0.079 |
| 2 | 0.071 | 0.074 | 0.075 | 0.076 |
| 3 | 0.074 | 0.076 | 0.077 | 0.078 |
| 4 | 0.066 | 0.067 | 0.067 | 0.070 |
| 5 | 0.068 | 0.070 | 0.070 | 0.072 |
| 6 | 0.065 | 0.066 | 0.067 | 0.069 |
| 7 | 0.071 | 0.074 | 0.075 | 0.076 |
| 8 | 0.074 | 0.075 | 0.076 | 0.078 |

**TABLE 9.5    The Average Overloading Time for IEEE 30-Bus System**

| Branch No. | AOT (h) | | | |
|---|---|---|---|---|
| | $a$ | $m1$ | $m2$ | $b$ |
| 1 | 1600 | 1640 | 1685 | 1725 |
| 2 | 1622 | 1655 | 1690 | 1750 |
| 3 | 1992 | 2048 | 2081 | 2123 |
| 4 | 1606 | 1640 | 1685 | 1725 |
| 5 | 1655 | 1690 | 1730 | 1780 |
| 6 | 1888 | 1922 | 1959 | 2027 |
| 7 | 1725 | 1750 | 1790 | 1834 |
| 8 | 2300 | 2365 | 2410 | 2470 |
| 9 | 1750 | 1800 | 1855 | 1888 |
| Others | 2300 | 2365 | 2410 | 2470 |

TABLE 9.6   The Branch Failure Probability for IEEE 30-Bus System

| Branch No. | $p_{ij}$ | | | |
|---|---|---|---|---|
| | $a$ | $m1$ | $m2$ | $b$ |
| 1 | 0.0833 | 0.0852 | 0.0874 | 0.0895 |
| 2 | 0.0821 | 0.0849 | 0.0866 | 0.0883 |
| 3 | 0.0680 | 0.0700 | 0.0700 | 0.0720 |
| 4 | 0.0833 | 0.0852 | 0.0874 | 0.0892 |
| 5 | 0.0808 | 0.0831 | 0.0849 | 0.0866 |
| 6 | 0.0710 | 0.0740 | 0.0750 | 0.0760 |
| 7 | 0.0790 | 0.0804 | 0.0821 | 0.0833 |
| 8 | 0.0589 | 0.0603 | 0.0615 | 0.0631 |
| 9 | 0.0760 | 0.0777 | 0.0800 | 0.0821 |
| Others | 0.0589 | 0.0603 | 0.0615 | 0.0631 |

TABLE 9.7   The Fuzzy Line Power Capacities for IEEE 30-Bus System

| Branch No. | $P_{ij\max}$ (p.u.) | $P_{ij\max}{}'$ (p.u.) |
|---|---|---|
| 1 | 1.30 | 1.30 |
| 2 | 1.30 | 1.30 |
| 3 | 0.65 | 0.80 |
| 4 | 1.30 | 1.30 |
| 5 | 1.30 | 1.30 |
| 6 | 0.60 | 0.80 |
| 7 | 0.90 | 1.20 |
| 8 | 0.70 | 1.00 |
| 9 | 1.30 | 1.50 |
| Others* | 0.65 | 0.80 |
| Others† | 0.32 | 0.50 |
| Others‡ | 0.16 | 0.25 |

*The power capacities of these lines are 0.65.

†The power capacities of these lines are 0.32.

‡The power capacities of these lines are 0.16.

It can be observed from Table 9.6 that the values of failure probability for branches 1, 2, 4, and 5 are higher than the *PI*. This means that the power capacity for these branches cannot be tapped by the potentialities. The fuzzy line power capacities of the 30-bus test system are listed in Table 9.7.

The calculating results are shown in Tables 9.8–9.13. Tables 9.8 and 9.11 provide the calculation results of security regions for the IEEE 6-bus and 30-bus systems when the $\lambda$-cuts of the fuzzy branch power capacity set $\mu(\theta_{ij})$ equal 0.0, 0.5, 0.6, and 1, respectively.

**TABLE 9.8   The Results for Security Regions on IEEE 6-Bus System (p.u.)**

| $\lambda$-cut | Regions | $P_{G4}$ | $P_{G5}$ | $P_{G6}$ |
|---|---|---|---|---|
| 1 | $P_{imax}$ | 3.9760 | 2.4240 | 3.8990 |
|   | $P_{imin}$ | 2.0250 | 0.4740 | 0.0000 |
| 0.6 | $P_{imax}$ | 3.9755 | 2.4245 | 4.5480 |
|   | $P_{imin}$ | 1.7010 | 0.1500 | 0.0000 |
| 0.5 | $P_{imax}$ | 3.9755 | 2.4245 | 4.7100 |
|   | $P_{imin}$ | 1.6200 | 0.0693 | 0.0000 |
| 0 | $P_{imax}$ | 3.9755 | 2.4245 | 5.1849 |
|   | $P_{imin}$ | 1.2151 | 0.000 | 0.0000 |

**TABLE 9.9   The Comparison of Security Region Results for IEEE 6-Bus System (p.u.)**

| Method | Regions | $P_{G4}$ | $P_{G5}$ | $P_{G6}$ |
|---|---|---|---|---|
| Method 1 | $P_{imax}$ | 3.9760 | 2.4240 | 3.8990 |
|   | $P_{imin}$ | 2.0250 | 0.4740 | 0.0000 |
| Method 2 | $P_{imax}$ | 3.7500 | 2.6490 | 2.5510 |
|   | $P_{imin}$ | 2.4490 | 1.4000 | 0.0000 |

Method 1: enhanced expanding method.

Method 2: traditional expanding method.

**TABLE 9.10   The Results for $N-1$ Security Regions on IEEE 6-Bus System**

| Gen. Node | Base Value $P^0$ | Security $P_{imin}$ | Regions $P_{imax}$ |
|---|---|---|---|
| $P_{G4}$ | 2.514 | 2.378 | 3.301 |
| $P_{G5}$ | 1.523 | 1.369 | 1.654 |
| $P_{G6}$ | 2.363 | 1.400 | 2.645 |

**TABLE 9.11   The Results for Security Regions on IEEE 30-Bus System (p.u.)**

| $\lambda$-cut | 1 | 0.6 | 0.5 | 0.0 |
|---|---|---|---|---|
| $P_{G2}$ $P_{imax}$ | 0.7350 | 0.7550 | 0.7600 | 0.7710 |
| $P_{imin}$ | 0.3712 | 0.3700 | 0.3656 | 0.3513 |
| $P_{G5}$ $P_{imax}$ | 0.4622 | 0.4733 | 0.4744 | 0.4895 |
| $P_{imin}$ | 0.1500 | 0.1500 | 0.1500 | 0.1500 |
| $P_{G8}$ $P_{imax}$ | 0.3500 | 0.3500 | 0.3500 | 0.3500 |
| $P_{imin}$ | 0.1110 | 0.1080 | 0.1000 | 0.1000 |
| $P_{G11}$ $P_{imax}$ | 0.3000 | 0.3000 | 0.3000 | 0.3000 |
| $P_{imin}$ | 0.1000 | 0.1000 | 0.1000 | 0.1000 |
| $P_{G13}$ $P_{imax}$ | 0.4000 | 0.4000 | 0.4000 | 0.4000 |
| $P_{imin}$ | 0.1200 | 0.1200 | 0.1200 | 0.1200 |

**TABLE 9.12  The Comparison of Security Region Results for IEEE 30-Bus System (p.u.)**

| Method | Method 1 | | Method 2 | |
|---|---|---|---|---|
| Regions | $P_{max}$ | $P_{min}$ | $P_{max}$ | $P_{min}$ |
| $P_{G2}$ | 0.7350 | 0.3712 | 0.7120 | 0.4280 |
| $P_{G5}$ | 0.4622 | 0.1500 | 0.4020 | 0.1500 |
| $P_{G8}$ | 0.3500 | 0.1110 | 0.3500 | 0.1480 |
| $P_{G11}$ | 0.3000 | 0.1000 | 0.3000 | 0.1000 |
| $P_{G13}$ | 0.4000 | 0.1200 | 0.4000 | 0.1770 |

Method 1: enhanced expanding method.

Method 2: traditional expanding method.

**TABLE 9.13  The Results for $N-1$ Security Regions on IEEE 30-Bus System (p.u.)**

| Gen. Node | Base Value $P_{Gi}^0$ | Lower Bound of Regions $P_{imin}$ | Upper Bound of Regions $P_{imax}$ |
|---|---|---|---|
| $P_{G2}$ | 0.566 | 0.2000 | 0.7350 |
| $P_{G5}$ | 0.293 | 0.1500 | 0.3500 |
| $P_{G8}$ | 0.306 | 0.1000 | 0.3500 |
| $P_{G11}$ | 0.154 | 0.1540 | 0.1600 |
| $P_{G13}$ | 0.295 | 0.2950 | 0.3000 |

It can be found from Tables 9.8 and 9.11 that the bigger the value of $\lambda$-cut of the fuzzy set $\mu(\theta_{ij})$, the higher will be the system reliability requirements and the smaller will be the acquired security regions. On the contrary, the smaller the value of $\lambda$-cut of the fuzzy set $\mu(\theta_{ij})$, the lower will be the system reliability requirements and the larger will be the acquired security regions. Therefore, it is very convenient to select the corresponding security regions to judge whether the power system operation is secure according to the given reliability requirements.

Tables 9.9 and 9.12 are the comparisons of results for the IEEE 6-bus and 30-bus systems with previous work. It can be observed from Tables 9.9 and 9.12 that steady-state security regions without the fuzzy line power flow capacity constraints (i.e., the value of $\lambda$-cut of the fuzzy set $\mu(\theta_{ij}) = 1$) computed by enhanced method are bigger than those computed by the general expanding method. Therefore, power security regions in this section are relatively less conservative than those of the previous work.

Tables 9.10 and 9.13 provide the calculation results of $N-1$ security regions for the IEEE 6-bus and 30-bus systems when the $\lambda$-cut of the fuzzy branch power capacity set $\mu(\theta_{ij})$ equals 1. It can be observed that the $N-1$ security regions for both 6-bus and 30-bus systems are far smaller than $N$ security regions. Especially for the IEEE 30-bus system, the range of expansion for generators 11 and 13 is almost zero in the calculation of $N - 1$ security regions. The reason is that the feasible region becomes narrow because of the introduction of $N - 1$ security constraints.

The results show that it is very important to calculate security regions with fuzzy line power flow constraints. It can provide more information for real-time security analysis and security operation in power systems compared with the previous methods. This is because different reliability requirements correspond to different security regions with fuzzy constraints, while only one reliability requirement corresponds to one security region in the previous method. Because of the adoption of the new expansion method, the computing time is also shorter than that of the traditional expansion method.

## 9.5 FUZZY SET AND LINEAR PROGRAMMING

### 9.5.1 Introduction

This section presents a new approach to construct the steady-state security regions of power systems, that is, the maximal security regions are directly computed using the optimization method [13,14]. First of all, the security regions model is converted into a linear programming (LP) optimization model, in which the upper and lower limits of each component forming a hyperbox are taken as unknown variables, and the objective is to maximize the sum of the generators' power adjustment ranges. The fuzzy branch power constraints and the $N - 1$ security constraints are also introduced into the optimization model of the steady-state security regions. The IEEE 6-bus and 30-bus systems are used as test examples.

### 9.5.2 Steady-State Security Regions Solved by Linear Programming

***Objective Function***  In the practical operation of power systems, it is desired to obtain each security region to cover as many points of operation as possible. This means that it is desired to make the volume of hyperbox as big as possible. However, it will become very complicated if the volume of hyperbox is directly taken as the objective function. In fact, there exists some approximately corresponding relation between the size of the hyperbox's volume and the sum of all sides of the hyperbox. Especially, for the practical operation of power systems, operators are mainly concerned about the secure and adjustable range of generator power output, rather than the volume of of $\Omega_P{}'$. Therefore, in the optimization model for $\Omega_P$, we do not directly select the volume of $\Omega_P$ as the objective function. The objective for optimization calculation $\Omega_P$ is to maximize the sum of the generators' power adjustment ranges, that is,

$$\max Z = \sum_{i=nd+1}^{n-1} W_i (P_{Gi}^M - P_{Gi}^m) \tag{9.72}$$

where $W_i$ is the weighting coefficient of $i$th generator.

$(P_{Gi}^M - P_{Gi}^m)$ is the secure and adjustable range of the $i$th generator power output. It is also the length of $i$th side of hyperbox. Obviously, it must satisfy the rated adjustable range of the $i$th generator power output $(P_{Gimax} - P_{Gimin})$, that is,

$$(P_{Gi}^M - P_{Gi}^m) \leq (P_{Gimax} - P_{Gimin}) \tag{9.73}$$

**Security Constraints** In the optimization calculation of hyperbox $\Omega_P$, the unknown variables are the upper and lower limits of each component in the hyperbox. This is different from the ordinary expanding method. Therefore, the constraints for constructing $\Omega_P$ need to be changed in the optimization method.

**(1)** Generation constraints

According to the definition of security regions, we get

$$P_{Gi} \geq P_{Gi}^m \geq P_{Gimin} \quad i = n_d + 1, \ldots\ldots, n - 1 \tag{9.74}$$

$$P_{Gi} \leq P_{Gi}^M \leq P_{Gimax} \quad i = n_d + 1, \ldots\ldots, n - 1 \tag{9.75}$$

For the slack generator, we have the following equations

$$\sum_{i=1}^{nd} P_i - \sum_{i=nd+1}^{n-1} P_{Gi}^M = P_{Gnm} \tag{9.76}$$

$$\sum_{i=1}^{nd} P_i - \sum_{i=nd+1}^{n-1} P_{Gi}^m = P_{GnM} \tag{9.77}$$

where $P_{Gnm}$ and $P_{GnM}$ are the lower and upper limits of the slack generator, respectively.

**(2)** Branch constraints

According to equation (9.39), the security constraints of branch $ij$ can be written as

$$\theta_{ijmin} \leq \sum_{k=nd+1}^{n-1} (A_{ik} - A_{jk})P_{Gk} \leq \theta_{ijmax} \tag{9.78}$$

For equation (9.78), the power injection of the $k$th generator $P_{Gk}$ can be replaced by $P_{Gk}^m$ and $P_{Gk}^M$ under the following conditions.

$$P_{Gk} = \begin{cases} P_{Gk}^m, & \text{when} \quad A_{ik} - A_{jk} \geq 0 \\ P_{Gk}^M, & \text{when} \quad A_{ik} - A_{jk} \leq 0 \end{cases} \tag{9.79}$$

In this way, the unknown variables in security constraints are all changed into $P_{Gk}^m$ and $P_{Gk}^M (k = n_d + 1, \ldots\ldots, n - 1)$.

### Linear Programming Model and Implementation

*Linear Programming Model for Computing* $\Omega_p$ According to equations (9.72)–(9.79), the optimization model for computing $\Omega_p$ is set up, that is, model $M - 1$.

$$\max Z = \sum_{k=nd+1}^{n-1} W_k (P_{Gk}^M - P_{Gk}^m) \tag{9.80}$$

subject to the constraints in equations (9.73)–(9.79)

Obviously, $M - 1$ is a linear programming model. It can be expressed by the standard form of linear programming, that is, model $M - 2$

$$\max Z = CX \tag{9.81}$$

such that

$$AX \leq B \tag{9.82}$$

$$X \geq 0 \tag{9.83}$$

The model $M - 2$ can be solved by the simplex method. The details of the LP algorithm are shown in the Appendix to this chapter.

*Calculation of Security Regions without Basic Operation Point* The steady-state security regions can be directly obtained through solving model $M - 1$ without a basic operation point. With this method, it is very convenient to judge whether there exists a security region under the given operation state. Meanwhile, it is easy to find the "security center point" of power system operation when the security region $\Omega_p$ is obtained. Therefore, this method can provide useful information for system operation.

*Calculation of Security Regions Considering Basic Operation Point* As described in the previous paragraph, the biggest hyperbox $\Omega_p$ can be acquired when the basic operation point has not been considered in the calculation of security regions. However, in some cases, it is possible that the obtained hyperbox $\Omega_p$ has not covered the basic operation point. Thus this $\Omega_p$ is not practical. For this reason, we introduce the following constraints into model $M - 1$, that is,

$$[P_G^M] \geq [P_{G0}] \tag{9.84}$$

$$[P_G^m] \leq [P_{G0}] \tag{9.85}$$

where $[P_{G0}]$ is the basic operation point.

Then we can obtain optimization model $M - 3$, which considers the basic operation point $[P_{G0}]$, that is,

$$\max Z = \sum_{k=nd+1}^{n-1} W_k (P_{Gk}^M - P_{Gk}^m) \tag{9.86}$$

subject to the constraints in equations (9.73)–(9.79), and (9.84), (9.85)

In this way, the hyperbox $\Omega_P$ obtained from model $M - 3$ certainly covers $[P_{G0}]$. If a solution does not exist in $M - 3$, then we can judge that the given operation point $[P_{G0}]$ is not secure.

It is noted that the optimal solution of the LP is certainly located at the summit on the feasible region. So, in some cases, it is possible that $[P_{G0}]$ will be located on some boundary of the hyperbox $\Omega_P$, although $\Omega_P$ contains $[P_{G0}]$. This means that the security-adjustable amount of the generator along some direction in $\Omega_P$ is zero in this situation. In other cases, although $[P_{G0}]$ is in $\Omega_P$ and is also not on the boundary of $\Omega_P$, it is possible that the security-adjustable amount of the generator along some direction in $\Omega_P$ is very small. Under the aforementioned cases, it is very difficult to judge whether the operation point is still secure when some perturbation occurs in the power system operation. For this reason, we adopt the following constraints to remedy this disadvantage.

$$[P_G^M] \geq [P_{G0}] + [\Delta P_{G0}] \tag{9.87}$$

$$[P_G^m] \leq [P_{G0}] + [\Delta P_{G0}] \tag{9.88}$$

where $[\Delta P_{G0}]$ is the vector of generation power deviation from the basic operation point $[P_{G0}]$. This is an estimated value and can be determined according to the requirement of system operation and experience of the operators.

Introducing constraints (9.87) and (9.88) into $M - 1$, the new optimization model $M - 4$ for computing $\Omega_P$ can be expressed as follows.

$$\max Z = \sum_{k=nd+1}^{n-1} W_k(P_{Gk}^M - P_{Gk}^m) \tag{9.89}$$

such that

$$(P_{Gi}^M - P_{Gi}^m) \leq (P_{Gimax} - P_{Gimin}) \tag{9.90}$$

$$P_{Gi} \geq P_{Gi}^m \geq P_{Gimin} \quad i = n_d + 1, \ldots\ldots, n - 1 \tag{9.91}$$

$$P_{Gi} \leq P_{Gi}^M \leq P_{Gimax} \quad i = n_d + 1, \ldots\ldots, n - 1 \tag{9.92}$$

$$\sum_{i=1}^{nd} P_i - \sum_{i=nd+1}^{n-1} P_{Gi}^M = P_{Gnm} \tag{9.93}$$

$$\sum_{i=1}^{nd} P_i - \sum_{i=nd+1}^{n-1} P_{Gi}^m = P_{GnM} \tag{9.94}$$

$$\theta_{ijmin} \leq \sum_{k=nd+1}^{n-1} (A_{ik} - A_{jk})P_{Gk} \leq \theta_{ijmax} \tag{9.95}$$

$$P_{Gk} = \begin{cases} P_{Gk}^m, & \text{when} \quad A_{ik} - A_{jk} \geq 0 \\ P_{Gk}^M, & \text{when} \quad A_{ik} - A_{jk} \leq 0 \end{cases} \tag{9.96}$$

$$[P_G^M] \geq [P_{G0}] + [\Delta P_{G0}] \tag{9.97}$$

$$[P_G^m] \leq [P_{G0}] + [\Delta P_{G0}] \tag{9.98}$$

The above model is a linear model, which can be solved by an LP algorithm.

### 9.5.3  Numerical Examples

***Comparison of Linear Programming and Expanding Method for $\Omega_P$***  The calculation of the maximal security region hyperbox $\Omega_P$ by the optimization method is examined with the IEEE 6-bus and 30-bus systems.

To assess or compare the size of $\Omega_P$ for different means, the following performance index is introduced:

$$PI = \frac{\displaystyle\sum_{i=nd+1}^{n-1} (P_{Gi}^M - P_{Gi}^m)}{\displaystyle\sum_{i=nd+1}^{n-1} (P_{Gimax} - P_{Gimin})} \tag{9.99}$$

or

$$PI_i = \frac{P_{Gi}^M - P_{Gi}^m}{P_{Gimax} - P_{Gimin}} \quad i = n_d + 1, \ldots\ldots, n - 1 \tag{9.100}$$

The calculation results for a steady-state security region are given in Tables 9.14 and 9.15, where the optimization approach for constructing the maximal security region is identified as method 1 and the expanding method is identified as method 2. Table 9.14 represents the results for security regions on the IEEE 6-bus system. Table 9.15 represents the results for security regions on the IEEE 30-bus system. For comparison, we also use the traditional expanding method to calculate the maximal security region for the IEEE 30-bus system under the same system parameters and conditions. The results are listed in Table 9.15.

**TABLE 9.14   The Comparison of Security Region Results for IEEE 6-Bus System**

| Methods | Security Regions | Gen. $P_{G4}$ | Gen. $P_{G5}$ | Total PI% |
|---------|------------------|---------------|---------------|-----------|
| Method 1 | $P_{Gi}^M$ | 4.200 | 2.200 | 71% |
|  | $P_{Gi}^m$ | 0.184 | 1.378 |  |
|  | $PI_i\%$ | 96% | 31% |  |
| Method 2 | $P_{Gi}^M$ | 3.750 | 2.649 | 37% |
|  | $P_{Gi}^m$ | 2.449 | 1.400 |  |
|  | $PI_i\%$ | 31% | 47% |  |

Method 1: optimization method.

Method 2: the expanding method.

**TABLE 9.15   The Comparison of Security Region Results for IEEE 30-Bus System**

| Methods | Security regions | Gen. $P_{G2}$ | Gen. $P_{G5}$ | Gen. $P_{G8}$ | Gen. $P_{G11}$ | Gen. $P_{G13}$ | Total PI% |
|---------|------------------|-----------|-----------|-----------|------------|------------|-----------|
| Method 1 | $P_{Gi}^M$ | 0.800 | 0.500 | 0.350 | 0.300 | 0.384 | 85% |
|  | $P_{Gi}^m$ | 0.439 | 0.150 | 0.100 | 0.100 | 0.120 |  |
|  | $PI_i\%$ | 80% | 100% | 100% | 100% | 94% |  |
| Method 2 | $P_{Gi}^M$ | 0.712 | 0.402 | 0.350 | 0.300 | 0.400 | 70% |
|  | $P_{Gi}^m$ | 0.428 | 0.150 | 0.148 | 0.100 | 0.177 |  |
|  | $PI_i\%$ | 47% | 72% | 81% | 100% | 80% |  |

Method 1 is optimization method.

Method 2 is the expanding method.

From Tables 9.14 and 9.15, we know that the security region $\Omega_P$ obtained by the optimization method in this section is far bigger than that obtained by the traditional expanding method described in Section 9.3. Therefore, the conservation of the maximal security regions computed on the basis of the optimization approach is relatively small. The computation time needed in this approach is also very short (only 1.1 s for the IEEE 6-bus system, and 4.37 s for the IEEE 30-bus system).

The calculation results and comparison show that the LP method is superior to the expanding method for computing security regions.

**Applying Linear Programming for $\Omega_P$ Considering Fuzzy Constraints**   The optimization computation of the steady-state security region with fuzzy constraints is examined with the IEEE 6-bus system. The parameters of the system including the fuzzy branch power capacities, the branch average contingency time ACTs, probability of branch temporary overload are the same as those in Section 9.3. Suppose the system operation time is 150 h. The performance index of branch failure probability PI is 0.085.

Table 9.16 provides the calculation results of security regions for the IEEE 6-bus system when the $\lambda$-cut of the fuzzy branch power capacity set $\mu(\theta_{ij})$ equals 0.0, 0.5, 0.6 and 1, respectively.

It can be observed from Table 9.16 that the bigger the value of the $\lambda$-cut of the fuzzy set $\mu(\theta_{ij})$, the higher will be the system reliability requirements and the smaller will be the acquired security regions. On the contrary, the smaller the value of the $\lambda$-cut of the fuzzy set $\mu(\theta_{ij})$, the lower will be the system reliability requirements and the larger will be the acquired security regions. Therefore, it is very convenient to select the corresponding security regions to judge whether the power system operation is secure according to the given reliability requirements.

Calculation of security regions with fuzzy line power flow constraints can provide more information for real-time security analysis and security operation in power system compared with the existing methods. Because of the adoption of the optimization method, the computing time of security regions is also shorter than that of the expanding methods.

**TABLE 9.16    The Results for Security Regions on IEEE 6-Bus System (p.u.)**

| $\lambda$-cut | Regions | $P_{G4}$ | $P_{G5}$ | $P_{G6}$ |
|---|---|---|---|---|
| 1 | $P_{Gi}^M$ | 4.2000 | 2.2240 | 3.8990 |
|  | $P_{Gi}^m$ | 0.1840 | 1.3700 | 0.0000 |
| 0.6 | $P_{Gi}^M$ | 4.0050 | 2.2245 | 4.5480 |
|  | $P_{Gi}^m$ | 0.1701 | 1.1500 | 0.0000 |
| 0.5 | $P_{Gi}^M$ | 4.0050 | 2.2245 | 4.7100 |
|  | $P_{Gi}^m$ | 0.1620 | 1.0693 | 0.0000 |
| 0 | $P_{Gi}^M$ | 3.9755 | 2.4245 | 5.1849 |
|  | $P_{Gi}^m$ | 0.1215 | 1.000 | 0.0000 |

# APPENDIX A: LINEAR PROGRAMMING

Linear programming (LP) is widely used in power system problems. Hence, we briefly describe the basic algorithm of LP [22–28].

## A.1    Standard Form of LP

Not all linear programming problems are easily solved. There may be many variables and many constraints. Some variables may be constrained to be nonnegative and others unconstrained. Some of the main constraints may be equalities and others, inequalities. However, two classes of problems, called here the standard maximum problem and the standard minimum problem, play a special role. In these problems, all variables are constrained to be nonnegative, and all main constraints are inequalities.

Given an $m$-vector, $b = (b_1, \ldots, b_m)^T$, an $n$-vector, $c = (c_1, \ldots, c_n)^T$, and an $m \times n$ matrix,

$$A = \begin{pmatrix} a_{11} & a_{12} & \cdots & a_{1n} \\ a_{21} & a_{22} & \cdots & a_{2n} \\ \vdots & \vdots & \ddots & \vdots \\ a_{m1} & a_{m1} & \cdots & a_{mn} \end{pmatrix}$$

The standard maximum problem of LP can be formulated as follows:

$$\text{maximize } c_1 x_1 + c_2 x_2 + \cdots + c_n x_n$$

$$\text{subject to } a_{11} x_1 + a_{12} x_2 + \cdots + a_{1n} x_n \le b_1$$

$$a_{21} x_1 + a_{22} x_2 + \cdots + a_{2n} x_n \le b_2$$

$$\cdots$$

$$a_{m1} x_1 + a_{m2} x_2 + \cdots + a_{mn} x_n \le b_m$$

$$x_1, x_2, \ldots x_n \ge 0$$

or

$$\max \quad c^T x$$

$$\text{s.t.} \quad Ax \leq b$$

$$x \geq 0$$

We shall always use $m$ to denote the number of constraints, and $n$ to denote the number of decision variables.

The standard minimum problem of the LP can be formulated as follows:

$$\text{Minimize } y_1 b_1 + y_2 b_2 + \cdots + y_m b_m$$

$$\text{subject to } y_1 a_{11} + y_2 a_{12} + \cdots + y_m a_{m1} \geq c_1$$

$$y_1 a_{12} + y_2 a_{22} + \cdots + y_m a_{m2} \geq c_2$$

$$\cdots$$

$$y_1 a_{1n} + y_2 a_{2n} + \cdots + y_m a_{mn} \geq c_n$$

$$y_1, y_2, \ldots y_m \geq 0$$

or

$$\min \quad y^T b$$

$$\text{s.t.} \quad y^T A \geq c$$

$$y \geq 0$$

The following terminologies are used in LP.

- The function to be maximized or minimized is called the objective function.
- A vector, $x$ for the standard maximum problem or $y$ for the standard minimum problem, is said to be feasible if it satisfies the corresponding constraints.
- The set of feasible vectors is called the constraint set.
- An LP problem is said to be feasible if the constraint set is nonempty; otherwise it is said to be infeasible.
- A feasible maximum (minimum) problem is said to be unbounded if the objective function can assume arbitrarily large positive (negative) values at feasible vectors; otherwise, it is said to be bounded. Thus there are three possibilities for a linear programming problem. It may be bounded feasible, it may be unbounded feasible, and it may be infeasible.
- The value of a bounded feasible maximum (minimum) problem is the maximum (minimum) value of the objective function as the variables range over the constraint set.

- A feasible vector at which the objective function achieves the value is said to be optimal.

***Example A.1:*** Consider the following LP problem:

$$\text{Maximize} \quad 7x_1 + 5x_2$$

$$\text{subject to} \quad x_1 + x_2 \leq 1$$

$$-3x_1 - 3x_2 \leq -15$$

$$x_1, x_2 \geq 0$$

Indeed, the second constraint implies that $x_1 + x_2 \geq 5.0$, which contradicts the first constraint. If a problem has no feasible solution, then the problem itself is called *infeasible*.

At the other extreme from infeasible problems, one finds unbounded problems. A problem is *unbounded* if it has feasible solutions with arbitrarily large objective values. For example, consider

$$\text{Maximize} \quad 3x_1 - 4x_2$$

$$\text{subject to} \quad -2x_1 + 3x_2 \leq -1$$

$$-x_1 - 2x_2 \leq -5$$

$$x_1, x_2 \geq 0$$

Here, we could set $x_2$ to zero and let $x_1$ be arbitrarily large. As long as $x_1$ is greater than 5 the solution will be feasible, and as it gets large the objective function does so too. Hence, the problem is unbounded. In addition to finding optimal solutions to linear programming problems, we shall also be interested in detecting when a problem is infeasible or unbounded.

An LP problem was defined as maximizing or minimizing a linear function subject to linear constraints. All such problems can be converted into the form of a standard maximum problem by the following techniques.

A minimum problem can be changed to a maximum problem by multiplying the objective function by $-1$. Similarly, constraints of the form $\sum_{j=1}^{n} a_{ij}x_j \geq b_i$ can be changed into the form $\sum_{j=1}^{n} (-a_{ij})x_j \leq -b_i$. Two other problems arise.

(1) Some constraints may be equalities. An equality constraint $\sum_{j=1}^{n} a_{ij}x_j = b_i$ may be removed, by solving this constraint for some $x_j$ for which $a_{ij} \neq 0$ and substituting this solution in the other constraints and in the objective function wherever $x_j$ appears. This removes one constraint and one variable from the problem.

(2) Some variables may not be restricted to be nonnegative. An unrestricted variable, $x_j$, may be replaced by the difference of two nonnegative variables, $x_j =$

$u_j - v_j$, where $u_j \geq 0$ and $v_j \geq 0$. This adds one variable and two nonnegativity constraints to the problem.

Any theory derived for problems in standard form is therefore applicable to general problems. However, from a computational point of view, the enlargement of the number of variables and constraints in (2) is undesirable and, as will be seen later, can be avoided.

## A.2   Duality

To every linear program there is a dual linear program with which it is intimately connected. We first state this duality for the standard programs.

Definition: The dual of the standard maximum problem

$$\text{maximize} \quad c^T x$$

$$\text{subject to the constraints } Ax \geq b$$

$$\text{and} \quad x \geq 0 \tag{9A.1}$$

is defined to be the standard minimum problem

$$\text{minimize} \quad y^T b$$

$$\text{subject to the constraints } y^T A \leq c^T$$

$$\text{and} \quad y \geq 0 \tag{9A.2}$$

***Example A.2:***   Find $x_1$ and $x_2$ to maximize $2x_1 + x_2$ subject to the constraints $x_1 \geq 0, x_2 \geq 0$, and

$$3x_1 + 2x_2 \leq 9$$

$$4x_1 + 3x_2 \leq 18$$

$$-x_1 + x_2 \leq 2$$

The dual of this standard maximum problem is therefore the standard minimum problem: Find $y_1$, $y_2$, and $y_3$ to minimize $9y_1 + 18y_2 + 2y_3$ subject to the constraints $y_1 \geq 0, y_2 \geq 0, y_3 \geq 0$, and

$$3y_1 + 4y_2 - y_3 \geq 2$$

$$2y_1 + 3y_2 + y_3 \geq 1$$

If the standard minimum problem (A2) is transformed into a standard maximum problem (by multiplying $A$, $b$, and $c$ by $-1$), its dual by the definition above is a standard minimum problem which, when transformed to a standard maximum problem (again by changing the signs of all coefficients) becomes exactly (A1). Therefore, the dual of the standard minimum problem (A2) is the standard maximum problem (A1). The problems (A1) and (A2) are said to be duals.

The general standard maximum problem and the dual standard minimum problem may be simultaneously exhibited in the display:

$$
\begin{array}{c|cccc|c}
 & x_1 & x_2 & \cdots & x_n & \\
\hline
y_1 & a_{11} & a_{12} & \cdots & a_{1n} & \leq b_1 \\
y_2 & a_{21} & a_{22} & \cdots & a_{2n} & \leq b_2 \\
\vdots & \vdots & \vdots & \ddots & \vdots & \vdots \\
y_m & a_{m1} & a_{m2} & \cdots & a_{mn} & \leq b_m \\
\hline
 & \geq c_1 & \geq c_2 & \cdots & \geq c_n &
\end{array}
\tag{9A.3}
$$

The relation between a standard problem and its dual is seen in the following theorem and its corollaries.

**Theorem 1** If $x$ is feasible for the standard maximum problem (A1) and if $y$ is feasible for its dual (A2), then

$$c^T x \leq y^T b \tag{9A.4}$$

**Proof.**

$$c^T x \leq y^T A x \leq y^T b$$

The first inequality follows from $x \geq 0$ and $c^T \leq y^T A$. The second inequality follows from $y \geq 0$ and $Ax \leq b$.

**Corollary 1** If a standard problem and its dual are both feasible, then both are bounded feasible.

**Proof.** If $y$ is feasible for the minimum problem, then (A4) shows that $y^T b$ is an upper bound for the values of $c^T x$ for $x$ feasible for the maximum problem. Similarly for the converse.

**Corollary 2** If there exists feasible $x^*$ and $y^*$ for a standard maximum problem (A1) and its dual (A2) such that $c^T x^* = y^{*T} b$, then both are optimal for their respective problems.

**Proof.** If $x$ is any feasible vector for (A1), then $c^T x \leq y^{*T} b = c^T x^*$, which shows that $x^*$ is optimal. A symmetric argument works for $y^*$.

The following fundamental theorem completes the relationship between a standard problem and its dual. It states that the hypotheses of Corollary 2 are always satisfied if one of the problems is bounded feasible.

### The Duality Theorem

If a standard linear programming problem is bounded feasible, then so is its dual, their values are equal, and there exist optimal vectors for both problems.

As a corollary of the duality theorem we have the equilibrium theorem. Let $x^*$ and $y^*$ be feasible vectors for a standard maximum problem (A1) and its dual (A2) respectively. Then $x^*$ and $y^*$ are optimal if, and only if,

$$y_i^* = 0 \quad \text{for all } i \text{ for which} \sum_{j=1}^{n} a_{ij}x_j^* < b_i \tag{9A.5}$$

and

$$x_j^* = 0 \quad \text{for all } j \text{ for which} \sum_{i=1}^{m} y_i^* a_{ij} > c_j \tag{9A.6}$$

Proof: For first part, "If"
If equation (9A.5) implies that $y_i^* = 0$ unless there is equality in $\sum_{j=1}^{n} a_{ij}x_j^* \le b_i$, thus

$$\sum_{i=1}^{m} y_i^* b_i = \sum_{i=1}^{m} y_i^* \sum_{j=1}^{n} a_{ij}x_j^* = \sum_{i=1}^{m}\sum_{j=1}^{n} y_i^* a_{ij}x_j^* \tag{9A.7}$$

Similarly, from equation (9A.6), we have

$$\sum_{i=1}^{m}\sum_{j=1}^{n} y_i^* a_{ij}x_j^* = \sum_{j=1}^{n} c_j x_j^* \tag{9A.8}$$

According to Corollary 2, the $x^*$ and $y^*$ are optimal.
For the second part, "Only If"
As in the first line of the proof of Theorem 9.1,

$$\sum_{j=1}^{n} c_j x_j^* \le \sum_{i=1}^{m}\sum_{j=1}^{n} y_i^* a_{ij}x_j^* \le \sum_{i=1}^{m} y_i^* b_i \tag{9A.9}$$

By the duality theorem, if $x^*$ and $y^*$ are optimal, the left side is equal to the right side so we get equality throughout. The equality of the first and second terms may be written as

$$\sum_{j=1}^{n}\left(c_j - \sum_{i=1}^{m} y_i^* a_{ij}\right)x_j^* = 0 \tag{9A.10}$$

Since $x^*$ and $y^*$ are feasible, each term in this sum is nonnegative. The sum can be zero only if each term is zero. Thus, if $\sum_{i=1}^{m} y_i^* a_{ij} > c_j$, then $x_j^* = 0$. A symmetric argument shows that if $\sum_{j=1}^{n} a_{ij} x_j^* < b_i$, then $y_i^* = 0$.

Equations (9A.5) and (9A.6) are sometimes called the complementary slackness conditions. They require that a strict inequality (a slackness) in a constraint in a standard problem implies that the complementary constraint in the dual be satisfied with equality.

## A.3   The Simplex Method

Before we present the simplex method for solving linear programming problems, look at the following example first to illustrate how the simplex method works.

*Example A.3:*

$$\text{Maximize}\quad 5x_1 + 4x_2 + 3x_3$$

$$\text{subject to}\quad 2x_1 + 3x_2 + x_3 \le 5$$

$$4x_1 + x_2 + 2x_3 \le 11$$

$$3x_1 + 4x_2 + 2x_3 \le 8$$

$$x_1, x_2, x_3 \ge 0$$

We start by adding so-called *slack variables*. For each of the less-than inequalities in the above problem we introduce a new variable that represents the difference between the right-hand side and the left-hand side. For example, for the first inequality,

$$2x_1 + 3x_2 + x_3 \le 5$$

we introduce the slack variable $w_1$ defined by

$$w_1 = 5 - 2x_1 - 3x_2 - x_3$$

so that the inequality becomes equality, that is

$$2x_1 + 3x_2 + x_3 + w_1 = 5$$

It is clear then that this definition of $w_1$, together with the stipulation that $w_1$ be nonnegative, is equivalent to the original constraint. We carry out this procedure for each of the less-than constraints to get an equivalent representation of the problem:

$$\text{Maximize}\quad y = 5x_1 + 4x_2 + 3x_3$$

$$\text{subject to}\quad w_1 = 5 - 2x_1 - 3x_2 - x_3$$

$$w_2 = 11 - 4x_1 - x_2 - 2x_3$$

$$w_3 = 8 - 3x_1 - 4x_2 - 2x_3$$

$$x_1, x_2, x_3, w_1, w_2, w_3 \geq 0 \qquad\qquad (9A.11)$$

Note that we have included a notation, $y$, for the value of the objective function, $5x_1 + 4x_2 + 3x_3$.

The simplex method is an iterative process in which we start with a solution $x_1, x_2, x_3, w_1, w_2, w_3$ that satisfies the equations and nonnegativities in the above equivalent problem and then look for a new solution $x_1', x_2', x_3', w_1', w_2', w_3'$, which is better in the sense that it has a larger objective function value

$$5x_1' + 4x_2' + 3x_3' > 5x_1 + 4x_2 + 3x_3$$

We continue this process until we arrive at a solution that cannot be improved. This final solution is then an optimal solution.

To start the iterative process, we need an initial feasible solution $x_1$, $x_2$, $x_3$, $w_1$, $w_2$, $w_3$. For our example, this is easy. We simply set all the original variables to zero and use the defining equations to determine the slack variables

$$x_1 = 0, \; x_2 = 0, \; x_3 = 0, \; w_1 = 5, \; w_2 = 11, \; w_3 = 8$$

The objective function value associated with this solution is $y = 0$.

We now ask whether this solution can be improved. Since the coefficient of $x_1$ is positive, if we increase the value of $x_1$ from zero to some positive value, we will increase $y$. But as we change its value, the values of the slack variables will also change. We must make sure that we do not let any of them become negative. Since $x_2 = x_3 = 0$, we see that $w_1 = 5 - 2x_1$, and so keeping $w_1$ nonnegative imposes the restriction that $x_1$ must not exceed $5/2$. Similarly, the nonnegativity of $w_2$ imposes the bound that $x_1 \leq 11/4$, and the nonnegativity of $w_3$ introduces the bound that $x_1 \leq 8/3$. Since all of these conditions must be met, we see that $x_1$ cannot be made larger than the smallest of these bounds: $x_1 \leq 5/2$. Our new, improved solution then is

$$x_1 = 5/2, \; x_2 = 0, \; x_3 = 0, \; w_1 = 0, \; w_2 = 1, \; w_3 = 1/2$$

This first step was straightforward. It is less obvious how to proceed. What made the first step easy was the fact that we had one group of variables that were initially zero and we had the rest explicitly expressed in terms of these. This property can be arranged even for our new solution. Indeed, we simply must rewrite the equations in (9A.11) in such a way that $x_1$, $w_2$, $w_3$, and $y$ are expressed as functions of $w_1$, $x_2$, and $x_3$, that is, the roles of $x_1$ and $w_1$ must be swapped. To this end, we use the equation for $w_1$ in (9A.11) to solve for $x_1$:

$$x_1 = \frac{5}{2} - \frac{1}{2}w_1 - \frac{3}{2}x_2 - \frac{1}{2}x_3$$

The equations for $w_2$, $w_3$, and $y$ must also be doctored so that $x_1$ does not appear on the right. The easiest way to accomplish this is to do so-called *row operations* on the equations in the equivalent problem. For example, if we take the equation for $w_2$ and subtract two times the equation for $w_1$ and then bring the $w_1$ term to the right-hand side, we get

$$w_2 = 1 + 2w_1 + 5x_2$$

Performing analogous row operations for $w_3$ and $\zeta$, we can rewrite the equations in (9A.11) as

$$y = 12.5 - 2.5w_1 - 3.5x_2 + 0.5x_3$$

$$x_1 = 2.5 - 0.5w_1 - 1.5x_2 - 0.5x_3$$

$$w_2 = 1 + 2w_1 + 5x_2$$

$$w_3 = 0.5 + 1.5w_1 + 0.5x_2 - 0.5x_3 \qquad (9A.12)$$

Note that we can recover our current solution by setting the "independent" variables to zero and using the equations to read off the values for the "dependent" variables.

Now we see that increasing $w_1$ or $x_2$ will bring about a *decrease* in the objective function value, so of $x_3$, being the only variable with a positive coefficient, is the only independent variable that we can increase to obtain a further increase in the objective function. Again, we need to determine how much this variable can be increased without violating the requirement that all the dependent variables remain nonnegative. This time we see that the equation for $w_2$ is not affected by changes in $x_3$, but the equations for $x_1$ and $w_3$ do impose bounds, namely, $x_3 \leq 5$ and $x_3 \leq 1$, respectively. The latter is the tighter bound, and so the new solution is

$$x_1 = 2, \qquad x_2 = 0, \quad x_3 = 1, \quad w_1 = 0, \qquad w_2 = 1, \ w_3 = 0$$

The corresponding objective function value is $y = 13$.

Once again, we must determine whether it is possible to increase the objective function further and, if so, how. Therefore, we need to write our equations with $y$, $x_1$, $w_2$, and $x_3$ written as functions of $w_1$, $x_2$, and $w_3$. Solving the last equation in (9A.12) for $x_3$, we get

$$x_3 = 1 + 3w_1 + x_2 - 2w_3$$

Also, performing the appropriate row operations, we can eliminate $x_3$ from the other equations. The result of these operations is

$$\zeta = 13 - w_1 - 3x_2 - w_3$$

$$x_1 = 2 - 2w_1 - 2x_2 + w_3$$

$$w_2 = 1 + 2w_1 + 5x_2$$

$$x_3 = 1 + 3w_1 + x_2 - 2w_3 \qquad (9A.13)$$

We are now ready to begin the third iteration. The first step is to identify an independent variable for which an increase in its value would produce a corresponding increase in $y$. But this time there is no such variable, as all the variables have negative coefficients in the expression for $\zeta$. This fact not only brings the simplex method to a standstill but also proves that the current solution is optimal. The reason is quite simple. Since the equations in (9A.13) are completely equivalent to those in (9A.11) and, as all the variables must be nonnegative, it follows that $y \le 13$ for every feasible solution. Since our current solution attains the value of 13, we see that it is indeed optimal.

Now for the standard maximum problem, the simplex method is presented as below.

First of all, we add the slack variables $w = b - Ax$. The problem becomes: Find $x$ and $w$ to maximize $c^T x$ subject to $x \ge 0$, $u \ge 0$, and $u = b - Ax$.

We may use the following table to solve this problem if we write the constraint, $w = b - Ax$ as $-w = Ax - b$.

$$
\begin{array}{c|cccc|c}
 & x_1 & x_2 & \cdots & x_n & -1 \\
\hline
-w_1 & a_{11} & a_{12} & \cdots & a_{1n} & b_1 \\
-w_2 & a_{21} & a_{22} & \cdots & a_{2n} & b_2 \\
\vdots & \vdots & \vdots & \ddots & \vdots & \vdots \\
-w_m & a_{m1} & a_{m2} & \cdots & a_{mn} & b_m \\
\hline
 & -c_1 & -c_2 & \cdots & -c_n & 0
\end{array}
$$

$$(9A.14)$$

We note as before that if $-c \ge 0$ and $b \ge 0$, then the solution is obvious: $x = 0$, $w = b$, and value equal to zero (as the problem is equivalent to minimizing $-c^T x$).

Suppose we want to pivot to interchange $w_1$ and $x_1$ and suppose $a_{11} = 0$. The equations

$$-w_1 = a_{11}x_1 + a_{12}x_2 + \cdots + a_{1n}x_n - b_1$$

$$-w_2 = a_{21}x_1 + a_{22}x_2 + \cdots + a_{2n}x_n - b_2$$

$$\cdots$$

$$-w_m = a_{m1}x_1 + a_{m2}x_2 + \cdots + a_{mn}\,x_n - b_m$$

become

$$-x_1 = \frac{1}{a_{11}}w_1 + \frac{a_{12}}{a_{11}}x_2 + \frac{a_{1n}}{a_{11}}x_n - \frac{b_1}{a_{11}}$$

$$-w_2 = -\frac{a_{21}}{a_{11}}w_1 + \left(a_{22} - \frac{a_{21}a_{12}}{a_{11}}\right)x_2 + \dots \text{ etc.}$$

In other words, for the same pivot rule, we apply

$$\begin{pmatrix} p & r \\ c & q \end{pmatrix} \Rightarrow \begin{pmatrix} 1/p & r/p \\ -c/p & q - (rc/p) \end{pmatrix}$$

If we pivot until the last row and column (exclusive of the corner) are non-negative, we can find the solution to the dual problem and the primal problem at the same time.

Let $x_{n+i} = w_i$, then we have $n + m$ variables $x$. Initially, we have $n$ nonbasic variables $N = \{1, \ 2, \ \dots, \ n\}$ (i.e., $x_1, \dots, x_n$) and $m$ basic variables $B = \{n + 1, \ n + 2, \dots, n + m\}$ (i.e., $x_{n=1}, \dots, x_{n+m}$).

Within each iteration of the simplex method, exactly one variable goes from nonbasic to basic and exactly one variable goes from basic to nonbasic. The variable that goes from nonbasic to basic is called the *entering variable*. It is chosen with the aim of increasing $y$; that is, one whose coefficient is positive: *pick $k$ from $\{j \in N :$ $c'_j > 0\}$*, where $N$ is the set of nonbasic variables. Note that if this set is empty, then the current solution is optimal. If the set consists of more than one element (as is normally the case), then we have a choice of which element to pick. There are several possible selection criteria. Generally, we pick an index $k$ having the largest coefficient (which again could leave us with a choice).

The variable that goes from basic to nonbasic is called the *leaving variable*. It is chosen to preserve nonnegativity of the current basic variables. Once we have decided that $x_k$ will be the entering variable, its value will be increased from zero to a positive value. This increase will change the values of the basic variables.

$$x_i = b'_i - a'_{ik} x_k, \quad i \in B$$

We must ensure that each of these variables remains nonnegative. Hence, we require that

$$b'_i - a'_{ik} x_k \geq 0, \quad i \in B$$

Of these expressions, the only ones that can go negative as $x_k$ increases are those for which $a'_{ik}$ is positive; the rest remain fixed or increase. Hence, we can restrict our attention to those $i$'s for which $a'_{ik}$ is positive. And for such an $i$, the value of $x_k$ at which the expression becomes zero is

$$x_k = \frac{b'_i}{a'_{ik}}$$

Since we do not want any of these to become negative, we must raise $x_k$ only to the smallest of all of these values

$$x_k = \min_i \left( \frac{b'_i}{a'_{ik}} \right), \quad i \in B, \ a'_{ik} > 0$$

Therefore, with a certain amount of latitude remaining, the rule for selecting the leaving variable is *pick l from* $\{i \in B : a'_{ik} > 0$ and $b'_i/a'_{ik}$ is minimal$\}$.

The rule just given for selecting a leaving variable describes exactly the process by which we use the rule in practice, that is, we look only at those variables for which $a'_{ik}$ is positive and among those we select one with the smallest value of the ratio $b'_i/a'_{ik}$.

This same "method" may be used to solve the dual problem—the standard minimum problem: Find $y$ to minimize $y^T b$ subject to $y \geq 0$ and $y^T A \geq c^T$.

Similarly, we convert the inequalities into equalities by adding slack variables $s^T = y^T A - c^T \geq 0$. The problem can be restated: Find $y$ and $s$ to minimize $y^T b$ subject to $y \geq 0$, $s \geq 0$ and $s^T = y^T A - c^T$.

We write this problem in a table to represent the linear equations $s^T = y^T A - c^T$.

| | $s_1$ | $s_2$ | $\cdots$ | $s_n$ | |
|---|---|---|---|---|---|
| $y_1$ | $a_{11}$ | $a_{12}$ | $\cdots$ | $a_{1n}$ | $b_1$ |
| $y_2$ | $a_{21}$ | $a_{22}$ | $\cdots$ | $a_{2n}$ | $b_2$ |
| $\vdots$ | $\vdots$ | $\vdots$ | $\ddots$ | $\vdots$ | $\vdots$ |
| $y_m$ | $a_{m1}$ | $a_{m2}$ | $\cdots$ | $a_{mn}$ | $b_m$ |
| $1$ | $-c_1$ | $-c_2$ | $\cdots$ | $-c_n$ | $0$ |

$$(9A.15)$$

The last column represents the vector whose inner product with $y$ we are trying to minimize.

If $-c \geq 0$ and $b \geq 0$, there is an obvious solution to the problem; namely, the minimum occurs at $y = 0$ and $s = -c$, and the minimum value is $y^T b = 0$. This is feasible because $y \geq 0$, $s \geq 0$, and $s^T = y^T A - c$, and yet $\Sigma y_i b_i$ cannot be made any smaller than 0, as $y \geq 0$, and $b \geq 0$.

Suppose then we cannot solve this problem so easily because there is at least one negative entry in the last column or last row. (exclusive of the corner). Let us pivot about $a_{11}$ (suppose $a_{11} \neq 0$), including the last column and last row in our pivot operations, we get

| | $y_1$ | $s_2$ | $\cdots$ | $s_n$ | |
|---|---|---|---|---|---|
| $s_1$ | $a'_{11}$ | $a'_{12}$ | $\cdots$ | $a'_{1n}$ | $b'_1$ |
| $y_2$ | $a'_{21}$ | $a'_{22}$ | $\cdots$ | $a'_{2n}$ | $b_2$ |
| $\vdots$ | $\vdots$ | $\vdots$ | $\ddots$ | $\vdots$ | $\vdots$ |
| $y_m$ | $a'_{m1}$ | $a'_{m2}$ | $\cdots$ | $a'_{mn}$ | $b_m$ |
| $1$ | $-c'_1$ | $-c'_2$ | $\cdots$ | $-c'_n$ | $v'$ |

$$(9A.16)$$

Let $r = (r_1, \ldots, r_n) = (y_1, s_2, \ldots, s_n)$ denote the variables on top, and let $t = (s_1, y_1, \ldots, y_m)$ denote the variables on the left. The set of equations are represented by the new table. Moreover, the objective function $y^T b$ may be written (replacing $y_1$ by its value in terms of $s_1$)

$$\sum_{i=1}^{m} y_i b_i = \frac{b_1}{a_{11}} s_1 + \left(b_2 - \frac{a_{21} b_1}{a_{11}}\right) y_2 + \ldots + \left(b_m - \frac{a_{m1} b_1}{a_{11}}\right) y_2 + \frac{c_1 b_1}{a_{11}}$$

$$= t^T b' + v'$$  (9A.17)

This is represented by the last column in the new table. We have transformed our problem into the following: Find vectors $y$ and $s$, to minimize $t^T b'$ subject to $y \geq 0$, $s \geq 0$ and $r = t^T A' - c'$ (where $t^T$ represents the vector $s_1, y_2, \ldots, y_m$ and $r^T$ represents the vector $y_1, s_2, \ldots, s_n$).

Again, if $-c' \geq 0$ and $b' \geq 0$, we have the obvious solution: $t = 0$ and $r = -c'$ with value $v'$.

Similar to the standard maximum problem solved by simplex method, this process will be continued until the optimal solution is obtained.

## PROBLEMS AND EXERCISES

1. What is the steady-state security region?

2. Explain the "Security Corridor"

3. What is the maximal security region?

4. State the differences of the traditional expanding method and new expanding method.

5. Do we need the stating points for LP-based security regions calculation? Why?

6. What does the hyperbox of the security region look like for a system with two generators? How about a system with three generators?

7. Can we find a hyperbox of the security region if a system has an infeasible constraints set? Why?

8. For a maximum problem as below, please write the dual LP problem.

$$\text{Maximize} \quad 5x_1 + 4x_2 + 3x_3$$

$$\text{subject to} \quad 2x_1 + 3x_2 + x_3 \leq 5$$

$$4x_1 + x_2 + 2x_3 \leq 11$$

$$3x_1 + 4x_2 + 2x_3 \leq 8$$

$$x_1, \ x_2, \ x_3 \geq 0$$

9. For a minimum problem as below, please write the dual LP problem.

$$\text{Minimize} \quad 8x_1 + 6x_2 + 2x_3$$

$$\text{subject to} \quad x_1 + x_2 + x_3 \geq 6$$

$$2x_1 + 3x_2 + x_3 \geq 10$$

$$x_1 + 4x_2 + x_3 \geq 15$$

$$x_1, \ x_2, \ x_3 \geq 0$$

**10.** A power system has two generators. The power output limits of two units and security constraint are

$$10 \leq P_{G1} \leq 50 \text{ MW}$$

$$15 \leq P_{G2} \leq 60 \text{ MW}$$

$$3P_{G1} + P_{G2} \leq 180 \text{ MW}$$

(1) If the initial operation point is $P_{G1} = 30, P_{G2} = 40$, use expanding method to compute the hyperbox of the steady-state security region.

(2) If the initial operation point is $P_{G1} = 25, P_{G2} = 30$, use expanding method to compute the hyperbox of the steady-state security region.

(3) Compare the sizes of the hyperboxes for the above cases.

**11.** A power system has two generators. The power output limits of two units and security constraint are

$$0 \leq P_{G1} \leq 55 \text{ MW}$$

$$10 \leq P_{G2} \leq 80 \text{ MW}$$

$$3P_{G1} + P_{G2} \leq 180 \text{ MW}$$

$$3P_{G1} - 2P_{G2} \leq 90 \text{ MW}$$

$$-4P_{G1} + P_{G2} \geq 20 \text{ MW}$$

$$2P_{G1} + P_{G2} \geq 40 \text{ MW}$$

(1) Draw the feasible constraints region.

(2) If the initial operation point is $P_{G1} = 30, P_{G2} = 40$, illustrate the hyperbox of the steady-state security region.

(3) If the initial operation point is $P_{G1} = 25, P_{G2} = 30$, illustrate the hyperbox of the steady-state security region.

(4) If the initial operation point is $P_{G1} = 10, P_{G2} = 10$, illustrate the hyperbox of the steady-state security region.

(5) If the initial operation point is $P_{G1} = 40, P_{G2} = 50$, illustrate the hyperbox of the steady-state security region.

(6) Are all initial points in the above cases in feasible regions?

**12.** A power system has three generators. The power output limits of three units and security constraints are

$$0 \leq P_{G1} \leq 55 \text{ MW}$$

$$10 \leq P_{G2} \leq 80 \text{ MW}$$

$$5 \le P_{G3} \le 100 \text{ MW}$$

$$P_{G1} + P_{G2} + P_{G3} \le 200 \text{ MW}$$

$$P_{G1} + P_{G2} + P_{G3} \ge 30 \text{ MW}$$

$$P_{G1} + P_{G2} \ge 15 \text{ MW}$$

**(1)** If the initial operation point is $P_{G1} = 30, P_{G2} = 40, P_{G3} = 40$, use expanding method to compute the hyperbox of the steady-state security region.

**(2)** If the initial operation point is $P_{G1} = 30, P_{G2} = 45, P_{G3} = 50$, use expanding method to compute the hyperbox of the steady-state security region.

**(3)** If the initial operation point is $P_{G1} = 20, P_{G2} = 35, P_{G3} = 50$, use expanding method to compute the hyperbox of the steady-state security region.

**(4)** Compare the sizes of the hyperboxes for the above cases.

# REFERENCES

1. Hnyilicza E, Lee STY , Schweppe FC. Steady-state security regions: set-theoretic approach, Proceedings of PICA Conference, pp. 347–355, 1975.
2. Fischl R, Ejebe GC, and DeMaio JA. Identification of power system steady-state security regions under load uncertainty, IEEE summer power meeting, Paper A76 495–2. *IEEE Trans on PAS* Vol. 95, No. 6, p. 1767, 1976
3. DeMaio JA, Fischl R. Fast identification of the steady-state security regions for power system security enhancement, IEEE winter power meeting, A.76076-0. *IEEE Trans. on PAS*, Vol. 95, No. 3, p. 758, 1976.
4. Galiana FD, Banakar MH. Approximation formula for dependent load flow variables, Paper F80 200–6, IEEE winter conference, New York, February 1980.
5. Banakar MH, Galiana FD. Power system security corridors concept and computation. IEEE Trans., PAS 1981;100:4524–4532.
6. Wu FF, Kumagai S. Steady-state security regions of power system. IEEE Trans., CAS 1982;29:703–711.
7. Liu CC. A new method for construction of maximal steady-state regions of Power Systems. IEEE Trans. PWRS. 1986;4:19–27.
8. Zhu JZ. A new expanding method to real power steady-state security regions of power system, Proceedings of Chinese Youth Excellent Science & Technology Papers, 1994, pp. 664-668, Science Press.
9. Zhu JZ, Xu GY. Application of network theory with fuzzy to real power steady-state security regions. Power Syst. Autom. 1994;5(3).
10. Fan RQ, Zhu JZ, Xu GY. Power steady-state security regions with loads change. Power Syst. Autom. 1994;5(4).
11. Li Y, Zhu JZ, Xu GY. Study of power system steady-state regions. Proc. Chinese Soc. Electr. Eng. 1993;13(2):15–22.
12. Zhu JZ, Chang CS. A new approach to steady-state security regions with N and N-1 security, 1997 Intern. Power Eng. Conf., IPEC'97, Singapore, May, 1997.
13. Zhu JZ, Fan RQ, Xu GY, Chang CS. Construction of maximal steady-state security regions of power systems using optimization method. Electr. Pow. Syst. Res. 1998;44:101–105.
14. Zhu JZ. Optimal power systems steady-state security regions with fuzzy constraints. Proceedings of IEEE Winter Meeting, New York, January 27–30, 2002, Paper No. 02WM033.
15. Stott B, Marinho JC. Linear programming for power system network security applications. IEEE Trans., PAS 1979;98:837–848.

16. Hobson E, Fletcher DL, Stadlin WO. Network flow linear programming techniques and their application to fuel scheduling and contingency analysis. IEEE Trans., PAS 1984;103:1684–1691.
17. Elacqua AJ, Corey SL. Security constrained dispatch at the New York power pool. IEEE Trans., PAS 1982;101:2876–2884.
18. Zhu JZ, Xu GY. A New Economic Power Dispatch Method with Security. Electr. Pow. Syst. Res. 1992;25:9–15.
19. Zhu JZ, Irving MR. Combined active and reactive dispatch with multiple objectives using an analytic hierarchical process, IEE Proceedings, Part C, Vol. 143, pp. 344–352, 1996
20. Zhu JZ, Xu GY. A Unified Model and Automatic Contingency Selection Algorithm for the P and Q Subproblems. Electr. Pow. Syst. Res. 1995;32:101–105.
21. Zhu JZ, Xu GY. Approach to automatic contingency selection by reactive type performance index, IEE Proceedings, Part C, Vol. 138, pp. 65–68, 1991
22. Ferguson TS. *Linear programming*. Academic Press; 1967.
23. Dantzig GB. *Linear Programming and Extensions*. Princeton University Press; 1963.
24. Vanderbei RJ. *Linear Programming: Foundations and Extensions*. Boston: Kluwer; 1996.
25. Luenberger DG. *Introduction to linear and nonlinear programming*. USA: Addison-wesley Publishing Company, Inc; 1973.
26. Hadley G. *Linear programming*. Reading, MA: Addison—Wesley; 1962.
27. Strayer JK. *StrayerLinear Programming and Applications*. Springer-Verlag; 1989.
28. Bazaraa M, Jarvis J, Sherali H. *Linear Programming and Network Flows*. 2nd ed. New York: Wiley; 1977.

CHAPTER **10**

# APPLICATION OF RENEWABLE ENERGY

## 10.1 INTRODUCTION

Renewable energy is energy that comes from natural resources such as sunlight, wind, rain, tides, and geothermal heat, which are renewable. Renewable energy sources differ from conventional sources in that, generally they cannot be scheduled, and they are often connected to the electricity distribution system rather than the transmission system.

The production and use of renewable fuels has grown more quickly in recent years as a result of higher prices for oil and natural gas, and faster development of all kinds of new technologies such as the distributed generation (DG) for use of renewable energy resources, as well as the development of the smart grid. Since the renewable energy resources are typically sited close to customer loads, they can help reduce the number of transmission and distribution lines that need to be upgraded or built. Obviously, they reduce transmission and distribution losses. However, owing to the introduction of renewable energy resources, the distribution network has multiple sources and it is possible to have power flow in the reverse direction, from renewable energy resources to the substations. Reverse power flow is the main problem in the integration of DG units in the smart grid. The smart grid including DG will be discussed in Chapter 14. This chapter focuses on the application of renewable energy in power systems [1–24].

## 10.2 RENEWABLE ENERGY RESOURCES

### 10.2.1 Solar Energy

Energy produced by sun is called solar energy. The light energy which we receive from the sun can be absorbed, stored, converted, and used for domestic purposes.

*Optimization of Power System Operation*, Second Edition. Jizhong Zhu.
© 2015 The Institute of Electrical and Electronics Engineers, Inc. Published 2015 by John Wiley & Sons, Inc.

Most renewable energy comes either directly or indirectly from the sun. Sunlight, or solar energy, can be used directly for heating and lighting homes and other buildings, for generating electricity, and for hot water heating, solar cooling, and a variety of commercial and industrial uses. One of most the commonly used solar technologies for electricity is the solar photovoltaic cell.

Solar cells, also called photovoltaic (PV) cells by scientists, convert sunlight directly into electricity. PV gets its name from the process of converting light (photons) to electricity (voltage), which is called the PV effect.

Solar panels used to power homes and businesses are typically made from solar cells combined into modules that hold about 40 cells. A typical home will use about 10–20 solar panels to power the home. The panels are mounted at a fixed angle facing south, or they can be mounted on a tracking device that follows the sun, allowing them to capture the maximum amount of sunlight. Many solar panels combined together to create one system is called a solar array. For a large electric utility or for industrial applications, hundreds of solar arrays are interconnected to form a large utility-scale PV system.

## 10.2.2 Wind Energy

Wind is a form of solar energy. It is a natural power source that can be economically used to generate electricity. The terms "wind energy" or "wind power" describe the process by which wind is used to generate mechanical power or electricity.

Wind turbines, such as aircraft propeller blades, turn in the moving air and power an electric generator that supplies an electric current. There are two different types of wind turbines that are currently in use. The first type originates from the vertical-axis design. The second type of wind turbine is based on the horizontal-axis design. These wind turbines are very much like the windmills found on farms used for daily chores like pumping water. Modern wind turbines, large in size, are created from the original horizontal-axis design. Wind turbines are often grouped together into a single wind power plant, also known as a wind farm, to generate bulk electrical power. Electricity from these turbines is fed into a utility grid and distributed to customers, just as with conventional power plants.

## 10.2.3 Hydropower

Flowing water creates energy that can be captured and turned into electricity. This is called hydroelectric power or hydropower.

There are several types of hydroelectric facilities; they are all powered by the kinetic energy of flowing water as it moves downstream. Turbines and generators convert the energy into electricity, which is then fed into the electrical grid to be used in homes, businesses, and by industry. Hydropower is currently the best known and most widely used source of renewable energy production, accounting for about 20% of present global energy production. The operation of hydropower was discussed in Chapter 4.

## 10.2.4   Biomass Energy

Biomass refers to relatively recently living organic material such as wood, leaves, paper, food waste, manure, and other items usually considered garbage. Biomass can be used to produce electricity, transportation fuels, or chemicals. The use of biomass for any of these purposes is called biomass energy or biomass power.

Bioenergy system technologies include direct-firing, co-firing, gasification, pyrolysis, and anaerobic digestion.

Most biopower plants use direct-fired systems. They burn bioenergy feedstocks directly to produce steam. This steam drives a turbine, which turns a generator that converts the power into electricity. In some biomass industries, the spent steam from the power plant is also used for manufacturing processes or to heat buildings. Such combined heat and power systems greatly increase overall energy efficiency.

Co-firing refers to mixing biomass with fossil fuels in conventional power plants. Coal-fired power plants can use co-firing systems to significantly reduce emissions, especially sulfur dioxide emissions. Gasification systems use high temperatures and an oxygen-starved environment to convert biomass into synthesis gas, a mixture of hydrogen and carbon monoxide. Gasification, anaerobic digestion, and other biomass power technologies can be used in small, modular systems with internal combustion or other generators. These could be helpful for providing electrical power to villages remote from the electrical grid.

## 10.2.5   Geothermal Energy

Geothermal energy is harnessed from the earth. Geothermal power plants harness the heat from the earth to produce electricity. There are three different ways that power plants process geothermal energy. They are the dry-steam, flash-steam and binary-cycle methods. All three methods use steam to power a turbine which drives a generator that produces electricity.

Dry-steam geothermal power plants use steam that is brought from below the earth's surface through pipes, directly to the power plant turbines.

Flash-steam geothermal power plants use hot water that is brought from below the earth's surface. The hot water is sprayed into a tank and creates steam.

Binary-cycle geothermal plants use moderate temperature water from a geothermal source and combine it with another chemical to create steam. The steam powers the turbine that drives the generator to create electricity.

# 10.3   OPERATION OF GRID-CONNECTED PV SYSTEM

## 10.3.1   Introduction

PV systems can be grouped into *stand-alone systems* (such as rural electrification, pumping water equipment, and industrial applications) and *grid-connected systems* (such as domestic systems and power plants). PV power supplied to the utility grid is gaining more and more visibility, because of the need for meeting the worldwide increase in the demand for electric power. The PV array normally uses a maximum power point tracking (MPPT) technique to continuously deliver the highest power to the load when there are variations in irradiation and temperature. The disadvantage of PV energy is that the PV output power depends on weather conditions and cell temperature, making it an uncontrollable source.

It is also not available during the night. In order to overcome these inherent drawbacks, grid-connected PV systems are widely adopted. The system includes PV panels (string and parallel connected to form PV arrays), on-grid inverters, and electricity-distributing devices. The following are the advantages of operating grid-connected PV systems:

- Reduction in the costs of the PV panels
- Reduction in transmission power losses
- No noise or pollution
- Less maintenance, simple structure
- Supply of power from the PV system to the grid, relieving the grid demand.

All PV systems interface the utility grid through a voltage source inverter and a boost converter. The introduction of a grid-connected PV system increases the voltage in its point of common coupling (PCC). The voltage level depends on the network configurations and the load conditions. It is proportional to the instantaneously produced power of the PV system. In this case, the structure of the distribution network changes from the single- to multipower sources, and the size and direction of the power flow in the feeder may change, leading to a change of the voltage profile in distribution feeders. Thus, the connection of a large PV system to utility grids may cause some operational problems for distribution networks. The severity of these problems directly depends on the percentage of PV penetration and the geography of the installation. Hence, knowing the possible impact of the grid-connected PV system on the distribution network can provide feasible solutions before real-time and practical implementation. The following sections introduce possible effects that PV system may impose on a distribution network.

## 10.3.2 Model of PV Array

Since the capacity of the PV cell is relatively small, output voltage being is less than 1 V and the peak output power being only around 1 W, a single PV cell cannot meet the load requirement; it is also inconvenient for installation and application. Therefore, a few dozens or even hundreds of PV cells are connected in parallel or series, according to the load requirement, to form a combined device and then they are encapsulated in a box made of transparent sheet with anode and cathode down-leads outside the box. Before and after the encapsulation, the combined device is called a PV module and PV panel, respectively. A few PV panels are connected in parallel or series to form a larger power supply, namely, a PV array.

A controlled current source is generally used for modeling a PV array. For a PV array with $N_S$ PV cells in series and $N_P$ PV cells in parallel, the terminal current $I_A$ can be expressed as follows [1]:

$$I_A = N_P \cdot I_L - N_P \cdot I_0 \cdot \left[ \exp\left( \frac{q \cdot (V_A + I_A \cdot R_{sa})}{N_S \cdot n \cdot m \cdot k \cdot T} \right) - 1 \right] \quad (10.1)$$

where $V_A$ and $I_A$ are terminal voltage and current in the PV array, respectively, $R_{sa}$ is the equivalent series resistance of the PV array.

## 10.3.3    Control of Three-Phase PV Inverter

The PV inverter is an important component of the PV system and is used to convert DC power from the PV array to AC power on the grid. Its performance determines the quality of PV system output power. With an increase in the types of inverters and the continuous development of control techniques, the PV system has been applied to all fields. For a high-performance PV inverter, the choice of circuit topology is very important, because the circuit topology concerns efficiency, cost, security, and reliability.

One of the control schemes of the grid-connected is the current-mode control scheme, which takes the output current as the controlled variables. The output current should be real-time controlled so that the output current of the inverter has the same phase and frequency as the grid voltage. The PV inverter will ensure that the output alternating current is the high-quality sine wave with the synchronized frequency. The goals of grid-connected control are to decouple control of the output power of PV array and to realize MPPT. The following is the analysis of the power-decoupling control of a three-phase PV inverter. The MPPT control will be introduced in the next section.

$$\begin{bmatrix} i_d \\ i_q \\ i_0 \end{bmatrix} = \frac{2}{3} \begin{bmatrix} \cos\theta & \cos(\theta - 120°) & \cos(\theta + 120°) \\ \sin\theta & \sin(\theta - 120°) & \sin(\theta + 120°) \\ \frac{1}{2} & \frac{1}{2} & \frac{1}{2} \end{bmatrix} \begin{bmatrix} i_a \\ i_b \\ i_c \end{bmatrix} \quad (10.2)$$

where $\theta$ is the phase that $d$-axis current lags behind $a$-axis current.

In the static $a - b - c$ coordinates, the speed of current regulation mode is rapid, but the frequency of inverter switching is not fixed, and the harmonic component of the output current is high. Therefore, the rotational $d - q - 0$ coordinates are generally applied to regulate the $q$- and $d$-axes currents. In this case, the output harmonics of voltage source inverter would be reduced. If the $q$-axis in rotational $d - q - 0$ coordinates lagged behind the $d$-axis by 90°, the three-phase current $i_a$, $i_b$, $i_c$ in the static $a - b - c$ can be transformed into $d$-, $q$- and 0-axes currents $i_d$, $i_q$ and $i_0$ through the Park transformation.

In a balanced three-phase system, the instantaneous active and reactive power could be described by $d$-, $q$-axes voltages $V_d$, $V_q$ and currents $I_d$, $I_q$.

$$P = \frac{3}{2} \cdot (V_d \cdot I_d + V_q \cdot I_q) \tag{10.3}$$

$$Q = \frac{3}{2} \cdot (V_d \cdot I_q - V_q \cdot I_d) \tag{10.4}$$

Here, $V_q$ is identical to the magnitude of the instantaneous voltage at the PV array terminal and $V_d$ is zero in the rotating $d - q - 0$ coordinates, so equations (10.3) and (10.4) may be contracted into the simpler equations (10.5) and (10.6).

$$P = \frac{3}{2} \cdot |V_o| \cdot I_q \tag{10.5}$$

$$Q = -\frac{3}{2} \cdot |V_o| \cdot I_d \tag{10.6}$$

where $|V_o|$ is the voltage magnitude of the instantaneous PV array. Since the voltage magnitude remains almost constant, the real and reactive power can be controlled by regulating the $q$- and $d$-axes currents ($I_q$ and $I_d$), respectively.

### 10.3.4 Maximum Power Point Tracking

In the actual PV power system, the changes in solar irradiance intensity and temperature are not controllable. To ensure that the PV arrays always work at maximum power operation point under certain light intensity and temperature, the MPPT controller must maintain the DC voltage of the PV array at the appropriate value all the time. The MPPT strategy requires real-time detection of the PV array output power, and applies some control algorithm to predict the possible maximum output power of the PV array under the current operating condition. Then it changes the current impedance to meet the requirements of maximum power output. Even if the rise of PV cells temperature causes a reduction in output power, the PV power generation system could still run in the optimum state under the current condition.

Figure 10.1 shows the P–V characteristic of a PV array. A different operating point of the PV array determines a different output power. The principle of MPPT is to seek the corresponding voltage of maximum power point under the specified sunshine and temperature conditions through detecting the output power at the different operating points. The MPPT algorithms mainly include the constant voltage tracking (CVT) method, current sweep method, perturbation and observation method, fractional open-circuit voltage method, and incremental conductance method [1].

### 10.3.5 Distribution Network with PV Plant

Figure 10.2 is a distribution feeder structure without PV plants, where the source is selected as the reference node with voltage $\dot{U}_0 = U_0 e^{j0}$. Figure 10.3 is a distribution feeder structure with a PV plant. If the PV power plant of node $k$ is in operation, the voltage drop of each node will be changed.

According to the model in Figure 10.2, the feeders have $n$ loads which are evenly distributed, and each load is assumed to have the same value $P_d + jQ_d$. Let $m$

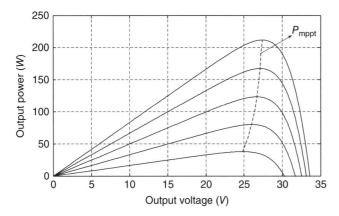

Figure 10.1    The maximum power curve of a PV array.

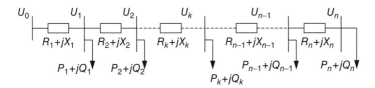

Figure 10.2    Distribution feeders without PV power plants.

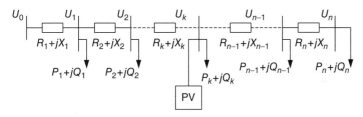

Figure 10.3    Distribution feeders with PV a power plant.

be any point on the feeder branch, then the active and reactive loads of point $m$ can be written as

$$P_{m-n} + jQ_{m-n} = (n - m + 1)P_d + j(n - m + 1)Q_d \qquad (10.7)$$

To simplify the calculation, the principle of superposition is applied to the voltage calculation of the feeders. This will consider the impact of both the main source and PV power plant on the distribution feeder. In this situation, the source of the distribution feeder is equivalent to a voltage source that is in short-circuit status, while the PV power plant is equivalent to a current source that is in open status.

## 10.4 VOLTAGE CALCULATION OF DISTRIBUTION NETWORK

Since the output power of the PV plant is affected by sunshine, temperature, and other weather factors, the PV output power has characteristics of fluctuations and intermittence that is prone to cause voltage fluctuations at the common connection point. The impact of PV power on the power system must be assessed in order to ensure that the increasing application of PV power does not bring negative consequences to the users. This section discusses the steady-state voltage distribution and dynamics of voltage fluctuation after PV power plants access the distribution network [8].

### 10.4.1 Voltage Calculation without PV Plant

Let us first analyze the voltage calculation of the traditional distribution system. From Figure 10.2, the voltage drop at any point $m$ on distribution line is

$$\Delta U_{sm} = \Delta U_{smf} + \Delta U_{sml} \tag{10.8}$$

where $\Delta U_{msf}$ is the voltage loss caused by the equivalent load after point $m$. $\Delta U_{msl}$ is the voltage loss caused by the load before point $m$. Assume that the line between two nodes has the same length, the voltage loss from point $m$ to the end of the feeder can be written as follows.

$$\Delta U_{sml} = m(n - m + 1)\frac{P_d R_1 + Q_d X_1}{U_N} \tag{10.9}$$

where $R_1$ and $X_1$ are the resistance and reactance of the line between two nodes, respectively. $U_N$ is the rated voltage.

Assuming that each load is evenly distributed at the midpoint of each line section, the voltage loss from the source to the point $m$ can be written as follows.

$$\Delta U_{sml} = \frac{m}{2}(m - 1)\frac{P_d R_1 + Q_d X_1}{U_N} \tag{10.10}$$

The total voltage loss at node $m$ is

$$\Delta U_{sm} = \frac{m}{2}(2n - m + 1)\frac{P_d R_1 + Q_d X_1}{U_N} \quad k \in [1, n] \tag{10.11}$$

Thus, the node voltage at any point $m$ on the distribution feeders without PV power plant can be calculated as follows.

$$U_m = U_0 - \frac{m}{2}(2n - m + 1)\frac{P_d R_1 + Q_d X_1}{U_N} \quad m \in [1, n] \tag{10.12}$$

## 10.4.2 Voltage Calculation with PV Plant Only

Now we discuss the voltage calculation of the distribution system with a PV plant. From Figure 10.3, if the feeder side of the main power source is a short circuit, the impedance of the circuit is small comparing to the loads on distribution feeders. Thus, the PV power plant has no direct impact on node voltage loss after node $k$ (between point $k$ and load $n$), but has an indirect impact because the voltage of node $k$ is improved as a result of the access of PV plant. At this point, where the grid-connected PV power plant provides sole injection power and the loss of line voltage is negative, the voltage loss of node $m$ can be expressed as follows:

$$\Delta U_{pv} = -m\frac{P_{pv}R_1 + Q_{pv}X_1}{U_N} \qquad m \in [1, k] \tag{10.13}$$

$$\Delta U_{pv} = -k\frac{P_{pv}R_1 + Q_{pv}X_1}{U_N} \qquad m \in [k+1, n] \tag{10.14}$$

## 10.4.3 Voltage Calculation of Distribution Feeders with PV Plant

By using the superposition theorem, the voltage loss of the distribution feeders with a PV power plants can be obtained by

$$\Delta U_m = \frac{m}{2}(2n - m + 1)\frac{P_dR_1 + Q_dX_1}{U_N} - m\frac{P_{pv}R_1 + Q_{pv}X_1}{U_N} \qquad m \in [1, k] \tag{10.15}$$

$$\Delta U_m = \frac{m}{2}(2n - m + 1)\frac{P_dR_1 + Q_dX_1}{U_N} - k\frac{P_{pv}R_1 + Q_{pv}X_1}{U_N} \qquad m \in [k+1, n] \tag{10.16}$$

Therefore, the node voltage at any point $m$ on the distribution feeders with the PV power plant can be calculated as follows.

$$U_m = U_0 - \frac{m}{2}(2n - m + 1)\frac{P_dR_1 + Q_dX_1}{U_N} + m\frac{P_{pv}R_1 + Q_{pv}X_1}{U_N} \qquad m \in [1, k] \tag{10.17}$$

$$U_m = U_0 - \frac{m}{2}(2n - m + 1)\frac{P_dR_1 + Q_dX_1}{U_N} + k\frac{P_{pv}R_1 + Q_{pv}X_1}{U_N} \qquad m \in [k+1, n]$$
$$\tag{10.18}$$

It can be observed from the above equations that the node voltages of the distribution network have been enhanced with the PV power plants in the network.

## 10.4.4 Voltage Impact of PV Plant in Distribution Network

It can be known from the above analysis that the reason for voltage fluctuation when PV plants connect to the network is the output power fluctuations of the plants. Figure 10.4 is the equivalent circuit when a PV plant is connected to a grid.

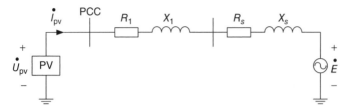

Figure 10.4    Equivalent circuit of PV power plant accessing the grid.

The node voltage of the PCC can be obtained from the voltage balance equation:

$$\dot{U}_{pv} = \dot{E} - (R_z + jX_z)\dot{I}_{pv} = \dot{E} - (R_z + jX_z)(I_{pv\_p} + jI_{pv\_q}) \tag{10.19}$$

where

$\dot{U}_{pv}$: the output voltage phasor of the PV plant

$\dot{E}$: the grid voltage phasor

$R_1$: the line resistance

$X_1$: the line reactance

$R_s$: the equivalent resistance

$X_s$: the equivalent reactance

$\dot{I}_{pv}$: the current phasor from the PV plant to the grid

$\dot{I}_{pv\_p}$: the active components of the injection current from the PV plant into the system

$\dot{I}_{pv\_q}$: the reactive components of the injection current from the PV plant into the system

In addition, $R_z = R_1 + R_s$, and $X_z = X_1 + X_s$ are the total impedance of the lines and the system, respectively.

When the injection power flows from the PV plant to the grid changes, the line current in the grid will change by $\Delta \dot{I}_{pv}$. Assuming the voltage of power grid $\dot{E}$ is a constant, the voltage change of PCC can be calculated as follows.

$$\Delta U_{PCC} = (R_z + jX_z) \cdot (\Delta I_{pv\_p} + j\Delta I_{pv\_q})$$

$$= |Z_z|(\cos \phi + j \sin \phi)|\Delta I|(\cos \theta + j \sin \theta)$$

$$= \frac{U^2}{S_K} \frac{\Delta S_{pv}}{U}[(\cos \phi \cos \theta - \sin \phi \sin \theta) + j(\sin \phi \cos \theta + \cos \phi \sin \theta)]$$

$$\tag{10.20}$$

$$Z_z = (R_1 + R_s) + j(X_1 + X_s) \tag{10.21}$$

$$\theta = \arctan \frac{\Delta I_{pv\_q}}{\Delta I_{pv\_P}} \tag{10.22}$$

where

$\Delta S_{pv}$: the power variation for the PV power plant injecting into the system

$\Delta I$: the current variation for the PV power plant injecting into the system

$S_k$: the short-circuit capacity of accessing point for the photovoltaic power station

$\phi$: the short-circuit impedance angle of PV plant accessing to the grid

$U$: the voltage of the public access point

$Z_z$: the equivalent impedance for the line and system

$\theta$: the power factor angle of the injected PV power change

According to previous analysis, the vertical component of voltage variation can be neglected. The horizontal component of voltage drop $\Delta U_{PCC}$ can be simplified as follows.

$$\Delta U_{PCC} = \frac{U^2}{S_K} \frac{\Delta S_{pv}}{U} (\cos\phi \cos\theta - \sin\phi \sin\theta)$$

$$= U \frac{\Delta S_{pv}}{S_K} \cos(\phi + \theta) \tag{10.23}$$

From the above equation, there are three factors that affect the voltage of PCC. They are the variation of the injection power, the short-circuit capacity of the system, and the power factor of the PV plant. Since the PV plant is often operated in the control mode of the unit power factor, the change in output power of the plant is equivalent to the total change in active output power.

$$\Delta U_{PCC} = U \frac{\Delta P_{pv}}{S_K} \cos(\phi) \tag{10.24}$$

It can be observed from the above equation that the fluctuation in the output power of PV systems is one of the main factors that may cause severe operational problems for the utility network. Power fluctuation occurs because of variations in (1) solar irradiance caused by the movement of clouds, which may continue for minutes or hours, and (2) the PV system topology. Power fluctuation may cause power swings in lines, over- and underloadings, unacceptable voltage fluctuations, and voltage flickers.

## 10.5 FREQUENCY IMPACT OF PV PLANT IN DISTRIBUTION NETWORK

Frequency is one of the more important factors in power quality. Any imbalance between the produced and the consumed power may lead to frequency change. When the generated power is less than the load power due to an accidental event, the electrical torque of generator is greater than the mechanical input torque. The speed of the generator will slow down, decreasing the frequency. Otherwise, when the generated power is greater than the load power, the speed of the generator will accelerate, increasing the frequency.

The small size of PV systems causes frequency fluctuation to be negligible compared with other renewable energy–based resources. However, this issue may become more severe with an increase in the penetration levels of the PV systems. Frequency fluctuation may change the winding speed in electro motors and may damage generators. Thus, it is necessary to analyze the frequency impact of grid-connected PV systems.

System frequency characteristic is the combined effect of load frequency characteristic, generator frequency characteristic, and voltage. Generally, the frequency characteristic can be classified as static and dynamic types. Static frequency characteristic is the relationship between power and frequency in the stable state (generation and consumption is balanced), which is beyond our scope. Dynamic frequency characteristics of power system refer to the time course when the frequency goes through transition from normal state to another stable state when the system's active power balance is destroyed. The process is complicated, and involves several factors. In order to analyze and calculate the dynamic characteristic, we do not consider the load changes with time and the role of the generator governor. At this point, the load can be expressed as a function of frequency,

$$P_L = P_{L0}\left(1 - \frac{K_L \Delta f}{f}\right) \tag{10.25}$$

The gain motion equation of generator rotor is as follows.

$$J\frac{d\omega}{dt} = T_m - T_e = \frac{P_{m0}}{\omega} - \frac{P_{L0}}{\omega}\left(1 - K_L\frac{\Delta\omega}{\omega_0}\right) \tag{10.26}$$

where

$P_{m0}$: the total active power output of the generator

   $J$: the total moment of inertia constant of the generator

$T_m$: the total input mechanical torque of the generator

$T_e$: the total electrical torque of the load

   $\omega$: the generator speed ($\omega = 2\pi f$)

During the relatively short period after a disturbance occurs, let $\omega = \omega_0 + \Delta\omega$, $T_m = T_{m0} + \Delta T_m$, $T_e = T_{e0} + \Delta T_e$, and considering $T_J = J\omega$, we get

$$T_J\frac{d}{dt}\frac{\Delta\omega}{\omega_0} + \frac{K_L}{\omega_0}\frac{\Delta\omega}{\omega_0} = \frac{P_{m0}}{\omega_0} - \frac{P_{L0}}{\omega_0} = T_a \tag{10.27}$$

where $T_a$ is the acceleration torque and $\Delta f/f_0 = \Delta\omega/\omega_0$.

Let $D_T = K_L/\omega_0$ be the total damping coefficient. The above equation can be written as

$$T_J\frac{d}{dt}\frac{\Delta f}{f_0} + D_T\frac{\Delta f}{f_0} = T_a \tag{10.28}$$

Let $T_a$ and $D_T$ remain unchanged within $\Delta t$. From equation (10.28), we get

$$\frac{\Delta f(t)}{f_0} = \frac{T_a}{D_T}\left(1 - e^{-\frac{D_T}{T_J}\Delta t}\right) \tag{10.29}$$

$$\Delta t = -\frac{T_J}{D_T}\ln\left(1 - \frac{D_T}{T_J}\frac{\Delta f}{f_0}\right) \tag{10.30}$$

In terms of the above two equations, $\Delta f(t)$ and $\Delta t$ can be estimated. It is required that the frequency deviation in the normal state should be less than 0.1–0.2 Hz.

In order to adjust power imbalance of system, there are mainly two methods: increasing the power input and load shedding. The level of primary spinning reserves is generally not less than 2% of all loads. Once there is a power short, the system's spinning reserve capacity will be activated as soon as possible to prevent a system crash.

***Example 10.1:***    Figure 10.5 is a distribution network with a PV plant, which is used to analyze the impact of the capacity of the PV plant on node voltage. Each load is $P_d + jQ_d = 0.42 + j\,0.24$.

The power factor of the output power of the PV inverter is selected as 0.9, and the data of the PV power plant capacity are given in Table 10.1. If the access point of the PV plant is located at node 4, the results of the voltage calculation for this distribution network are shown in Figure 10.6. It can be observed from Figure 10.6 that the line voltage loss reduces and the feeder voltage is gradually increases with an increase in the capacity of the PV power plant. The feeder voltage will be higher than

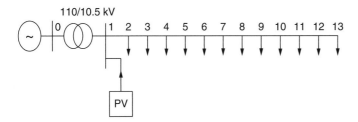

Figure 10.5    A substation with a PV plant.

**TABLE 10.1    Capacity Changes of PV the Power Plant**

| Curve No. | 1 | 2 | 3 | 4 | 5 | 6 |
|---|---|---|---|---|---|---|
| PV power (MVA) | $1 + j0.48$ | $3 + j1.45$ | $5 + j2.42$ | $7 + j3.39$ | $10 + j4.84$ | $15 + j7.26$ |
| $S_{pv}/S_{load}$ | 20% | 60% | 100% | 140% | 200% | 300% |

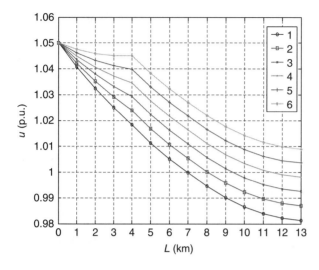

Figure 10.6　Voltage changes with different PV power plant capacity at node 4.

the standard deviation of the voltage if the capacity of the PV power plant exceeds a certain level.

## 10.6　OPERATION OF WIND ENERGY [1,10–16]

### 10.6.1　Introduction

Energy from the wind has been harnessed for thousands of years, making wind power one of the oldest forms of renewable energy. Compared with the other energy sources, there are many advantages of using wind energy:

- Wind energy relies on the renewable power of the wind, which cannot be used up.
- Wind energy is fueled by the wind, so it is a clean fuel source. It does not pollute the airlike power plants that rely on combustion of fossil fuels, such as coal or natural gas. Wind turbines do not produce atmospheric emissions that cause acid rain or greenhouse gases.
- Wind energy is one of the lowest-priced renewable energy technologies available today.
- Wind turbines can be built on farms or ranches, thus benefiting the economy in rural areas, where most of the best wind sites are found. Farmers and ranchers can continue to work the land because the wind turbines use only a fraction of the land.

At present, wind power generation has become a new energy source with great market potential. According to the estimation of the International Energy Agency, by 2030, wind power generation would provide 9% of the world's power demand. With the increasing of capacity of wind fields, many important issues need to be studied and resolved:

**(1)** Voltage fluctuation and flicker governance. Wind speed changes and the shadow effect of wind turbines will cause fluctuation in their power and voltage flicker problem of the power grid.

**(2)** Voltage stability. Voltage stability can be affected and widespread in local areas. At present, to reduce costs and simplify operation and management, the grid-connected wind generator often uses the squirrel-cage asynchronous generator. When the capacity of wind field is enlarged, the reactive power characteristics of the squirrel-cage asynchronous generator will affect the voltage stability; this would need external reactive power compensation to counteract the effect.

**(3)** Frequency stability. With large-capacity wind fields integrating into the grid, dynamic response ability of power system should be capable of tracking high-frequency fluctuation in wind power.

## 10.6.2    Operation Principles of Wind Energy

Wind energy is generated by converting kinetic energy through friction process into useful forms such as electricity and mechanical energy. These two energy sources are put to use by humans to achieve various purposes. Wind turbines use wind energy to produce electricity. Wind turbines are machines that have a rotor with three propeller blades. These blades are specifically arranged in a horizontal manner to propel wind for generating electricity. Wind turbines are placed in areas that have high speeds of wind, to spin the blades much faster for the rotor to transmit the electricity produced to a generator. Thereafter, the electricity produced is supplied to different stations through the grid. According to the rule, the higher you go, the cooler it becomes and more air is circulated. This rule is applied by constructing turbines at high altitudes, to use the increased air circulation at high altitudes to propel the turbines much faster.

A wind energy plant normally consists of many wind turbines each of length 30–50 m. The plant needs to maintain a certain distance during layout of the wind power machines. If the space interval among the wind power machines is too large, area covered by a single machine will be increased; this will reduce the number of wind power machines within the same area of the wind field, or expand the area of the wind field under the same installed capacity. Consequently, the transmission distance, investment, and operating loss will increase. Usually, the space interval among wind power machines in the dominant wind direction is about 8–12 times the diameter of the wind wheel and about 2–4 times in the vertical direction of the dominant wind direction.

## 10.6.3    Types and Operating Characteristics of the Wind Turbine

A wind farm is a group of wind turbines in the same location used for production of electric power. Individual turbines are interconnected with a medium voltage (usually

34.5 kV) power collection system and communications network. A simple relationship exists relating the power generated by a wind turbine and the wind parameters:

$$P_W = 0.5\rho A V_W^3 C_P \tag{10.31}$$

where

$P_W$: the wind power
$P$: the air density
$A$: the fan blade sweep for the wind section
$V_W$: the wind speed
$C_P$: the wind energy utilization factor

It can be seen from (10.31) that as fan power and the output are proportional to the cube of wind speed, the smaller changes in wind speed would cause a larger change in wind power.

The angle between wind flows and the strings section of the fan blades is called the pulp angle, recorded as $\beta$. At the same time, the ratio of cutting-edge rotation of the fan blades to the wind velocity is defined as the tip-speed ratio (TSR), recorded as $\lambda$. According to the dynamic characteristics of fan blades, fan conversion efficiency is a function of the pulp angle and TSR, that is, $C_P = f(\beta, \lambda)$.

At a certain angle $\beta$, the characteristics of the wind turbine can be expressed by the wind energy conversion efficiency curve $(C_P - \lambda)$. The relationship between conversion efficiency and the TSR $\lambda$ is shown in Figure 10.7.

For a different pulp angle $\beta$, a group of curves $C_P - \lambda$ can be obtained as shown in Figure 10.8. It can be seen from the figure that, for the same $C_P$, the wind machine has two operating points $A$ and $B$, which correspond respectively to the high-speed

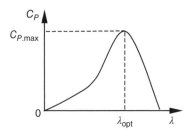

Figure 10.7   Relationship between $C_p$ and $\lambda$.

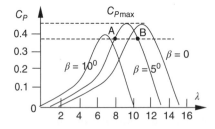

Figure 10.8   Typical curves of $C_p = f(\beta, \lambda)$.

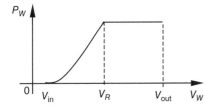

Figure 10.9    Characteristic curve figure of ideal wind turbine power.

operation area of the wind machines and the low-speed running area. With the wind speed changing, the wind turbine operation point also changes. The ideal power and wind speed curve of a wind turbine, which is shown in Figure 10.9, is achieved through controlling the blade angle $\beta$ and rotor speed.

In Figure 10.9, $P_W$ is the wind power (W), $V_W$ is the wind speed (m/s), $V_{in}$ is the starting wind speed of the fans (m/s), $V_R$ is the rated wind speed (m/s), and $V_{out}$ is cutout wind speed (m/s).

Owing to a larger moment of inertia of the blades and hub, the conversion from wind energy to mechanical energy has a certain time lag. The one-order inertia link can be used to simulate the process of analysis of the dynamic characteristics of wind power systems; this is shown in Figure 10.10.

In the Figure 10.10, $P_m$ is the wind turbine output mechanical power; $T_W$ is the wind turbine inertia constant.

According to whether the pitch angle $\beta$ is adjustable, there are two types of wind machines: (1) rated-blade pitch wind turbine (or pitch-rated wind turbine), with fixed blade angle ($\beta$ = const.); (2) variable-blade pitch wind turbine (or pitch-regulated wind turbine), with adjustable blade angle.

For the rated blade pitch wind power, stall regulation can limit the energy capture at a specified value, which relies on the blade's aerodynamic shape (the leaves twist angle). Under the condition of rated wind speed, the air flows along (close to) the surface of the stable leaf. The wind energy absorbed by the leaves is proportional to the wind speed.

If the wind speed is over the rated value, the air flows at the back of leaves and separates from the leaves, causing the efficiency of the plant to absorb the wind to drop when the wind speed increases. In this case, the power absorbed by leaves is slightly lower than the rated power. This kind of wind turbine has a simple structure, but it bears the loss of torque, enabling the leaves to bear a larger force.

The variable-blade pitch wind turbine is able to maintain constant output power through adjustment of the pitch angle $\beta$, which changes the blade angle between the windward side and the longitudinal axis of rotation to affect the impact force, and thus

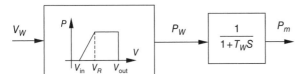

Figure 10.10    Mathematic model of a wind turbine.

to regulate the output power of the fans. If the wind speed is below the rated value, the controller will set the power angle of the blade close to zero, which is equivalent to pitch-rated regulation. If the wind speed is over the rated value, the variable-blade pitch wind turbine adjusts the power angle blade and controls the output power around the rated value.

Compared to the pitch-rated wind turbine, the variable-blade pitch wind turbine has the following advantages:

**(1)** By adjusting the pitch angle, the pitch-regulated wind turbine has higher wind energy conversion efficiency than pitch-rated wind turbine at a low wind speed. Therefore, there is greater energy output, and the starting wind speed is also higher, which is more suitable for the regions with low average wind speed installation.

**(2)** The variable-blade pitch wind turbine is much less impacted by force than the pitch-rated wind turbine. This can reduce material usage and lower the overall weight of the turbine.

**(3)** When the wind speed exceeds a certain value, the pitch-rated wind turbine must be shut down, while the blades of the pitch-regulated wind turbine can be adjusted to the no-load position without shutting down the turbine, which is the launch mode of the entire wing.

Because of the above advantages, variable-blade pitch wind turbines can increase annual generating capacity of wind power rather than pitch-rated wind turbines.

### 10.6.4   Generators Used in Wind Power

There are three types of generators used in wind power: (1) synchronous generator; (2) squirrel-cage induction generator; (3) doubly fed generator.

The synchronous generator is an AC generator that was first used in wind power. Owing to its excitation systems with complex structure, high cost, and the high probability of failure, most of them have been replaced by squirrel-cage induction generators after 1990s.

The rotor of the squirrel-cage induction generator is a short-circuit winding. Its equivalent circuit is shown in Figure 10.11.

In the Figure 10.11, $I_m$ is the generator exciting current; $I_r$ is the rotor current; $R_2$ is the rotor resistance; $X_m$ is the magnetizing reactance; $X_\sigma$ is the leakage reactance.

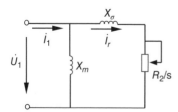

Figure 10.11   Simplified equivalent circuit of an induction generator.

Since the rotor relies on the excitation of the stator, the squirrel-cage induction generator outputs active power and absorbs reactive power. The power equation of the squirrel-cage induction generator can be obtained from Figure 10.11.

$$P_e = \frac{(sU)^2}{(sX_\sigma)^2 + R_r^2} \frac{R_r}{s} \tag{10.32}$$

$$Q = Q_m + Q_\sigma = \frac{U^2}{X_m} + \frac{(sU)^2}{(sX_\sigma)^2 + R_r^2} X_\sigma \tag{10.33}$$

where

$P_e$: the output average power of the generator

$Q$: the reactive power absorption of the generator

$Q_m$: the magnetizing reactive component

$Q_\sigma$: the reactive component due to the leakage reactance

$s$: the generator slip, which is defined as

$$s = \frac{\omega_r - \omega_0}{\omega_0} \tag{10.34}$$

where, $\omega_0$ is the angular velocity of the stator flux; $\omega_r$ is the rotor angular velocity; $s > 0$ is the condition for power generation.

It can be seen from equation (10.32) that the output active power of electro-magnetic induction generator is a function of the slip $s$. Figure 10.12 shows the characteristic curves $P_e - s$ of the induction generator with two voltage levels, where $U_1 < U_2$.

In Figure 10.12, $P_{e.max}$ stands for the maximum active power, and $s_m$ is the slip corresponding to $P_{e.max}$. These two factors can be calculated by the following equation.

$$\left. \begin{array}{l} s_m = \dfrac{R_r}{X_\sigma} \\[2mm] P_{e.max} = \dfrac{U^2}{2X_\sigma} \end{array} \right\} \tag{10.35}$$

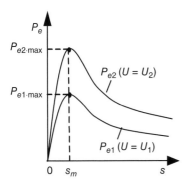

Figure 10.12  $P_e - s$ characteristics of an asynchronous generator.

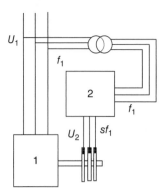

Figure 10.13   Structure of a doubly fed generator.

The structure of the doubly fed generator is shown in Figure 10.13.

In Figure 10.13, box 1 contains stators and rotors, which are the generators of three-phase AC windings. Box 2 is the AC–AC frequency converter.

Through the AC–AC frequency converter, the rotor will be excited by the doubly fed generator. If the frequency of the excited current decreases to zero, that is, the excited current is a DC current, the rotor at this moment will operate at the synchronous speed, and its characteristics will be exactly same as the characteristics of traditional synchronous generator. Meanwhile, the excited current of the rotor will be completely determined by $U_2$, which is the output voltage of the AC–AC frequency converter. If the frequency of the excited current is not zero, which means the excited current is an AC current, the rotor will not operate at the synchronous speed. As a result, the excited current of the rotor will be determined by both the voltage of the AC–AC frequency converter and the voltage excited by rotor winds cutting the magnetic field of the stator. Therefore, it has the double characteristics of a traditional synchronous machine and a squirrel-cage induction generator.

## 10.7   VOLTAGE ANALYSIS IN POWER SYSTEM WITH WIND ENERGY

### 10.7.1   Introduction

Voltage stability refers to the ability of a power system to maintain steady voltages at all buses in the system after being subject to a disturbance from a given initial operating condition. It depends on the ability to maintain/restore equilibrium between load demand and load supply from the power system. Instability that may result occurs in the form of a progressive fall or rise of voltages of some buses. A possible outcome of voltage instability is loss of load in an area or tripping of transmission lines and other elements by their protective systems, leading to cascading outage [19]. Generally, voltage stability can be classified into two subcategories: large-disturbance voltage stability and small-disturbance voltage stability. The former refers to the system's ability to maintain steady voltages following large disturbances such as system fault, loss of generation, or circuit contingencies, and the latter refers to system's ability to maintain steady voltages when subjected to small perturbations such as incremental changes in system load [19,20].

The impact of the wind power on voltage distribution levels has been addressed in the literature. The majority of these works deals with the determination of the maximum active and reactive power that is possible to be connected on a system load bus, until the voltage at that bus reaches the voltage collapse point. It is done by the traditional methods of PV curves reported in many references [21–24]. These studies handle small-disturbance voltage stability.

Wind power generation can affect the large-disturbance voltage stability of power systems because a fault in the network results in a reduction in the supply voltage to a wind generator for a short period of time (voltage dip) and subsequently to generator tripping due to minimum voltage protections.

## 10.7.2 Voltage Dip

The distribution network with installation of wind energy will experience voltage dips due to faults at distribution voltage levels. An important characteristic of those dips is that they are associated with a jump in characteristic phase angle, that is, the characteristic voltage $V$ becomes complex:

$$V^* = V e^{j\varphi} \tag{10.36}$$

where $V$ now stands for the absolute value of the characteristic voltage (the characteristic retained voltage) and $j$ for its phase angle (the characteristic phase-angle jump).

The relation between the retained voltage and the phase-angle jump can be described on the basis of the following simple distribution network (Figure 10.14):

If a two or three-phase fault location is at the terminal of the distribution network, the complex voltage at the PCC is found from

$$U_{\text{PCC}} = U \frac{Z_f}{Z_f + Z_s} \tag{10.37}$$

where

$U$: the pre-fault voltage
$Z_f$: the impedance between the PCC and the fault location
$Z_s$: the source impedance at the PCC.

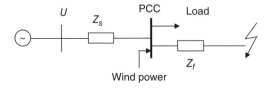

Figure 10.14   A distribution network with wind power.

As both impedances are complex numbers, the resulting voltage $U_{PCC}$ will show a magnitude drop and a change in phase angle compared to the pre-fault voltage $U$. In transmission systems, both $Z_f$ and $Z_S$ are formed mainly by transmission lines and so the phase angle jump will be small. In distribution systems, $Z_S$ is typically formed by a transformer with a rather large X/R ratio, whereas $Z_f$ is formed by lines or cables with a much smaller X/R ratio. This will lead to a significant change in phase angle, especially for cable faults.

Equation (10.37) can be rewritten as follows:

$$\frac{U_{PCC}}{U} = \frac{\lambda e^{j\alpha}}{1 + \lambda e^{j\alpha}} \tag{10.38}$$

$$\frac{Z_f}{Z_s} = \lambda e^{j\alpha} \tag{10.39}$$

The value of $\lambda$ depends on the distance to the fault. $\alpha$ is the angle between source impedance and the feeder impedance, which is constant for any feeder/source combination. This angle is referred to as the "impedance angle" in [18]. The phase-angle jump is the argument (angle) of (10.39):

$$\Delta\phi = \arg\left(\frac{\lambda e^{j\alpha}}{1 + \lambda e^{j\alpha}}\right) \tag{10.40}$$

For any given impedance angle, there is a unique relation between voltage dip magnitude and phase-angle jump.

The above expressions only hold for voltage dips due to two- and three-phase faults, not for voltage dips due to single-phase faults. The effect of a single-phase fault depends on the type of system earthing used. In a high-impedance earthed system, single-phase faults do not cause any significant voltage dips at all. They do cause a zero sequence voltage but this does not affect end-user equipment so there is no need to consider them as voltage dips. In solidly earthed system single-phase faults do lead to voltage dips but less severe ones than those due to the other types of faults. Even retained voltage is not higher; in addition, the phase-angle jump is less severe [17].

From the point of view of a sensitive wind power installation, what matters is the expected number of severe dips (i.e., retained voltages below the critical voltages) because all these dips will expose the installation to a potential disconnection or mal-operation. A suitable way to present this information is the cumulative histogram of retained voltages (dips).

### 10.7.3 Simulation Results

Figure 10.15 is a practical power system with wind power installation, which is used to analyze the voltage impact of the system with wind power penetration. Bus 0 is the system slack bus that corresponds to the equivalent power network. Buses 1–6 stand for 330 kV substations. Buses 7–9 correspond to three wind power fields

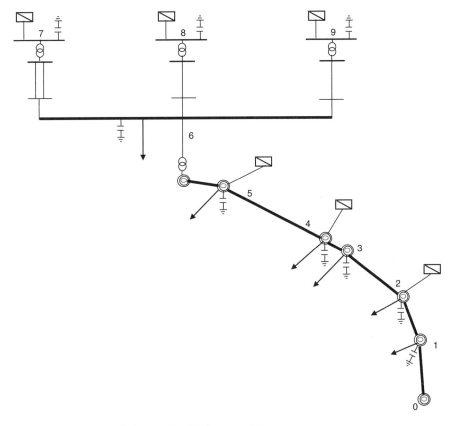

Figure 10.15    A practical network with three wind farms.

that have 100 MW, 50 MW, 50 MW capacities, respectively. The active power supply, load demand, and the capacity of shunt reactive power compensation of each bus are summarized in Table 10.2. The reactance of each branch is summarized in Table 10.3.

The equivalent impedances of asynchronous generator are rotor resistance $R_r =$ 0.055 p.u., leakage reactance $X_\sigma = 0.2875$ p.u., and magnetizing reactance $X_m =$ 3.3 p.u.

The following cases are simulated and analyzed for the aforementioned practical power system with wind power penetration.

**Case 1    A three-phase fault occurs at the 10 kV incoming feeder in the substation of the wind power field (bus 7). The short-circuit period is 1.0 s.**
*Fault starting time is set to t = 10s, the fault-clearing time is set to t = 10.5s, and the simulation duration is 30 s. The corresponding simulation results, which are the maximum slips in the process of the fault for each wind power field, are shown in Table 10.4.*

**TABLE 10.2  Power Supply and Demand of Each Bus**

| Bus Number | Capacity of Power Supply (MW) | Active Power of Load (MW) | Capacity of Shunt Capacitor (Mvar) |
|---|---|---|---|
| 1 | 0 | 252 | 200.4 |
| 2 | 100 | 640 | 196.86 |
| 3 | 0 | 100 | 64.6 |
| 4 | 600 | 372 | 111.72 |
| 5 | 300 | 406 | 111.72 |
| 6 | 0 | 40 | 54.5 |
| 7 | 100 | 0 | 75 |
| 8 | 50 | 0 | 37.5 |
| 9 | 50 | 0 | 37.5 |

**TABLE 10.3  Parameter Value of Inductive Reactance**

| Inductive Reactance of Ranch | Per unit Value | Inductive Reactance of Ranch | Per Unit Value |
|---|---|---|---|
| $X_{01}$ | 0.0250 | $X_{56}$ | 0.0850 |
| $X_{12}$ | 0.0125 | $X_{67}$ | 0.0554 |
| $X_{23}$ | 0.0139 | $X_{68}$ | 0.1282 |
| $X_{34}$ | 0.0104 | $X_{69}$ | 0.1235 |
| $X_{45}$ | 0.0290 | — | — |

**TABLE 10.4  Maximum Slip in the Process of the Fault for Case 1**

| Bus number | 7 | 8 | 9 |
|---|---|---|---|
| Voltage at $t = 30s$ | 1.0287 | 1.0275 | 1.0278 |
| Slip at $t = 30s$ | −0.0554 | −0.0555 | −0.0555 |
| Maximum slip | 0.1866 | 0.1589 | 0.1585 |

*The simulation results show that the rotor speed of each wind power generator increases rapidly and the corresponding node voltages drop a lot in the process of the fault. In addition, the bus voltage at bus 6 also drops rapidly. But after fault clearing, the active power, the reactive power, the wind turbine power, and the voltage of each node recover to normal states and thus, power system retains stability.*

## Case 2  A three-phase fault close to the 10 kV bus in the wind power field (bus 8); the fault-clearing time is 1.0 s

*The other conditions or simulation parameters are the same as in Case 1. The simulation results-the maximum slip in the process of fault and slip at $t = 30s$ of each wind power field-are summarized in Table 10.5.*

TABLE 10.5   Maximum Slip in the Process of the Fault for Case 2

| Node number | 6 | 7 | 8 | 9 |
|---|---|---|---|---|
| Voltage at $t = 30s$ | – | 1.028968 | 1.027434 | 1.028027 |
| Slip at $t = 30s$ | – | −0.056292 | −0.056465 | −0.056417 |
| Maximum slip | – | −0.161821 | −0.291627 | −0.163487 |

*Because the capacity of wind turbines at a fault point is small, the increment of absorbing reactive power of asynchronous generators is small in the process of a fault. The fault at wind power field bus 8 or bus 9 has small effect on the voltage at bus 6, as well as the other two normal operation wind power fields.*

**Case 3   A three-phase fault occurs at the 110 kV outgoing line in a substation at bus 6; the fault-clearing time is 0.25 s**
*With the same simulation parameters as Case 1, the maximum slip in the process of fault and slip at t = 30s of the three wind power fields are summarized in Table 10.6.*
*From Table 10.6, when a three-phase fault occurs on the 110 kV outgoing line in the substation (bus 6), the system can keep stable because of fast operating of protection and short fault duration.*

**Case 4   A three-phase fault occurs on 110 kV outgoing line in substation (bus 6); fault-clearing time is 0.5 s**
*With the same simulation parameters as in Case 1, the voltage and slip of the three wind power fields at t = 30s are summarized in Table 10.7.*
*It can be seen from Table 10.7 that the power system has lost stability. Compared with the faults occurring in the wind power field, the fault close to the 110 kV substation (bus 6) can cause the problem of voltage stability. The reason is that the fault at bus 6 causes the speedup of generator groups in three wind power fields and makes the 200 MW asynchronous generator absorb the huge reactive power. In*

TABLE 10.6   Maximum Slip in the Process of the Fault for Case 3

| Node number | 6 | 7 | 8 | 9 |
|---|---|---|---|---|
| Voltage at $t = 30s$ | 1.037247 | 1.028828 | 1.027509 | 1.027814 |
| Slip at $t = 30s$ | – | −0.056292 | −0.056465 | −0.056417 |
| Maximum slip | – | −0.185367 | −0.192433 | −0.190475 |

TABLE 10.7   Voltage and Slip Value at $t = 30s$ for Case 4

| Node number | 6 | 7 | 8 | 9 |
|---|---|---|---|---|
| Voltage | 0.737900 | 0.62869 | 0.614298 | 0.618134 |
| Slip | – | −0.836934 | −0.848868 | −0.845768 |

*addition, three wind power fields are located at the terminal of the power network, thus, the voltage is difficult to recover after fault clearing. Consequently, voltage stability is destroyed.*

*Through simulation calculations, we can conclude that the capacity of the wind power unit, the fault location, and fault clearing time are very important factors of voltage stability of a power system. If the fault point is close to a large-capacity wind power field, it is easier for the system to lose stability. If the fault point is closer to a public access point of wind power field, the system is also easier to lose stability. If the fault duration is longer, the system is also easier to lose stabilility.*

## PROBLEMS AND EXERCISES

1. What is renewable energy?

2. What are the purposes for which we use renewable energy sources?

3. What is MPPT?

4. What is PCC?

5. What are the advantages of the grid-connected PV system?

6. What advantages does the variable blade pitch wind turbine have?

7. State "True" or "False"

  7.1 Hydropower is a renewable energy resource.

  7.2 Energy takes many forms.

  7.3 Using biomass as an energy source does not pollute the environment.

  7.4 Using hydropower does not impact the environment.

  7.5 Electricity is a nonrenewable resource.

  7.6 There is no voltage stability problem for a system with wind power penetration.

8. Which of the following is a renewable source of energy?

  A. Coal

  B. Hydropower

  C. Natural gas

  D. Petroleum

9. Of the following choices, which best describes or defines *biomass*?

  A. Massive living things

  B. Inorganic matter that can be converted to fuel

  C. Organic matter that can be converted to fuel

  D. Petroleum

10. Of the following choices, which best describes or defines *geothermal energy*?

  A. Heat energy from volcanic eruptions

  B. Heat energy from hot springs

    **C.** Heat energy from inside the earth

    **D.** Heat energy from rocks on Earth's surface

**11.** Which of the following is *not* a renewable source of energy?

    **A.** Geothermal

    **B.** Propane

    **C.** Solar

    **D.** Wind

**12.** Which of the following is not a fossil fuel?

    **A.** Biomass

    **B.** Coal

    **C.** Natural gas

    **D.** Petroleum

**13.** New renewable energy resources are

    **A.** Solar, wind, geothermal

    **B.** Wind, wood, alcohol

    **C.** Hydro, biomass

    **D.** Coal, natural gas

**14.** At present, the fastest growing source of electricity generation using a new renewable source

    **A.** Solar

    **B.** Wind

    **C.** Hydro

    **D.** Natural gas

**15.** A major disadvantage of solar power is

    **A.** its cost effectiveness compared to other types of power

    **B.** its efficiency level compared to other types of power

    **C.** the variation in sunshine around the world

    **D.** the lack of knowledge on long-term economic impact

**16.** Windmill towers are generally more productive if they are

    **A.** higher, to minimize turbulence and maximize wind speed

    **B.** lower, to minimize turbulence and maximize wind speed

    **C.** higher, to minimize the number of birds that interfere with blade turning

    **D.** lower, to increase heat convection from the ground

**17.** A major disadvantage in using wind to produce electricity is

    **A.** the emissions it produces once in place

    **B.** its energy efficiency compared to that of conventional power sources

    **C.** Wind Turbines Kill Birds

    **D.** the initial startup cost

**18.** The largest problem with adopting the new technology of renewable resources is

    **A.** in evaluating the scientific and economic impact

    **B.** the high start-up costs

    **C.** higher long-term maintenance costs than those for fossil fuels

    **D.** energy production facilities not being located near consumers

# REFERENCES

1. Zhu JZ. *Renewable Energy Applications in Power Systems*. New York: Nova Press; 2012.
2. Zhu JZ, Cheung K. Summary of environment impact of renewable energy resources. Adv. Mat. Res. 2013;616–618:1133–1136.
3. Hamrouni N, Chérif A. Modelling and control of a grid connected photovoltaic system. Int. J. Elec. and Power Eng. 2007;1(3):307–313.
4. Babu BP, Reddy IP. Operation and control of grid connected PV-FC hybrid power system. Int. J. Eng. Sci. 2013;2(9):52–63.
5. Chenni R, Makhlouf M, Kerbache T, et al. A detailed modeling method for photovoltaic cells. Energy 2007;32(9):1724–1730.
6. Kim SK, Jeon JH, Cho CH, et al. Modeling and simulation of a grid-connected PV generation system for electromagnetic transient analysis. Sol. Energ. 2009;83(5):664–678.
7. Zhu JZ, Cheung K. Voltage impact of photovoltaic plant in distributed network. 2012 IEEE APPEEC Conference, Shanghai, China, March 27–29, 2012.
8. Zhu JZ, Xiong XF, Cheung K. Research and development of photovoltaic power systems in China, IEEE PES General Meeting, Detroit, MI, July 25–28, 2011.
9. Zhou NC, Zhu JZ. Voltage assessment in distributed network with photovoltaic plan. ISRN Renew. Energ., 2011.
10. Ackerman T. *Wind Power in Power Systems*. Wiley; 2005.
11. Hatziargyriou N, Zervos A. Wind power development in Europe. Proc. IEEE 2001;89(12): 1765–1782.
12. Wiser R, Bolinger M. Annual report on US wind power installation, cost, and performance trends: 2007, *Energy Efficiency and Renewable Energy*, US Department of Energy, Washington, DC; 2008.
13. Zhu JZ, Cheung K. Analysis of regulating wind power for power system, IEEE PES General Meeting, Calgary, Canada, July 26–30, 2009.
14. Zhu JZ, Cheung K. Selection of wind farm location based on fuzzy set theory. 2010 IEEE PES General Meeting, Minneapolis, MN, July, 2010.
15. Lin L, Zhou N, Zhu JZ. Analysis of voltage stability in a practical power system with wind power. Electr. Pow. Compo. Sys. 2010;38(7):753–766.
16. Lin L, Guo W, Wang J, Zhu JZ. Real-time voltage control model with power and voltage characteristics in the distribution substation. Int. J. Elec. Power. 2013;33(1):8–14.
17. Bollen MHJ, Olguin G, Martins M. Voltage dips at the terminals of wind power installations, NORDIC Wind Power Conference, March 1–2, 2004, Chalmers University of Technology.
18. Bollen MHJ. *Understanding Power Quality—Voltage sags and Interruptions*. New York: IEEE Press; 2000.
19. Van Cutsem T, Vournas C. *Voltage Stability of Electric Power Systems*. Norwell, MA: Kluwer; 1998.
20. Fan YF, Chao Q. Modeling and simulation of wind asynchronous-generator. Comput. Simulat. 2002;19(5):56–58.
21. Arai J, Yokoyama R, Iba K, Zhou Y, Nakanishi Y. Voltage deviation of power generation due to wind velocity change. Int. J. Energ. 2007;2(1).
22. Chuong TT. Voltage stability investigation of grid connected wind farm, Proceedings of World Academy of Science Engineering and Technology, Vol. 32, 2008.

23. El-Kashlan SA, Abdel-Rahman M, El-Desouki H, Mansour MM. *Voltage Stability of Wind Power Systems Using Bifurcation Analysis. European Power and Energy Systems*; 2005.

24. Singh B, Singh SN. Voltage stability assessment of grid-connected offshore wind farms. Wind Energy 2009;12(2):157–169.

# OPTIMAL LOAD SHEDDING

When all available controls are unable to maintain the security of system operation during a disturbance or contingency, optimal load shedding is used as the last resort to make the loss of blackout minimum. This chapter first introduces the traditional load-shedding methods such as under-frequency or under-voltage load shedding, and then studies optimal power system load-shedding methods. These include intelligent load shedding (ILS), distributed interruptible load shedding, Everett optimization, analytic hierarchical process (AHP), and network flow programming (NFP). The related topic on congestion management is also introduced in this chapter.

## 11.1 INTRODUCTION

The security and stability of electrical power systems have always been among the central and fundamental issues of concern in network planning and operation. Serving users of electricity is the duty of power systems that generate, transmit, and distribute electrical energy. Therefore, system operation, network growth and expansion are highly user dependent and the system should be able to satisfy their needs and requirements. Central requirements include reliability, quality of energy, and continued load capacity. Network designers and operation managers should continuously pay attention to these requirements and take the necessary steps to fulfill these requirements and maintain the desired qualities. Especially, in the United States, the electricity market is in the midst of major changes designed to promote competition. Vertical integration with guaranteed customers and suppliers is no longer there. Electricity generators and distributors have to compete to sell and buy electricity. The stable utilities of the past find themselves in a highly competitive environment [1–3]. In this new competitive power environment, buy/sell decision support systems are to find economic ways to serve critical loads with limited sources under various uncertainties. Decision making is significantly affected by limited energy sources, generation cost, and network-available transfer capacity. Generally, system congestion or system overloading can be reduced through some control strategy such as a

*Optimization of Power System Operation*, Second Edition. Jizhong Zhu.
© 2015 The Institute of Electrical and Electronics Engineers, Inc. Published 2015 by John Wiley & Sons, Inc.

generation-rescheduling scheme, obtaining power support from a neighboring utility as well as optimal load shedding [4–7]. In the particular case of power shortage, load shedding cannot be avoided. This, in turn, requires that the load demand be as determinate as possible so that each watt can be allocated.

In general, load shedding can be defined as the amount of load that must almost instantly be removed from a power system to keep the remaining portion of the system operational. This load reduction is in response to a system disturbance (and consequent possible additional disturbances) that results in a generation-deficiency condition or network-overloading situation. Common disturbances that can cause these conditions to occur include transmission line or transformer faults, loss of generation, switching errors, lightning strikes, etc. When a power system is exposed to a disturbance, its dynamics and transient responses are mainly controlled through two major dynamic loops. One is the excitation (including AVR) loop that controls the generator reactive power and system voltage. The other is the prime-mover loop, which controls the generator active power and system frequency.

## 11.2 CONVENTIONAL LOAD SHEDDING

Load shedding by frequency relays is the most commonly used method for controlling the frequency of power networks within set limits and maintaining network stability under critical conditions. In conventional load-shedding methods, when frequency drops below the operational plan's set point, the frequency relays of the system issue commands for a stepwise disconnection of parts of the electrical power load, thereby preventing further frequency drop and its consequential effects [8].

Frequency is the main criteria of system quality and security because it is

- a global variable of interconnected networks that has the same value in all parts of the network;
- an indicator of the balance between supply and demand;
- a critically important factor for smooth operation of all users, particularly manufacturing and industries.

One of the main problems of all interconnected networks is a total blackout because of frequency drop as a consequence of some power station failure or transmission line breakage. At present, in power generation and transmission systems of the world, the most appropriate way of preventing a total or partial blackout that is triggered by frequency drop is quick and automatic load shedding.

To study situations of imbalance between power supply and demand, and the resulting frequency variations under the circumstances of severe and major disorders, a simplified model of the steady state for systems that consist mainly of thermal units is used [8–10], which is shown in Figure 11.1.

The expression of the model is as follows.

$$\Delta\omega = \frac{P_a}{D}\left(1 - e^{-\frac{D}{2H}t}\right) \tag{11.1}$$

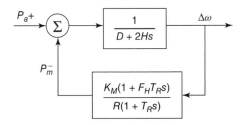

Figure 11.1    Steady-state frequency-response model.

where

$H$: system's inertial constant

$D$: load damping coefficient

$K_m$: frequency control loop gain

$F_H$: high pressure re-warmed turbines' power portion

$TR$: re-warming time constant

$Pm$: mechanical power of the turbine (per unit)

$P_a$: accelerator's power

$\Delta\omega$: speed change (per unit).

Equation (11.1) models the system under the initial conditions of major disorder when the governor's effect is lifted off because during the first seconds of the disorder, due to the governor's response delay and its operating time constant, it cannot play a role in prevention of the frequency drop [9].

According to equation (11.1), the main factors and parameters that control the behavior of frequency and overloading are the amount of overloading and the $D$ and $H$ parameters. The effect of these two parameters should be definitely be considered in any load-shedding scheme.

The load damping coefficient ($D$) is an effective parameter that represents the relation between the load and the frequency. It cannot be ignored in planning for load-shedding schemes. In planning for load shedding, the load damping coefficient is normally expressed per unit as shown in the following formula:

$$D = \frac{F}{P}\frac{\Delta P}{\Delta F} \tag{11.2}$$

The value of $D$ varies from 0 to 7 and, for each system, it is to be determined once and used in all cases of planning. The latest studies have shown $D = 3.3$ for the sample network [8].

The effect of $D$ on the frequency-drop gradient is quite visible as an increase in $D$ causes a decrease in the frequency-drop gradient. For any specified overloading, systems with a higher value of $D$ will have a higher stability and the final system frequency will be stabilized at a higher level. Figure 11.2 clearly shows the effect of $D$ on the frequency-drop curve.

In commonly used stepwise methods, the load-shedding scheme has little relation to the degree of overload. Any overload triggers the same strategy of load

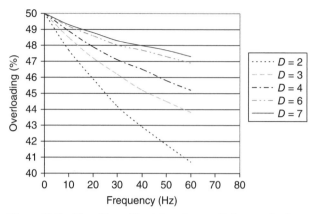

Figure 11.2   The effect of load damping coefficient on the frequency-drop curve (system stability curves for various overloading).

shedding, as the degree of overload does not determine the number or quantity of the load shedding.

This kind of scheme greatly simplifies the task of harmonizing the relays and the steps of load shedding, as simple calculations and a process of trial-and-error would suffice. It is one of the obvious advantages of this kind of scheme. Once the steps of load shedding are specified, if at any step the frequency continues to drop (with regard to the specified delay times), then the next step will be automatically activated until the frequency stops dropping. In such strategies, increasing the number of steps can increase the costs and allow a more precise harmony and a minimized blackout area. Nevertheless, in almost all countries, only three to five steps are planned, with rare cases of more steps.

In such strategies or plans, the first step of load shedding is regulated in such a way that with any frequency drop below the set point, this step is activated to operate within its specific time delay. The time duration for frequency to drop from normal to below the set point is not taken into consideration, despite the fact that we know that the gradient of frequency drop is directly proportional to the amount of overload and severity of the case; therefore, it can be a basis to decide on whether only one step is adequate.

## 11.3   INTELLIGENT LOAD SHEDDING

### 11.3.1   Description of Intelligent Load Shedding

Conventional load-shedding systems that rely solely on frequency-measuring systems cannot be programmed with the knowledge gained by the power system designers. The system engineer must perform numerous system studies that include all the conceivable system operating conditions and configurations to correctly design the power system. Unfortunately, the engineer's knowledge of the system, which is gained through the studies, is not utilized fully. In addition, most data and study results are simply lost. This nonavailability of information for future changes and enhancement of the system will significantly reduce the system protection performance.

The state-of-the-art load-shedding system uses real-time, systemwide data acquisition that continually updates a computer-based real-time system model. This system produces the optimum solution for system preservation by shedding only the necessary amount of load and is called intelligent load shedding [11].

This system must have the following capabilities:

- to map a very complex and nonlinear power system with a limited number of data collection points to a finite space;
- to automatically remember the system configuration and operation conditions as load is added or removed, and the system response to disturbances with all the system configurations;
- to recognize different system patterns in order to predict system response for different disturbances;
- to utilize a built-in knowledge base trainable by user-defined cases;
- to make use of adaptive self-learning and automatic training of the system knowledge base obtained as a result of system changes;
- to make fast, correct, and reliable decisions on load-shedding priority based on the actual loading status of each breaker;
- to shed the minimum amount of load to maintain system stability and nominal frequency;
- to shed the optimal combinations of load breakers with complete knowledge of system dependencies.

In addition to having the above list of capabilities, the ILS system must have a dynamic knowledge base. For the knowledge base to be effective, it must be able to capture the key system parameters that have a direct impact on the system frequency response following disturbances. These parameters include the following:

- power exchanged between the system and the grid both before and after disturbance;
- generation available before and after disturbances;
- on-site generator dynamics;
- updated status and actual loading of each sheddable load;
- the dynamic characteristics of the system loads which this include rotating machines, constant impedance loads, constant current loads, constant power loads, frequency-dependent loads, or other types of loads.

Some additional requirements must be met during the designing and tuning of an ILS scheme:

- carefully selected and configured knowledge base cases;
- ability to prepare and generate sufficient training cases for the system knowledge base to ensure accuracy and completeness;

- ability to ensure that the system knowledge base is complete, correct, and tested;
- ability to add user-defined logics;
- ability to add system dependencies;
- to have an online monitoring system that is able to coherently acquire real-time system data;
- ability to run in a preventive and predictive mode so that it can generate a dynamic load-shedding table that corresponds to the system configuration changes and prespecified disturbances (triggering);
- a centralized distributed local control system for the power system that the ILS system supervises.

## 11.3.2  Function Block Diagram of the ILS

In Figure 11.3, the system knowledge base is pretrained by using carefully selected input and output databases from offline system studies and simulations. System dynamic responses, including frequency variation, are among the outputs of the knowledge base.

The trained knowledge base runs in the background of an advanced monitoring system, which constantly monitors all the system operating conditions. The network models and the knowledge base provide power system topology, connection information, and electric properties of the system component for ILS. The disturbance list is prepared for all prespecified system disturbances (triggers). On the basis of input data and system updates, the knowledge base periodically sends requests to the ILS

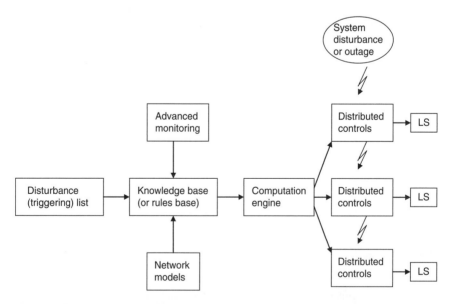

Figure 11.3  Function block diagram of the ILS scheme.

computation engine to update the load-shedding tables, thus ensuring that the optimum load will be shed when a disturbance occurs. The load-shedding tables, in turn, are downloaded to the distributed controls that are located close to each sheddable load. When a disturbance occurs, fast load-shedding action can be taken.

## 11.4  FORMULATION OF OPTIMAL LOAD SHEDDING

In a competitive resource allocation environment, buy/sell decision support systems are needed to find economic ways to serve critical loads with limited sources under different uncertainties. Therefore, a value-driven load-shedding approach is proposed for this purpose. The mathematical model of load shedding is expressed as follows.

### 11.4.1  Objective Function — Maximization of Benefit Function

$$\text{Max } H_i = \sum_{j=1}^{ND(K)} w_{ij} v_{ij} x_{ij}$$

or

$$\text{Min } (-H_i) \tag{11.3}$$

where

$x_{ij}$: decision variable (it equals 0 or 1) on load bus $j$ at the $i$th time stage

$ND(K)$: total number of load sites in load center $K$

$w_{ij}$: load priority to indicate the importance of the $j$th load site of the $i$th time stage

$v_{ij}$: independent load values (or costs) in a specific load bus $j$ at the $i$th time stage ($/kW or $/MW)

$H$: benefit function

In the objective function (11.3), decision variable $x_{ij}$ equals 1 if load demand $P_{ij}$ is satisfied; otherwise it equals 0 if the load demand is not satisfied, that is, load shedding appears on the $j$th load site at the $i$th time stage. There are several different kinds of loads in a power system, such as critical load, important load, unimportant load, etc., and $w_{ij}$ can reflect the relative importance of the different kinds of loads. The more important the load site is (e.g., first important load), the larger the $w_{ij}$ of the load site will be. In addition, each specific load has its independent load value (cost) $v_{ij}$, which is the value/cost per kilowatt load at this location. Therefore, the unit of $v_{ij}$ is $/kW.

### 11.4.2  Constraints of Load Curtailment

The constraints of load curtailment reflect the system congestion case. These constraints include limited capacity in each load center and the whole system, as well as available transfer capacity of the key line (e.g., the tie line connecting different load centers or the source), which can be expressed as follows:

$$\sum_{j\in K} P_{ij}x_{ij} \leq P_{iK} \qquad (11.4)$$

$$\sum_{j=1}^{ND} P_{ij}x_{ij} \leq P_D \qquad (11.5)$$

$$\sum_{j\in K} P_{ij}x_{ij} = P_{SK} \leq P_{SK\,\text{ATC}} \qquad (11.6)$$

where

$P_{ij}$: load demand of the $j$th load site of the $i$th time stage

$P_{iK}$: total amount of load center $K$ available at the $i$th time stage

$P_D$: total amount of system load available at the $i$th time stage

$P_{SK}$: transmission power on the line connecting the load center $K$

$P_{SK\,\text{ATC}}$: available transfer capacity of the line connecting the load center $K$.

It is noted that the power flow equation or Kirchhoff's current law must be satisfied during the load shedding, that is,

$$\sum_{G\to\omega} P_{iG} + \sum_{T\to\omega} P_{iT} + \sum_{j\to\omega} x_{ij}P_{ij} = 0 \quad \omega \in n \qquad (11.7)$$

$$-P_{iT\max} \leq P_{iT} \leq P_{iT\max} \qquad (11.8)$$

where $n$ is the total node number in the system; $G \to \omega$ indicates that generator $G$ is adjacent to node $\omega$; $T \to \omega$ indicates that transmission line $T$ is adjacent to node $\omega$; $j \to \omega$ indicates that load $j$ is adjacent to node $\omega$.

The direction of power flow is specified as positive when power enters the node, while it is negative when it leaves from the node. Equation (11.8) gives the system network security constraints.

## 11.5 OPTIMAL LOAD SHEDDING WITH NETWORK CONSTRAINTS

### 11.5.1 Calculation of Weighting Factors by AHP

It is very difficult to compute exactly the weighting factor of each load in equation (11.3). The reason is that the relative importance of these loads is not the same, which is related to the power market operation condition. According to the principle of AHP described in Chapter 7, the weighting factors of the loads can be determined through the ranking computation of a judgment matrix, which reflects the judgment and comparison of a series of pairs of factors. The hierarchical model for computing the load weighting factors is shown in Figure 11.4, in which $PI$ is the performance index of load center $K$.

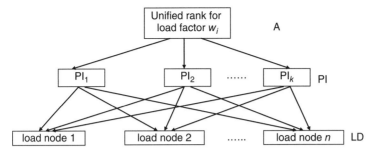

Figure 11.4 Hierarchy model of load weighting factor rank.

The judgment matrix $A - LD$ of the load-shedding problem can be written as follows.

$$A - LD = \begin{bmatrix} w_{D1}/w_{D1} & w_{D1}/w_{D2} \cdots\cdots & w_{D1}/w_{Dn} \\ w_{D2}/w_{D1} & w_{D2}/w_{D2} \cdots\cdots & w_{D2}/w_{Dn} \\ \vdots & \vdots \\ w_{Dn}/w_{D1} & w_{Dn}/w_{D2} \cdots\cdots & w_{Dn}/w_{Dn} \end{bmatrix} \quad (11.9)$$

where $w_{Di}$, which is just what we need, is unknown. $w_{Di}/w_{Dj}$, which is the element of the judgment matrix $A - LD$, represents the relative importance of the $i$th load compared with the $j$th load. The value of $w_{Di}/w_{Dj}$ can be obtained according to the experiences of electrical engineers or system operators using some ratio scale methods. For example, a "1–9" scale method from Chapter 7 can be used.

Similarly, the judgment matrix $A - PI$ can be written as follows.

$$A - PI = \begin{bmatrix} w_{K1}/w_{K1} & w_{K1}/w_{K2} \cdots\cdots & w_{K1}/w_{Kn} \\ w_{K2}/w_{K1} & w_{K2}/w_{K2} \cdots\cdots & w_{K2}/w_{Kn} \\ \vdots & \vdots \\ w_{Kn}/w_{K1} & w_{Kn}/w_{K2} \cdots\cdots & w_{Kn}/w_{Kn} \end{bmatrix} \quad (11.10)$$

where $w_{Ki}$ is unknown. $w_{Ki}/w_{Kj}$, which is the element of judgment matrix $A - PI$, represents the relative importance of the $i$th load center compared with the $j$th load center. The value of $w_{Ki}/w_{Kj}$ can also be obtained according to the experiences of electrical engineers or system operators using some ratio scale methods [12, 13].

Therefore, the unified weighting factor of the load $w_i$ can be obtained from the following equation.

$$w_i = w_{Kj} \times w_{Di} \quad Di \in Kj \quad (11.11)$$

where $Di \in Kj$ means load $Di$ is located in load center $Kj$.

## 11.5.2 Network Flow Model

After the weighting factors are computed by AHP, the above optimization model of load shedding corresponds to a network flow problem and can be solved by NFP.

According to Chapter 5, the general NFP model can be written as

$$\text{Min } F = \sum C_{ij} f_{ij} \tag{11.12}$$

such that

$$\sum (f_{ij} - f_{ji}) = r \tag{11.13}$$

$$0 \le f_{ij} \le U_{ij} \tag{11.14}$$

However, there exist three disadvantages in the general NFP algorithm [14], that is,

(a) the initial arc flows must be feasible;

(b) the lower bound of flows should be 0;

(c) all flow variables must be nonnegative.

Because of these disadvantages, it is difficult to solve the optimal load-shedding problem effectively by using the general NFP algorithm. A special NFP algorithm – "out-of-kilter algorithm" (OKA), which is analyzed in Chapter 5, is adopted. The mathematical representation of the OKA network can be written as follows.

$$\text{Min } F = \sum C_{ij} f_{ij} \tag{11.15}$$

such that

$$\sum (f_{ij} - f_{ji}) = 0 \tag{11.16}$$

$$L_{ij} \le f_{ij} \le U_{ij} \tag{11.17}$$

Obviously, the optimal load-shedding model that is mentioned in Section 11.4 can be transformed into the OKA model shown in equations (11.15)–(11.17) and solved by OKA. The details of the OKA model and algorithm can be found in Chapter 5.

## 11.5.3 Implementation and Simulation

The simulation system for load shedding is the IEEE 30-bus system. The capacity of the generator is given in Table 11.1. The daily load data including the independent load value/cost at each load site are listed in Table 11.2, in which the loads are divided

**TABLE 11.1 Capacity of Generators for IEEE 30-Bus System**

| Gen. | PG1 | PG2 | PG5 | PG8 | PG11 | PG13 |
|------|------|------|------|------|------|------|
| $P_{G\text{max}}$ (MW) | 200.00 | 80.00 | 50.00 | 35.00 | 30.00 | 30.00 |
| $P_{G\text{min}}$ (MW) | 50.00 | 12.00 | 10.00 | 10.00 | 10.00 | 10.00 |

**TABLE 11.2  Load Data for IEEE 30-Bus System**

| Load Center | Load Node | $v_{ij}$ ($/kW) | Load t1 0.00–4.00 (MW) | Load t2 4.01–8.00 (MW) | Load t3 8.01–12.00 (MW) | Load t4 12.01–16.00 (MW) | Load t5 16.01–20.00 (MW) | Load t6 20.01–24.00 (MW) |
|---|---|---|---|---|---|---|---|---|
| CK1 | PD2  | 300.0 | 15.15 | 19.53 | 21.7 | 19.62 | 19.53 | 17.36 |
| CK1 | PD3  | 300.0 | 1.89  | 2.43  | 2.7  | 2.57  | 2.43  | 2.16  |
| CK1 | PD4  | 300.0 | 5.46  | 6.86  | 7.8  | 7.41  | 6.86  | 6.24  |
| CK1 | PD6  | 280.0 | 65.94 | 84.78 | 94.2 | 85.49 | 84.78 | 75.36 |
| CK1 | PD7  | 280.0 | 15.96 | 20.52 | 22.8 | 21.66 | 20.52 | 18.24 |
| CK1 | PD8  | 300.0 | 21.00 | 27.00 | 30.0 | 27.50 | 27.00 | 24.00 |
| CK1 | PD10 | 300.0 | 4.06  | 5.22  | 5.8  | 5.51  | 5.22  | 4.64  |
| CK1 | PD12 | 280.0 | 7.84  | 10.08 | 11.2 | 10.64 | 10.08 | 8.96  |
| CK1 | PD14 | 280.0 | 4.34  | 5.58  | 6.2  | 5.89  | 5.58  | 4.96  |
| CK2 | PD15 | 245.0 | 5.74  | 7.38  | 8.2  | 7.79  | 7.38  | 6.56  |
| CK2 | PD16 | 220.0 | 2.45  | 3.15  | 3.5  | 3.33  | 3.15  | 2.80  |
| CK2 | PD17 | 280.0 | 6.30  | 8.10  | 9.0  | 8.55  | 8.10  | 7.20  |
| CK2 | PD18 | 220.0 | 2.24  | 2.82  | 3.2  | 3.04  | 2.82  | 2.56  |
| CK2 | PD19 | 245.0 | 6.65  | 8.65  | 9.5  | 9.03  | 8.65  | 7.60  |
| CK3 | PD20 | 280.0 | 1.54  | 1.98  | 2.2  | 2.09  | 1.98  | 1.76  |
| CK3 | PD21 | 280.0 | 12.25 | 15.75 | 17.5 | 16.63 | 15.75 | 14.00 |
| CK3 | PD23 | 220.0 | 2.24  | 2.82  | 3.2  | 3.04  | 2.82  | 2.56  |
| CK3 | PD24 | 220.0 | 6.09  | 7.83  | 8.7  | 8.27  | 7.83  | 6.96  |
| CK3 | PD26 | 300.0 | 2.45  | 3.15  | 3.5  | 3.33  | 3.15  | 2.80  |
| CK3 | PD29 | 220.0 | 1.68  | 2.16  | 2.4  | 2.28  | 2.16  | 1.92  |
| CK3 | PD30 | 245.0 | 7.42  | 9.54  | 10.6 | 10.07 | 9.54  | 8.48  |

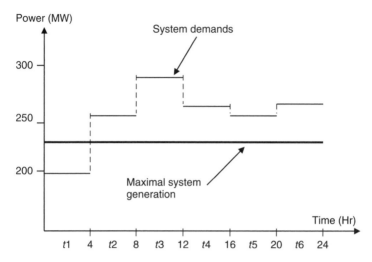

Figure 11.5   Total system generation and load demands.

**TABLE 11.3   Judgment Matrix A – PI**

| PI   | CK1 | CK2 | CK3 |
|------|-----|-----|-----|
| CK1  | 1   | 2   | 5   |
| CK2  | 1/2 | 1   | 1/2 |
| CK3  | 1/5 | 2   | 1   |

into three load centers. Suppose generator G1 is out of service. The total source power is only 225.0 MW. This, in turn, leads to the power shortage for IEEE 30-bus system, that is; the power supply is limited at some time stages. The total system generation resources and load demands are shown in Figure 11.5.

The judgment matrix $A - LD$ and $A - PI$ are provided in Tables 11.3 and 11.4, respectively. The weighting factors that reflect the relative importance of each load or each load center are computed by AHP. The results of the weighting factors are listed in Table 11.5. The optimal load-shedding schemes are computed and obtained by the proposed approach. The calculation results are shown in Tables 11.6 and 11.7.

In Table 11.6, the decision variable $x = 1$ means that this load is committed, and $x = 0$ means that this load is curtailed. It can be known from Tables 11.6 and 11.7 that load curtailment appeared at time stage $t2 \sim t6$. Load 15, 16, 18, 19, 29 and 30 are curtailed at time stage $t2 \sim t5$. Load 24 is curtailed at time stage $t2 \sim t6$. Load 21 is curtailed at time stage $t3$ and $t4$, and Load 20 is curtailed at time stage $t3$. The total load curtailments at each time stage are summarized in Table 11.7. It is noted that network security constraints are satisfied at any time period by using the proposed approach.

**TABLE 11.4  Judgment Matrix A – LD**

**(1)**

| LD | 2 | 3 | 4 | 6 | 7 | 8 | 10 | 12 | 14 | 15 |
|----|-----|-----|-----|-----|-----|-----|-----|-----|-----|-----|
| 2 | 1 | 2 | 2 | 1/3 | 1/5 | 2 | 1/2 | 2 | 2 | 3 |
| 3 | 1/2 | 1 | 1/2 | 1/4 | 2 | 1/2 | 1 | 2 | 2 | 3 |
| 4 | 1/2 | 2 | 1 | 1/2 | 2 | 1/3 | 2 | 2 | 3 | 2 |
| 6 | 3 | 4 | 2 | 1 | 4 | 2 | 3 | 3 | 3 | 3 |
| 7 | 5 | 1/2 | 1/2 | 1/4 | 1 | 1/2 | 2 | 2 | 2 | 3 |
| 8 | 1/2 | 2 | 3 | 1/2 | 2 | 1 | 3 | 2 | 2 | 4 |
| 10 | 2 | 1 | 1/2 | 1/3 | 1/2 | 1/3 | 1 | 2 | 3 | 3 |
| 12 | 1/2 | 1/2 | 1/2 | 1/3 | 1/2 | 1/2 | 1/2 | 1 | 1 | 2 |
| 14 | 1/2 | 1/2 | 1/3 | 1/3 | 1/2 | 1/2 | 1/3 | 1 | 1 | 2 |
| 15 | 1/3 | 1/3 | 1/2 | 1/3 | 1/3 | 1/4 | 1/3 | 1/2 | 1/2 | 1 |
| 16 | 1/3 | 1/2 | 1/3 | 1/4 | 1/3 | 1/4 | 1/3 | 1/2 | 1/3 | 1/2 |
| 17 | 1/2 | 2 | 1/2 | 1/2 | 1/3 | 1/2 | 2 | 1/2 | 1/2 | 3 |
| 18 | 1/3 | 1 | 1/2 | 1/3 | 1/3 | 1/3 | 1/2 | 1/2 | 1/3 | 1/2 |
| 19 | 1/3 | 1/2 | 1/2 | 1/3 | 1/3 | 1/3 | 1/3 | 1/2 | 1/3 | 1/2 |
| 20 | 1/3 | 1/2 | 1/3 | 1/3 | 1/3 | 1/3 | 1/2 | 1/3 | 1/2 | 5 |
| 21 | 1/3 | 1/3 | 1/2 | 1/3 | 1/4 | 1/4 | 1/3 | 1/3 | 1/2 | 5 |
| 23 | 2 | 3 | 1/2 | 1/2 | 1/2 | 1/2 | 1/2 | 1/2 | 1/3 | 3 |
| 24 | 1/3 | 1/3 | 1/2 | 1/3 | 1/3 | 1/2 | 1/3 | 1/3 | 1/3 | 1/3 |
| 26 | 1/3 | 1/3 | 1/2 | 1/3 | 1/2 | 1/3 | 1/3 | 1/2 | 1/2 | 3 |
| 29 | 1/3 | 1/3 | 1/3 | 1/3 | 1/3 | 1/2 | 1/3 | 1/3 | 1/3 | 1/2 |
| 30 | 1/3 | 1/3 | 1/2 | 1/3 | 1/3 | 1/3 | 1/2 | 1/3 | 1/3 | 2 |

**(2)**

| LD | 16 | 17 | 18 | 19 | 20 | 21 | 23 | 24 | 26 | 29 | 30 |
|----|-----|-----|-----|-----|-----|-----|-----|-----|-----|-----|-----|
| 2 | 3 | 2 | 3 | 3 | 3 | 3 | 1/2 | 3 | 3 | 3 | 3 |
| 3 | 2 | 1/2 | 1 | 2 | 2 | 3 | 1/3 | 3 | 3 | 3 | 3 |
| 4 | 3 | 2 | 2 | 2 | 3 | 2 | 2 | 2 | 2 | 3 | 2 |
| 6 | 4 | 2 | 3 | 3 | 3 | 3 | 2 | 3 | 3 | 3 | 3 |
| 7 | 3 | 3 | 3 | 3 | 3 | 4 | 2 | 3 | 2 | 3 | 3 |
| 8 | 4 | 2 | 3 | 3 | 3 | 4 | 2 | 2 | 3 | 2 | 3 |
| 10 | 3 | 1/2 | 2 | 3 | 2 | 3 | 2 | 3 | 3 | 3 | 2 |
| 12 | 2 | 2 | 2 | 2 | 3 | 3 | 2 | 3 | 2 | 3 | 3 |
| 14 | 3 | 2 | 3 | 3 | 2 | 2 | 3 | 3 | 2 | 3 | 3 |
| 15 | 2 | 1/3 | 2 | 2 | 1/5 | 1/5 | 1/3 | 3 | 1/3 | 2 | 1/2 |
| 16 | 1 | 1/3 | 2 | 3 | 1/2 | 1/2 | 1/3 | 3 | 1/2 | 2 | 1/2 |
| 17 | 3 | 1 | 2 | 2 | 3 | 3 | 2 | 2 | 2 | 3 | 3 |
| 18 | 1/2 | 1/2 | 1 | 1/2 | 2 | 2 | 1/2 | 3 | 1/3 | 2 | 1/2 |
| 19 | 1/3 | 1/2 | 2 | 1 | 2 | 3 | 1/3 | 2 | 1/2 | 3 | 1/2 |
| 20 | 2 | 1/3 | 1/2 | 1/2 | 1 | 3 | 1/2 | 2 | 1/3 | 2 | 4 |

(*continued*)

**TABLE 11.4** *(Continued).*

| 21 | 2 | 1/3 | 1/2 | 1/3 | 1/3 | 1 | 1/3 | 2 | 1/2 | 3 | 4 |
|----|----|-----|-----|-----|-----|-----|-----|-----|-----|-----|-----|
| 23 | 3 | 1/2 | 2 | 3 | 2 | 3 | 1 | 3 | 2 | 3 | 3 |
| 24 | 1/3 | 1/2 | 1/3 | 1/2 | 1/2 | 1/2 | 1/3 | 1 | 1/2 | 1/2 | 1/3 |
| 26 | 2 | 1/2 | 3 | 2 | 3 | 2 | 1/2 | 2 | 1 | 4 | 3 |
| 29 | 1/2 | 1/3 | 1/2 | 1/3 | 1/2 | 1/3 | 1/3 | 2 | 1/4 | 1 | 1/2 |
| 30 | 2 | 1/3 | 2 | 2 | 1/4 | 1/4 | 1/3 | 3 | 1/3 | 2 | 1 |

**TABLE 11.5** **Weighting Factors Computed By AHP**

| Load Center | Weighting Factor $w_{Kj}$ | Load Node | $v_{ij}$ ($/kW) | Weighting Factor $w_{Di}$ | Unified Weighting Factor $w_i$ |
|-------------|---------------------------|-----------|-----------------|---------------------------|--------------------------------|
| CK1 | 0.61185 | PD2 | 300.0 | 0.07007 | 0.042872 |
| CK1 | 0.61185 | PD3 | 300.0 | 0.05425 | 0.033193 |
| CK1 | 0.61185 | PD4 | 300.0 | 0.06824 | 0.041753 |
| CK1 | 0.61185 | PD6 | 280.0 | 0.11115 | 0.068007 |
| CK1 | 0.61185 | PD7 | 280.0 | 0.08006 | 0.048985 |
| CK1 | 0.61185 | PD8 | 300.0 | 0.08616 | 0.052717 |
| CK1 | 0.61185 | PD10 | 300.0 | 0.06148 | 0.037617 |
| CK1 | 0.61185 | PD12 | 280.0 | 0.04999 | 0.030586 |
| CK1 | 0.61185 | PD14 | 280.0 | 0.05201 | 0.031822 |
| CK2 | 0.17891 | PD15 | 245.0 | 0.02356 | 0.004215 |
| CK2 | 0.17891 | PD16 | 220.0 | 0.02340 | 0.004186 |
| CK2 | 0.17891 | PD17 | 280.0 | 0.05430 | 0.009715 |
| CK2 | 0.17891 | PD18 | 220.0 | 0.02601 | 0.004653 |
| CK2 | 0.17891 | PD19 | 245.0 | 0.02701 | 0.004832 |
| CK3 | 0.20925 | PD20 | 280.0 | 0.03219 | 0.006736 |
| CK3 | 0.20925 | PD21 | 280.0 | 0.02843 | 0.005949 |
| CK3 | 0.20925 | PD23 | 220.0 | 0.05438 | 0.011379 |
| CK3 | 0.20925 | PD24 | 220.0 | 0.01677 | 0.003509 |
| CK3 | 0.20925 | PD26 | 300.0 | 0.03848 | 0.008052 |
| CK3 | 0.20925 | PD29 | 220.0 | 0.01686 | 0.003528 |
| CK3 | 0.20925 | PD30 | 245.0 | 0.02521 | 0.005275 |

To further verify the AHP-based NFP approach, linear programming (LP) is used to solve the same load-shedding problem without load priority factors $w_{ij}$ that are determined by AHP. The corresponding results are compared with those obtained by AHP-based NFP method and also listed in the Tables 11.6 and 11.7 (Figures 11.6 and 11.7). In the LP method, the loads with small MW demands and small costs are first considered for curtailment. The LP method also cannot handle or consider the relative importance of the load locations. The result comparison shows that the

**TABLE 11.6   Optimal Load-Shedding Schemes and Comparison for IEEE 30-Bus System**

| Methods | AHP | LP | AHP | LP | AHP | LP | AHP | LP | AHP | LP | AHP | LP |
|---|---|---|---|---|---|---|---|---|---|---|---|---|
| Time stage | $t1$ | $t1$ | $t2$ | $t2$ | $t3$ | $t3$ | $t4$ | $t4$ | $t5$ | $t5$ | $t6$ | $t6$ |
| $X2$ | 1 | 1 | 1 | 1 | 1 | 1 | 1 | 1 | 1 | 1 | 1 | 1 |
| $X3$ | 1 | 1 | 1 | 1 | 1 | 1 | 1 | 1 | 1 | 1 | 1 | 1 |
| $X4$ | 1 | 1 | 1 | 1 | 1 | 1 | 1 | 1 | 1 | 1 | 1 | 1 |
| $X6$ | 1 | 1 | 1 | 1 | 1 | 1 | 1 | 1 | 1 | 1 | 1 | 1 |
| $X7$ | 1 | 1 | 1 | 1 | 1 | 1 | 1 | 1 | 1 | 1 | 1 | 1 |
| $X8$ | 1 | 1 | 1 | 1 | 1 | 1 | 1 | 1 | 1 | 1 | 1 | 1 |
| $X10$ | 1 | 1 | 1 | 1 | 1 | 1 | 1 | 1 | 1 | 1 | 1 | 1 |
| $X12$ | 1 | 1 | 1 | 1 | 1 | 1 | 1 | 1 | 1 | 1 | 1 | 1 |
| $X14$ | 1 | 1 | 1 | 1 | 1 | 0 | 1 | 1 | 1 | 1 | 1 | 1 |
| $X15$ | 1 | 1 | 0 | 0 | 0 | 0 | 0 | 0 | 0 | 0 | 1 | 1 |
| $X16$ | 1 | 1 | 0 | 0 | 0 | 0 | 0 | 0 | 0 | 0 | 1 | 1 |
| $X17$ | 1 | 1 | 1 | 1 | 1 | 0 | 1 | 1 | 1 | 1 | 1 | 1 |
| $X18$ | 1 | 1 | 0 | 0 | 0 | 0 | 0 | 0 | 0 | 0 | 1 | 0 |
| $X19$ | 1 | 1 | 0 | 0 | 0 | 0 | 0 | 0 | 0 | 0 | 1 | 1 |
| $X20$ | 1 | 1 | 1 | 1 | 0 | 0 | 1 | 0 | 1 | 1 | 1 | 1 |
| $X21$ | 1 | 1 | 1 | 1 | 0 | 1 | 1 | 1 | 1 | 1 | 1 | 1 |
| $X23$ | 1 | 1 | 1 | 0 | 1 | 0 | 1 | 0 | 1 | 0 | 1 | 0 |
| $X24$ | 1 | 1 | 0 | 0 | 0 | 0 | 0 | 0 | 0 | 0 | 0 | 1 |
| $X26$ | 1 | 1 | 1 | 1 | 1 | 1 | 1 | 1 | 1 | 1 | 1 | 1 |
| $X29$ | 1 | 1 | 0 | 0 | 0 | 0 | 0 | 0 | 0 | 0 | 1 | 0 |
| $X30$ | 1 | 1 | 0 | 0 | 0 | 0 | 0 | 0 | 0 | 0 | 1 | 1 |

AHP-based NFP approach is truly optimal. It not only has maximal load benefits but also considers the relative importance of the load sites. For example, load site 23, which is always curtailed in the LP method when system generation is limited, is not curtailed in the AHP-based NFP method although it has the minimal load cost (220$/kW) and small MW load demands.

# 11.6   OPTIMAL LOAD SHEDDING WITHOUT NETWORK CONSTRAINTS

## 11.6.1   Everett Method

If the network constraints are neglected, the load-shedding problem in equations (11.3)–(11.6) can be easily solved by the Everett optimization technique, a generalized Lagrange multiplier [15–17]. The problem of load shedding can be

TABLE 11.7  Summary and Comparison of Optimal Load Shedding for IEEE 30-Bus System

| Methods | AHP | LP | AHP | LP | AHP | LP | AHP | LP | AHP | LP | AHP | LP |
|---|---|---|---|---|---|---|---|---|---|---|---|---|
| Time Stage | t1 | t1 | t2 | T2 | t3 | t3 | t4 | t4 | t5 | t5 | t6 | t6 |
| Max. system gen. (MW) | 225.0 | 225.0 | 225.0 | 225.0 | 225.0 | 225.0 | 225.0 | 225.0 | 225.0 | 225.0 | 225.0 | 225.0 |
| System demands (MW) | 198.38 | 198.38 | 255.06 | 255.06 | 283.4 | 283.4 | 263.23 | 263.23 | 255.06 | 255.06 | 266.72 | 226.72 |
| Committed loads (MW) | 198.38 | 198.38 | 213.53 | 210.71 | 217.6 | 216.7 | 219.42 | 216.38 | 213.53 | 210.71 | 219.76 | 219.68 |
| Total load shedding (MW) | 0.0 | 0.0 | 41.53 | 44.35 | 65.80 | 66.7 | 43.81 | 46.85 | 41.53 | 44.35 | 6.96 | 7.04 |
| Objective Hi | 130.87 | – | 123.87 | – | 120.32 | – | 120.32 | – | 120.32 | – | 130.10 | – |
| Benefit $\Sigma\ v_{ij}P_{ij}\,(\times10^3)$\$ | 55058 | 55058 | 60979 | 60358 | 62306 | 62246 | 62717 | 61463 | 60979 | 60358 | 61406 | 61388 |
| Network security satisfied | Yes | Yes | Yes | Yes | Yes | Yes | Yes | Yes | Yes | Yes | Yes | Yes |

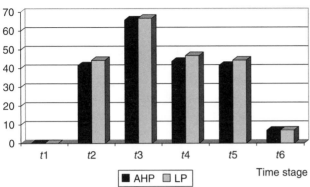

Figure 11.6   Comparison of optimal load-shedding results.

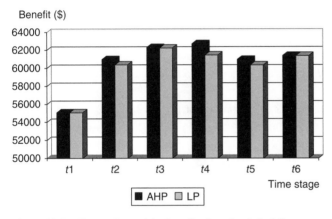

Figure 11.7   Comparison of the benefits from load shedding.

represented as follows:

$$\text{Max } H_i = \sum_{i=1}^{m} H_i(x_i) \quad x_i \in s \tag{11.18}$$

such that

$$\sum_{i=1}^{m} C_i^k(x_i) \leq c^k \text{ for all } k \tag{11.19}$$

where

$x_i$: a 0-1 integer variable

$S$: set that is interpreted as the set of possible strategies or actions

$H(x)$: benefit that accrues from employing the strategies $x \in S$

$C^k$: resource function.

This load-shedding model is a 0-1 integer optimization problem. It is possible to solve problem (11.18) and (11.19) with integer-based optimization techniques. But this will be a variable dimension problem in the large-scale power systems. Everett [14] showed that the Lagrange multiplier can be used to solve the maximization problem with many variables without any restrictions on continuity or differentiability of the function being maximized. The aim of the generalized Lagrange multiplier is maximization rather than the location of stationary points as with the traditional Lagrange multipliers. This technique is discussed as follows.

The main theorem of the generalized Lagrange multiplier is as follows.

**_Theorem 1 [15]_** If (1) $\lambda^k (k = 1, 2, \ldots \ldots n)$ are nonnegative real numbers,

(2) $x^* \in S$ maximizes the function

$$H(x) - \sum_{k=1}^{n} \lambda^k C^k(x) \quad x \in S \tag{11.20}$$

then (3) $x^*$ maximizes $H(x)$ over all of those $x \in S$ such that $C^k \leq C^k(x^*)$ for all $k$.

**_Proof._** By assumptions 1 and 2 of Theorem 1, $\lambda^k (k = 1, 2, \ldots, n)$ are nonnegative real numbers, and $x^* \in S$ maximizes

$$H(x) - \sum_{k=1}^{n} \lambda^k C^k(x) \tag{11.21}$$

Over all $x \in S$. This means that, for all $x \in S$,

$$H(x^*) - \sum_{k=1}^{n} \lambda^k C^k(x^*) \geq H(x) - \sum_{k=1}^{n} \lambda^k C^k(x) \tag{11.22}$$

and hence that

$$H(x^*) \geq H(x) + \sum_{k=1}^{n} \lambda^k [C^k(x^*) - C^k(x)] \tag{11.23}$$

for all $x \in S$. However, if the latter inequality is true for all $x \in S$, it is necessarily true for any subset of $S$ and hence true on that subset $S^*$ of $S$ for which the resources never exceed the resources $C^k(x^*)$, that is, $C^k \leq C^k(x^*)$, $x \in S^*$ for all $k$. Thus on the subset $S^*$ the term

$$\sum_{k=1}^{n} \lambda^k [C^k(x^*) - C^k(x)] \tag{11.24}$$

is nonnegative by definition of the subset and the nonnegativity of $\lambda^k$. Consequently, the inequality equation (11.23) reduces to

$$H(x^*) \geq H(x) \tag{11.25}$$

For all $x \in S^*$, and the theorem is proved.

In accordance with Theorem 1, for any choice of nonnegative $\lambda^k (k = 1, 2, \ldots \ldots n)$, if an unconstrained maximum of the new Lagrange function [eq. (11.20)] can be found (where $x^*$, e.g., is a strategy that produces the maximization), then this solution is a solution to that constrained maximization problem whose constraints are, in fact, the amount of each resource expended in achieving the unconstrained solution. Therefore, if $x^*$ produces the unconstrained maximum and the required resources $C^k(x^*)$, then $x^*$ itself produces the greatest benefit that can be achieved without using additional resource allocation.

With the Everett method, the problem of load shedding is changed into an unconstrained maximization. The key to solve this problem is choosing the Lagrange multipliers that correspond to the trial prices in the new competitive power market. In general, different choices of the trial prices $\lambda^k$ lead to different schemes to resources provided and demands of customers to achieve the maximal benefit.

## 11.6.2   Calculation of the Independent Load Values

Suppose $v_i$ is the independent load value in a specific load bus. It reflects the value of supplement unit capacity generator for eliminating the load curtailment at node $i$ (in $/kW). However, load shedding is time dependent. Different time stages correspond to different levels of load shedding. Thus a load-shedding study should be performed on the basis of hourly load and the corresponding independent load values converted into hourly values.

The annual equipment value method, which is a dynamic assessment method, converts the cost of the operational lifetime to an annual cost. According to this method,

The value $v_i^t$ per hour can be calculated as follows:

$$v_i^t = \frac{\beta v_i \times 10^3}{365 \times 24} (\$/MW/hr) \tag{11.26}$$

$$\beta = \frac{r(1 + r)^n}{(1 + r)^n - 1} \tag{11.27}$$

where

$v_i$: the independent load value in a specific load bus (\$/kW)
$v_i^t$: the per hour independent load value in a specific load bus (\$/MW/h)
$r$: the interest rate

$n$: the capital recovery years

$\beta$: the capital recovery factor (CRF), which is an important factor in economic analysis. It is supposed that 1 year = 365 days in equation (11.26).

***Example 11.1:*** The testing system is shown in Figure 11.8, which is taken from reference [16], but with modified data. It consists of two generators, and five loads at buses 3, 4, 5, 8, and 9, where loads 3, 4, and 5 are located in load center 1, and the others are located in load center 2. The weight factors reflecting the relative values of load centers are $w_1 = 0.58$, and $w_2 = 0.42$. The independent load values $v$ in a specific load bus, the absolute load priority $\alpha$ to indicate the importance of each load bus and the load demand for each load bus are given in Table 11.8. The capacities of generator 1 and generator 2 are $P_{G1} = 0.90$ and $P_{G2} = 0.6$ p.u., respectively. The available transfer capacities of the key lines are $P_{1-6\,max} = 0.60$ p.u., $P_{2-7\,max} = 0.58$ p.u., $P_{1-7\,max} = 0.5$ p.u., respectively.

There are two test cases:

Case 1: two generators are in operating, tie line 1-7 is in outage.

Case 2: generator 2 is in outage. No line outage.

First of all, we assume that the capital recovery years of investing in the generators $n = 10$ years, and that interest rate is 6%. According to equation (11.27), we get the capital recovery factor (CRF) $\beta = 1.3587$. Then according to

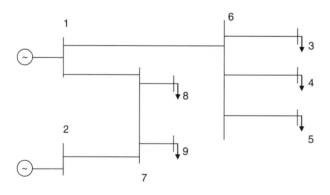

Figure 11.8　A simple network.

**TABLE 11.8　The Values of Load Buses**

| Values | Load 3 | Load 4 | Load 5 | Load 8 | Load 9 |
|---|---|---|---|---|---|
| $v_i$ (S/kW) | 150 | 200 | 180 | 190 | 220 |
| $\alpha_i$ | 1.14 | 1.25 | 1.30 | 1.10 | 1.22 |
| Demand $P_D$ (p.u.) | .270 | .280 | .260 | .305 | .310 |

**TABLE 11.9   The Hourly Independent Values of Load Buses**

| Values | Load 3 | Load 4 | Load 5 | Load 8 | Load 9 |
|---|---|---|---|---|---|
| $v_i$ (S/kW) | 150 | 200 | 180 | 190 | 220 |
| $v_i$ (S/MW/hr) | 23.26 | 31.02 | 27.92 | 29.47 | 34.12 |
| $v_i$ (S/p.u. MW/hr) | 2326 | 3102 | 2792 | 2947 | 3412 |

equation (11.26), we get the hourly independent load values, which are shown in Table 11.9.

For case 1, we can get the following objective function and constraints.

$$H = \sum_i \alpha_i v_i x_i$$

and constraints

$$P_{D3} x_3 + P_{D4} x_4 + P_{D5} x_5 \le P_{G1}$$

$$P_{D3} x_3 + P_{D4} x_4 + P_{D5} x_5 \le P_{1-6\,\text{max}}$$

$$P_{D8} x_8 + P_{D9} x_9 \le P_{G2}$$

$$P_{D8} x_8 + P_{D9} x_9 \le P_{2-7\text{max}}$$

Since tie line $1-7$ is in outage, the system becomes two subsystems, each of them has one generator. Thus we can solve two subproblems separately.

For subproblem 1:

Objective

$$H_1 = \sum_i \alpha_i v_i x_i = \alpha_3 v_3 x_3 + \alpha_4 v_4 x_4 + \alpha_5 v_5 x_5$$

$$= 1.14 \times 2326 x_3 + 1.25 \times 3120 x_4 + 1.30 \times 2792 x_5$$

$$= 2651.64 x_3 + 3900 x_4 + 3629.6 x_5$$

subject to

$$P_{D3} x_3 + P_{D4} x_4 + P_{D5} x_5 \le \min\{P_{G1}, P_{1-6\,\text{max}}\}$$

that is,

$$0.27 x_3 + 0.28 x_4 + 0.26 x_5 \le \min\{0.90, 0.60\} = 0.60$$

Compared with equations (11.18) and (11.19), the above load-shedding problem is a linear model, that is,

$$\max H(x) = \sum_i H_i x_i \tag{11.28}$$

such that

$$\sum_i P_i x_i \le C \tag{11.29}$$

According to the generalized Lagrange multiplier technique, the Everett model for the load-shedding problem can be written as follows.

$$\text{Max } L = H(x) - \sum_{k=1}^{n} \lambda^k C^k(x)$$

$$= \sum_i \{H_i x_i - \lambda[P_i x_i - C]\} = \sum_i \delta_i x_i + \lambda C \tag{11.30}$$

where

$$\delta_i = H_i - \lambda P_i \tag{11.31}$$

Thus, we have

$$L = 2651.64 x_3 + 3900 x_4 + 3629.6 x_5 - \lambda(0.20 x_3 + 0.22 x_4 + 0.28 x_5 - 0.60)$$

$$= (2651.64 - 0.27\lambda) x_3 + (3900 - 0.28\lambda) x_4 + (3629.6 - 0.26\lambda) x_5 + 0.60\lambda$$

If all $x_i = 1$, $\sum_i P_i x_i = 0.81 > C$, which equals 0.60. Thus some load should be curtailed. It can be observed from the above Lagrange function that shedding load 3 will have the maximum benefit no matter what the value of the trial price $\lambda$ is.

For subproblem 2:

Objective

$$H_2 = \sum_i \alpha_i v_i x_i = \alpha_8 v_8 x_8 + \alpha_9 v_9 x_9$$

$$= 1.1 \times 2947 x_8 + 1.22 \times 3412 x_9$$

$$= 3241.7 x_8 + 4162.64 x_9$$

subject to

$$P_{D8} x_8 + P_{D9} x_9 \le \min\{P_{G2}, P_{2-7\max}\}$$

that is,

$$0.305 x_8 + 0.310 x_9 \le \min\{0.70, 0.58\} = 0.58$$

Then, we have

$$L = 3241.7 x_8 + 4162.64 x_9 - \lambda(0.305 x_8 + 0.310 x_9 - 0.58)$$

$$= (3241.7 - 0.305\lambda) x_8 + (4162.64 - 0.310\lambda) x_9 + 0.58\lambda$$

If all $x_i = 1$, $\sum_i P_i x_i = 0.61 > C$, which equals 0.58. Thus, some load should be curtailed. It can be observed from the above Lagrange function that shedding load 8 will have the maximum benefit no matter what the value of the trial price $\lambda$ is.

For case 2, one generator will supply two load centers because generator 2 is in outage. We have the following objective function and constraints.

Objective

$$h = w_1 H_1 + w_2 H_2 = w_1(\alpha_3 v_3 x_3 + \alpha_4 v_4 x_4 + \alpha_5 v_5 x_5) + w_2(\alpha_8 v_8 x_8 + \alpha_9 v_9 x_9)$$

$$= 0.58(1.14 \times 2326 x_3 + 1.25 \times 3120 x_4 + 1.30 \times 2792 x_5)$$

$$+ 0.42(1.1 \times 2947 x_8 + 1.22 \times 3412 x_9)$$

$$= 1537.95 x_3 + 2262 x_4 + 2105.17 x_5 + 1361.51 x_8 + 1748.31 x_9$$

subject to

(1)  $0.27 x_3 + 0.28 x_4 + 0.26 x_5 \leq P_{1-6\text{max}} = 0.60$
(2)  $0.305 x_8 + 0.31 x_9 \leq P_{1-7\text{max}} = 0.50$
(3)  $0.27 x_3 + 0.28 x_4 + 0.26 x_5 + 0.305 x_8 + 0.31 x_9 \leq P_{G1} = 0.90$

Then we have the following Lagrange function for case 2.

$$L = 1537.95 x_3 + 2262 x_4 + 2105.17 x_5 + 1361.51 x_8 + 1748.31 x_9$$

$$- \lambda_1(0.27 x_3 + 0.28 x_4 + 0.26 x_5 - 0.60)$$

$$- \lambda_2(0.305 x_8 + 0.310 x_9 - 0.50)$$

$$- \lambda_3(0.27 x_3 + 0.28 x_4 + 0.26 x_5 + 0.305 x_8 + 0.310 x_9 - 0.90)$$

$$= (1537.95 - 0.27\lambda_1 - 0.27\lambda_3)x_3 + (2262 - 0.28\lambda_1 - 0.28\lambda_3)x_4$$

$$+ (2105.17 - 0.26\lambda_1 - 0.26\lambda_3)x_5 + (1361.51 - 0.305\lambda_2 - 0.305\lambda_3)x_8$$

$$+ (1748.31 - 0.31\lambda_2 - 0.31\lambda_3)x_9 + 0.60\lambda_1 + 0.50\lambda_2 + 0.90\lambda_3$$

If $\lambda_1 = \lambda_2 = \lambda_3 = 2000$ \$/p.u. MW/h, and assume there is no load shedding, we get the following results according to equations (11.28)–(11.31).

| Load $i$ | $\delta_i$ | $x_i$ | $H_i x_i$ | $P_i x_i$ | $\delta_i x_i$ | Rank ($x_i$) |
|---|---|---|---|---|---|---|
| Load 3 | 457.50 | 1 | 1537.95 | 0.27 | 457.50 | 4 |
| Load 4 | 1142.00 | 1 | 2262.00 | 0.26 | 1142.00 | 1 |
| Load 5 | 1065.70 | 1 | 2105.17 | 0.28 | 1065.70 | 2 |
| Load 8 | 141.51 | 1 | 1361.51 | 0.305 | 141.51 | 5 |
| Load 9 | 508.31 | 1 | 1748.31 | 0.31 | 508.31 | 3 |

However, constraints (1)–(3) are not satisfied. According to the above table, the optimal load-shedding scheme is that load 8 and load 3 are curtailed, and the maximum benefit for this case is $H = 6115.27$.

If $\lambda_1 = \lambda_2 = \lambda_3 = 2500$ \$/p.u. MW/h, we get the following results.

| Load $i$ | $\delta_i$ | $x_i$ | $H_i x_i$ | $P_i x_i$ | $\delta_i x_i$ | Rank ($x_i$) |
|---|---|---|---|---|---|---|
| Load 3 | 187.95 | 1 | 1537.95 | 0.27 | 187.95 | 4 |
| Load 4 | 862.00 | 1 | 2262.00 | 0.26 | 862.00 | 1 |
| Load 5 | 805.70 | 1 | 2105.17 | 0.28 | 805.70 | 2 |
| Load 8 | −138.49 | 1 | 1361.51 | 0.305 | −138.49 | 5 |
| Load 9 | 198.31 | 1 | 1748.31 | 0.31 | 198.31 | 3 |

According to the above table, the same optimal load-shedding scheme is obtained, that is, load 8 and load 3 are curtailed, and the maximum benefit for this case is $H = 6115.27$.

However, if $\lambda_1 = \lambda_2 = \lambda_3 = 2700$ \$/p.u. MW/h, we get the following results.

| Load $i$ | $\delta_i$ | $x_i$ | $H_i x_i$ | $P_i x_i$ | $\delta_i x_i$ | Rank ($x_i$) |
|---|---|---|---|---|---|---|
| Load 3 | 79.95 | 1 | 1537.95 | 0.27 | 79.95 | 3 |
| Load 4 | 750.00 | 1 | 2262.00 | 0.26 | 750.00 | 1 |
| Load 5 | 701.17 | 1 | 2105.17 | 0.28 | 701.17 | 2 |
| Load 8 | −285.49 | 1 | 1361.51 | 0.305 | −285.49 | 5 |
| Load 9 | 74.31 | 1 | 1748.31 | 0.31 | 74.31 | 4 |

According to the above table, a different load-shedding scheme is obtained, that is, load 8 and load 9 are curtailed, and the maximum benefit for this case is $H = 5905.12$.

Obviously, the trial price $\lambda_i$ affects the results of load shedding. Further calculations show that the optimal load-shedding scheme will be that loads 8 and 3 are curtailed if $\lambda_1 = \lambda_2 = \lambda_3 \le 2629.52700$ \$/p.u. MW/h, and the optimal load-shedding scheme will be that loads 8 and 9 are curtailed if $\lambda_1 = \lambda_2 = \lambda_3 > 2629.52700$ \$/p.u. MW/h.

## 11.7 DISTRIBUTED INTERRUPTIBLE LOAD SHEDDING (DILS)

### 11.7.1 Introduction

Blackouts are becoming more frequent in industrial countries because of network deficiencies and continuous load growing. One possible solution to prevent blackouts is load curtailment. Both *Demand side management* (*DSM*) and *load shedding* (*LS*) have been used to provide reliable power system operation under normal and emergency conditions. DSM is specifically devoted to peak demand shaving [18] and to encourage efficient use of energy. LS is still a methodology used worldwide to prevent power system degradation to blackouts [19–21] and it acts in a repressive way.

To perform the LS program, it could be necessary to increase the number of interruptible customers and distribute them over the entire system. Considering such small percentage values of load shedding, if the number of interruptible customers increased, the impact on users would be negligible. Instead of detaching all the interruptible loads, only a part of the load could be disconnected from the network, in particular the part that can be interrupted or controlled (such as the lighting system, air conditioning, devices under UPS, pumps dedicated to tanks filling, etc.). This method is called *distributed interruptible load shedding* (*DILS*) program [18].

Generally speaking, at least the following three levels of action should be assumed so that a customer can participate in the DILS, allowing the network manager to control the peak power withdrawal or to act during the periods of network dysfunctions:

- the financing of technologies that enable the implementation of DILS (electronic power meters, domestic and similar appliances, etc.);
- incentives aimed at changing the behavior of some categories of end users;
- definition of *ad hoc* instruments for particular classes of consumers such as Public Administration, Data Centers, etc.

In addition, the customers could find it convenient to participate in the day-ahead market. Users with reducible power above a minimal threshold could present offers in the previous day market that, if accepted because they are competitive, could take part in the dispatch services market.

This way, the load curtailment would be paid according to the actual recorded interruption. Moreover, there would be more market efficiency, created by the competition between both the interruptible services themselves and between these and the generation.

## 11.7.2 DILS Methods

To participate in the DILS program with interest, a user must have an economic profit and/or be less sensitive to dysfunctions. There are two different DILS techniques that can be adopted in automation sceneries only, for obtaining the desired load relief during criticalities:

1. The first technique increases the cost of electric energy for all the users [8]. One can assume to know the response of the users statistically, in particular, as to the way they change the subdivision between interruptible loads (which would become disconnectable) and uninterruptible loads depending on the cost of energy. In this case, the transmission of a price signal via the electronic power meter could be sufficient to avoid the loss.

2. The second technique is based on the transmission of an interruptible load percentage reduction signal $p$ to every customer participating in the DILS program.

The duration of the reduction might be contractually determined. Because of the uncertainty on how much power each single interruptible customer is actually drawing, the value of $p$ will be larger than the fraction of the expected interruptible load, giving the wanted load relief.

Since it is more easily adoptable in practice by the distributor and the end user, the second DILS technique is analyzed here.

***Analysis of Interruptible Load*** The interruptible load of a customer can be considered as an essentially continuous random variable. This ensures that every percentage $p$ of load reduction is actually achievable (possibly with low probability for some values of $p$). We denote by $Y_{I,k}(t)$ the random value of the interruptible load power of the single customer $k$ of a given sector at time $t$ and build its probability distribution at a fixed time, so that we omit the time argument temporarily and write $Y_{I,k}$ only.

The load $Y_{I,k}$ is composed of various combinations of continuous adjustable and step-adjustable interruptible loads, which we write as

$$Y_{I,k} = Y_{CAI,k} + Y_{SAI,k} \tag{11.32}$$

where

$Y_{CAI,k}$: the interruptible continuous adjustable loads
$Y_{SAI,k}$: the interruptible step adjustable loads.

The combinations of step-adjustable loads give rise to, say, $m$ possible well-separated load levels of $Y_{SAI,k}$, denoted by $l_1, \ldots, l_m$. Each level is taken with a different probability, so we introduce the probabilities $w(1), \ldots \ldots, w(m)$, which sum up to 1, giving the probability distribution $w(\cdot)$ of $Y_{SAI,k}$. On the other hand, $Y_{CAI,k}$ has an absolutely continuous probability distribution with density $f_{CAI}(\bullet)$ on the range $(0, L_{CAI})$, where $L_{CAI}$ is the maximum power of the interruptible continuous adjustable load.

Assuming $Y_{SAI,k}$ and $Y_{CAI,k}$ are independent, the distribution of $Y_{I,k}$ is the mixture density resulting from the convolution of $w(\cdot)$ and $f_{CAI}(\bullet)$, that is,

$$f_Y(y) = \sum_{i=1}^{m} f_{CAI}(y - l_i) \bullet w(i) \tag{11.33}$$

where

$Y$: a random variable
$y$: a particular value that $Y$ can take with ranging in $(0, l_m + L_{CAI})$.

Since $f_{CAI}(\cdot)$ is a density, the mixture density $f_Y(\bullet)$ is never 0 in $(0, l_m + L_{CAI})$ provided $L_{CAI}$ is greater than the largest difference between consecutive step-adjustable load levels. This makes every load level within this interval actually achievable.

The argument we are making here ensures a smooth transition to lower load levels following reduction signals sent to customers. This is important if DILS is applied to few customers, but it becomes less and less important as the number of customers increases.

Suppose a load point has $N$ users connected to it. We now analyze the effect of a load-shedding signal $p$ sent to a given number of customers at time $t$ to be carried out at time $(t + u)$. Let $n$, the number of customers participating in the DILS program, be less than $N$. If we know the probabilistic characterization of the load of a typical customer at any time $t$, and its subdivision into interruptible and uninterruptible, which will be the tool to assess the probability of reaching the desired load relief. For expository purposes, we take all the $N$ users belonging to the same class (e.g., all residential).

The total load of a single user can be written as

$$Y_k(t) = Y_{I,k}(t) + Y_{U,k}(t) \tag{11.34}$$

where $Y_{I,k}(t)$ and $Y_{U,k}(t)$ are the interruptible and the uninterruptible part of the load respectively. Obviously $Y_{I,k}(t)$ is 0 for uninterruptible customers. Let us consider NA appliances (such as refrigerators, washing machines, dishwashers, etc.), and let the percentages of customers who possess each appliance be given by $p_1, \dots, p_{NA}$. Finally, the indicator function $I$ ($i$ has $j$), takes a value of 1 if customer has the appliance $j$ and 0 otherwise. Then we can write

$$Y_k(t) = \sum_{j=1}^{NA} I(i \text{ has } j) \bullet w_j(t) \tag{11.35}$$

where $w_j(t)$ is the (random) power absorbed by the appliance $j$ at time $t$. $I(\bullet)$ is the indicator function of a statement.

Let

$$\mu_j(t) = E(w_j(t)) \tag{11.36}$$

$$\sigma_j^2(t) = \text{Var}(w_j(t)) \tag{11.37}$$

We can derive the expected value and the variance of the load absorbed by a customer picked at random, under the hypothesis that the appliances are used independently of each other:

$$\mu_T(t) = \sum_{j=1}^{NA} p_j \mu_j(t) \tag{11.38}$$

$$\sigma_T^2(t) = \sum_{j=1}^{NA} p_j [\sigma_j^2(t) + (1 - p_j \mu_j^2(t))] \tag{11.39}$$

If the appliances are not independent, equation (11.38) is unchanged, whereas equation (11.39) is modified by adding twice the sum of all the covariances between pairs of products of random variables $I$ ($i$ has $j$) $w_j(t)$ and $I$ ($i$ has $j'$) $w_{j'}(t)$.

The mean and variance in equations (11.38) and (11.39), are sufficient to approximate the probability distribution of the load with a Gaussian by the central limit theorem, provided the total number $N$ of customers connected to a given load point is large enough, so that we can state that the total power $S(t)$ absorbed at time $t$ has a Gaussian distribution, with mean $N\mu_T(t)$ and variance $N\sigma_T^2(t)$, as follows:

$$S(t) = S_I(t) + S_U(t) = \sum_{k=1}^{N} Y_{I,k}(t) + \sum_{k=1}^{N} Y_{U,k}(t)$$

$$= \sum_{k=1}^{N} Y_k(t) \sim N(N\mu_T(t), N\sigma_T^2(t)) \tag{11.40}$$

By indexing from 1 to $n$ those customers who take part in the DILS program, we can write the share of the total load actually available for curtailment as

$$S_{I,n}(t) = \sum_{k=1}^{n} Y_{I,k}(t) \tag{11.41}$$

Suppose now that we possess a load-forecasting method, which is precise enough to consider $s(t + u)$ as known when data is available up to time $t$. Certainly, $S_{I,n}(t)$ remains unobserved (we can only measure the total power taken by all the $N$ customers), but the precisely forecasted $s(t + u)$ gives us some information about $S_{I,n}(t + u)$. This information is summarized by the conditional distribution $P(S_{I,n}(t + u)|S(t + u) = s(t + u))$. Let $\mu_I(t)$ and $\sigma_I^2(t)$ be the mean and variance of the load drawn by the interruptible appliances of a customer picked at random. By normal approximation, this conditional distribution is still Gaussian with mean

$$n\left\{\mu_I(t + u) + \frac{1}{N}\frac{\sigma_I^2(t + u)}{\sigma_T^2(t + u)}[s(t + u) - N\mu_T(t + u)]\right\} = n\mu \tag{11.42}$$

and variance

$$n\left[\sigma_I^2(t + u)\left(1 - \frac{n}{N}\frac{\sigma_I^2(t + u)}{\sigma_T^2(t + u)}\right)\right] = n\sigma^2 \tag{11.43}$$

This conditional Gaussian distribution will be the main ingredient for the determination of the optimal value of $p$.

If the customers connected to the same load point are not homogeneous, they can be split into homogeneous groups. If these groups are large enough, then the Gaussian approximation still applies for each group so that $S(t)$ will be Gaussian distributed and the conditional distribution of the interruptible load can be found in a similar way as above.

The effectiveness of the central limit theorem depends on both the shape of the individual load probability distribution and the degree of statistical correlation among customers' loads. A recent study [22] on the probability distribution of the aggregated residential load for extra-urban areas, based on a bottom-up approach, shows that the Gamma distribution exhibits the best goodness of fit among a set of candidate distributions, but that the Gaussian approximation still passes the test for a reasonably large number of users. If strong stochastic dependence among customers persists, for example, due to spatial autocorrelation (the means $\mu_T(t)$ depend on time only), the Gaussian distribution could be inappropriate, and further study would be necessary to model the specific situation correctly.

***Load Shedding via the Probability of Failure***   A load-shedding request $p$, sent to customer $k$, implies a load relief of $pY_{I,k}$ kW. The customer can attain the new load level $(1 - p)Y_{I,k} + Y_{U,k}$. Overall, the load relief obtained when $p$ is applied to the $n$ customers is

$$p\sum_{k=1}^{n} Y_{I,k} = pS_{I,n} \tag{11.44}$$

Then we must set up a decision criterion to set $p$ in such a way that we are confident that the requested load relief of $r$ kW is achieved. We can formalize this by stating that $p$ must be such that

$$P(pS_{I,n} < r) \le \alpha \tag{11.45}$$

where $\alpha$ is an acceptable probability that the desired load relief is not attained. In principle $\alpha$ can be zero, if the interruptible load is greater than $(r/p)$ with probability 1 for some $p$. In some situations, when the absorbed load is very high and a small load relief is requested, this condition can be met.

Let $F$ denote the cumulative conditional distribution function of $S_{I,n}$. Then the decision criterion for $p$ is written as

$$F\left(\frac{r}{p}\right) \le \alpha \tag{11.46}$$

and is satisfied if

$$\frac{r}{p} = F^{-1}(\alpha) = q_\alpha \Rightarrow p = \frac{r}{q_\alpha} \tag{11.47}$$

The condition $r < q_\alpha$ is required for this to have an admissible solution.

In general, there will be no closed-form expression for $F$. But we may employ the central limit theorem approximation introduced above with the appropriate conditional mean and variance of the single customer's load indicated by $\mu$ and $\sigma^2$. Then

$$F\left(\frac{r}{p}\right) \cong \Phi\left(\frac{\frac{r}{np} - \mu}{\frac{\sigma}{\sqrt{n}}}\right) \tag{11.48}$$

where $\Phi$ is the standard Gaussian cumulative density function, and the solution to equation (11.46) is

$$p = \frac{\frac{r}{n}}{\mu + z_\alpha \frac{\sigma}{\sqrt{n}}} \qquad (11.49)$$

where $z_\alpha$ is the $\alpha$-level percentage of the standard Gaussian distribution.

The probability level $\alpha$ can be chosen if a measure of the cost of not achieving the desired load relief is available, say, $c_0$. Then the expected cost of not attaining the load relief is given by $\alpha c_0$ and $\alpha$ can be increased from 0 up to a value $c_A/c_0$, where $c_A$ is the maximum acceptable cost (which would be lower than $c_0$).

**Load Shedding via the General Cost Function**    A more sophisticated decision criterion of load shedding can be based on a cost function which increases with the actual load relief distance from the target, such as

$$c(p, S_{I,n}) = c_1 p S_{I,n} I(p S_{I,n} > r) + c_2 S_{I,n} I(p S_{I,n} < r) \qquad (11.50)$$

As mentioned before, $I(\cdot)$ is the indicator function of a statement and $s$ is the total load at the time of the shedding. The two addenda account for the cost of an overshooting and an undershooting, respectively. The cost constants $c_1$ and $c_2$ can include per-kWh costs on the distributor's (energy not sold) and on the customer's side (energy not available), because of a blackout or of an excessive curtailment (as we are talking about energy and the cost function depends on power, we are implying a fixed duration of the shedding intervention). One should note that for the network operator, which manages the shedding action, it will be difficult to give a fair assessment of costs not incurred by itself. Considering the costs of the energy not sold only, given $c_2$, the order of magnitude of $c_1$ should be $c_2$, one possible choice being $c_1 = c_2$.

The load-shedding problem becomes a search for the minimization of the expected value of the following cost function.

$$c(p) = E(c(p, S_{I,n})) = c_1 p \int_{r/p}^{\infty} s' f(s') ds' + c_2 s F\left(\frac{r}{p}\right) \qquad (11.51)$$

where $f$ is the density function associated with $F$.

By using the Gaussian approximation, we get

$$c(p) \cong c_1 p \left\{ n\mu \left[ 1 - \Phi\left(\frac{\frac{r}{p} - n\mu}{\sqrt{n}\sigma}\right) \right] + \sqrt{n}\sigma\phi\left(\frac{\frac{r}{p} - n\mu}{\sqrt{n}\sigma}\right) \right\}$$

$$+ c_2 s \Phi\left(\frac{\frac{r}{p} - n\mu}{\sqrt{n}\sigma}\right) \qquad (11.52)$$

where $\varphi$ is the standard Gaussian density function.

This decision criterion based on the conditional Gaussian is an instance of Bayesian expected loss minimization [23]. The loss is represented by $c(p, S_{I,n})$ and the expectation is taken with respect to the posterior distribution of an unobservable quantity $(S_{I,n})$ conditionally on another observed quantity $(s)$, through which the prior information on the former is updated.

## 11.8    UNDERVOLTAGE LOAD SHEDDING

### 11.8.1    Introduction

We discuss the load-shedding problem from the view of the voltage stability in this section. Load shedding is the ultimate countermeasure to save a voltage unstable system, when there is no other alternative to stop an approaching collapse [24–29]. This countermeasure is cost effective in the sense that it can stop voltage instability triggered by large disturbances, against which preventive actions would not be economically justified (if at all possible) in view of the low probability of occurrence [26]. Load shedding is also needed when the system undergoes an initial voltage drop that is too pronounced to be corrected by generators (because of their limited range of allowed voltages) or load tap changers (because of their relatively slow movements and also limited control range).

In the practical system, this kind of load shedding belongs to the family of system protection schemes (also referred to as special protections scheme) (SPS) against long-term voltage instability. An SPS is a protection designed to detect abnormal system conditions and take predetermined corrective actions (other than the isolation of the faulted elements) to preserve system integrity as far as possible and regain acceptable performance [27].

The following SPS design has been chosen [29]:

- *Response-based*: load shedding will rely on voltage measurements which reflect the initiating disturbance (without identifying it) and the actions taken so far by the SPS and by other controllers. On the contrary, an event-based SPS would react to the occurrence of specific events [28].

- *Rule-based*: load shedding will rely on a combination of rules of the type:

$$\text{If } V < V_{\text{threshold}} \text{ during } t \text{ seconds, shed } \Delta P \text{ MW} \qquad (11.53)$$

where $V$ is measured voltage, and $V_{\text{threshold}}$ is the corresponding threshold value.

- *Closed-loop operation*: an essential feature of the scheme considered here is the ability to activate the rule equation (11.53) several times, based on the measured result of the previous activations. This closed-loop feature allows the load-shedding controllers to adapt their actions to the severity of the disturbance. Furthermore, it increases the robustness with respect to operation failures as well as system behavior uncertainties [30]. This is particularly important in voltage instability, where load plays a central role but its composition varies with time and its behavior under large voltage drops may not be known accurately;

- A *distributed* scheme is proposed for its ability to adjust to the disturbance location.

  It is well-known that time, location, and amount are three important and closely related aspects of load shedding against voltage instability [31]. The time available for shedding is limited by the necessity to avoid [25]

  o reaching the collapse point corresponding to generator loss of synchronism or motor stalling;

  o further system degradation due to undervoltage tripping of field current–limited generators, or line tripping by protections;

  o the nuisance for customers of sustained low voltages. This requires fast action even in the case of long-term voltage instability, if the disturbance has a strong initial impact [30].

As far as long-term voltage instability is concerned, if none of the above factors is limiting, one can show that there is a maximum delay beyond which shedding later requires shedding more [25]. On the other hand, it may be appropriate to activate other emergency controls first so that the amount of load shedding is reduced [30].

The shedding location matters a lot when dealing with voltage instability: Shedding at a less appropriate place requires shedding more. In practice, the region prone to voltage instability is well known beforehand. However, within this region, the best location for load shedding may vary significantly with the disturbance and system topology.

### 11.8.2 Undervoltage Load Shedding Using Distributed Controllers

This undervoltage load-shedding scheme relies on a set of controllers distributed over the region prone to voltage instability [30]. Each controller monitors the voltage $V$ at a transmission bus and acts on a set of loads located at distribution level and having influence on $V$. Each controller operates as follows:

- It acts when its monitored voltage $V$ falls below some threshold $V_{threshold}$.

- It can act repeatedly, until $V$ recovers above $V_{threshold}$. This yields the already mentioned closed-loop behavior.

- It waits in between two sheddings, in order to assess the effect of the actions taken both by itself and by the other controllers.

- The delay between successive sheddings varies with the severity of the situation;

- The same holds true for the amount shed.

***Individual Controller Design***   As long as $V$ remains above the specified threshold, the controller is idle, while it is starts as soon as a (severe) disturbance causes $V$ to drop below $V_{threshold}$. Let $t_0$ be the time when this change takes place. The controller remains started until either the voltage recovers, or a time $\tau$ is elapsed since

$t_0$. In the latter case, the controller sheds a power $\Delta P^{sh}$ and returns to either idle (if $V$ recovers above $V_{threshold}$) or started state (if $V$ remains smaller than $V_{threshold}$). In the second case, the current time is taken as the new value of and the controller is ready to act again (provided of course that there remains load to shed).

The delay $\tau$ depends on the time evolution of $\tau$ as follows. A block of load is shed at a time such that

$$\int_{t_0}^{t_0+\tau} (V_{threshold} - V(t))dt = C \tag{11.54}$$

where $C$ is a constant to be adjusted. This control law yields an inverse-time characteristic: The deeper the voltage drops, the less time it takes to reach the value $C$ and, hence, the faster the shedding. The larger the value $C$ is, the more time it takes for the integral to reach this value and hence, the slower the action.

Furthermore, the delay $\tau$ is lower bounded:

$$\tau_{min} \leq \tau \tag{11.55}$$

to prevent the controller from reacting on a nearby fault. Indeed, in normal situations, time must be left for the protections to clear the fault and the voltage to recover to normal values.

The amount of load shedding depends on the voltage drop at the time period, that is

$$\Delta P^{sh} = K\Delta V^d \tag{11.56}$$

where $K$ is another constant to be adjusted, and $\Delta V^d$ is the average voltage drop over the time period $\tau$, that is,

$$\Delta V^d = \frac{1}{\tau} \int_{t_0}^{t_0+\tau} (V_{threshold} - V(t))dt \tag{11.57}$$

The controller acts by opening distribution circuit breakers and may disconnect interruptible loads only. Hence, the minimum load shedding corresponds to the smallest load whose breaker can be opened, while the maximum shedding corresponds to opening all the maneuverable breakers. Furthermore, to prevent unacceptable transients, it may be appropriate to limit the power disconnected in a single step to some value $\Delta P^{sh}_{tr}$, which can be written as

$$\min_k P_k \leq \Delta P^{sh} \leq \Delta P^{sh}_{max} \tag{11.58}$$

with

$$\Delta P^{sh}_{max} = \min \left( \sum_k P_k, P^{sh}_{tr} \right) \tag{11.59}$$

where $P_k$ denotes the individual load power behind the $k$th circuit breaker under control, and the minimum in equation (11.58) and the sum in equation (11.59) extend over all maneuverable breakers.

The control logic focuses on active power but load reactive power is obviously reduced together with active power. In the absence of more detailed information, we assume that both powers vary in the same proportion.

***Cooperation Between Controllers***   In this section, we discuss the interaction of the various controllers used in load shedding.

Let us consider two close controllers: $C_i$ monitoring bus $i$ and $C_j$ monitoring bus $j$. Let us assume that both controllers are started by a disturbance. When $C_i$ sheds some load, it causes the voltages to increase not only at bus $i$ but also at neighboring buses including the monitoring bus $j$. Since $V_i$ increases, the integral $\int (V_{\text{threshold}} - V_j(t)) dt$ decreases. It can be observed from equations (11.56) and (11.57) that the $\Delta V^d$ decreases; consequently, the amount of load shedding will be reduced for the controller $j$. If $V_i$ is increased and is larger than $V_{\text{threshold}}$, the controller $j$ will return to idle. Thus, when one controller sheds load, it slows down or inhibits the other controllers to restore voltages in the same area. This cooperation avoids excessive load shedding.

Obviously, the whole system will tend to automatically trigger the controller to shed the load first where voltages drop the most at the location of the controller. It means that operating the controllers in a fully distributed way, each controller using local information and taking local actions, as underfrequency load-shedding controllers do, which we discussed in Section 11.2.

Another way to implement the load-shedding scheme in a centralized way is by collecting all voltage measurements at a central point, running the computations involved in equations (11.54)–(11.59) in a single processor, and sending back load-shedding orders (with some communication delays being taken into account). In this case, additional information exchanges and interactions between controllers may be envisaged without further penalizing the scheme. To protect the SPS against erroneous measurements, it is desirable for each controller to rely on several voltage measurements, taken at closely located buses. Some filtering can remove outliers from the measurements, and the average value of the valid ones can be used as $V$ in equations (11.54) and (11.57). If all data are dubious, the controller should not be started; other controllers will take over.

***Tuning the Parameters of the Controller***   Obviously, the parameters of the controller affect the response of the controller as well as the scheme of load shedding. The tuning of the controllers should rely on a set of scenarios combining different operating conditions and disturbances, as typically considered when planning SPS [30,31].

The following are the basic requirements:

**(1)** Protection security: The SPS does not act in a scenario with acceptable post-disturbance system response. This is normally the case following any contingency.

**(2)** Protection dependability: All unacceptable post-disturbance system responses are saved by the SPS, possibly in conjunction with other available controls.

**(3)** Protection selectivity: In the latter case, the minimum load power possible is interrupted.

The tuning mainly consists of choosing the best values for $V_{\text{threshold}}, C, K, \Delta P_{\text{tr}}^{\text{sh}}$, and $\tau_{\text{min}}$. It is noted that the voltage threshold should be set high enough to avoid excessive shedding delays, which in turn would require to shed more and/or cause low load voltages. On the other hand, it should be low enough to obey requirement (1) above. It should thus be set a little below the lowest voltage value reached during any of the acceptable post-disturbance evolutions.

As for $C$ and $K$, they should be selected so that, for all scenarios,

- the protection sheds the minimum load possible and
- some security margin is left with respect to values causing protection failure.

Certainly, using the same $C$ and $K$ values for all controllers makes the design definitely simpler.

## 11.8.3  Optimal Location for Installing Controller

We know that the location of the controller affects not only the improvement of the voltage profile, but also the economy of the system operation. Thus the location of installing the controller or SPS is very important. The following conditions should be satisfied at the optimal location of the installing controller:

**(1)** There is considerable improvement in voltage at the location.

**(2)** The probability of the outage at the location is high.

**(3)** There is considerable reduction in system loss.

**(4)** The load at the location is of low importance.

**(5)** The load center that the load is located at is of low importance.

For item 1, the performance index to evaluate the voltage improvement by load shedding can be computed as follows:

$$PI_{\text{LSV}}^{j} = \frac{V_j(\Delta P_j^{\text{sh}}) - V_j(0)}{\Delta P_j^{\text{sh}}} \qquad (11.60)$$

where

$V_j(0)$: the voltage at bus $j$ before the load shedding

$V_j(\Delta P_j^{\text{sh}})$: the voltage at bus $j$ after the load shedding

$\Delta P_j^{\text{sh}}$: the amount of the load shedding at bus $j$

$PI_{\text{LSV}}^{j}$: the performance index to assess the voltage improvement at bus $j$.

The probability of the outage for each location can be obtained according to analysis of the historical outage or disturbance data in the system.

For item 3, the performance index to evaluate the loss reduction by load shedding can be computed as follows:

$$PI^j_{LSPL} = \frac{P_L(\Delta P^{sh}_j) - P_L(0)}{\Delta P^{sh}_j} \tag{11.61}$$

where

$P_L(0)$: the system loss before the load shedding at bus $j$

$P_L(\Delta P^{sh}_j)$: the system loss after the load shedding at bus $j$

$PI^j_{LSPL}$: the performance index to assess the loss reduction at bus $j$.

Actually, the performance index to evaluate the loss reduction by load shedding can also be obtained using loss sensitivity of load that was discussed in Chapter 3.

For items 4 and 5, which are related to the less important of the loads, we can use one performance $PI^j_{LSKEY}$ to express them. In Section 11.5, we computed the unified weighting factors $w_i$ of loads on the basis of their importance. Obviously, the less important performance index $PI^j_{LSKEY}$ will be

$$PI^j_{LSKEY} = \frac{1}{w_i} \tag{11.62}$$

Therefore, the hierarchical model for computing the optimal location for installing the controller can be constructed as in Figure 11.9.

For the lower layers in the hierarchy model (Figure 11.9), the performance indices for evaluating the individual load location can be computed on the basis of equations (11.60)–(11.62). But for the upper layer in the hierarchy model, the relationship among the all kinds of performance indices cannot be computed exactly. It can be only obtained on the basis of system operation cases and the judgment of the engineer or operators. According to AHP, the judgment matrix $A - PI$ can be

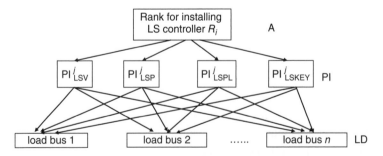

Figure 11.9  Hierarchy model of optimal location for installing LS controller.

written as follows:

$$A - PI = \begin{bmatrix} w_{PI1}/w_{PI1} & w_{PI1}/w_{PI2} \cdots\cdots & w_{PI1}/w_{PIn} \\ w_{PI2}/w_{PI1} & w_{PI2}/w_{PI2} \cdots\cdots & w_{PI2}/w_{PIn} \\ \vdots & \vdots \\ w_{PIn}/w_{PI1} & w_{PIn}/w_{PI2} \cdots\cdots & w_{PIn}/w_{PIn} \end{bmatrix} \qquad (11.63)$$

where, $w_{PIi}$ is unknown. $w_{PIi}/w_{PIj}$, which is the element of judgment matrix $A - PI$, represents the relative importance of the $i$th performance index compared with the $j$th performance index. Here, there are only four performance indices for selecting the location of the controller. Thus, $n = 4$ in equation (11.63).

According to the hierarchy model in Figure 11.9 and AHP approach, we can get the unified rank for all the locations for installing the LS controller. The number one in the rank list of locations will be first selected to install the LS controller. If there are $K$ controllers, they will be installed in the system where the locations are the top $K$ in the rank list.

## 11.9  CONGESTION MANAGEMENT

### 11.9.1  Introduction

Transmission congestion occurs when there is insufficient transmission capacity to meet the demands of all customers.

Congestion can be reduced by the following methods [32]:

**(1)** Generation re-dispatch

**(2)** Load shedding

**(3)** Using VAR support

**(4)** Expansion of transmission lines.

Obviously, expansion of transmission lines involves a large number of factors such as financial, time, environment, etc., and it is not realistic to solve the current congestion problem. Several previous chapters have analyzed generation dispatch and re-dispatch issues. The congestion may be reduced by modification of generating schedules, but not for every situation. In heavily congested conditions, transmission congestion can only be relieved by curtailing a portion of non-firm transactions. Thus we focus on the load-shedding method for analyzing congestion management in this section.

### 11.9.2  Congestion Management in US Power Industry

In the United States, the congestion management is implemented in the various ISOs such as the Pennsylvania–New Jersey–Maryland Interconnection (PJM), Electric

Reliability Council of Texas (ERCOT), and New York Independent System Operator (NYISO).

**PJM** PJM Interconnection is a regional transmission organization that ensures the reliability of the electric power supply system in 13 states. PJM operates the wholesale electricity market and manages regional electric transmission planning to maintain the reliability of the power system.

Different methods to mitigate transmission emergencies due to overloads and excess transfers in transmission lines are adopted in PJM [33]. They are:

- Generator active power adjustment—raise/lower MW
- Phase angle regulator adjustment—increase/decrease phase angle
- Interchange schedule adjustment—import/export MW
- Transmission line switching—selected line switching
- Circuit breaker switching—change network topology
- Customer load shedding—internal procedure and NERC transmission loading relief procedure.

Load shedding is the last option when the congestion cannot be alleviated through the remaining transmission emergency methods. Flow limits are further distinguished into normal limits, short-term emergency (STE) limits, and load dump (115% of STE). Violations may occur under actual (precontingency) or contingency (postcontingency) conditions.

PJM curtails loads that contribute to the overload before redispatching the generators if the transmission customers have indicated that they are not willing to pay transmission congestion charges. If overload persists, even after redispatching the system, PJM will implement the NERC transmission loading relief procedure (TLR) [34]. The steps of TLR are:

1. Notification of reliability coordinator
2. Hold interchange transactions
3. Reallocate firm transmission service
4. Reallocate non-firm transmission service
5. Curtail non-firm load
6. Redispatch generation
7. Curtail firm load
8. Implement emergency procedures.

**ERCOT** ERCOT directs and ensures the reliable and cost-effective operation of its electric grid and enables fair and efficient market-driven solutions to meet customer's electric needs [35]. The following issues are addressed:

(1) Ensures the grid can accommodate the scheduled energy transfers.

**(2)** Ensures grid reliability.

**(3)** Oversees retail transactions.

ERCOT develops four types of action plans to respond to electric system congestions.

- Precontingency action plan—used ahead of the contingency because it is not feasible once the contingency occurs.
- Remedial action plan—used after contingency occurs.
- Mitigation plan—similar to remedial action plan but only used after all available generation redispatch is exhausted. After the precontingency and remedial action plans are executed and if relief is still needed, this method is appropriate
- Special protection plan—automatic actions using special protection systems

The Emergency Electric Curtailment Plan (EECP) [36,37] was developed to respond to short-supply situations and restore responsive reserve to required levels. This procedure will direct the system operator to declare an emergency notice for frequency restoration purposes.

*NYISO*   The NYISO manages New York's electricity transmission grid, a network of high-voltage lines that carries electricity throughout the state, and oversees the wholesale electricity market. NYISO addresses the following issues:

**(1)** Maintains and enhances regional reliability

**(2)** Promotes and operates a fair and competitive electric wholesale market

**(3)** Provides quality customer service

**(4)** Tries to achieve these objectives in a cost-effective manner.

Severe system disturbances generally result in critically loaded transmission facilities, critical frequency deviations, high or low voltage conditions, or stability problems. The following operating states are defined for the state of New York [38]:

**(1)** Warning

**(2)** Alert

**(3)** Major emergency

**(4)** Restoration

The NYISO schedule coordinator, or the NYISO shift supervisor forecasts the likelihood of the occurrence of states other than the Normal State in advance. If it is predicted that load relief either by voltage reduction or load shedding may be necessary during a future period, then the NYISO shift supervisor notifies all transmission owners and arranges corrective measures including

- Curtailment of interruptible load
- Manual voltage reduction
- Curtailment of nonessential market participant load

- Voluntary curtailment of large load-serving entities (LSEs).
- Public appeals

NYISO reduces transmission flows that may cause thermal, voltage, and stability violations to properly allocate the reduction of transmission flows to relieve violations. When there are security violations that require reductions in transmission flow, NYISO takes action in the following sequence and to the extent possible, when system conditions and time permit:

1. Implement all routine actions using phase angle regulator tap positions, where possible.
2. Request all overgeneration suppliers that are contributing to the problem to adjust their generation to match their schedules.
3. Request voluntary shifts on generation either below minimum dispatchable levels or above normal maximum levels to help relieve the violation.
4. Request reduction or cancellation of all transactions that contribute to the violation. Applicable transactions shall be curtailed in accordance with curtailment procedures described in the NYISO Transmission and Dispatching Operations Manual [39].

## 11.9.3   Congestion Management Method

The previous sections presented several approaches for optimal load shedding, which are can be used for congestion management. Here, we present simple-load shedding or load-management methods for congestion management. They are

- TLR Sensitivities-Based Load Curtailment
- Economic Load Management for Congestion Relief

***TLR Sensitivities-Based Load Curtailment***   We discussed power transfer distribution factors (PTDFs) in Chapter 3. The transmission line relief (TLR) sensitivities can be considered as the inverse of the PTDF. Both TLR and PTDF can measure the sensitivity of the flow on a line-to-load curtailment. PTDFs determine the sensitivity of the flow on an element such as transmission line to a single power transfer. TLR sensitivities determine the sensitivity of the flow on the single monitored element such as a transmission line to many different transactions in the system. In other words, TLR sensitivities gauge the sensitivity of a single monitored element to many different power transfers.

The TLR sensitivity values at all the load buses for the most overloaded lines are considered and used for calculating the necessary load curtailment for the alleviation of the transmission congestion. The TLR sensitivity at a bus $k$ for a congested line $ij$ is $S_{ij}^k$, and is calculated by [32]

$$S_{ij}^k = \frac{\overline{\Delta P_{ij}}}{\Delta P_K} \tag{11.64}$$

The excess power flow on transmission line $ij$ is given by

$$\overline{\Delta P_{ij}} = P_{ij} - P_{ij\text{max}} \tag{11.65}$$

where

$P_{ij}$: the actual power flow through transmission line $ij$
$P_{ij\text{max}}$: the flow limit of transmission line $ij$.

The new load $P_k^{\text{new}}$ at the bus $k$ can be calculated by

$$P_k^{\text{new}} = P_k - \frac{S_{ij}^k}{\sum_{l=1}^{ND} S_{ij}^l} \overline{\Delta P_{ij}} \tag{11.66}$$

where

$P_k^{\text{new}}$: the load after curtailment at bus $k$
$P_k$: the load before curtailment at bus $k$
$S_{ij}^l$: the sensitivity of power flow on line $ij$ due to load change at bus $k$
$ND$: the total number of load buses.

The higher the TLR sensitivity the more the effect of a single MW power transfer at any bus. So, on the basis of the TLR sensitivity values, the loads are curtailed in required amounts at the load buses in order to eliminate transmission congestion on the congested line $ij$.

This method can be implemented for systems where load curtailment is a necessary option for obtaining $(N-1)$ secure configurations.

It is noted that the sensitivity computed here is based on perturbation, which is discussed in Chapter 3—Sensitivity Calculation. A limitation exists for this approach, that is, the sensitivity results are not stable. They are affected by the power flow solution, including the selection of initial operation points. The more precise method for sensitivity calculation is based on matrix operation, which is purely related to network topology, and will not be affected by the solution of power flow. The details are described in Chapter 3.

***Economic Load Management for Congestion Relief***   Another possible solution for congestion management is to find customers who will volunteer to lower their consumption when transmission congestion occurs. By lowering the consumption, the congestion will "disappear" resulting in a significant reduction in bus marginal costs. A strategy to decide how much load should be curtailed for which customer is discussed here.

The anticipated effect of this congestion relief solution is to encourage consumers to be elastic against high prices of electricity. Hence, this congestion relief procedure could eventually protect all customers from high electricity prices in a deregulated environment [40].

The following three factors can be considered for the analysis of load management:

**(1)** Power flow effect through sensitivity index

**(2)** Economic factor for LMP index

**(3)** Load reduction preference for customer load curtailment index.

The possible methods for these load management are presented in the following.

*Sensitivity Index*  In Chapter 3, we discussed load-distributed sensitivity, which can be used to rank load sensitivity. The sensitivity of the congested line $ij$ with respect to load bus $k$ is $S_{ij}^k$. We can convert it to the new sensitivity with the load distribution reference.

$$S_{ij}^{k\,new} = S_{ij}^k - S_{ldref}^k \quad k = 1, \ldots\ldots, ND \tag{11.67}$$

where

$S_{ldref}^k$: the sensitivity of load distribution reference for the constraint $ij$, that is,

$$S_{ldref}^k = \frac{\sum_{k=1}^{ND}(S_{ij}^k * P_{dk})}{\sum_{k=1}^{ND} P_{dk}} \tag{11.68}$$

The load shedding can be performed on the basis of the ranking of the distributed load reference-based sensitivity $S_{ij}^{k\,new}$. The load with high $S_{ij}^{k\,new}$ value will be curtailed first as it is more efficient to relieve the congestion than in the load with low $S_{ij}^{k\,new}$ value.

*LMP Index*  High electricity price or locational marginal price (LMP) is an incentive to reduce load. The following index measures the level of customer incentive to cut down on electricity consumption.

$$LMP^{k\,new} = LMP^k - LMP_{ldref}^k \quad k = 1, \ldots\ldots, ND \tag{11.69}$$

where

$LMP^k$: the electricity price of the load bus $k$ without considering the load factor.
$LMP^{k\,new}$: the electricity price of the load bus $k$ considering load factor.
$LMP_{ldref}^k$: the electricity price of the load bus $k$ based on load distribution reference, that is,

$$LMP_{ldref}^k = \frac{\sum_{k=1}^{ND}(LMP^k * P_{dk})}{\sum_{k=1}^{ND} P_{dk}} \tag{11.70}$$

The load shedding can be performed on the basis of the ranking of the distributed load reference-based electricity price $LMP^{k\,new}$. The load with high $LMP^{k\,new}$

value will be curtailed first as it provides greater incentive for customer to cut down on electricity consumption than the load with low $\text{LMP}^{k\,\text{new}}$ value. This is especially for customers with high load amounts.

*Customer Load Curtailment Index*  If the required reduction of the power flow on the congested branch is given by $\Delta P_{ijc}$, the required amount of adjustment $\Delta P_k$ at bus $k$ will be given by

$$\Delta P_K = \frac{\Delta P_{ijc}}{S_{ij}^k} \tag{11.71}$$

Generally, the higher the sensitivity, the smaller the amount of curtailment needed. The customer is supposed to express the acceptable range of curtailment by $\Delta P^{\max}$ and $\Delta P^{\min}$ at bus $k$, and the curtailment acceptance level is measured by

$$\mu_{LK} = \frac{\Delta P^{\max} - \Delta P_k}{\Delta P^{\max} - \Delta P^{\min}} \tag{11.72}$$

If the index $\mu_{Lk}$ is between 0 and 1, then the required amount of load reduction is in the acceptable range of the customer, and if $\mu_{Lk}$ is less than 0 or greater than 1, then the required amount of load curtailment is more than the acceptable range.

*Comprehensive Index for Congestion Relief*  We can comprehensively consider the three indices mentioned above. First of all, normalize each of them as follows:

$$\text{CR}_{\text{SI}}^k = \frac{S_{ij}^{k\,\text{new}}}{\sum_{k=1}^{ND} S_{ij}^{k\,\text{new}}} \quad k = 1, \dots\dots, ND \tag{11.73}$$

$$\text{CR}_{\text{LMP}}^k = \frac{\text{LMP}^{k\,\text{new}}}{\sum_{k=1}^{ND} \text{LMP}^{k\,\text{new}}} \quad k = 1, \dots\dots, ND \tag{11.74}$$

$$\text{CR}_{\text{LCI}}^k = \frac{\mu_{LK}}{\sum_{k=1}^{ND} \mu_{LK}} \quad k = 1, \dots\dots, ND \tag{11.75}$$

where
  $\text{CR}_{\text{SI}}^k$: the normalized sensitivity index
  $\text{CR}_{\text{LMP}}^k$: the normalized LMP index
  $\text{CR}_{\text{LCI}}^k$: the normalized customer load curtailment index.

Then we compute comprehensive index for congestion relieve (CICR) using following expression.

$$\text{CICR}^k = W_{\text{SI}}\text{CR}_{\text{SI}}^k + W_{\text{LMP}}\text{CR}_{\text{LMP}}^k + W_{\text{LCI}}\text{CR}_{\text{LCI}}^k, \quad k = 1, \dots\dots, ND \tag{11.76}$$

where

$W_{SI}$: the weight for the normalized sensitivity index
$W_{LMP}$: the weight for the normalized LMP index
$W_{LCI}$: the weight for the normalized customer load curtailment index
$CICR^k$: the comprehensive index for congestion relief.

The weight factors can be determined according to the practical system operation status. If they cannot be easily obtained, the AHP method can be used. Their sum should be 1.0, that is,

$$W_{SI} + W_{LMP} + W_{LCI} = 1 \qquad (11.77)$$

## PROBLEMS AND EXERCISES

1. What is underfrequency load shedding?

2. What is ILS? State the capabilities of the ILS.

3. Describe the effect of the load damping coefficient on the frequency drop.

4. What is the DILS method?

5. What is undervoltage load shedding?

6. List several important methods to reduce network congestion.

7. What is SPS?

8. List several proper locations to install voltage controller or SPS.

9. State the function of TLR.

10. The system shown in Figure 11.8 consists of two generators and two load centers. The weight factors reflecting the relative values of the load centers are $w_1 = 0.6$, and $w_2 = 0.4$. The independent load values $v$ in a specific load bus, the absolute load priority $\alpha$ to indicate the importance of each load bus, and the load demand for each load bus are given in Table 11.10. The capacities of generator 1 and generator 2 are $P_{G1} = 0.95$ and $P_{G2} = 0.65$ p.u., respectively. The available transfer capacity of key lines is $P_{1-6max} = 0.65$ p.u., $P_{2-7max} = 0.6$ p.u., $P_{1-7max} = 0.55$ p.u., respectively.

TABLE 11.10 The Values of Load Buses for Exercise 10

| Values | Load 3 | Load 4 | Load 5 | Load 8 | Load 9 |
|---|---|---|---|---|---|
| $v_i$ (S/kW) | 150 | 200 | 180 | 190 | 220 |
| $\alpha_i$ | 1.14 | 1.25 | 1.30 | 1.10 | 1.22 |
| Demand $P_D$ (p.u.) | .280 | .290 | .270 | .31 | .315 |

Compute load-shedding schemes for the following two test cases:

Case 1: two generators are in operation, tie line 1−7 is in outage.

Case 2: generator 2 is in outage, no line outage.

# REFERENCES

1. Richter CW, Sheble GB. A profit-based unit commitment GA for the competitive environment. IEEE Trans. on Power Syst. 2000;15(2):715–721.
2. Tenenbaum B, Lock R, Barker J. Electricity privatization: structural, competitive and regulatory options. Energy Policy 1992;12(12):1134–1160.
3. Rudnick H, Palma R, Fernandez JE. Marginal pricing and supplement cost allocation in transmission open access. IEEE Trans. on Power Syst. 1995;10(2):1125–1142.
4. Zhu JZ. Optimal load shedding using AHP and OKA. Int. J. Power Energ. Syst. 2005;25(1):40–49.
5. Lee TH, Thorne DH, Hill EF. A transportation method for economic dispatching—application and comparison. IEEE Trans. on Power Syst. 1980;99(5):2372–2385.
6. Hobson E, Fletcher DL, Stadlin WO. Network flow linear programming techniques and their application to fuel scheduling and contingency analysis. IEEE Trans. on Power Syst. 1984;103(4):1684–1691.
7. Zhu JZ, Xu GY. Application of out-of-kilter algorithm to automatic contingency selection and ranking, Proceedings of International Symposium on Eng. Mathematics and Application, ISEMA-88, 1988, pp. 301–305.
8. Ameli MT, Moslehpour S, Rahimikhoshmakani H. The role of effective parameters in automatic load-shedding regarding deficit of active power in a power system. Int. J. Mod. Eng. 2006;7(1).
9. Anderson P, Mirheydar M. A low order system frequency response model. IEEE Trans. Power Syst. 1990;5(3).
10. Chelgardee S, Doozbakhshian M. The designs for under-frequency relays used in load–shedding dispatching department, G 2162, 1990.
11. Shokooh F, Dai JJ, Shokooh S, Tastet J, Castro H, Khandelwal T, Donner G. An Intelligent Load Shedding (ILS) System Application in a Large Industrial Facility. Ind. Appl. Conf. 2005;1:417–425.
12. Saaty TL. The Analytic Hierarchy Process. New York: McGraw Hill, Inc.; 1980.
13. Zhu JZ, Irving MR, Xu GY. Automatic contingency selection and ranking using an analytic hierarchical process. Electr. Mach. Power Syst. 1998;16(4):389–398.
14. Smith DK. Network Optimization Practice. Chichester, UK: Ellis Horwood; 1982.
15. Everett H. Oper. Res. 1963;11:399–417.
16. Momoh JA, Zhu JZ, Dolce JL. Optimal allocation with network limitation for autonomous space power system. AIAA Journal—J. Propul. Power 2000;16(6):1112–1117.
17. Momoh JA, Zhu JZ, Dolce JL. Aerospace power system automation—using Everett method, AIP Conference Proceedings – January 22, 1999, Vol. 458, pp. 1653–1658.
18. Faranda R, Pievatolo A, Tironi E. Load shedding: a new proposal. IEEE Trans. on Power Syst. 2007;22(4):2086–2093.
19. Concordia C, Fink LH, Poullikkas G. Load shedding on an isolated system. IEEE Trans. Power Syst. 1995;10(3):1467–1472.
20. Delfino B, Massucco S, Morini A, Scalera P, Silvestro F. Implementation and comparison of different under frequency load- shedding schemes, presented at the IEEE summer meeting 2001, Vancouver , BC, Canada, July 15–19, 2001.
21. Xu D, Girgis AA. Optimal load shedding strategy in power systems with distributed generation, presented at the IEEE Power Engineering Society winter meeting, Columbus, OH, February 2001.
22. Carpaneto E Chicco G. Probability distributions of the aggregated residential loads, presented at the 9th International Conference Probabilistic Methods Applied to Power Systems, Stockholm, Sweden, June 11–15, 2006.
23. Berger JO. Statistical Decision Theory and Bayesian Analysis. 2nd ed. New York: Springer-Verlag; 1985.
24. Taylor CW. Power System Voltage Stability. EPRI Power Engineering Series. New York: McGraw-Hill; 1994.
25. Van Cutsem T, Vournas C. Voltage Stability of Electric Power Systems. Boston, MA: Kluwer; 1998.
26. Taylor CW. Concepts of undervoltage load shedding for voltage stability. IEEE T. Power Deliver. 1992;7(2):480–488.
27. Karlsson DH. System protection schemes in power networks, final report of CIGRE task force 38.02.19.

28. Nikolaidis VC, Vournas CD, Fotopoulos GA, Christoforidis GP, Kalfaoglou E, Koronides A. Automatic load shedding schemes against voltage stability in the Hellenic system, IEEE PES General Meeting, Tampa, FL, June 2007.

29. Otomega B, Cutsem TV. Undervoltage load shedding using distributed controllers. IEEE Trans. on Power Syst. 2007;22(4):1898–1907.

30. Lefebvre D, Moors C, Van Cutsem T. "Design of an undervoltage load shedding scheme for the Hydro-Québec system," IEEE PES General Meeting, Toronto, ON, Canada, July 2003.

31. Arnborg S, Andersson G, Hill DJ, Hiskens IA. On under-voltage load shedding in power systems. Int. J. Electr. Power Energ. Syst. 1997;19:141–149.

32. Parnandi S. Power Market Analysis Tool for Congestion Management, Master Thesis, West Virginia University, 2007.

33. PJM Transmission and Voltage Emergicies Online: http://www.pjm.com/services/training/downloads/ops101-transemer.pdf

34. NERC Reliability Coordination Transmission Loading Relief, ftp://www.nerc.com/pub/sys/all updl/standards/rs/IRO-006-1.pdf

35. ERCOT, Online: http://www.ercot.com/about/index.html

36. ERCOT Emergency Electric Curtailment Plan, Online: http://nodal.ercot.com/protocols/nprr/040/keydocs/040NPRR-01 Synchronization of Emergency Electric Curtailment.doc

37. ERCOT Mitigation, Online: http://www.ercot.com/meetings/tac/keydocs/2004/1202/TAC12022004-19.doc

38. NewYork ISO Emergency Operations, Online: http://www.nyiso.com/public/webdocs/documents/manuals/operations/em op mnl.pdf

39. New York ISO Transmission and Dispatching Manual, Online: http://www.nyiso.com/public/webdocs/documents/manuals/operations/trans disp.pdf

40. Niimura T. Niu Y, Transmission congestion relief by economic load management, IEEE Power Engineering Society summer meeting, Vol. 3, pp. 1645–1649, 2002.

# CHAPTER 12

# OPTIMAL RECONFIGURATION OF ELECTRICAL DISTRIBUTION NETWORK

The reconfiguration of the distribution network is also part of power system operation. This chapter sums up several major methods used to date in optimal reconfiguration of electric distribution network. These are the simple branch exchange method, the optimal flow pattern, the rule-based comprehensive approach, mixed-integer linear programming, the genetic algorithm (GA) with matroid theory, and multiobjective evolution programming (EP).

## 12.1    INTRODUCTION

Distribution networks are the most extensive part of the electrical power system. They produce a large number of power losses because of the low voltage level of the distribution system. The goal of reconfiguration of the distribution network is to find a radial operating structure that minimizes the power losses of the distribution system under the normal operation conditions. Generally, distribution networks are built as interconnected networks, while in operation they are arranged into a radial tree structure. This means that distribution systems are divided into subsystems of radial feeders, which contain a number of normally closed switches and a number of normally open switches. According to graph theory, a distribution network can be represented with a graph $G(N, B)$ that contains a set of nodes $N$ and a set of branches $B$. Every node represents either a source node (supply transformer) or a sink node (customer load point), whereas a branch represents a feeder section that can either be loaded (switch closed) or unloaded (switch open). The network is radial, so that the feeder sections form a set of trees where each sink node is supplied from exactly one source node. Therefore, the distribution network reconfiguration (DNRC) problem is to find a radial operating structure that minimizes the system power loss while satisfying operating constraints [1]. In fact, DNRC can be viewed as a problem of determining an optimal tree of a given graph. Many algorithms have been used to solve the reconfiguration problem: heuristic methods [2–10], expert system,

*Optimization of Power System Operation*, Second Edition. Jizhong Zhu.
© 2015 The Institute of Electrical and Electronics Engineers, Inc. Published 2015 by John Wiley & Sons, Inc.

combinatorial optimization with discrete branch and bound methods [11–17], and EP or GA [1,18–21].

Merlin and Back first proposed the discrete branch and bound method to reduce losses in a distribution network [3]. Because of the combinatorial nature of the problem, it requires checking a great number of configurations for a real-sized system. Shirmohammadi and Hong [8] used the same heuristic procedure mentioned in [3]. Castro et al. [4] proposed heuristic search techniques to restore the service and load balance of the feeders. Castro and Franca [6] proposed modified heuristic algorithms to restore the service and load balance. The operation constraints are checked through a load flow solved by means of modified fast decoupled Newton–Raphson method. Baran and Wu [5] presented a heuristic reconfiguration methodology based on the method of branch exchange to reduce losses and balance loads in the feeders. To assist in the search, two approximated load flows for radial networks with different degrees of accuracy are used. Also they propose an algebraic expression that allows estimating the loss reduction for a given topological change. Liu et al [14] proposed an expert system to solve the problem of restoration and loss reduction in distribution systems. The model for the reconfiguration problem is a combinatorial nonlinear optimization problem. To find the optimal solution, it is necessary to consider all the possible trees generated owing to the opening and closing of the switches existing in the network.

Nahman et al presented another heuristic approach in [10]. The algorithm starts from a completely empty network, with all switches open and all loads disconnected. Load points are connected one by one by switching branches onto the current subtree. The search technique also does not necessarily guarantee global optima.

Zhu et al [22] proposed a rule-based comprehensive approach to study DNRC. The DNRC model with line power constraints is set up, in which the objective is to minimize the system power loss. Unlike the traditional branch exchange–based heuristic method, the switching branches are divided into three types. The rules that are used to select the optimal reconfiguration of the distribution network are formed on the basis of system operation experiences and the types of switching branches [23].

Recently, new methods based on GA have been used in DNRC [1,18–20]. GA-based methods are better than traditional heuristic algorithms in the aspect of obtaining the global optima.

## 12.2 MATHEMATICAL MODEL OF DNRC

The mathematical model of DNRC can be expressed by either branch current or branch power.

**(1)** Use of Current Variable

$$\min f = \sum_{l=1}^{NL} k_l R_l I_l^2 \quad l \in NL \tag{12.1}$$

such that

$$k_l/I_l/ \leq I_{l\max} \quad l \in NL \tag{12.2}$$

$$V_{i\min} \leq V_i \leq V_{i\max} \quad i \in N \tag{12.3}$$

$$g_i(I,k) = 0 \tag{12.4}$$

$$g_i(V,k) = 0 \tag{12.5}$$

$$\varphi(k) = 0 \tag{12.6}$$

where

$I_l$: the plural current in branch $l$
$R_l$: the resistance of branch $l$
$V_i$: the node voltage at node $i$
$K_l$: the topological status of the branches—$k_l = 1$ if the branch $l$ is closed and $k_l = 0$ if the branch $l$ is open
$N$: the set of nodes
$NL$: the set of branches.

In the above model, equation (12.2) stands for the branch current constraints. Equation (12.3) stands for the node voltage constraints. Equation (12.4) represents Kirchhoff's first law (KCL) and equation (12.5) represents Kirchhoff's second law (KVL). Equation (12.6) stands for topological constraints that ensure radial structure of each candidate topology. It consists of two structural constraints:

**(a)** Feasibility: all nodes in the network must be connected by some branches, that is, there is no isolated node.

**(b)** Radiality: the number of branches in the network must be smaller than the number of nodes by one unit ($k_l^* NL = N - 1$).

Therefore, the final network configuration must be radial and all loads must remain connected.

**(2)** Use of Power Variable

$$\min f = \sum_{l=1}^{NL} k_l R_l \left( \frac{P_l^2 + Q_l^2}{V_l^2} \right) \quad l \in NL \tag{12.7}$$

such that

$$k_l/P_l/ \leq P_{l\max} \quad l \in NL \tag{12.8}$$

$$k_l/Q_l/ \leq Q_{l\text{max}} \quad l \in NL \tag{12.9}$$

$$V_{i\text{min}} \leq V_i \leq V_{i\text{max}} \quad i \in N \tag{12.3}$$

$$g_i(P, k) = 0 \tag{12.10}$$

$$g_i(Q, k) = 0 \tag{12.11}$$

$$g_i(V, k) = 0 \tag{12.5}$$

$$\varphi(k) = 0 \tag{12.6}$$

where,

$P_l$: the real power in branch $l$
$Q_l$: the reactive power in branch $l$

The objective function in equation (12.7) is power losses. If voltage magnitudes are assumed to be 1.0 p.u. and reactive power losses are ignored in the objective function, equation (12.7) may be simplified as

$$\min f = \sum_{l=1}^{NL} k_l R_l P_l^2 \quad l \in NL \tag{12.12}$$

In the above model, equations (12.8) and (12.9) stand for the branch real power and reactive power constraints. Equations (12.10), (12.11) represent Kirchhoff's first law.

Obviously, both DNRC models, whether with branch current expression or power expression, have the same function.

## 12.3 HEURISTIC METHODS

### 12.3.1 Simple Branch Exchange Method

The basic idea of the heuristic branch exchange method is to compute the change of power losses through operating a pair of switches (close one and open another one at the same time). The goal is to reduce power losses. The advantage of this method is simple and easily understood. The following are the disadvantages:

(1) The final configuration depends on the initial network configuration.
(2) The solution is a local optima, rather than global optima.
(3) Selecting and operating each pair of switches as well as computing the corresponding radial network load flow is time consuming.

## 12.3.2    Optimal Flow Pattern

If the impedances of all branches in the loop network are replaced by the corresponding branch resistances, the load flow distribution that satisfies the KCL and KVL is called an optimal flow pattern. When the load flow distribution in a loop is an optimal flow, the corresponding network power losses will be minimal. Thus the basic idea of the optimal flow pattern is to open the switch of the branch that has a minimal current value in the loop [8]. The steps of the heuristic algorithm based on optimal flow pattern are as follows:

(1) Compute load flow of initial radial network.

(2) Close all normal open switches to produce loop networks.

(3) Compute the equivalent injection current at all nodes in loops through the injecting current method.

(4) Replace branch impedance by the corresponding branch resistance in the loop and then compute the optimal flow.

(5) Open a switch of the branch that has a minimal current value in the loop. Recompute the load flow for the remaining part of the network.

(6) Open the next branch switch, and repeat step (5) until the network becomes a radial.

The advantages of this method are that (i) the final network configuration will not depend on the initial network topology; (ii) the computing speed is much quicker than that in the simple branch exchange method; (iii) the complicated combination problem of switch operation becomes a heuristic problem by opening one switch each time.

However, there are some disadvantages because all normally open switches are closed in the initial network, that is,

(1) If there are many normal open switches in a network, it means the calculation of optimal flow involves a number of loops. The final solution may not be optimal because of the mutual effects among the loops.

(2) When load flow is solved by the equivalent injection current method, it needs to compute the impedance matrix of the Thevenin equivalent network with multiports. This will increase the calculation burden.

(3) The loop network load flow needs to be computed twice for each switch operation (before and after one switch is opened).

## 12.3.3    Enhanced Optimal Flow Pattern

The enhanced optimal flow pattern combines the advantages of the two heuristic algorithms mentioned in Sections 12.3.1 and 12.3.2, that is, the approach is based on optimal flow pattern but does not close all normally open switches (it only closes one switch and opens another switch each time). In addition, this method ignores the accuracy of network losses. It only focuses on the change in losses that are caused by the operation of the switches. The calculation steps of the enhanced optimal flow pattern are as follows.

**(1)** Open all normally open switches in the network so that the initial network is a tree structure.

**(2)** Close any one switch. In this way, there is only one loop in the network.

**(3)** Compute the load flow for the single loop network and get the equivalent injection current for all nodes in the loop.

**(4)** Change the single loop network into a pure resistance network, and compute the optimal flow to find the branch with the minimal current value. Open the switch on this branch.

**(5)** Compute the load flow for this new radial network, and proceed with the calculation of the next switch operation as in steps (2)–(4).

**(6)** The algorithm will be stopped after we go through all the open switches.

The enhanced optimal flow pattern has eliminated the effect among multiple loops. Although the convergence process is related to the initial network, the final solution is stable and not related to the order of operation of the switches [9]. The disadvantages of this method are as follows:

**(1)** It needs twice load flow calculations for operation of each pair of switches.

**(2)** The convergence process and speed are affected by the order of the switches operation.

## 12.4 RULE-BASED COMPREHENSIVE APPROACH

This section uses a rule-based comprehensive approach to study DNRC. The algorithm consists of a modified heuristic solution methodology and the rules base. It determines the switching actions on the basis of a search by branch exchange to reduce the network's losses as well as to balance the load of the system.

### 12.4.1 Radial Distribution Network Load Flow

In order to get a precise expression for system power loss, the branch power will be computed through a radial distribution network load flow (RDNLF) method in the study. It is well known that in the distribution network, the ratio of $R/X$ (resistance/reactance) is relatively big, even bigger than 1.0 for some transmission lines. In this case, P–Q decoupled load flow is invalid for distribution network load-flow calculation. It will also be complicated and time consuming to use the Newton–Raphson load flow because the distribution network is only a simple radial tree structure. Therefore, the power summation–based radial distribution

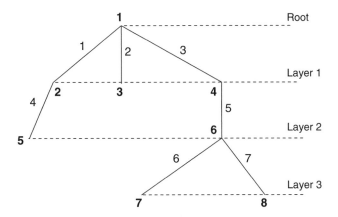

Figure 12.1    Example of optimal node order.

network load flow (PSRDNLF) method is presented in this section. The PSRDNLF calculation consists of three parts:

**(1)** Conduct the optimal node order calculation for all redial network based on graph theory. Consequently, the branches are divided into different layers according to the distant between the ordered node and "root of a tree" node. Figure 12.1 is an example on how to make optimal node order.

The rules of node order are as follows:

**(a)** Start from the root node.

**(b)** The nodes that connect to the root belong to first layer.

**(c)** The nodes that connect to the nodes in the first layer become second layer, and so on.

**(d)** The node number in layer $n$ must be greater than the node number in layer $(n-1)$. The node numbers in the same layer may be arbitrary.

**(e)** For the branch number, for example, connecting to layer $n$ and layer $n-1$, the start node of the branch is the node in layer $n-1$, and the end node is the node in layer $n$.

**(2)** Calculate the branch real power and reactive power from the "top of a tree" node to the "root of a tree" node, that is, from the last layer to the first layer.

**(3)** Compute the node voltage from the "root of a tree" node to the "top of a tree node," that is, from the first layer to the last layer.

The initial conditions are the given voltage vectors at root nodes as well as real and reactive power at load nodes. Finally, the deviation of injection power at all nodes can be computed. The iteration calculation will cease if the deviation is less than the given permissive error.

If there are multiple generation sources in the distribution network, one source will be selected as a reference/slack source and others can be handled as negative loads.

## 12.4.2 Description of Rule-Based Comprehensive Method

Unlike the traditional branch exchange–based heuristic method, the rule-based comprehensive method combined the traditional branch exchange approach with the set of rules. The rules that are used to select the optimal reconfiguration of the distribution network are formed on the basis of the system operation experiences.

In the rule-based comprehensive method, the switching branches are divided into three types:

(1) Type I: the switching branches that are planned for maintenance in a short period according to the equipment maintenance schedule.

(2) Type II: the power flows of the switching branches that almost reach their maximal power limits (e.g., 90%).

(3) Type III: the other switching branches that have enough available transfer capacity under the system operation conditions.

Thus the following rules will be used for the modified heuristic approach according to the practical system operation experiences of the engineers.

(1) If the switching branches lead to an increase in system power losses, do not switch them.

(2) If the switching branches lead to a reduction in system power losses but cause system overload, do not switch them.

(3) If the switching branches belong to type I mentioned above and also can lead to system power losses reduction, then select one that results in maximal reduction in power losses, $\Delta \mathrm{PL_I}$.

(4) If the switching branches belong to type II mentioned above, and also can lead to a reduction insystem power losses, select one that results in maximal reduction in power losses, $\Delta \mathrm{PL_{II}}$.

(5) If the switching branches belong to type III mentioned above, and also can lead to reduction in system power losses, select one that makes maximal power losses reduction, $\Delta \mathrm{PL_{III}}$.

(6) From (3)–(5), use the following formula to determine the branch that will be switched.

$$\mathrm{PI}_{\mathrm{SW}i} = \frac{W_i \Delta \mathrm{PL}_i}{W_\mathrm{I} \Delta \mathrm{PL_I} + W_\mathrm{II} \Delta \mathrm{PL_{II}} + W_\mathrm{III} \Delta \mathrm{PL_{III}}} \quad i = \mathrm{I, II, III} \quad (12.13)$$

where

$\Delta \mathrm{PL}_i$: the change of system power losses before and after the branch switch

$W$: the weighting coefficient of the different types of switching branches. According to the experiences of the engineers, the weighting factors of the three types of switches may be 1.0, 0.6, and 0.3, respectively. They may also be adjusted according the practical system operation situations.

$PI_{SWi}$: the performance index of the switching branch $i$. The largest $PI_{SWi}$ of each switching loop will be switched.

### 12.4.3 Numerical Examples

The rule-based comprehensive approach for DNRC is tested on 14-bus and 33-bus distribution systems as shown in Figures 12.2 and 12.3, respectively. The system data and parameters of the 14-bus system are listed in Tables 12.1 and 12.2.

The 14-bus test system contains two source transformers and 12 load nodes. The three initially open switches are "4–9," "14–11," and "6–3." The initial system power loss is 0.0086463 MW.

The results of the optimal configuration for the 14-bus distribution network are shown in Tables 12.3–12.5. Table 12.3 is the node voltage results comparison between the initial network and final configuration. Table 12.4 is the load flow results of the optimal configuration for the 14-bus system. Table 12.5 is the optimal open switches of the final network and the corresponding system losses, from which we can know that the system losses reduction is 0.0003765 MW, that is, 4.354%.

The system data and parameters of the 33-bus system are listed in Tables 12.6 and 12.7. The 33-bus test system consists of one source transformer and 32 load nodes. The five initially open switches are "33," "34," "35," "36," and "37." The total system load is 3.715 MW, while the initial system power loss is 0.202674 MW. The system base is $V = 12.66$ kV and $S = 10$ MVA.

The calculation results of the final configuration of the 33-bus system are shown in Table 12.8. It can be observed that the same results are obtained as in reference [8].

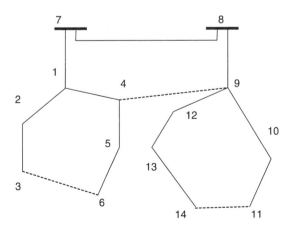

Figure 12.2  A 14-bus distribution system.

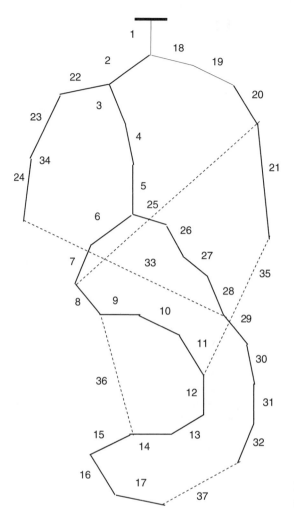

Figure 12.3   A 33-bus distribution system.

## 12.5   MIXED-INTEGER LINEAR-PROGRAMMING APPROACH

Because of the magnitude of the DNRC problem and its nonlinear nature, the use of a blend of optimization and heuristic techniques is one choice as in Section 12.4. The linearization of DNRC is another choice. Through performing a linearization of both the objective function and constraints, the DNRC is changed to a mixed-integer linear optimization problem [17].

**TABLE 12.1    System Load Demand for 14-Bus System**

| Node | Load $P$ (MW) | Load $Q$ (MVAR) |
|------|------|------|
| 1 | 0.0000 | 0.0000 |
| 2 | 0.9000 | 0.7000 |
| 3 | 0.7000 | 0.5500 |
| 4 | 0.0000 | 0.0000 |
| 5 | 0.9000 | 0.7600 |
| 6 | 0.4000 | 0.3000 |
| 7 | 0.0000 | 0.0000 |
| 8 | −2.2000 | −1.0800 |
| 9 | 0.3000 | 0.2000 |
| 10 | 0.6000 | 0.4500 |
| 11 | 0.9000 | 0.7500 |
| 12 | 0.0000 | 0.0000 |
| 13 | 0.8000 | 0.6500 |
| 14 | 0.3000 | 0.2200 |

**TABLE 12.2    System Branch Parameters for 14-Bus Distribution Network**

| Line No. | From Node $i$ | To Node $j$ | Resistance $R(\Omega)$ | Reactance $X(\Omega)$ |
|------|------|------|------|------|
| 1 | 7 | 1 | 0.00575 | 0.00893 |
| 2 | 1 | 2 | 0.02076 | 0.03567 |
| 3 | 2 | 3 | 0.01284 | 0.01663 |
| 4 | 1 | 4 | 0.01023 | 0.01567 |
| 5 | 9 | 12 | 0.01023 | 0.01976 |
| 6 | 4 | 5 | 0.09385 | 0.11457 |
| 7 | 5 | 6 | 0.03220 | 0.04985 |
| 8 | 8 | 9 | 0.00575 | 0.00793 |
| 9 | 9 | 10 | 0.03076 | 0.04567 |
| 10 | 10 | 11 | 0.02284 | 0.03163 |
| 11 | 12 | 13 | 0.09385 | 0.11457 |
| 12 | 13 | 14 | 0.02810 | 0.04085 |
| 13 | 7 | 8 | 0.02420 | 0.42985 |
| 14 | 14 | 11 | 0.02500 | 0.04885 |
| 15 | 4 | 9 | 0.02300 | 0.04158 |
| 16 | 6 | 3 | 0.02105 | 0.04885 |

**TABLE 12.3   Initial and Final Node Voltages for 14-Bus Distribution Network**

| Node | Initial $V$ (p.u.) | Initial $\Theta$ | Final $V$ (p.u.) | Final $\theta$ |
|------|-----------|-----------|-----------|-----------|
| 1 | 1.04964 | −0.00656 | 1.04951 | −0.00625 |
| 2 | 1.04890 | −0.02275 | 1.04858 | −0.02664 |
| 3 | 1.04873 | −0.02514 | 1.04831 | −0.03048 |
| 4 | 1.04936 | −0.01157 | 1.04906 | −0.00899 |
| 5 | 1.04704 | −0.03738 | 1.04742 | −0.02557 |
| 6 | 1.04678 | −0.04275 | 1.04809 | −0.03738 |
| 7 | 1.05000 | 0.00000 | 1.05000 | 0.00000 |
| 8 | 1.04927 | −0.00317 | 1.04863 | −0.00405 |
| 9 | 1.04894 | −0.00834 | 1.04843 | −0.00990 |
| 10 | 1.04798 | −0.02480 | 1.04729 | −0.02999 |
| 11 | 1.04756 | −0.03072 | 1.04673 | −0.03824 |
| 12 | 1.04867 | −0.01503 | 1.04823 | −0.01467 |
| 13 | 1.04674 | −0.03819 | 1.04681 | −0.03067 |
| 14 | 1.04657 | −0.04136 | 1.04656 | −0.04302 |

**TABLE 12.4   Load Flow of Optimal Configuration for 14-Bus Distribution Network**

| Line No. | From Node $i$ | To Node $j$ | Real power $P$ (MW) | Reactive power $Q$ (MVAR) |
|------|------|------|-----------|-----------|
| 1 | 7 | 1 | 4.30930 | 1.92709 |
| 2 | 1 | 2 | 3.10318 | 1.40266 |
| 3 | 1 | 4 | 1.20496 | 0.52344 |
| 4 | 2 | 3 | 2.20100 | 1.00101 |
| 5 | 4 | 5 | 0.90083 | 0.40074 |
| 6 | 4 | 9 | 0.30398 | 0.12245 |
| 7 | 3 | 6 | 1.00032 | 0.30039 |
| 8 | 9 | 12 | 0.80069 | 0.30062 |
| 9 | 9 | 8 | −3.19940 | −1.07969 |
| 10 | 9 | 10 | 2.40266 | 0.80148 |
| 11 | 12 | 13 | 0.80062 | 0.30056 |
| 12 | 10 | 11 | 1.80087 | 0.60057 |
| 13 | 11 | 14 | 0.70013 | 0.20020 |

**TABLE 12.5   Optimal Configuration Results for 14-Bus Distribution Network**

| Radial Network | Initial Network | Optimal Configuration |
|------|------|------|
| Open switches | Switch 4−9 | Switch 7−8 |
|  | Switch 14−11 | Switch 13−14 |
|  | Switch 6−3 | Switch 5−6 |
| Power loss (MW) | 0.008646 | 0.008270 |

**TABLE 12.6   System Data and Parameters for 33-Bus Distribution Network**

| Line No. | Node $i$ | Node $J$ | Resistance $R(\Omega)$ | Reactance $X(\Omega)$ |
|---|---|---|---|---|
| 1 | 1 | 2 | 0.0922 | 0.0470 |
| 2 | 2 | 3 | 0.4930 | 0.2512 |
| 3 | 3 | 4 | 0.3661 | 0.1864 |
| 4 | 4 | 5 | 0.3811 | 0.1941 |
| 5 | 5 | 6 | 0.8190 | 0.7070 |
| 6 | 6 | 7 | 0.1872 | 0.6188 |
| 7 | 7 | 8 | 0.7115 | 0.2351 |
| 8 | 8 | 9 | 1.0299 | 0.7400 |
| 9 | 9 | 10 | 1.0440 | 0.7400 |
| 10 | 10 | 11 | 0.1967 | 0.0651 |
| 11 | 11 | 12 | 0.3744 | 0.1298 |
| 12 | 12 | 13 | 1.4680 | 1.1549 |
| 13 | 13 | 14 | 0.5416 | 0.7129 |
| 14 | 14 | 15 | 0.5909 | 0.5260 |
| 15 | 15 | 16 | 0.7462 | 0.5449 |
| 16 | 16 | 17 | 1.2889 | 1.7210 |
| 17 | 17 | 18 | 0.7320 | 0.5739 |
| 18 | 2 | 19 | 0.1640 | 0.1565 |
| 19 | 19 | 20 | 1.5042 | 1.3555 |
| 20 | 20 | 21 | 0.4095 | 0.4784 |
| 21 | 21 | 22 | 0.7089 | 0.9373 |
| 22 | 3 | 23 | 0.4512 | 0.3084 |
| 23 | 23 | 24 | 0.8980 | 0.7091 |
| 24 | 24 | 25 | 0.8959 | 0.7071 |
| 25 | 6 | 26 | 0.2031 | 0.1034 |
| 26 | 26 | 27 | 0.2842 | 0.1447 |
| 27 | 27 | 28 | 1.0589 | 0.9338 |
| 28 | 28 | 29 | 0.8043 | 0.7006 |
| 29 | 29 | 30 | 0.5074 | 0.2585 |
| 30 | 30 | 31 | 0.9745 | 0.9629 |
| 31 | 31 | 32 | 0.3105 | 0.3619 |
| 32 | 32 | 33 | 0.3411 | 0.5302 |
| 34 | 8 | 21 | 2.0000 | 2.0000 |
| 36 | 9 | 15 | 2.0000 | 2.0000 |
| 35 | 12 | 22 | 2.0000 | 2.0000 |
| 37 | 18 | 33 | 0.5000 | 0.5000 |
| 33 | 25 | 29 | 0.5000 | 0.5000 |

**TABLE 12.7    System Load Demand for 33-Bus Distribution Network**

| Node No. | Real Power Load $P$ (MW) | Reactive Power Load $Q$ (MVAr) |
|---|---|---|
| 2 | 100.0 | 60.0 |
| 3 | 90.0 | 40.0 |
| 4 | 120.0 | 80.0 |
| 5 | 60.0 | 30.0 |
| 6 | 60.0 | 20.0 |
| 7 | 200.0 | 100.0 |
| 8 | 200.0 | 100.0 |
| 9 | 60.0 | 20.0 |
| 10 | 60.0 | 20.0 |
| 11 | 45.0 | 30.0 |
| 12 | 60.0 | 35.0 |
| 13 | 60.0 | 35.0 |
| 14 | 120.0 | 80.0 |
| 15 | 60.0 | 10.0 |
| 16 | 60.0 | 20.0 |
| 17 | 60.0 | 20.0 |
| 18 | 90.0 | 40.0 |
| 19 | 90.0 | 40.0 |
| 20 | 90.0 | 40.0 |
| 21 | 90.0 | 40.0 |
| 22 | 90.0 | 40.0 |
| 23 | 90.0 | 50.0 |
| 24 | 420.0 | 200.0 |
| 25 | 420.0 | 200.0 |
| 26 | 60.0 | 25.0 |
| 27 | 60.0 | 25.0 |
| 28 | 60.0 | 20.0 |
| 29 | 120.0 | 70.0 |
| 30 | 200.0 | 100.0 |
| 31 | 150.0 | 70.0 |
| 32 | 210.0 | 100.0 |
| 33 | 60.0 | 40.0 |

## 12.5.1    Selection of Candidate Subnetworks

The simplest way of modeling the topology of an electrical network is by means of the branch-to-node incidence matrix $A$, in which as many rows as connected components are omitted to assure linear independence of the remaining rows. Given a single-component meshed network with $N + 1$ buses, a well-known theorem states that a set of $N$ branches is a spanning tree if, and only if, the respective columns

**TABLE 12.8    Optimal Configuration Results for 33-Bus Distribution Network**

| Radial Network | Initial Network | Final Configuration | Results in Ref. [8] |
|---|---|---|---|
| Open switches | Switch 33 | Switch 7 | Switch 7 |
| | Switch 34 | Switch 10 | Switch 10 |
| | Switch 35 | Switch 14 | Switch 14 |
| | Switch 36 | Switch 33 | Switch 33 |
| | Switch 37 | Switch 37 | Switch 37 |
| Power loss (MW) | 0.202674 | 0.141541 | 0.141541 |

of constitute a full rank submatrix [27]. Thus graph-based algorithms are usually adopted to select the candidate subnetworks. Given the undirected graph of a single-component network, determining whether a candidate set of $N$ branches constitutes a spanning tree reduces to checking whether they form a single connected component. Alternatively, instead of checking for radiality, an a posteriori, straightforward algorithm is available to generate radial subnetworks, either from scratch or by performing branch exchanges on existing radial networks.

For a meshed network, there are, in general, several alternative paths connecting a given bus to the substation, whereas in a radial network, each bus is connected to the substation by a single unique path. Furthermore, the union of all node paths gives rise to the entire system. The connectivity of a meshed network, as well as that of its radial subnetworks, can then be represented by means of paths. Let $\pi_n^i$ be the set of paths associated to bus $i$

$$\Pi_n^i = \{\pi_1^i, \ \dots \ , \pi_p^i, \ \dots \ , \pi_n^i\} \tag{12.14}$$

where each path is a set of branches connecting the bus to the substation. As noted above, any radial network is characterized by only one of those paths being active for each bus. Therefore, there is a need to represent the status of each path, for which the following binary variable is defined:

$$K_p^i = \begin{cases} 1, & \text{if } \pi_p^i \text{ is the active path for bus } i \\ 0, & \text{otherwise} \end{cases} \tag{12.15}$$

A candidate subnetwork is both connected and radial if the following constraints are satisfied:

**(1)** Every node has at most one active path, that is,

$$\sum_{p \in \Pi_n^i} K_p^i = 1, \quad \forall \text{ node } i. \tag{12.16}$$

**(2)** If $\pi_p^i$ is active, then any path contained in $\pi_p^i$ must be also active, which can be written as follows:

$$K_p^i \leq K_l^j, \quad \forall \pi_l^j \subset \pi_p^i \tag{12.17}$$

Figure 12.4 is a simple electrical network with one source node and three load nodes. Table 12.9 presents all possible paths for this network [17].

It is worth noting that, for computational efficiency, not all of the possible paths shown in Table 12.9 should be considered in practice. For example, assuming the branch lengths represented in Figure 12.4 are proportional to their resistance, it is clear that paths $\pi_3^A$ and $\pi_3^B$ can be discarded, as they involve much greater electrical distance than that of alternative paths for nodes $A$ and $B$, respectively. Hence, for each node, only those paths whose total resistance does not exceed a previously defined threshold times the lowest node path resistance are considered. This significantly reduces the number of relevant candidate paths for realistic networks.

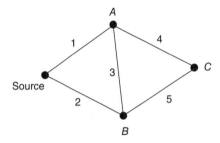

Figure 12.4   Simple electrical network with one source.

**TABLE 12.9   Node Paths for the Example of Figure 12. 4**

| Node | Path | Path Branches |
|------|------|---------------|
| $A$ | $\pi_1^A$ | 1 |
|  | $\pi_2^A$ | 2, 3 |
|  | $\pi_3^A$ | 2, 4, 5 |
| $B$ | $\pi_1^B$ | 2 |
|  | $\pi_2^B$ | 1, 3 |
|  | $\pi_3^B$ | 1, 4, 5 |
| $C$ | $\pi_1^C$ | 1, 4 |
|  | $\pi_2^C$ | 2, 5 |
|  | $\pi_3^C$ | 1, 3, 5 |
|  | $\pi_4^C$ | 2, 3, 4 |

The inequality constraint in equation (12.17) can be better understood with the help of this example. We can easily check that the following inequalities hold (paths $\pi_3^A$, $\pi_3^B$ are discarded).

$$
\left.
\begin{aligned}
W_3^C &\le W_2^B \le W_1^A \\[4pt]
W_1^C &\le W_1^A \\[4pt]
W_4^C &\le W_2^A \le W_1^B \\[4pt]
W_2^C &\le W_1^B
\end{aligned}
\right\}
$$

Although the concepts and variables presented above suffice for modeling the network radial structure, in order to handle other branch-related electrical constraints a second set of paths is introduced:

$$\Pi_b^j = \{\text{set of node paths sharing branch } j\}$$

Table 12.10 shows the set $\Pi_b^j$ for every branch in the sample system of Figure 12.4. A graph-based effective procedure is as follows.

Before describe the graph-based procedure, we assume that the meshed network connectivity is conveniently represented by a sparse structure allowing fast access to the set of buses adjacent to a given bus. The main idea consists of building an auxiliary tree, named a *mother tree*, by a breadth-first search, which contains all the feasible paths for the network under study. The system shown in Figure 12.4, whose *mother tree* is presented in Figure 12.5, will be used to illustrate this concept.

Every node $N_L$ in the *mother tree* corresponds to a possible path for the related bus $L$. In this case, according to Table 12.9, the four-bus system will be translated into a *mother tree* with 8 nodes, assuming paths $\pi_3^A$ and $\pi_3^B$ are discarded. For example, bus $A$ is associated to nodes $1_A$ and $5_A$ in the *mother tree*, corresponding to paths $\pi_1^A$ and $\pi_2^A$ (see Table 12.9).

**TABLE 12.10    Sets $\Pi_b^j$ for the Example of Figure 12.4**

| Branch $j$ | $\Pi_b^j$ |
|---|---|
| 1 | $\Pi_b^1 = \{\pi_1^A, \pi_2^B, \pi_3^B, \pi_1^C, \pi_3^C\}$ |
| 2 | $\Pi_b^2 = \{\pi_2^A, \pi_3^A, \pi_1^B, \pi_2^C, \pi_4^C\}$ |
| 3 | $\Pi_b^3 = \{\pi_2^A, \pi_2^B, \pi_3^C, \pi_4^C\}$ |
| 4 | $\Pi_b^4 = \{\pi_3^A, \pi_3^B, \pi_1^C, \pi_4^C\}$ |
| 5 | $\Pi_b^5 = \{\pi_3^A, \pi_3^B, \pi_2^C, \pi_3^C\}$ |

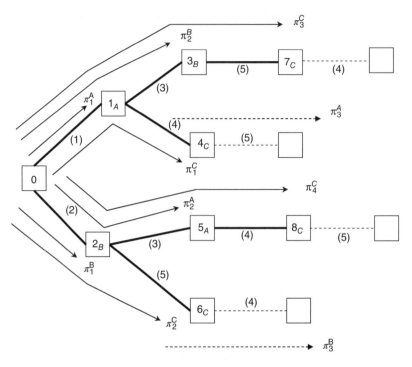

Figure 12.5 *Mother tree* for the example of Figure 12.4.

When building the *mother tree*, the following rules are taken into account.

A. Before a new node $N_L$ is added to the *mother tree*, two conditions are checked:

   (1) A node $N_L$, associated with the same bus $L$, is not located upstream in the tree. Returning to the example, a new node, say $9_A$, is not appended to node $7_C$ through branch (4) because bus $A$ already appears upstream in the tree (node $1_A$). These dead ends are shown in Figure 12.5 by dashed lines.

   (2) The impedance of the total path from the substation to the new node $N_L$ does not exceed a threshold times the impedance of the electrically shortest path for bus $L$. In Figure 12.5, paths $\pi_3^A$ and $\pi_3^B$ of Table 12.9 are not considered for this reason. These cases are represented in Figure 12.5 by dashed arrows.

B. The *mother tree* is only swept two times, first downstream and then upstream. During the downstream sweep, both the *mother tree* and associated paths are obtained simultaneously. When the two rules described above preclude the addition of new nodes, the resulting *mother tree* is swept upstream in order to define the inequality constraints among the paths represented by (2), as well as the minimum and maximum power flows through every branch in the system.

For the radiality and electrical constraints to be easily expressed in the standard matrix–vector form, sets $\Pi_n^i$ and $\Pi_b^j$ are stored as sparse linked lists.

## 12.5.2 Simplified Mathematical Model

The mathematical model of DNRC can be written as follows.

$$\min f = \sum_{l=1}^{NL} R_l \left( \frac{P_l^2 + Q_l^2}{V_l^2} \right) \quad l \in NL \tag{12.18}$$

such that

$$\sum_{j \to i} P_{Gj} + \sum_{l \to i} P_l + \sum_{k \to i} P_{Dk} = 0 \tag{12.19}$$

$$\sum_{j \to i} Q_{Gj} + \sum_{l \to i} Q_l + \sum_{k \to i} Q_{Dk} = 0 \tag{12.20}$$

$$P_l^2 + Q_l^2 \le S_{lmax}^2 \tag{12.21}$$

$$\Delta V_l \le \Delta V_{lmax} \tag{12.22}$$

$$\sum_{p \in \Pi_n^i} K_p^i = 1, \quad \forall \ \text{node} \ i. \tag{12.23}$$

$$K_p^i \le K_{l}^j, \quad \forall \pi_l^j \subset \pi_p^i \tag{12.24}$$

If voltage magnitudes are assumed to be 1.0 p.u. the objective function becomes

$$\min f = \sum_{l=1}^{NL} R_l (P_l^2 + Q_l^2) \quad l \in NL \tag{12.25}$$

The power flow $P_l$ and $Q_l$ comprise the total real and reactive load demanded downstream from node $j$ plus the real and reactive losses of the respective branches. For simplification, the latter component of $P_l$ and $Q_l$ are omitted as system losses are much smaller than power loads. Therefore, the real and reactive power flows equal the sum of real and reactive power loads located downstream from the node, that is,

$$P_l = \sum_{p \in \Pi_{NL}^l} K_p^i P_{Di} \tag{12.26}$$

$$Q_l = \sum_{p \in \Pi_{NL}^l} K_p^i Q_{Di} \tag{12.27}$$

These are equivalent to the node real and reactive balance equations without considering the branch loss.

Substituting them into objective function, we get

$$\min f = \sum_{l=1}^{NL} R_l \left[ \left( \sum_{p \in \Pi_{NL}^l} K_p^i P_{Di} \right)^2 + \left( \sum_{p \in \Pi_{NL}^l} K_p^i Q_{Di} \right)^2 \right] \quad l \in NL \qquad (12.28)$$

The network connectivity is incorporated through the binary variables $K_p^i$. When this is simplified, the computed power losses will be smaller than the actual losses.

Substituting equations (12.26), (12.27) in equation (12.21), we get

$$\left( \sum_{p \in \Pi_{NL}^l} K_p^i P_{Di} \right)^2 + \left( \sum_{p \in \Pi_{NL}^l} K_p^i Q_{Di} \right)^2 \leq S_{lmax}^2 \quad l \in NL \qquad (12.29)$$

According to [5,20], the voltage drop without considering power losses can be expressed as

$$V_i^2 - V_l^2 \approx 2(R_l P_l + X_l Q_l) \qquad (12.30)$$

Then the total quadratic voltage drop through a path $\pi_p^i$ reaching bus $i$ is approximated by

$$V_s^2 - V_i^2 \approx 2 \sum_{l \in \Pi_p^i} (R_l P_l + X_l Q_l) \qquad (12.31)$$

The voltage constraint can be expressed as

$$2 \sum_{l \in \Pi_p^i} \left[ R_l \left( \sum_{p \in \Pi_{NL}^l} K_p^i P_{Di} \right) + X_l \left( \sum_{p \in \Pi_{NL}^l} K_p^i Q_{Di} \right) \right] \leq \Delta V_{max} \qquad (12.32)$$

### 12.5.3   Mixed-Integer Linear Model

In Section 12.5.2, the DNRC model is simplified, in which the branch losses are ignored and bus complex voltages are removed from the model. Thus load flow calculation is not required during the solution process. However, the resulting optimization problem is still quadratic in the binary variables $K_p^i$ (path statuses). A piecewise linear function is used to replace approximately the quadratic branch power flows. In this way, the DNRC model is converted into a standard mixed-integer linear model:

$$\min f = \sum_{l=1}^{NL} R_l \left[ \left( \sum_{t \in tp} C_p^{p(t)} \sum_{p \in \Pi_{NL}^l} K_p^i P_{Di} \right) + \left( \sum_{t \in tq} C_p^{q(t)} \sum_{p \in \Pi_{NL}^l} K_p^i Q_{Di} \right) \right] \quad l \in NL \quad (12.33)$$

such that

$$P_l^{(t)} = \sum_{p \in \Pi_{NL}^l} K_p^i P_{Di}, \quad t \in tp \tag{12.34}$$

$$Q_l^{(t)} = \sum_{p \in \Pi_{NL}^l} K_p^i Q_{Di}, \quad t \in tq \tag{12.35}$$

$$\left( \sum_{t \in tp} C_p^{p(t)} \sum_{p \in \Pi_{NL}^l} K_p^i P_{Di} \right) + \left( \sum_{t \in tq} C_p^{q(t)} \sum_{p \in \Pi_{NL}^l} K_p^i Q_{Di} \right) \leq S_{l\max}^2 \tag{12.36}$$

$$\begin{cases} 0 \leq P_l^{(1)} \leq \overline{P}_l^{(1)} \\[2mm] 0 \leq P_l^{(2)} \leq (\overline{P}_l^{(2)} - \overline{P}_l^{(1)}) \\[2mm] \quad \cdots \cdots \qquad\qquad\qquad l \in NL \\[2mm] 0 \leq P_l^{(tp)} \leq (\overline{P}_l^{(tp)} - \overline{P}_l^{(tp-1)}) \end{cases} \tag{12.37}$$

$$\begin{cases} 0 \leq Q_l^{(1)} \leq \overline{Q}_l^{(1)} \\[2mm] 0 \leq Q_l^{(2)} \leq (\overline{Q}_l^{(2)} - \overline{Q}_l^{(1)}) \\[2mm] \quad \cdots \cdots \qquad\qquad\qquad l \in NL \\[2mm] 0 \leq Q_l^{(tq)} \leq (\overline{Q}_l^{(tq)} - \overline{Q}_l^{(tq-1)}) \end{cases} \tag{12.38}$$

To reduce the problem size and to speed up the calculation, some additional features are considered.

- As noted earlier, those paths whose electrical length exceeds a certain threshold times the shortest distance to the substation for that node are discarded.
- If the set of paths $\Pi_b^j$ associated with branch $j$ comprises a single element $\pi_l^i$, then the respective flow $P_j$ is constant and equal to $P_{Li}$, provided $W_l^i = 1$.
- If the set of paths $\Pi_n^j$ associated with bus $i$ comprises a single element, $\pi_l^i$, then $W_l^i = 1$.

After the final reconfiguration is obtained by solving the mixed-integer linear model of DNRC, the exact losses as well as node voltage and branch flow may be computed by solving the radial load flow.

## 12.6 APPLICATION OF GA TO DNRC

### 12.6.1 Introduction

Chapter 4 discussed the application of GAs to the economic dispatch problem. GAs are considered when conventional techniques have not achieved the desired speed, accuracy, or efficiency [24–26].

The basic steps of general GA are as follows.

(**1**) Initialization

For the given control variables $X$, randomly select a variable population $\{X_0^1, X_0^2, \ldots, X_0^p\}$, where each individual $X_0^i$ is represented by a binary code string. Each string consists of some binary codes and each code is either 0 or 1. Then each individual corresponds to a fitness $f(X_0^i)$, and the population corresponds to a set of fitness $\{f(X_0^1), f(X_0^2), \ldots, f(X_0^p)\}$. Let generation be zero (i.e. $k = 0$) go to the next step.

(**2**) Selection

Select a pair of individuals from the population as a parent. Generally, the individual with higher fitness has higher probability of being selected.

(**3**) Crossover

The crossover is an important operation in the GA. The purpose of the crossover is to exchange fully information among individuals. There are many crossover methods such as one-point crossover and multipoint crossover.

(**a**) One-point crossover. Select randomly a truncation point in the parent strings and divide them into two parts. Then exchange the tail parts of the parent strings. The example of a one-point crossover is given in the following.

| Parent generation | | Child generation |
|---|---|---|
| 100110 : 01101 | One point crossover | 100110 : 10000 |
| 111011 : 10000 | → | 111011 : 01101 |

(**b**) Multipoint crossover. Select randomly several truncation points in the parent strings and divide them into several parts. Then exchange some parts of the parent strings. The examples of two- and three-point crossovers are given in the following.

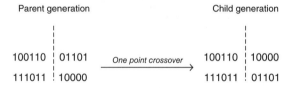

| Parent generation | | Child generation |
|---|---|---|
| 100 : 11001 : 101 | Two points crossover | 100 : 01110 : 101 |
| 111 : 01110 : 000 | → | 111 : 11001 : 000 |

| Parent generation | | | | | Child generation | | | |
|---|---|---|---|---|---|---|---|---|
| 100 | 110 | 01 | 101 | *Three points crossover* → | 100 | 011 | 01 | 000 |
| 111 | 011 | 10 | 000 | | 111 | 110 | 10 | 101 |

**(4)** Mutation

Mutation is another important operation in GA. A good mutation will be kept and a bad mutation will be discarded. Generally, the individual with lower fitness has a greater mutation probability. Similar to crossover, there are one-point mutations and multipoint mutations.

**(a)** One-point mutation. Select randomly a binary code in the parent string and reverse the value of the binary code. The example of one-point mutation is below.

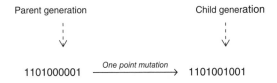

Parent generation → Child generation

1101000001 — *One point mutation* → 1101001001

**(b)** Multi-point mutation. Select randomly several truncation points in the parent strings and divide them into several parts. Then reverse the value of the binary code in some parts. The examples of two- and three-point mutations are given in the following.

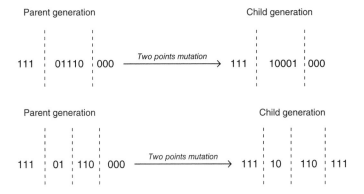

| Parent generation | | | | Child generation | | |
|---|---|---|---|---|---|---|
| 111 | 01110 | 000 | *Two points mutation* → | 111 | 10001 | 000 |

| Parent generation | | | | | Child generation | | | |
|---|---|---|---|---|---|---|---|---|
| 111 | 01 | 110 | 000 | *Two points mutation* → | 111 | 10 | 110 | 111 |

**(5)** Through steps 2–4, a new generation population is produced. Replace the parent generation with the new population and discard some bad individuals. In this way, the new parent population is formed. The calculation will be stopped if the convergence condition is satisfied. Otherwise, go back to step 2.

## 12.6.2   Refined GA Approach to DNRC Problem

GA has shown to be an effective and useful approach for the DNRC problem [1,18]. Some refinements of the approach are described in this section.

***Genetic String***   In the early application of GA to DNRC, the string structure is expressed by "arc no. ($i$)" and "SW no. ($i$)" for each switch $i$. The "arc no. ($i$)" identifies the arc (branch) number which contains the $i$th open switch, and "SW. no. ($i$)" identifies the switch which is normally open on arc no.($i$). For large distribution networks, it is not efficient to represent every arc in the string, as its length will be very long. In fact, the number of open switch positions is identical to keep the system radial once the topology of the distribution networks is fixed, even if the open switch positions are changed. Therefore, to memorize the radial configuration, it is enough to number only the open switch positions. Figure 12.6 shows a simple distribution network with five switches that are normally open.

In Figure 12.6 (a), positions of the five initially open switches 5, 8, 10, 13, and 14 determine a radial topology. In Figure 12.6 (b), positions of the five initially open switches 1, 4, 7, 9, and 10 determine another radial topology. Therefore, to represent a network topology, only positions of the open switches in the distribution network need to be known. Suppose the number of normally open switches is $N_o$, the length of a genetic string depends on the number of open switches $N_o$. Genetic strings for Figure 12.6 (a) and (b) are represented as follows:

| 0 1 0 1 | 1 0 0 0 | 1 0 1 0 | 1 1 0 1 | 1 1 1 0 |
|---------|---------|---------|---------|---------|

switch 5;   switch 8;   switch 10;   switch 13;   switch 14

Genetic string for Figure 12.6 (a)

| 0 0 0 1 | 0 1 0 0 | 0 1 1 1 | 1 0 0 1 | 1 0 1 0 |
|---------|---------|---------|---------|---------|

switch 1;   switch 4;   switch 7;   switch 9;   switch 10

Genetic string for Figure 12.6 (b)

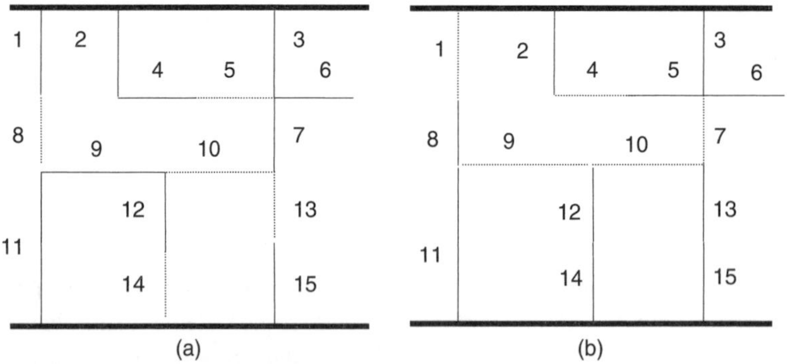

Figure 12.6   A simple distribution network.

***Fitness Function***   GAs are essentially unconstrained search procedures within a given represented space. Therefore, it is very important to construct an accurate fitness function as its value is the only information available to guide the search. In this section, the fitness function is formed by combining the object function and the penalty function, that is,

$$\max f = 1/L \tag{12.39}$$

where

$$L = \sum_i |I_i|^2 k_i R_i + \beta_1 \max\{0, (|I_i| - I_{i\max})^2\}$$

$$+ \beta_2 \max\{0, (V_{i\min} - V_i)^2\}$$

$$+ \beta_3 \max\{0, (V_i - V_{i\max})^2\} \tag{12.40}$$

where $\beta_i$ ($i = 1, 2, 3$) is a large constant.

Suppose $m$ is the population size, the values of the maximum fitness, the minimum fitness, sum of fitness, and average fitness of a generation are calculated as follows.

$$f_{\max} = \{f_i / f_i \geq f_j \quad \forall f_j, \quad j = 1, \ldots \ldots, m\} \tag{12.41}$$

$$f_{\min} = \{f_i / f_i \leq f_j \quad \forall f_j, \quad j = 1, \ldots \ldots, m\} \tag{12.42}$$

$$f_\Sigma = \sum_i f_i, \quad i = 1, \ldots \ldots, m \tag{12.43}$$

$$f_{\mathrm{av}} = f_\Sigma / m \tag{12.44}$$

The strings are sorted according to their fitness and are then ranked accordingly.

***Selection***   To obtain and maintain good performance of the fittest individuals, it is important to keep the selection competitive enough. It is no doubt that the fittest individuals have higher chances of being selected. In this chapter, the "roulette wheel selection" scheme is used, in which each string occupies an area of the wheel that is equal to the string's share of the total fitness, that is, $f_i / f_\Sigma$ .

***Crossover and Mutation***   Crossover takes random pairs from the mating pool and produces two new strings, each being made of one part of the parent string. Mutation provides a way to introduce new information into the knowledge base. With this operator, individual genetic representations are changed according to some probabilistic rules. In general, the GA mutation probability is fixed throughout the whole search process. However, in practical application of DNRC, a small fixed mutation probability can only result in a premature convergence. Here, an adaptive mutation process is used to change the mutation probability, that is,

$$p(k+1) = \begin{cases} p(k), & \text{if } f_{min}(k) \text{ unchanged} \\ p(k) - p_{step}, & \text{if } f_{min}(k) \text{ decreased} \\ p_{final}, & \text{if } p(k) - p_{step} < p_{final} \end{cases} \quad (12.45)$$

$$p(0) = p_{init} = 1.0 \quad (12.46)$$

$$p_{step} = 0.001 \quad (12.47)$$

$$p_{final} = 0.05 \quad (12.48)$$

where $k$ is the generation number; and $p$ is the mutation probability.

The mutation scale will decrease as the process continues. The minimum mutation probability in this study is given as 0.05. This adaptive mutation not only prevents premature convergence but also leads to a smooth convergence.

### 12.6.3 Numerical Examples

The modified GA approach for DNRC is tested on the 16-bus and 33-bus distribution systems. System data and parameters of the 16-bus system are listed in Table 12.11. The 16-bus test system, which is shown in Figure 12.7, contains three source transformers and 13 load nodes. The three initially open switches are "4," "11," and "13." The total system load is 23.7 MW, while the initial system power loss is 0.5114 MW. The 33-bus test system consists of one source transformer and 32 load points.

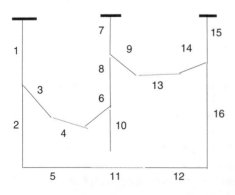

Source transformer busbars

Closed switches

Open switches

Sink nodes (load nodes)

Figure 12.7   A 16-bus distribution system.

**TABLE 12.11 System Data and Parameters for 16-bus Distribution Network**

| Line No. | Node $i$ | Node $j$ | Resistance $R(\Omega)$ | Reactance $X(\Omega)$ | Receiving Node $j$ $P$ (MW) | $Q$ (MVAr) | Receiving Node $j$ Voltage (p.u.) |
|---|---|---|---|---|---|---|---|
| 1 | 1 | 4 | 0.0750 | 0.1000 | 2.0 | 1.6 | $0.9907\angle-.3968$ |
| 3 | 4 | 5 | 0.0800 | 0.1100 | 3.0 | 0.4 | $0.9878\angle-.5443$ |
| 2 | 4 | 6 | 0.0900 | 0.1800 | 2.0 | −0.4 | $0.9860\angle-.6972$ |
| 5 | 6 | 7 | 0.0400 | 0.0400 | 1.5 | 1.2 | $0.9849\angle-.7043$ |
| 7 | 2 | 8 | 0.1100 | 0.1100 | 4.0 | 2.7 | $0.9791\angle-.7635$ |
| 8 | 8 | 9 | 0.0800 | 0.1100 | 5.0 | 1.8 | $0.9711\angle-1.452$ |
| 9 | 8 | 10 | 0.1100 | 0.1100 | 1.0 | 0.9 | $0.9769\angle-.7701$ |
| 6 | 9 | 11 | 0.1100 | 0.1100 | 0.6 | −0.5 | $0.9710\angle-1.526$ |
| 10 | 9 | 12 | 0.0800 | 0.1100 | 4.5 | −1.7 | $0.9693\angle-1.837$ |
| 15 | 3 | 13 | 0.1100 | 0.1100 | 1.0 | 0.9 | $0.9944\angle-.3293$ |
| 14 | 13 | 14 | 0.0900 | 0.1200 | 1.0 | −1.1 | $0.9948\angle-.4562$ |
| 16 | 13 | 15 | 0.0800 | 0.1100 | 1.0 | 0.9 | $0.9918\angle-.5228$ |
| 12 | 15 | 16 | 0.0400 | 0.0400 | 2.1 | −0.8 | $0.9913\angle-.5904$ |
| 4 | 5 | 11 | 0.0400 | 0.0400 | | | |
| 13 | 10 | 14 | 0.0400 | 0.0400 | | | |
| 11 | 7 | 16 | 0.0900 | 0.1200 | | | |

**TABLE 12.12  DNRC Results for 16-Bus Test System**

| Radial Network | Initial Network | Refined GA Method |
|---|---|---|
| Open switches | Switch 4 | Switch 6 |
| | Switch 11 | Switch 9 |
| | Switch 13 | Switch 11 |
| Power loss (MW) | 0.5114 | 0.4661 |

**TABLE 12.13  Comparison of DNRC Results for 33-Bus Test System**

| Radial Network | Initial Network | Method in Ref. [8] | Refined GA Method |
|---|---|---|---|
| Open switches | Switch 33 | Switch 7 | Switch 7 |
| | Switch 34 | Switch 10 | Switch 9 |
| | Switch 35 | Switch 14 | Switch 14 |
| | Switch 36 | Switch 33 | Switch 32 |
| | Switch 37 | Switch 37 | Switch 33 |
| Power loss (MW) | 0.202674 | 0.141541 | 0.139532 |

The five initially open switches are "33", "34," "35," "36," and "37." The total system load is 3.715 MW, while the initial system power loss is 0.202674 MW. The system base is $V = 12.66$ kV and $S = 10$ MVA.

Results on the two systems are listed in Tables 12.12 and 12.13. By comparing results with reference [8], it can be seen that global optima have been found by the refined GA.

# 12.7  MULTIOBJECTIVE EVOLUTION PROGRAMMING TO DNRC

Reducing the real power loss is the primary aim of network reconfiguration. Thus power loss is generally selected as the objective function of DNRC. If we handle some power and voltage constraints as objective functions, the DNRC will become a constrained multiobjective optimization problem.

## 12.7.1  Multiobjective Optimization Model

Three objective functions are considered here; they are minimization of power losses, minimizing the deviation of node voltage, and maximizing the branch capacity margin, which are expressed as follows.

**1.** *Minimization of power losses*

$$\min f_1 = \sum_{l=1}^{NL} k_l R_l \left( \frac{P_l^2 + Q_l^2}{V_l^2} \right) \quad l \in NL \tag{12.49}$$

**2.** *Minimizing the deviation of node voltages*

$$\min f_2 = \min|V_i - V_{irate}| \quad i \in N \tag{12.50}$$

where

$V_{irate}$: the rated voltage at node $i$.

$f_2$: the maximal deviation of node voltage in the network.

Obviously, lower $f_2$ values indicate a higher quality voltage profile and better security of the considered network configuration.

**3.** *Branch capacity margin*

$$\min f_3 = 1 - \max_l \left[ \frac{S_{lmax} - S_l}{S_{lmax}} \right] \quad l \in NL \tag{12.51}$$

where

$S_{lmax}$: the megavolt amperes (MVA) capacity of the branch $l$

$S_l$: the actual megavolt amperes (MVA) loading of the branch $l$

$f_3$: the relative value of the margin between the capacity and the actual megavolt amperes (MVA) loading of the branch.

Obviously, a lower $f_3$ indicates a greater MVA reserve in the branches, implying that the considered network configuration is more secure.

Since the node voltages and branch flows are reflected in the objective functions, the corresponding constraints are omitted. The remaining constraints will be governed by KCL and KVL laws, as well as the network topological constraints in equation (12.6).

## 12.7.2  EP-Based Multiobjective Optimization Approach

***Multiobjective Optimization Algorithm [28,29]***  The aforementioned multiobjective DNRC problem can be expressed in the following form:

$$\min f_i(x), \quad i \in N_o \tag{12.52}$$

subject to

$$g(x) = 0 \tag{12.53}$$

$$h(x) \leq 0 \tag{12.54}$$

where $N_o$ is number of objective functions, and $x$ is the decision vector.

These three objective functions compete with each other, no point $X$ simultaneously minimizes all of the objective functions. This multiobjective optimization problem can be solved using the concept of noninferiority.

***Definition*** The *feasible region of the constraints*, $\Omega$, in the decision vector space $X$ is the set of all decision vectors $x$ that satisfy the constraints, such that

$$\Omega = \{x | g(x) = 0, \; h \leq (x) = 0\} \tag{12.55}$$

The *feasible region of objective functions*, $\psi$, in the objective function space $F$ is the image of $f$ of the feasible region $\Omega$ in the decision vector space

$$\psi = \{f | f = f(x), \; x \in \Omega\} \tag{12.56}$$

A point $\hat{x} \in \Omega$ is a *local noninferior point* if, and only if, for some neighborhood of $\hat{x}$, there does not exist a $\Delta x$ such that

$$\hat{x} + \Delta x \in \Omega \tag{12.57}$$

and

$$f_i(x + \Delta x) \leq f_i(\hat{x}), \quad i = 1, 2, \ldots, N_o \tag{12.58}$$

$$f_j(x + \Delta x) < f_j(\hat{x}), \quad \text{for some } j \in N_o \tag{12.59}$$

A point $\hat{x} \in \Omega$ is a *global noninferior point* if and only if no other point $x \in \Omega$ exists there such that

$$f_i(x) \leq f_i(\hat{x}), \quad i = 1, 2, \ldots, N_o \tag{12.60}$$

$$f_j(x) < f_j(\hat{x}), \quad \text{for some } j \in N_o \tag{12.61}$$

Thus a global noninferior solution of the multiobjective problem is one where any improvement of one objective function can be achieved only at the expense of at least one of the other objectives. Typically, an infinite number of noninferior points exist in a given multiobjective problem. A noninferior point is the same as the intuitive notion of an optimum trade-off solution, as a design is noninferior if it improves an objective that requires degradation in at least one of the other objectives. Clearly, if a decision-maker were able, he or she would not want to choose an inferior design.

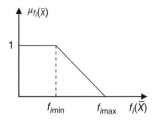

Figure 12.8   Fuzzy membership model.

Thus the decision-maker attempts to generate noninferior solutions to a multiobjective problem when trying to obtain a final design.

The decision-maker combines subjective judgment with the quantitative analysis, as the noninferior optimal solutions generally consist of an infinite number of points. This section introduces the interactive fuzzy satisfying algorithm based on EP to determine the optimal noninferior solution for the decision-maker.

### EP Algorithm with Fuzzy Objective Functions

*Fuzzy Objective Function*   A fuzzy set is typically represented by a membership function. A higher membership function implies greater satisfaction with the solution. One of the typical membership functions is triangle, which is shown in Figure 12.8.

Here, we use triangle model for representing fuzzy objective functions. The triangle membership function consists of lower and upper boundaries, together with a strictly monotonically decreasing membership function, which can be expressed as follows.

$$\mu_{f_i(\overline{X})} = \begin{cases} 1, & \text{if } f_i \leq f_{i\min} \\ \dfrac{f_{i\max} - f_i}{f_{i\max} - f_{i\min}}, & \text{if } f_{i\min} \leq f_i \leq f_{i\max} \\ 0, & \text{if } f_i \geq f_{i\max} \end{cases} \qquad (12.62)$$

*Evolution Programming [21]*   The state variable $\overline{X}$ represents a chromosome of which each gene represents an open switch to the network reconfiguration problem. The fitness function of $\overline{X}$ can be defined as

$$C(\overline{X}) = \frac{1}{1 + F(\overline{X})} \qquad (12.63)$$

where

$$F(\overline{X}) = \min_{X \in \Omega} \left\{ \max_{i=1,2,\dots N_o} \left[ \overline{\mu_{f_i}} - \mu_{f_i}(\overline{X}) \right] \right\} \qquad (12.64)$$

$\overline{\mu_{f_i}}$: the expected values of objective function
$\mu_{f_i(\overline{X})}$: the actual values of objective function
$C(\overline{X})$: the fitness function.

The function $F(\overline{X})$ is to minimize the objective with a maximum distance away from its expected value among the multiple objective functions. For a given $\overline{\mu_{f_i}}$, the solution reaches the optimum as the fitness value increases.

The steps of EP are detailed as follows.

Step 1:   Input parameters.
Input the parameters of EP, such as the length of the individual and the population size $N_P$.

Step 2:   Initialization.
The initial population is determined by selecting $P_j$ from the set of the original switches and their derivatives according to the mutation rules. $P_j$ is an individual, $j = 1, 2, \ldots, N_P$, with $N_S$ dimensions, where $N_S$ is the total number of switches.

Step 3:   Scoring.
Calculate the fitness value of an individual by equations (12.63) and (12.64).

Step 4:   Mutation.
In the network reconfiguration problem, the radial structure must be retained for each new structure and power must be supplied to each loading demand. Consequently, each $P_j$ is mutated and assigned to $P_{j+N_P}$. The number of offspring $n_j$ for each individual $P_j$ is given by

$$n_j = G\left(N_P \times \frac{C_j}{\sum_{j=1}^{N} C_j}\right) \qquad (12.65)$$

Where $G(x)$ is a function that rounds the element of $x$ to the nearest integer number. More offspring are generated from the individual with a greater fitness. A combined population is formed from the old generation and the new generation is mutated from the old generation.

Step 5:   Competition.
Each individual $P_j$ in the combined population has to compete with some other individuals to have the opportunity to be transcribed to the next generation. All individuals of the combined population are ranked in descending order of their corresponding fitness values. Then, the first $N_P$ individuals are transcribed to the next generation.

Step 6:   Stop criterion.
Convergence is achieved when either the number of generations reaches the maximum number of generations or the sampled mean fitness function values do not change noticeably throughout several consecutive generations. The process will stop if one of these conditions is met, otherwise returns to the mutation step.

**Optimization Approach**   For using the fuzzy objective function, the values of expected membership functions will be selected to generate a candidate solution of the multiobjective problem. The expected value is a real number in [0, 1], and represents the importance of each objective function. The afore mentioned min–max problem is solved to generate the optimal solution. The optimization technique can now be described as follows.

Step (1)    Input the data and set the interactive pointer $p = 0$.

Step (2)    Determine the upper and lower bounds for every objective function $f_{i\max}$ and $f_{i\min}$, as well as fuzzy membership $\mu_{f_i(X)}$.

Step (3)    Set the initial expected value of each objective function $\overline{\mu_{f_i(0)}}$ for $i = 1, 2, \ldots, N_o$.

Step (4)    Apply EP to solve the min–max problem (12.64).

Step (5)    Calculate the values of $\overline{X}$, $f_i(\overline{X})$, and $\mu_{f_i(\overline{X})}$. Go to the next step if they are satisfactory. Otherwise, set the interactive pointer $p = p + 1$ and choose a new expected value $\overline{\mu_{f_i(p)}}$, $i = 1, 2, \ldots, N_o$, Then go to step 4.

Step (6)    Output the most satisfactory feasible solution $X^*$, $f_i(X^*)$, and $\mu_{f_i(X^*)}$.

# 12.8    GENETIC ALGORITHM BASED ON MATROID THEORY

Section 12.5 analyzed the application of GAs to solve the DNRC problem in equations (12.1)–(12.6). To accelerate the GA convergence, a GA based on network matroid theory [30] is used to solve the same DNRC problem in this section.

## 12.8.1    Network Topology Coding Method

The distribution network topology coding method is fundamental for GA convergence. On the one hand, a complex strategy could increase considerably the convergence time. On the other hand, a simple strategy does not allow an effective exploration of the research field. Various coding strategies are detailed in this section and the GA operator's mechanisms are explained. Finally, their advantages and drawbacks are discussed.

***Different Topology Coding Strategies***    The most simple topology representation for the GA is to consider a topology string formed by the binary status (closed/open) of each network branch [31] or at least each network switch. In [18] the arc (a branch or a series of branches) number and the switch position in each branch are considered for the radial topology representation. In [1,32], only the positions of open switches are stored in the topology string.

Reference [17] proposes an efficient modeling method for the distribution networks connectivity. The path (a set of branches to the source) is determined for each node of the network. For a radial configuration, only a path to the source $S$ is considered for each node. This method is discussed in Section 12.5. For example, in the

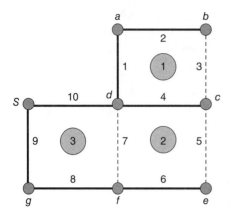

Figure 12.9   A simple meshed topology.

simple topology in Figure 12.9, paths from each node to the source $S$ are

$a$:  $\pi_1^a = [1, 10]$,  $\pi_2^a = [2, 3, 4, 10]$,  $\pi_3^a = [7, 8, 9]$

  $\pi_4^a = [2, 3, 5, 6, 8, 9]$,  $\pi_5^a = [2, 3, 5, 6, 7, 10]$,  $\pi_6^a = [2, 3, 4, 7, 8, 9]$

$b$:  $\pi_1^b = [3, 4, 10]$,  $\pi_2^b = [2, 1, 10]$,  $\pi_3^b = [3, 4, 6, 8, 9]$

  $\pi_4^b = [3, 5, 6, 7, 10]$,  $\pi_5^b = [2, 1, 7, 8, 9]$,  $\pi_6^b = [3, 4, 7, 8, 9]$

$g$:  $\pi_1^g = [8, 7, 10]$,  $\pi_2^g = [9]$,  $\pi_3^g = [8, 6, 5, 4, 10]$,  $\pi_4^g = [8, 6, 5, 3, 2, 1, 10]$

As mentioned in Section 12.5, $\pi_i^j$ is the path number $i$ between node $j$ and source $S$.

The general structure of the topology string for the simple topology Figure 12.9 can be handled as follows: for node $a$, only one of four paths is represented by the bit 1, the rest are represented by the bit 0. The same procedure is used for the other nodes.

***The GA Operators***   As we discussed above, the GA operators are mutation, selection, and crossover. The crossover is the most important operator of the GA. The traditional crossover process randomly selects two parents (chromosomes) for a gene exchange with a given crossover rate. This operator aims at mixing up genetic information coming from the two parents, to create new individuals.

The coding diagram is very important for the success of the crossover operation. A binary coding method cannot allow a high efficiency of the crossover process. Furthermore, *mesh checks* have to be performed in order to validate each resulting topology (to detect any loop in the network or any non-energized node).

The mutation operator can allow GA to avoid local optima. This operator randomly changes one gene in the string, and is applied with a probability that has been set in the initial phase. As in the crossover process, the topology coding strategy is very important for a fast and effective mutation operation.

## 12.8.2   GA with Matroid Theory

The reconfiguration problem tries to find out the optimal spanning tree among all the spanning trees of the DN graph for a given objective. In the first part of this section, an interesting property of the graph-spanning trees is discussed. In the second part, it is shown that this can be generalized using some properties proved for the matroid theory. The GA operators are then explained on the basis of this new theoretical approach.

***The Kruskal Lemma for the Graph-Spanning Trees***   For the graphs, the spanning trees exchange property has been proved by Kruskal [33]:

*Let U and T be two spanning trees of the graph G, let $a \in U$, $a \notin T$, then there exists $b \in T$, such that $U - a + b$ is also a spanning tree in the graph G.*

For the graph represented in the Figure 12.9, two spanning trees are drawn in Figure 12.10. Consider the edge $a = 6(a \in T)$ in the $U$ spanning tree. One edge $b$ that replaces $a = 6$ in $T$ in order to form another spanning tree can be found. Edge $b$ can be selected in the loop formed by $T \cup a(= 6)$. In Figure 12.10, this loop is formed by branches 4, 5, 6, and 7 (dotted arrow). Only edges 5 and 7 can replace edge 6 in $U$.

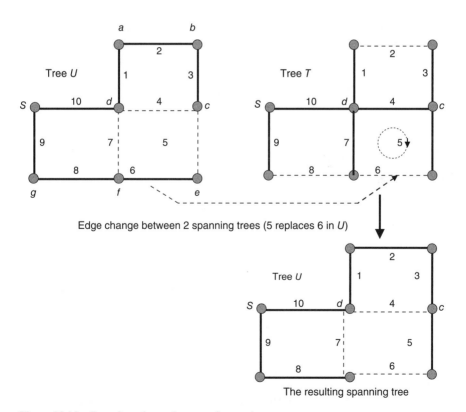

Figure 12.10   Branch exchange between 2 spanning trees.

Finally, edge 5 is chosen to replace edge 6 in $U$ and a new spanning tree is obtained (see the resulting tree in Figure 12.10).

The matroid theory abstracts the important characteristics of matrix theory and graph theory. A matroid is defined by axioms of independent sets [34].

A pair $(S, \ T)$ is called a matroid if $S$ is a finite set and $T$ is a nonempty collection of subsets of $S$:

if $I \in T$ and $J \subseteq I$ then $J \in T$,

if $I, \ J \in T$ and $I \leq J$, then $I + z \in T$ for some $z \in J \backslash I$.

The *base* concept has to be introduced. For $U \subseteq S$, a subset $B$ of $U$ is called a *base* of $U$ if $B$ is an inclusionwise maximal independent subset of $U$ [34], that is, $B \in T$ and there is no $Z \in T$ with $B \subset Z \subseteq U$. A subset of $S$ is called

*spanning* if it contains a *base* like a subset, so *bases* are just the inclusionwise minimal and independent spanning sets.

One of the matroid classes is the *graphic matroids*. Let $G = (V, E)$ be a graph (with $V$ the vertices set and $E$ the edges set). Let $T$ be the collection of all subsets of $E$ that form a *forest* (a graph in which any two vertices are connected by only one path), then $M = (E, T)$ is a matroid. The matroid $M$ is called the *cycle matroid* of graph $G$, denoted $M(G)$. The *bases* of $M(G)$ are exactly the inclusionwise maximal forests of $G$. So if the graph $G$ is connected, the *bases* are spanning trees (the forest equivalent for a connected graph or radial configurations for a DN).

In order to link these theoretical aspects with the problem of spanning trees (radial topologies), the exchange property of *bases*, given in [34], is considered.

Let $M = (S, T)$ be a matroid. Let $B_1$ and $B_2$ be bases and let $x \in B_1 \backslash B_2$. Then there exists an element $y \in B_2 \backslash B_1$ such that both $B_1 - x + y$ and $B_2 - y + x$ are *bases*.

***Application to the DN Topology Modeling for GA***   According to graph theory, the group (*the set of graph edges, the collection of all spanning trees*) is a matroid. The spanning tree of a graph is a *base*. A branch exchange between two spanning trees of the same graph is always possible. New spanning trees for the same graph are obtained.

Furthermore, on the basis of the matroid theory approach, not just one spanning tree is obtained (as shown in Figure 12.10), but two spanning trees are obtained. Moreover, in order to find easily which edge in a spanning tree can replace another in the other spanning tree, the loop formed by adding the edge to the other spanning tree has to be determined.

From an electrical point of view, the branch exchange between two spanning trees can be seen as a load transfer between two supply points or between two paths to the same supply point.

The matroid approach allows the use of GA operators without checking the DN graph planarity. Besides, on the basis of this approach, the GA operator success is always guaranteed, without a supplementary *mesh check* and extra computation time. An example is given in the next subsection.

**GA Operators Based on the Matroid Approach**    The examples for the mutation and crossover operators and initial population are given using the matroid approach [30] and applied to the graph illustrated in Figure 12.9.

*Crossover*    The crossover operator represents a gene exchange between two chromosomes. One or multiple crossover points can be randomly chosen. For the reconfiguration problem, this operation means one or several edges are exchanged between two spanning trees for a given DN graph.

In Figure 12.11, the first step for a crossover operation is shown between two chromosomes. Each chromosome represents two spanning trees for the graph illustrated in Figure 12.9. Only the open branches are considered here. In the graph theory this is called the *co-tree* concept (the branches missing from the tree). The theoretical approach given in the previous paragraph can be reformulated for the *co-trees*: a bidirectional branch exchange can be performed in order to obtain new *co-trees*. A crossover point is randomly chosen between the first and the second gene of the upper *co-tree* (see Figure 12.11). In the corresponding *co-tree* represented in [30], the genes (branches) 7 and 5 have to be exchanged with branches in the second *co-tree*.

Firstly, branch 7 is replaced. In order to identify rapidly what are the branches of the second *co-tree* that could replace the branch 7, the loop formed by closing

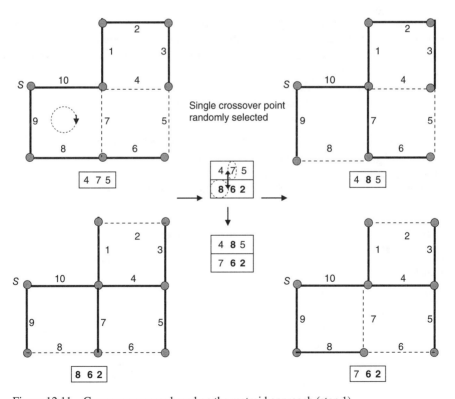

Figure 12.11    Crossover process based on the matroid approach (step 1).

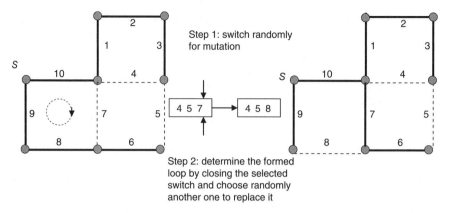

Figure 12.12   Mutation process based on the matroid approach.

branch 7 in the upper *tree* is determined (see the dotted arrow in Figure 12.11). For this purpose, a *depth-first graph search* algorithm was used [34]. This loop is formed by branches 7, 8, 9, and 10. Only the branch 8 is in the lower *co-tree* and it can then replace branch 7. The same procedure is employed in the second step.

*Mutation*   The mutation process is shown in Figure 12.12. After random selection of one (or multiple) branches in the chosen *co-tree* to be mutated, the corresponding loop is determined with a *depth-first graph search* algorithm (see the interrupted arrow in Figure 12.12).

A new branch is randomly chosen in this loop, in order to replace the one first selected. No other test is necessary in order to validate the new radial configuration.

*Initial Population Generation*   Even if this step is performed once in the GA process, the random creation of the initial population can be time consuming. The initial population is generated using the mutation process shown in Figure 12.12. An initial feasible chromosome (*co-tree*) is randomly generated. The mutation process

**TABLE 12.14   The DNRC Results by GA Based on Matroid Theory and Comparison for 33-Bus Test System**

| Radial Network | Initial Network | Branch Exchange Method | Refined GA Method | GA Based on Matroid Theory |
|---|---|---|---|---|
| Open switches | Switch 33 | Switch 7 | Switch 7 | Switch 7 |
| | Switch 34 | Switch 10 | Switch 9 | Switch 9 |
| | Switch 35 | Switch 14 | Switch 14 | Switch 14 |
| | Switch 36 | Switch 33 | Switch 32 | Switch 32 |
| | Switch 37 | Switch 37 | Switch 33 | Switch 37 |
| Power loss (MW) | 0.202674 | 0.141541 | 0.139532 | 0.136420 |

is used to randomly change each initial *co-tree* branch. The new chromosome feasibility is also implicitly guaranteed. The process is progressively repeated in order to perform the initial population.

The same 33-bus system used in Section 12.4 is adopted for DNRC test. The results and comparison are listed in Table 12.14, where the results based on refined GA and matroid theory–based GA are better than those based on the branch exchange method.

# APPENDIX A: EVOLUTIONARY ALGORITHM OF MULTIOBJECTIVE OPTIMIZATION

In power system optimization, some objective functions are noncommensurable and often competing objectives. Multiobjective optimization with such objective functions gives rise to a set of optimal solutions, instead of one optimal solution. The reason for the optimality of many solutions is that no one can be considered to be better than any other with respect to all objective functions. These optimal solutions are known as *Pareto-optimal* solutions.

A general multiobjective optimization problem consists of a number of objectives to be optimized simultaneously and is associated with a number of equality and inequality constraints. It can be formulated as follows:

$$\text{Minimize } f_i(x) \quad i = 1, \dots, N_{\text{obj}} \tag{12A.1}$$

subject to

$$g_j(x, u) = 0 \quad j = 1, \dots, M \tag{12A.2}$$

$$h_k(x, u) \leq 0 \quad k = 1, \dots, K \tag{12A.3}$$

where $f_i$ is the $i$th objective functions, $x$ is a decision vector that represents a solution, and $N_{\text{obj}}$ is the number of objectives.

For a multiobjective optimization problem, any two solutions $x^1$ and $x^2$ can have one of two possibilities: one covers or dominates the other or none dominates the other. In a minimization problem, without loss of generality, a solution $x^1$ dominates $x^2$ if the following two conditions are satisfied [35]:

1. $\forall i \in \{1, 2, \dots, N_{\text{obj}}\} : f_i(x^1) \leq f_i(x^2)$

2. $\exists j \in \{1, 2, \dots, N_{\text{obj}}\} : f_j(x^1) < f_j(x^2)$

If any of the above conditions is violated, the solution $x^1$ does not dominate the solution $x^2$. The solutions that are nondominated within the entire search space are denoted as *Pareto-optimal* and constitute the *Pareto-optimal set* or *Pareto-optimal front*.

There are some difficulties for the classic methods to solve such multiobjective optimization problems:

- An algorithm has to be applied many times to find multiple Pareto-optimal solutions.
- Most algorithms demand some knowledge about the problem being solved.
- Some algorithms are sensitive to the shape of the Pareto-optimal front.
- The spread of Pareto-optimal solutions depends on efficiency of the single objective optimizer.

As we analyzed in the book, AHP can be used to solve the mentioned multiobjective optimization problem. Here, we use another method—the strength Pareto evolutionary algorithm (SPEA) to solve it.

The SPEA-based approach has the following features [36]:

- It stores externally those individuals that represent a nondominated front among all solutions considered so far.
- It uses the concept of Pareto dominance in order to assign scalar fitness values to individuals.
- It performs clustering to reduce the number of individuals externally stored without destroying the characteristics of the trade-off front.

Generally, the algorithm can be described in the following steps.

Step 1    (*Initialization*): Generate an initial population and create an empty external Pareto-optimal set.

Step 2    (*External set updating*): The external Pareto-optimal set is updated as follows.

(a) Search the population for the nondominated individuals and copy them to the external Pareto set.

(b) Search the external Pareto set for the nondominated individuals and remove all dominated solutions from the set.

(c) If the number of the individuals externally stored in the Pareto set exceeds the prespecified maximum size, reduce the set by clustering.

Step 3    (*Fitness assignment*): Calculate the fitness values of individuals in both external Pareto set and the population as follows.

(a) Assign a real value $r \in [0, 1)$ called strength for each individual in the Pareto optimal set. The strength of an individual is proportional to the number of individuals covered by it. The strength of a Pareto solution is at the same time its fitness.

(b) The fitness of each individual in the population is the sum of the strengths of all external Pareto solutions by which it is covered. In order to guarantee that Pareto solutions are most likely to be produced, a small positive number is added to the resulting value.

Step 4  (*Selection*): Combine the population and the external set individuals. Select two individuals at random and compare their fitness. Select the better one and copy it to the mating pool.

Step 5  (*Crossover and Mutation*): Perform the crossover and mutation operations according to their probabilities to generate the new population.

Step 6  (*Termination*): Check for stopping criteria. If any one is satisfied *then* stop *else* copy new population to the old population and go to Step 2. In this study, the search will be stopped if the generation counter exceeds its maximum number.

In some problems, the Pareto optimal set can be extremely large. In this case, reducing the set of nondominated solutions without destroying the characteristics of the trade-off front is desirable from the decision-maker's point of view. An average linkage-based hierarchical clustering algorithm [37] is employed to reduce the Pareto set to manageable size. It works iteratively by joining the adjacent clusters until the required number of groups is obtained. It can be described as follows: given a set $P$ the size of which exceeds the maximum allowable size $N$, it is required to form a subset $P^*$ with size $N$. The algorithm is illustrated in the following steps.

Step 1:  Initialize cluster set $C$; each individual $i \in P$ constitutes a distinct cluster.

Step 2:  If the number of clusters $\leq N$, then go to Step 5, else go to Step 3.

Step 3:  Calculate the distances between all possible pairs of clusters.
The distance $d_c$ between two clusters $c1$ and $c2 \in C$ is given as the average distance between pairs of individuals across the two clusters

$$d_c = \frac{1}{n_1 n_2} \sum_{i_1 \in c_1, i_2 \in c_2} d(i_1, i_2) \tag{12A.4}$$

where $n_1$ and $n_2$ are the number of individuals in the clusters $c_1$ and $c_2$ respectively. The function $d$ reflects the distance in the objective space between individuals $i_1$ and $i_2$.

Step 4:  Determine two clusters with minimal distance $d_c$ between them. Combine them into a larger cluster. Go to Step 2.

Step 5:  Find the centroid of each cluster. Select the nearest individual in this cluster to the centroid as a representative individual and remove all other individuals from the cluster.

Step 6:  Compute the reduced nondominated set $P^*$ by uniting the representatives of the clusters.

Upon having the Pareto-optimal set of the nondominated solution, we can obtain one solution to the decision-maker as the best compromise solution. Owing to imprecise nature of the decision-maker's judgment, the $i$th objective function $F_i$

is represented by a membership function $\mu_i$ defined as

$$
\mu_i = \begin{cases} 1 & F_i \leq F_i^{\min} \\[2mm] \dfrac{F_i^{\max} - F_i}{F_i^{\max} - F_i^{\min}} & F_i^{\min} < F_i < F_i^{\max} \\[2mm] 0 & F_i \geq F_i^{\max} \end{cases} \tag{12A.5}
$$

where $F_i^{\min}$ and $F_i^{\max}$ are the minimum and maximum values of the $i$th objective function among all nondominated solutions.

For each nondominated solution $k$, the normalized membership function $\mu^k$ is calculated as

$$
\mu^k = \frac{\sum_{i=1}^{N_{\text{obj}}} \mu_i^k}{\sum_{k=1}^{M} \sum_{i=1}^{N_{\text{obj}}} \mu_i^k} \tag{12A.6}
$$

where $M$ is the number of nondominated solutions. The best compromise solution is that having the maximum value of $\mu^k$.

The following modifications have been incorporated in the basic SPEA algorithm [35].

(1) A procedure is imposed to check the feasibility of the initial population of individuals and the generated children through GA operations. This ensures the feasibility of Pareto-optimal nondominated solutions.

(2) In every generation, the nondominated solutions in the first front are combined with the existing Pareto-optimal set. The augmented set is processed to extract its nondominated solutions that represent the updated Pareto-optimal set.

(3) A fuzzy-based mechanism is employed to extract the best compromise solution over the trade-off curve.

# PROBLEMS AND EXERCISES

1. State the purpose of distribution network reconfiguration.

2. List several major methods that are used in DNRC.

3. Why do we not use the P–Q decouple power flow or Newton power flow methods to compute the flow of the distribution network?

4. What is the topological constraint in traditional DNRC calculation?

5. Is optimal flow pattern a heuristic algorithm in DNRC? Why?

6. Describe the power summation–based radial distribution network load-flow (PSRDNLF) method

7. Crossover is an important operation in GA. For the given parent strings,

(a) Use the one-point crossover to get the child generation.

(b) Use the two-point crossover to get the child generation.

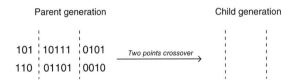

(c) Use the three-point crossover to get the child generation.

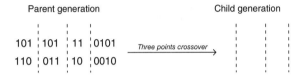

8. Mutation is another important operation in GA. For the given parent string,

   (a) use the one-point mutation to get the child generation.

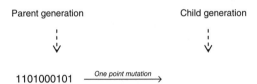

   (b) use the two-point mutation to get the child generation.

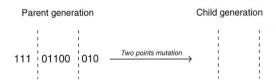

   (c) use the three-point mutation to get the child generation.

9. A 16-bus distribution system is shown in Figure 12.7. Use a GA string to express the initial open switches 4, 11, and 13.

**10.** A simple distribution system is shown in Figure 12.1. The loads are $P_{D2} = 0.6 + j0.3$, $P_{D3} = 0.9 + j0.6$, $P_{D4} = 0.6 + j0.4$, $P_{D5} = 0.4 + j0.2$, $P_{D7} = 0.3 + j0.1$, $P_{D8} = 0.2 + j0.1$; the branch resistances are $R_1 = 0.006$, $R_2 = 0.005$, $R_3 = 0.055$, $R_4 = 0.0045$, $R_5 = 0.003$, $R_6 = 0.0036$, $R_7 = 0.0038$; the voltage at source 1 is 1.05 (all data are p.u.). Compute the flow of this radial network.

# REFERENCES

1. Zhu JZ. Optimal reconfiguration of electrical distribution network using the refined genetic algorithm. Electr. Pow. Syst. Res. 2002;62(1):37–42.
2. Wojciechowski JM. An approach formula for counting trees in a graph. IEEE Trans. Circuits Syst. 1985;32(4):382–385.
3. Merlin A, Back H. Search for minimum-loss operating spanning tree configuration in an urban power distribution system, Proc. 5th Power System Computation Conference; Cambridge, 1975 Paper 1.2/6.
4. Castro CH, Bunchand JB, Topka TM. Generalized algorithms for distribution feeder deployment and sectionalizing. IEEE T. Power Ap. Syst. 1980;99(2):549–557.
5. Baran ME, Wu F. Network Reconfiguration in distribution systems for loss reduction and load balancing. IEEE T Power Deliver. 1989;4(2):1401–1407.
6. Castro CH, Franca ALM. Automatic power distribution reconfiguration algorithm including operating constraints. IFAC Symposium on Planning and Operation of Electric Energy Systems; Rio de Janeiro 1985. p. 181–186.
7. Civanlar S et al. Distribution feeder reconfiguration for loss reduction. IEEE T Power Deliver. 1988;13(3):1217–1223.
8. Shirmohammadi D, Hong HW. Reconfiguration of electric distribution networks for resistive line losses reduction. IEEE Trans. PWRD 1989;4(2):1492–1498.
9. Goswami SK. A new algorithm for the reconfiguration of distribution feeders for loss minimization. IEEE T Power Deliver. 1992;17(3):1484–1491.
10. Nahman J, Strbac G. A New Algorithm for Service Restoration in Large-scale Urban Distribution Systems. Electric Power Systems Research 1994;29:181–192.
11. Glamocanin V. Optimal loss reduction of distribution networks. IEEE Trans. Power Systems 1990;5(3):774–782.
12. Ross DW, Patton J, Cohen AI, Carson M. New methods for evaluating distribution automation and control system benefits. IEEE Trans. PAS 1981;100:2978–2986.
13. Strbac G, Nahman J. Reliability aspects in structuring of large scale urban distribution systems, IEE Conf. on Reliability of Transmission and Distribution Equipment; March 1995. p. 151–156.
14. Liu CC, Lee SJ, Venkata SS. An expert system operational aid for restoration and loss reduction of distribution systems. IEEE Transaction on Power Systems 1988;3(2):619–626.
15. Kendrew TJ, Marks JA. Automated distribution comes of age. IEEE Computer Applications in Power 1989;2(1):7–10.
16. Chiang HD, Jumeau RJ. Optimal network reconfiguration in distribution systems, part 1: a new formulation and a solution methodology. IEEE Trans. Power Delivery 1990;5(4):1902–1909.
17. Ramos ER, Expósito AG, Santos JR, Iborra FL. Path-based distribution network modeling: application to reconfiguration for loss reduction. IEEE Trans. Power Systems 2005;20(2):556–564.
18. Nara K, Shiose A, Kitagawa M, Ishihara T. Implementation of genetic algorithm for distribution system loss minimum reconfiguration. IEEE Trans. Power Systems 1992;7(3):1044–1051.
19. Souza BA, Braz HDM , Alves HN. genetic algorithm for optimal feeders configuration. Proceedings of the Brazilian Symposium of Intelligent Automation—SBAI; Bauru, Brazil; 2003.
20. Souza BA, Alves HN, Ferreira HA. Microgenetic algorithms and fuzzy logic applied to the optimal placement of capacitor banks in distribution networks. IEEE Trans. Power Systems 2004;19(2):942–947.
21. Hsiao YT. Multiobjective evolution programming method for feeder reconfiguration. IEEE Trans. Power Systems 2004;19(1):594–599.

22. Zhu JZ, Xiong XF, Hwang D, Sadjadpour A. A comprehensive method for reconfiguration of electrical distribution network, *IEEE/PES 2007 General Meeting*; Tampa; USA; June 24–28; 2007

23. Zhu JZ. *Application of Network Flow Techniques to Power Systems*. First ed. WA: Tianya Press, Technology; 2005.

24. Zhu JZ, Chang CS, Xu GY, Xiong XF. Optimal load frequency control using genetic algorithm, Proceedings of 1996 International Conference on Electrical Engineering, ICEE'96; Beijing; China; August 12–15; 1996; p. 1103–1107.

25. Holland JH. *Adaptation in Nature and Artificial Systems*. The University of Michigan Press; 1975.

26. Goldberg DE. *Genetic Algorithms in Search, Optimization and Machine Learning*. Reading, MA: Addision-Wesley; 1989.

27. Chua LO, Desoer CA, Kuh ES. *Linear and Nonlinear Circuits*. New York: McGraw-Hill; 1987.

28. Chen JBC, Zhang ZJ. The interactive step trade-off method for multi-objective optimization. IEEE Trans. Syst., Man, Cybern. 1990;SMC-20:688–694.

29. Lin JG. Multiple-objective problems: pareto-optimal solutions by method of proper equality constraints. IEEE Trans. Automat. Contr. 1976;AC-21:641–650.

30. Enacheanu B, Raison B, Caire R, Devaux O, Bienia W, HadjSaid N. Radial network reconfiguration using genetic algorithm based on the matroid theory. IEEE Trans. Power Systems 2007;22.

31. Hsiao YT, Chien CY. Implementation of genetic algorithm for distribution systems loss minimum re-configuration. IEEE Trans. Power Syst. 2000;15(4):1394–1400.

32. Zhu JZ, Chang CS. "Refined genetic algorithm for minimum-loss reconfiguration of electrical distribution network", 1998 Intern. Conf. On EMPD; Singapore; 1998.

33. Kruskal JB Jr. On the shortest spanning subtree of a graph and the traveling salesman problem. Proceedings of the American Mathematical Society 1956;7(1):48–50.

34. Schrijver A. *Combinatorial Optimization—Polyhedra and Efficiency*. Berlin: Springer-Verlag; 2003. p 651–671.

35. Abido MA. "Multiobjective Optimal VAR Dispatch Using Strength Pareto Evolutionary Algorithm," 2006 IEEE Congress on Evolutionary Computation, Sheraton Vancouver; Wall Centre Hotel; Vancouver, BC, Canada; July 16–21, 2006

36. Zitzler E, Thiele L. "An evolutionary algorithm for multiobjective optimization: the strength pareto approach," Swiss Federal Institute of Technology, TIK-Report, No. 43, 1998.

37. Morse N. Reducing the size of nondominated set: pruning by clustering. Computers and Operations Research 1980;7(1–2):55–66.

# CHAPTER 13

# UNCERTAINTY ANALYSIS IN POWER SYSTEMS

In most cases in the first 12 chapters, the variables and parameters have been deterministic. Actual power systems exhibit numerous parameters and phenomena that are either nondeterministic or so complex and dependent on so many diverse processes that they may readily be regarded as nondeterministic or uncertain. This chapter comprehensively deals with various uncertain problems in power system operation such as uncertainty load analysis, probabilistic power flow, fuzzy power flow, economic dispatch with uncertainties, fuzzy economic dispatch, hydrothermal system operation with uncertainty, unit commitment with uncertainties, VAR optimization with uncertain reactive load, and probabilistic optimal power flow (P-OPF).

## 13.1  INTRODUCTION

The planning process of the regulated utilities does not capture the uncertainties in the operation and planning of power systems. In particular, the factors of uncertainties increase as the utility industry undergoes restructuring. Because of restructuring under the pressure of various driving forces, we can foresee that those changes will become even greater in the near future. This is mainly because of the impact on this industry of the many uncertainty factors as also external factors related to the environment. Modern power systems are thus facing many new challenges, owing to environment and market pressures, as well as other uncertainties or/and inaccuracies [1–11]. Environment pressure implies more loaded networks, market pressure increases competition, while uncertainty and inaccuracy increase the complexity of operation and planning. Consequently, these new challenges have huge and direct impact on the operation and planning of modern power systems. They also demand some high requirements for modern power systems operation, such as,

(a) a stronger expectation from customers for higher reliability and quality of supply owing to the uncertainty factors as well as the increase of the share of electrical power in their overall energy consumption;

*Optimization of Power System Operation*, Second Edition. Jizhong Zhu.
© 2015 The Institute of Electrical and Electronics Engineers, Inc. Published 2015 by John Wiley & Sons, Inc.

**(b)** more electricity exchanges across large geographical areas resulting from a greater cooperation in the electricity market and greater competition in the energy market, resulting in a number of uncertainties in both the electricity market and the energy markets;

**(c)** the need for low production fuel cost and low price of electricity in order to achieve competitive strength in the energy market.

Furthermore, we can state only one thing with absolute certainty with regard to the electrical power industry today: we are living and working with many unknowns [2]. Especially in modern power system operation, the several inaccuracies and uncertainties will lead to deviation from operation and planning. These are on the one hand the inaccuracies and uncertainties in the input information needed by the operation and planning, and on the other hand, the modeling and solution inaccuracies. Therefore, it is very important to analyze the uncertainty in operation of modern power systems and to use the available controls to ensure their security and reliability.

## 13.2   DEFINITION OF UNCERTAINTY

Generally speaking, there are two kinds of uncertainties in power systems operation and planning [4]:

**(1)** uncertainty in a mathematical sense, which means the difference between measured, estimated values and true values, including errors in observation or calculation;

**(2)** sources of uncertainty, including transmission capacity, generation availability, load requirements, unplanned outages, market rules, fuel price, energy price, market forces, weather and other interruption, etc.

These uncertainties will affect power systems planning and operation in the following aspects:

- Entry of new energy producing/trading participants
- Increases in regional and intraregional power transactions
- Increasingly sensitive loads
- New types and numbers of generation resources.

## 13.3   UNCERTAINTY LOAD ANALYSIS

Power loads especially residential loads are variable and their data are uncertain. For example, the variability of the electricity consumption of a single residential customer generally depends on the presence at home of the family members and on the time of use of a few high-power appliances with relatively short duration of use during the day, and is subject to very high uncertainty. Probabilistic analysis and fuzzy theory can be used to analyze the uncertainty load.

### 13.3.1 Probability Representation of Uncertainty Load

Different probability distribution functions may be selected for the different kinds of uncertainty loads. The following probability distribution functions are often used [12]:

**Normal Distribution**   The general formula for the probability density function of the normal distribution for uncertain load $P_D$ is

$$f(P_D) = \frac{e^{-\frac{(P_D-\mu)^2}{2\sigma^2}}}{\sigma\sqrt{2\pi}} \tag{13.1}$$

$$-\infty \leq P_D \leq \infty$$

$$\sigma > 0 \tag{13.2}$$

where

$P_D$: the uncertain load.

$\mu$: the mean value of the uncertain load. It is also called the location parameter.

$\sigma$: the standard deviation of the uncertain load. It is also called the scale parameter.

The shape of plot of the normal probability density function is shown in Figure 13.1.

**Lognormal Distribution**   Many probability distributions are not a single distribution, but are in fact a family of distributions. This is due to the distribution having one or more shape parameters.

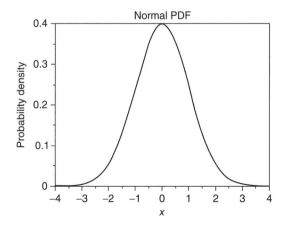

Figure 13.1   The plot of the normal probability density function.

Shape parameters allow a distribution to take on a variety of shapes, depending on the value of the shape parameter. These distributions are particularly useful in modeling applications because they are flexible enough to model a variety of uncertainty load data sets. The following is the equation of the lognormal distribution for uncertain load $P_D$.

$$f(P_D) = \frac{e^{-\frac{\left(\ln\left(\frac{(P_D-\mu)}{m}\right)\right)^2}{2a^2}}}{\sigma(P_D - \mu)\sqrt{2\pi}} \tag{13.3}$$

$$P_D \geq \mu$$

$$\sigma > 0 \tag{13.4}$$

where

$m$: the scale parameter.
ln: the natural logarithm.

Figure 13.2 is an example of the shape for the plot of the lognormal probability density function for four values of $\sigma$.

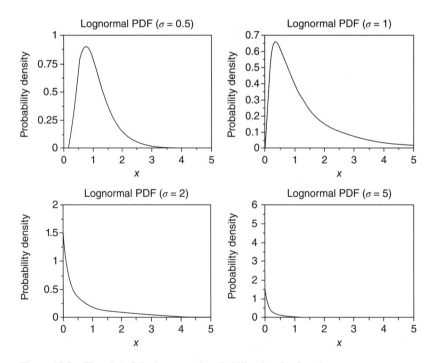

Figure 13.2   The plot of the lognormal probability density function.

***Exponential Distribution*** The formula for the probability density function of the exponential distribution for uncertain load $P_D$ is

$$f(P_D) = \frac{e^{-\frac{P_D - \mu}{b}}}{b} \tag{13.5}$$

$$P_D \geq \mu$$

$$b > 0 \tag{13.6}$$

where

$b$: the scale parameter.

Figure 13.3 is an example of the shape for the plot of the exponential probability density function.

***Beta Distribution*** The general formula for the probability density function of the beta distribution for uncertain load $P_D$ is

$$f(P_D) = \frac{(P_D - d)^{a-1}(c - P_D)^{b-1}}{B(a,b)(c - d)^{a+b-1}} \tag{13.7}$$

$$= \frac{\Gamma(a+b)(P_D - d)^{a-1}(c - P_D)^{b-1}}{\Gamma(a)\Gamma(b)(c - d)^{a+b-1}}$$

$$d \leq P_D \leq c$$

$$a > 0 \tag{13.8}$$

$$b > 0$$

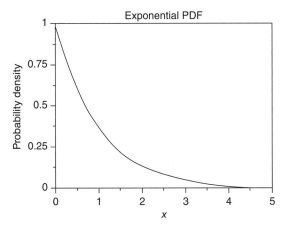

Figure 13.3 The plot of the exponential probability density function.

where

    $a, b$: the shape parameters.

      $c$: the upper bound.

      $d$: the lower bound.

$B(a, b)$: the beta function

Typically, we define the general form of a distribution in terms of location and scale parameters. The beta distribution is different in that we define the general distribution in terms of the lower and upper bounds. However, the location and scale parameters can be defined in terms of the lower and upper limits as follows:

$$\text{location} = d$$

$$\text{scale} = c - d$$

Figure 13.4 is an example of the shape for the plot of the beta probability density function for four different values of the shape parameters.

**Gamma Distribution**   The general formula for the probability density function of the gamma distribution for uncertain load $P_D$ is

$$f(P_D) = \frac{(P_D - \mu)^{a-1}}{b^a \Gamma(a)} e^{-\left(\frac{P_D - \mu}{b}\right)} \tag{13.9}$$

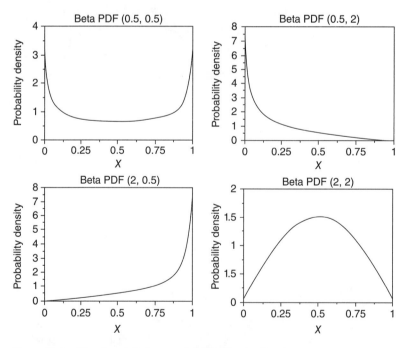

Figure 13.4   The plot of the beta probability density function.

$$P_D \geq \mu$$

$$a > 0$$

$$b > 0 \tag{13.10}$$

where $a$ is the shape parameter, $\mu$ is the location parameter, $b$ is the scale parameter, and $\Gamma$ is the gamma function, which has the formula

$$\Gamma(a) = \int_0^\infty t^{a-1} e^{-t} dt \tag{13.11}$$

Figure 13.5 is an example of the shape for the plot of the gamma probability density function.

***Gumbel Distribution*** The Gumbel distribution is also referred to as the extreme-alue type I distribution. The extreme-value type I distribution has two forms. One is based on the smallest extreme and the other is based on the largest extreme. We call these the minimum and maximum cases, respectively. Formulas and plots for both cases are given.

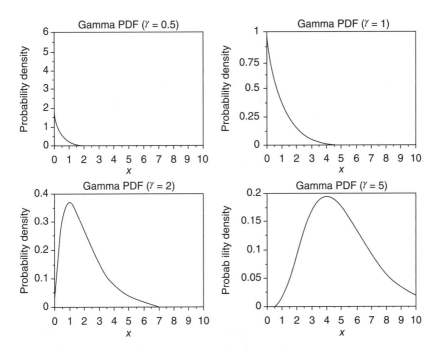

Figure 13.5    The plot of the gamma probability density function.

The general formula for the probability density function of the Gumbel (maximum) distribution for uncertain load $P_D$ is

$$f(P_D) = \frac{1}{b}e^{\left(\frac{\mu - P_D}{b}\right)}e^{-e^{\left(\frac{\mu - P_D}{b}\right)}} \tag{13.12}$$

$$-\infty \leq P_D \leq \infty$$

$$b > 0 \tag{13.13}$$

where $\mu$ is the location parameter and $b$ is the scale parameter.

Figure 13.6 is an example of the shape for the plot of the Gumbel probability density function for the maximum case.

**_Chi-Square Distribution_** The chi-square distribution results when $v$ independent variables with standard normal distributions are squared and summed. The formula for the probability density function of the chi-square distribution for uncertain load $P_D$ is

$$f(P_D) = \frac{P_D^{\frac{v}{2}-1}}{2^{\frac{v}{2}}\Gamma\left(\frac{v}{2}\right)}e^{-\left(\frac{P_D}{2}\right)} \tag{13.14}$$

$$P_D \geq 0 \tag{13.15}$$

where $v$ is the shape parameter and $\Gamma$ is the gamma function.

Figure 13.7 is an example of the shape for the plot of the chi-square probability density function for four different values of the shape parameter.

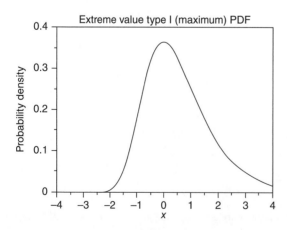

Figure 13.6 The plot of the Gumbel probability density function.

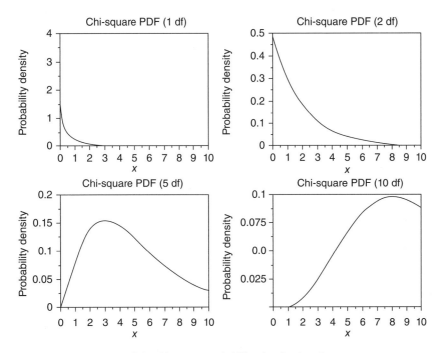

Figure 13.7   The plot of the chi-square probability density function.

**Weibull Distribution**   The formula for the probability density function of the Weibull distribution for uncertain load $P_D$ is

$$f(P_D) = \frac{a(P_D - \mu)^{a-1}}{b^a} e^{-\left(\frac{P_D-\mu}{b}\right)^a} \qquad (13.16)$$

$$P_D \geq \mu$$

$$a > 0$$

$$b > 0 \qquad (13.17)$$

where $a$ is the shape parameter, $\mu$ is the location parameter, and $b$ is the scale parameter.

Figure 13.8 is an example of the shape for the plot of the Weibull probability density function.

## 13.3.2   Fuzzy Set Representation of Uncertainty Load

The uncertainty load $P_D$ can also be represented by fuzzy sets, which are defined in the number set $R$ and satisfy the normality and boundary conditions that are designed by fuzzy numbers. The membership function of a fuzzy number for the uncertainty load $P_D$ corresponds to:

$$\mu_{P_D(x)} : R \in [0, 1] \qquad (13.18)$$

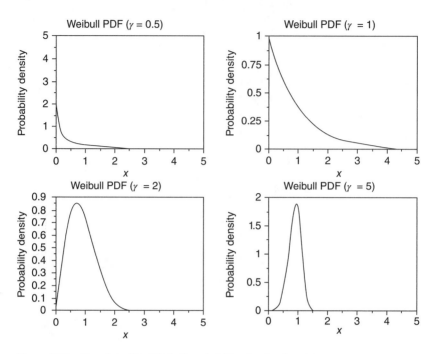

Figure 13.8    The plot of the Weibull probability density function.

The easiest way to express the fuzzy number is the LR fuzzy number. The uncertainty load $P_D$ is said to be an LR-type fuzzy number if

$$
\mu_{P_D(x)} =
\begin{cases}
L\left(\dfrac{m-x}{a}\right), & x \le m, \ a > 0 \\
R\left(\dfrac{x-m}{b}\right), & x \ge m, \ b > 0
\end{cases}
\tag{13.19}
$$

where $m$ is the mean value of load $P_D$.

The left-right (LR) type fuzzy number of the uncertainty load $P_D$ can be written as

$$
P_D = (m, a, b)_{LR}
\tag{13.20}
$$

One of the common LR fuzzy numbers is the triangular fuzzy number, which is shown in Figure 13.9.

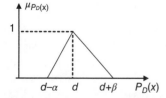

Figure 13.9    Uncertainty load with triangular fuzzy number.

The membership function of the fuzzy load in Figure 13.9 can be expressed as

$$
\mu_{P_D(x)} = \begin{cases} \dfrac{x - (d - \alpha)}{\alpha}, & \text{if } x \in [(d - \alpha), d] \\[3mm] \dfrac{(d + \beta) - x}{\beta}, & \text{if } x \in [d, (d + \beta)] \\[3mm] 0, & \text{otherwise} \end{cases} \tag{13.21}
$$

where

$d$: the model value of uncertainty load.
$\alpha$: the inferior dispersion of uncertainty load.
$\beta$: the superior dispersion of uncertainty load.

The principle of fuzzy numbers can be used to handle the uncertainty load. For example, for getting the sum of two uncertainty loads with a positive triangular fuzzy number, the following fuzzy operation is used.
Let uncertainty load 1

$$
P_{D1} = (d1, \alpha1, \beta1)_{\text{LR}} \tag{13.22}
$$

and uncertainty load 2

$$
P_{D2} = (d2, \alpha2, \beta2)_{\text{LR}} \tag{13.23}
$$

The sum of the two uncertainty loads will be

$$
(d1, \alpha1, \beta1)_{\text{LR}} \oplus (d2, \alpha2, \beta2)_{\text{LR}} = (d1 + d2, \alpha1 + \alpha2, \beta1 + \beta2)_{\text{LR}} \tag{13.24}
$$

Sometimes, a simple way to represent the uncertainty load is using an interval format of fuzzy number, which is based on $\gamma$-cuts of the fuzzy number. The values of $\gamma$ are between 0 and 1. Applying the $\gamma$-cuts, the uncertainty load $P_D$ can be represented as

$$
P_D^\gamma = [\gamma\alpha + (d - \alpha), (d + \beta) - \gamma\beta] \tag{13.25}
$$

or

$$
P_D^\gamma = \left[ P_{D\min}^\gamma, P_{D\max}^\gamma \right] \tag{13.26}
$$

$$
P_{D\min}^\gamma = \gamma\alpha + (d - \alpha) \tag{13.27}
$$

$$
P_{D\max}^\gamma = (d + \beta) - \gamma\beta \tag{13.28}
$$

For two different $\gamma$-cuts ($\gamma 1 < \gamma 2$), the relationship between two interval values of uncertainty load $P_D$ is

$$\left[P_{Dmin}^{\gamma 2}, P_{Dmax}^{\gamma 2}\right] \subset \left[P_{Dmin}^{\gamma 1}, P_{Dmax}^{\gamma 1}\right] \tag{13.29}$$

Let $P_{D1}$ and $P_{D2}$ be two uncertainty loads. Then

$$P_{D1} = \left[P_{D1min}, P_{D1max}\right] \tag{13.30}$$

$$P_{D2} = \left[P_{D2min}, P_{D2max}\right] \tag{13.31}$$

Addition, subtraction, multiplication, and division of the two uncertainty loads are defined as

$$P_{D1} + P_{D2} = \left[P_{D1min}, P_{D1max}\right] + \left[P_{D2min}, P_{D2max}\right]$$

$$= \left[P_{D1min} + P_{D2min}, P_{D1max} + P_{D2max}\right] \tag{13.32}$$

$$P_{D1} - P_{D2} = \left[P_{D1min}, P_{D1max}\right] - \left[P_{D2min}, P_{D2max}\right]$$

$$= \left[P_{D1min} - P_{D2max}, P_{D1max} - P_{D2min}\right] \tag{13.33}$$

$$P_{D1} \times P_{D2} = \left[P_{D1min}, P_{D1max}\right] \times \left[P_{D2min}, P_{D2max}\right]$$

$$= \big[\min\left(P_{D1min} \times P_{D2min}, P_{D1min} \times P_{D2min}, P_{D1max} \times P_{D2min}, P_{D1max} \times P_{D2max}\right),$$

$$\max\left(P_{D1min} \times P_{D2min}, P_{D1min} \times P_{D2min}, P_{D1max} \times P_{D2min}, P_{D1max} \times P_{D2max}\right)\big]$$

$$\tag{13.34}$$

$$P_{D1}/P_{D2} = \left[P_{D1min}, P_{D1max}\right] / \left[P_{D2min}, P_{D2max}\right]$$

$$= \left[P_{D1min}, P_{D1max}\right] \left[1/P_{D2max}, 1/P_{D2min}\right] \quad \text{if } 0 \notin \left[P_{D2min}, P_{D2max}\right] \tag{13.35}$$

Some of the algebraic laws valid for real numbers remain valid for intervals of fuzzy numbers. Intervals addition and multiplication are associative and commutative:

(a) Commutative:

$$P_{D1} + P_{D2} = P_{D2} + P_{D1} \tag{13.36}$$

$$P_{D1} \times P_{D2} = P_{D2} \times P_{D1} \tag{13.37}$$

**(b)** Associative:

$$(P_{D1} + P_{D2}) \pm P_{D3} = P_{D1} + (P_{D2} \pm P_{D3}) \qquad (13.38)$$

$$(P_{D1} \times P_{D2})P_{D3} = P_{D1}(P_{D2} \times P_{D3}) \qquad (13.39)$$

**(c)** Neutral element:

$$P_{D1} + 0 = 0 + P_{D1} = P_{D1} \qquad (13.40)$$

$$1 \times P_{D1} = P_{D1} \times 1 = P_{D1} \qquad (13.41)$$

***Example 13.1:*** There are two uncertainty loads, $P_{D1} = (20, 3, 5)_{LR}$ and $P_{D2} = (23, 8, 5)_{LR}$, which are shown in Figure 13.10.

The corresponding fuzzy membership functions can be presented as below.

$$\mu_{P_{D1}(x)} = \begin{cases} \dfrac{x - 17}{3}, & \text{if } x \in [17, 20] \\[2mm] \dfrac{25 - x}{5}, & \text{if } x \in [20, 25] \\[2mm] 0, & \text{otherwise} \end{cases}$$

$$\mu_{P_{D2}(x)} = \begin{cases} \dfrac{x - 15}{8}, & \text{if } x \in [15, 23] \\[2mm] \dfrac{28 - x}{5}, & \text{if } x \in [23, 28] \\[2mm] 0, & \text{otherwise} \end{cases}$$

The sum of the two uncertainty loads will be

$$(20, 3, 5)_{LR} \oplus (23, 8, 5)_{LR} = (43, 11, 10)_{LR}$$

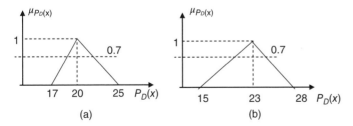

Figure 13.10   Two uncertainty loads with triangular fuzzy number.

Let us represent two uncertainty load by using an interval format of fuzzy number, and 0.7-cut of the fuzzy number. The uncertainty loads $P_{D1}$ and $P_{D2}$ can be represented as:

$$P_{D1}^{0.7} = [0.7 \times 3 + (20 - 3), (20 + 5) - 0.7 \times 5] = [19.1, 21.5]$$

$$P_{D2}^{0.7} = [0.7 \times 8 + (23 - 8), (23 + 5) - 0.7 \times 5] = [20.6, 24.5]$$

The sum of two uncertainty loads in interval format is computed as

$$
\begin{aligned}
P_{D1}^{0.7} + P_{D2}^{0.7} &= \left[ P_{D1\min}^{0.7}, P_{D1\max}^{0.7} \right] + \left[ P_{D2\min}^{0.7}, P_{D2\max}^{0.7} \right] \\
&= \left[ P_{D1\min}^{0.7} + P_{D2\min}^{0.7}, P_{D1\max}^{0.7} + P_{D2\max}^{0.7} \right] \\
&= [19.1 + 20.6, 21.5 + 24.5] \\
&= [39.7, 46]
\end{aligned}
$$

The same result can be obtained by using the sum of two uncertainty loads $P_{D\,\mathrm{sum}} = (43, 11, 10)_{\mathrm{LR}}$ and a 0.7-cut of fuzzy number, that is,

$$P_{D\,\mathrm{sum}}^{0.7} = [0.7 \times 11 + (43 - 11), (43 + 10) - 0.7 \times 10] = [39.7, 46]$$

## 13.4   UNCERTAINTY POWER FLOW ANALYSIS

In general power flow analysis, the input variables to the power flow problem are assumed to be deterministically known. The practical operation conditions with uncertainty factors are not considered. Consequently, the power flow results may not reflect the real status of system operation. This limitation will be overcome if a probabilistic approach or a fuzzy approach is applied.

### 13.4.1   Probabilistic Power Flow

From Chapter 2, the standard form of the load flow equations is

$$P_i = P_{Gi} - P_{Di} = \sum_j Y_{ij} V_i V_j \cos(\theta_i - \theta_j - \delta_{ij}) \tag{13.42}$$

$$Q_i = Q_{Gi} - Q_{Di} = \sum_j Y_{ij} V_i V_j \sin(\theta_i - \theta_j - \delta_{ij}) \tag{13.43}$$

where

$i, j$: the bus number.
$P_i$: the net real power injection.
$Q_i$: the net reactive power injection.

$V$: the magnitude of the bus voltage.

$\theta$: the phase angle of the bus voltage.

$Y_{ij}$: the magnitude of the $i$-$j$th element of the admittance matrix.

$\delta_{ij}$: the angle of the $i$-$j$th element of the admittance matrix.

The power flow problem can be expressed as two sets of nonlinear equations as follows:

$$Y = g(X) \tag{13.44}$$

$$Z = h(X) \tag{13.45}$$

where $X$ is the vector of unknown state variables (voltage magnitudes and angles at PQ buses; and voltage angles and reactive power outputs at PV buses); $Y$ is the vector of predefined input variables (real and reactive power at PQ buses; and voltage magnitudes and real power at PV buses); $Z$ is the vector of unknown output variables (real and reactive flows in the network elements); $g$ and $h$ are the power flow functions.

As mentioned in Section 13.3, the input variables such as power loads are uncertain, and can be expressed with probabilistic distributions. Probabilistic power flow models input data (generation and loads) in a probabilistic way and calculate the probability distribution functions of line flows.

We assume the input data has the nature of a normal distribution, and the mean values and variances of input variables $Y$ are $\overline{Y}$ and $\sigma_Y^2$, respectively. With the mean values $\overline{Y}$, the mean values of the state variables and output variables can be computed with the conventional power flow methods. Then the variances of state variables and branch power flows can be computed with the following formulas:

$$\sigma_X^2 = \mathrm{diag}(J^t \Lambda^{-1} J)^{-1} \tag{13.46}$$

$$\sigma_Z^2 = \mathrm{diag}(D(J^t \Lambda^{-1} J)^{-1} D^t) \tag{13.47}$$

where

$\sigma_X^2$: the variances of state variables $X$.

$\sigma_Z^2$: the variances of branch power flows $Z$.

$J$: the Jacobian matrix of the power flow equations.

$\Lambda$: the diagonal matrix of variances of the injected power $\sigma_Y^2$.

$D$: the first order matrix from the Taylor series expansion of $g(x)$.

With mean values and variances of the state variables and output variables, the probabilistic distribution of power flow is obtained.

The probabilistic power flow provides the complete spectrum of all probable values of output variables, such as bus voltages and flows, with their respective probabilities, taking into account generation unit unavailability, load uncertainty, dispatching criteria effects, and topological variations.

## 13.4.2 Fuzzy Power Flow

Fuzzy power flow analysis is needed if the input data such as load and generation power are given as fuzzy numbers.

Section 13.3.2 analyzes uncertain load by using fuzzy numbers. Other input data with uncertainty in power flow calculation can be handled as the same way. If we use the interval format of fuzzy numbers to deal with the uncertain input data, the fuzzy power flow can be computed using interval arithmetic method.

Power flow problems are nonlinear equations $F(x)$. One of the iteration operators for the solution of interval nonlinear equations is the Newton operator [13–16]:

$$N(x,\widetilde{x}) := \widetilde{x} - F'(x)^{-1}F(\widetilde{x}) \tag{13.48}$$

where

$F'(x)$: the interval Jacobian matrix.

$N(x,\widetilde{x})$: the Newton operator.

$\widetilde{x}$: the midpoint of the interval $[x_{min}, x_{max}]$, defined as:

$$\widetilde{x} := \frac{(x_{min} + x_{max})}{2} \tag{13.49}$$

For each iteration, we need to solve the following interval linear equations for $\Delta x$:

$$F\prime(x)\Delta x = F(\widetilde{x}) \tag{13.50}$$

Therefore, the solution of nonlinear equations reduces to the solutions of linear equation, but using interval arithmetic. It is noted that the solution of interval linear equations, which is at the heart of the nonlinear iterative solution, is a different proposition from the solution of ordinary linear equations. The solution set of the interval linear equations has a very complex nonconvex structure. The hull of the solution set is used, which is defined as the smallest interval vector that contains the solution set. Generally, the hull contains, in addition to the entire solution set, many nonsolutions. Therefore, solving interval linear equations involves obtaining the hull of the solution set. There are several methods to solve interval linear equations such as

(1) Krawczyk's method [11]
(2) Interval Gauss–Seidel iteration [14]
(3) LDU Decomposition.

The most widely used method to solve interval linear equations is the Gauss–Seidel iteration. The purpose of Gauss–Seidel iterations here is not to solve the power flow problems, but to solve the linear equations that result from Newton's method.

In short, the fuzzy power flow problem can be solved by using interval arithmetic through linearizing the problem. However, the resulting linear equations must be solved by a Gauss–Seidel iterative process instead of by direct LDU factorization. The solution obtained is conservative in that it contains all solution points, but may also contain many nonsolutions.

## 13.5    ECONOMIC DISPATCH WITH UNCERTAINTIES

### 13.5.1    Min–Max Optimal Method

Chapters 4 and 5 discussed the economic dispatch problem, where uncertain factors are not included. However, the economy of short-term operation of thermal power systems is influenced by approximations in the operation planning methods and by the inaccuracies and uncertainties of input data. There are two major uncertain factors in economic dispatch.

***Uncertain loads***   The forecast loads are important input information, which are characterized by uncertainty and inaccuracy because of the stochastic nature of the loads, as discussed in Section 13.3.

Let the load duration curve $P_D(t)$ be given in the form of intervals

$$P_{D\min}(t) \leq P_D(t) \leq P_{D\max}(t), \quad 0 \leq t \leq T \tag{13.51}$$

where $T$ is the time period.

***Inaccuracy Fuel Cost Function***

- Inaccuracy in the process of measuring or forecasting of input data
- Change of unit performance during the period between measuring and operation.

The inaccuracies in the cost functions for steady-state operation are caused by the limited accuracy of the determination of the thermal dynamic performance, changing cooling water temperatures, changing calorific values and contamination, and erosion and attrition in the boiler and turbine. These deviations lead to inaccurate values for heat inputs and fuel prices.

Similar to the uncertain load, the cost functions of generating units are also expressed in the form of intervals.

$$F_{\min}(P_{Gi}) \leq F(P_{Gi}) \leq F_{\max}(P_{Gi}), \quad i \in NG \tag{13.52}$$

where

$$P_{Gi\min} \leq P_{Gi} \leq P_{Gi\max}, \quad i \in NG \tag{13.53}$$

The most well-founded criterion for optimal scheduling of real power in a power system under uncertainty is the criterion of min–max risk [17,18] or possible losses caused by uncertainty of information. The risk function can be written as

$$R(\overline{P}_{Gi}(t), \widetilde{U}(t)) = F_{\Sigma} - F_{\Sigma\min} \tag{13.54}$$

where

$F_{\Sigma}$: the actual total fuel cost of the generators, which is expressed as

$$F_\Sigma = \sum_{i=1}^{NG} F_i(\overline{P}_{Gi}(t), \widetilde{U}(t)) \tag{13.55}$$

$F_{\Sigma \min}$: the minimal total fuel cost of the generators if we could obtain the deterministic information about the uncertainty factors, which is expressed as

$$F_{\Sigma \min} = \min \sum_{i=1}^{NG} F_i(P_{Gi}(t), \widetilde{U}(t)) \tag{13.56}$$

$\widetilde{U}(t)$: The uncertain factors.
$\overline{P}_{Gi}(t)$: The planned or expected power duration curve of units for the time period $T$.

The operator min max $R$ means the minimization of maximum risk caused by uncertainty factors, that is,

$$\min_{\overline{P}_{Gi}(t)} \max_{\widetilde{U}(t)} \int_0^T R(\overline{P}_{Gi}(t), \widetilde{U}(t)) dt \tag{13.57}$$

The optimality conditions of the min−max problem arise from the main theorem of game theory and can be expressed as follows:
If $\overline{P}_{Gi}^0(t)$ is the optimal plan for min max $R$ criterion, then

$$R(\overline{P}_{Gi}^0(t), U^-(t)) = R(\overline{P}_{Gi}^0(t), U^+(t)) \tag{13.58}$$

Let $E$ be the expected value of risk $R$, and $\Omega$ be a set of mixed strategy of uncertain factors. The minimal-maximal problem can be expressed as follows:

$$\min_{\overline{P}_{Gi}(t)} \max_{\Omega} \int_0^T E(R(\overline{P}_{Gi}(t), \widetilde{U}(t))) dt \tag{13.59}$$

It is possible to compose the deterministic equivalent of min-max problem on the basis of the conditions given above. This requires finding the min−max load demand curves and cost functions of generating units. If we replace the deterministic curves by the min−max curves, we can use the initial deterministic model for calculating the min−max optimal results.

## 13.5.2    Stochastic Model Method

In this section, we present another approach to handle the uncertainty of fuel cost of the generator units by use of the stochastic model.

A method of obtaining a stochastic model is to take a deterministic model and transform it into a stochastic model by (1) introducing random variables as inputs or as coefficients or as both; and (2) introducing equation errors as disturbances. Since this type of model is only an approximation, what is important in this approach is to make the randomness reflect a real situation.

From Chapter 4, the economic dispatch model can be expressed as follows.

$$\min F = \sum_{i=1}^{N} F_i(P_{Gi}) \tag{13.60}$$

such that

$$\sum_{i=1}^{N} P_{Gi} = P_D + P_L \tag{13.61}$$

$$P_{Gi\min} \leq P_{Gi} \leq P_{Gi\max} \tag{13.62}$$

Suppose the fuel cost is a quadratic function, that is,

$$F_i = a_i P_{Gi}^2 + b_i P_{Gi} + c_i \tag{13.63}$$

A stochastic model of the function $F_1$ is formulated by taking the deterministic fuel cost coefficients $a_2$, $b$, $c$ and the generator real power $P_{Gi}$ as random variables. Any possible deviation of the operating cost coefficients from their expected values is manipulated through the randomness of generator power output $P_{Gi}$. The randomness of $P_{Gi}$ implies that the power balance equation (13.61) is not a rigid constraint to be satisfied.

A simple way of converting a stochastic model to a deterministic one is to take its expected value [19]; therefore, the expected value of the operating cost becomes

$$\overline{F} = E\left[ \sum_{i=1}^{N} \left( a_i P_{Gi}^2 + b_i P_{Gi} + c_i \right) \right]$$

$$= \sum_{i=1}^{N} \left[ E\left( a_i \right) E(P_{Gi}^2) + E(b_i)E(P_{Gi}) + E(c_i) \right]$$

$$= \sum_{i=1}^{N} \left[ \overline{a}_i \left( varP_{Gi} + \overline{P}_{Gi}^2 \right) + \overline{b}_i \overline{P}_{Gi} + \overline{c}_i \right]$$

$$= \sum_{i=1}^{N} \left[ \overline{a}_i v \overline{P}_{Gi}^2 + \overline{a} \overline{P}_{Gi}^2 + \overline{b}_i \overline{P}_{Gi} + \overline{c}_i \right]$$

$$= \sum_{i=1}^{N} \left[ \overline{a}_i \overline{P}_{Gi}^2 (v + 1) + \overline{b}_i \overline{P}_{Gi} + \overline{c}_i \right] \tag{13.64}$$

where $v$ is the coefficient of variation of the random variable $P_{Gi}$. It is the ratio of standard deviation to the mean and is a measure of relative dispersion or uncertainty in the random variable. If $v = 0$, it implies no randomness or, in other words, complete certainty about the value of the random variable.

If we use the $B$ coefficient to compute the system network losses, we get

$$P_L = \sum_i \sum_j P_{Gi} B_{ij} P_{Gj} \tag{13.65}$$

Then the expected value of the network power losses is

$$\bar{P}_L = E\left[\sum_i \sum_j P_{Gi} B_{ij} P_{Gj}\right] = \sum_i \sum_j \bar{P}_{Gi} B_{ij} \bar{P}_{Gj} + \sum_i B_{ii} var P_{Gi}$$

$$\approx \sum_i \sum_j \bar{P}_{Gi} B_{ij} \bar{P}_{Gj} \tag{13.66}$$

where, the variance of network loss has been neglected as it is usually small.

In addition, the expected value of the load can be expressed as

$$\bar{P}_D = E[P_D] \tag{13.67}$$

The stochastic model of economic dispatch can be written as follows:

$$\min \bar{F} = \sum_{i=1}^{N} [\bar{a}_i \bar{P}_{Gi}^2(v + 1) + \bar{b}_i \bar{P}_{Gi} + \bar{c}_i] \tag{13.68}$$

such that

$$\sum_{i=1}^{N} \bar{P}_{Gi} = \bar{P}_D + \bar{P}_L \tag{13.69}$$

$$\bar{P}_{Gi\,min} \leq \bar{P}_{Gi} \leq \bar{P}_{Gi\,max} \tag{13.70}$$

Since there is a stochastic error for the stochastic model, the expected value associated with deficit or surplus of generation can be treated as the deviation proportional to the expectation of the square of power mismatch.

$$\delta = E\left[\left(\bar{P}_D + \bar{P}_L - \sum_{i=1}^{N} P_{Gi}\right)^2\right] = \sum_{i=1}^{N} E[\bar{P}_{Gi} - P_{Gi}]^2 = \sum_{i=1}^{N} var P_{Gi} \tag{13.71}$$

Using the Lagrange multiplier method to solve the above model, we get

$$L = \sum_{i=1}^{N} [\bar{a}_i \bar{P}_{Gi}^2(v + 1) + \bar{b}_i \bar{P}_{Gi} + \bar{c}_i] + \lambda\left(\bar{P}_D + \bar{P}_L - \sum_{i=1}^{N} P_{Gi}\right) + \mu \sum_{i=1}^{N} var P_{Gi} \tag{13.72}$$

According to optimality condition $\frac{\partial L}{\partial P_{Gi}} = 0$, we have

$$2\bar{a}_i\overline{P}_{Gi} + \bar{b}_i + \lambda\left(\sum_j 2B_{ij}\overline{P}_{Gj}\right) + 2(\bar{a}_i + \mu)v\overline{P}_{Gi} = 0 \qquad (13.73)$$

Solving the above equation, the stochastic optimal results of the economic dispatch can be obtained.

### 13.5.3 Fuzzy ED Algorithm

***Fuzzy ED Model***   Section 13.3 discusses the real load that can be modeled as fuzzy. Assume the load is a trapezoidal possibility distribution as shown in Figure 13.11. There are four break points: $P_D^{(1)}$, $P_D^{(2)}$, $P_D^{(3)}$ and $P_D^{(4)}$. The possibility distribution of each load refers to the mapping of a fuzzy variable on the [0,1] interval, which is expected to be between $P_D^{(1)}$ and $P_D^{(4)}$, however it is more likely to be between $P_D^{(2)}$ and $P_D^{(3)}$.

Similarly, the corresponding real power generation can also be modeled as fuzzy. Therefore, the economic dispatch with fuzzy loads can be expressed as follows.

$$\min F = \sum_{i=1}^{NG} F_i(\widetilde{P}_{Gi}) \qquad (13.74)$$

such that

$$\sum_{i=1}^{NG} \widetilde{P}_{Gi} = \sum_{j=1}^{ND} \widetilde{P}_{Dj} + \widetilde{P}_L \qquad (13.75)$$

$$P_{Gi\,\min} \le \widetilde{P}_{Gi} \le P_{Gi\,\max} \qquad (13.76)$$

where

$\widetilde{P}_{Gi}$: the fuzzy real power generation.
$\widetilde{P}_{Dj}$: the fuzzy real power load demand.
$\widetilde{P}_L$: the fuzzy real power losses.

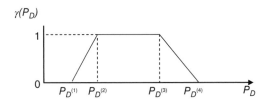

Figure 13.11    Uncertainty load with trapezoidal possibility distribution.

For simplifying the fuzzy economic dispatch problem, neglecting the network losses, and assuming the fuel cost is a linear function, that is,

$$F_i = c_i \widetilde{P}_{Gi} \tag{13.77}$$

then the minimization of cost function is equivalent to the minimization of fuzzy variable $\widetilde{P}_{Gi}$, which can be translated to the minimization of its distance from the $\gamma(P_G)$ axis.

According to Figure 13.2, the distance of the fuzzy variable $\widetilde{P}_{Gi}$ is given as [20,21].

$$d = \frac{A_1 + (A_1 + A_2)}{2} \tag{13.78}$$

where, $A_1$ and $A_2$ are the areas shown in Figure 13.12. They can be computed as follows.

$$A_1 = \frac{P_{Gi}^{(1)} + P_{Gi}^{(2)}}{2} \tag{13.79}$$

$$A_2 = \frac{(P_{Gi}^{(3)} - P_{Gi}^{(2)}) + (P_{Gi}^{(4)} - P_{Gi}^{(1)})}{2} \tag{13.80}$$

Substituting equations (13.79) and (13.80) in equation (13.78), we get

$$d = \frac{P_{Gi}^{(1)} + P_{Gi}^{(2)} + P_{Gi}^{(3)} + P_{Gi}^{(4)}}{4} = \sum_{k=1}^{4} \frac{P_{Gi}^{(k)}}{4} \tag{13.81}$$

Thus the aforementioned fuzzy economic dispatch problem can be written as follows.

$$\min F = \sum_{i=1}^{NG} \sum_{k=1}^{4} c_i \frac{P_{Gi}^{(k)}}{4} \tag{13.82}$$

such that

$$\sum_{i=1}^{NG} P_{Gi}^{(k)} = \sum_{j=1}^{ND} P_{Di}^{(k)}, \quad k = 1, \dots, 4 \tag{13.83}$$

$$P_{Gi\,min} \le P_{Gi}^{(1)} \le P_{Gi}^{(2)} \le P_{Gi}^{(3)} \le P_{Gi}^{(4)} \le P_{Gi\,max} \quad i = 1, \dots, NG \tag{13.84}$$

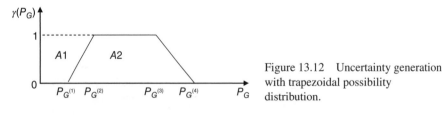

Figure 13.12   Uncertainty generation with trapezoidal possibility distribution.

***Fuzzy Line Constraint*** The above fuzzy representation of real loads will result in fuzzy line flows with trapezoidal possibility distributions. Since DC flow is considered in fuzzy ED analysis, the fuzzy line flow can be expressed as follows.

$$\widetilde{P}_l = \sum_{m=1}^{NB} S_{lm}\widetilde{P}_m, \quad l = 1,\ldots,NL \tag{13.85}$$

where

$\widetilde{P}_m$: the fuzzy bus real power injection.
$\widetilde{P}_l$: the fuzzy line real power flow.
$S$: the DC-based sensitivity matrix.

A contingency analysis is used to detect most severe outages, and contingency constraints are augmented to the base case to assure a preventive control. According to Chapter 5, the contingency constraints are represented similarly to equation (13.85) except that the sensitivity coefficients are adjusted for the contingency under consideration, that is,

$$\widetilde{P}'_l = \sum_{m=1}^{NB} S'_{lm}\widetilde{P}_m, \quad l = 1,\ldots,NL \tag{13.86}$$

where

$\widetilde{P}'_l$: the fuzzy line real power flow under the contingency situation.
$S'$: the DC-based sensitivity matrix under the contingency situation.

If the phase shifter is considered, we represent phase shifters in terms of equivalent injected power. If a phase shifter is located on line $t$ which connects buses $i$ and $j$, the equivalent injected power at buses $i$ and $j$ and phase shifter angle can be simplified as

$$P_{\phi i} = b_t\phi_t = -\frac{\phi_t}{x_t} \tag{13.87}$$

$$P_{\phi j} = -b_t\phi_t = \frac{\phi_t}{x_t} \tag{13.88}$$

where

$P_{\phi i}$: the bus real power injection due to a phase shifter.
$\phi_t$: the phase shifter angle located on line $t$.
$x_t$: the reactance of line $t$.
$b_t$: the susceptance of line $t$.

Thus, the constraint related to the phase shifter angle in the fuzzy case can be written as

$$\phi_{i\min} \le x_t\widetilde{P}_{\phi i} \le \phi_{i\max} \tag{13.89}$$

The fuzzy line flow with phase shifter can be expressed as follows:

$$\widetilde{P}_l = \sum_{m=1}^{NB} S_{lm}(\widetilde{P}_m + \widetilde{P}_{\varphi m}), \quad l = 1, \ldots, NL \tag{13.90}$$

$$\widetilde{P}'_l = \sum_{m=1}^{NB} S'_{lm}(\widetilde{P}_m + \widetilde{P}_{\phi m}), \quad l = 1, \ldots, NL \tag{13.91}$$

Therefore, the fuzzy economic dispatch model with the line constraints is written as

$$\min F = \sum_{i=1}^{NG} \sum_{k=1}^{4} c_i \frac{P_{Gi}^{(k)}}{4} \tag{13.92}$$

s.t.

$$\sum_{i=1}^{NG} P_{Gi}^{(k)} = \sum_{j=1}^{ND} P_{Di}^{(k)}, \quad k = 1, \ldots, 4 \tag{13.93}$$

$$P_{l\min} \le \sum_{m=1}^{NB} S_{lm}(\widetilde{P}_m + \widetilde{P}_{\phi m}) \le P_{l\max}, \quad l = 1, \ldots, NL \tag{13.94}$$

$$P_{l\min} \le \sum_{m=1}^{NB} S'_{lm}(\widetilde{P}_m + \widetilde{P}_{\phi m}) \le P_{l\max}, \quad l = 1, \ldots, NL \tag{13.95}$$

$$P_{Gi\min} \le P_{Gi}^{(1)} \le P_{Gi}^{(2)} \le P_{Gi}^{(3)} \le P_{Gi}^{(4)} \le P_{Gi\max} \quad i = 1, \ldots, NG \tag{13.96}$$

$$\frac{\phi_{i\min}}{x_t} \le P_{\phi i}^{(1)} \le P_{\phi i}^{(2)} \le P_{\phi i}^{(3)} \le P_{\phi i}^{(4)} \le \frac{\phi_{i\max}}{x_t} \quad t = 1, \ldots, NP \tag{13.97}$$

where,

$NP$: Number of phase shifters.
$NB$: Number of buses.
$NL$: Number of lines.

Since we use four sets of variables each describing one break point of the possibility distributions, Dantzig–Wolf decomposition (DWD) is applied to decompose the problem into four subproblems coupled by the constraints in equations (13.96) and (13.97). The dimension of the master problem is equal to the number of coupling constraints plus the number of subproblems, while each subproblem has a dimension equal to the number of constraints corresponding to each break point. The solution of the master problem generates new simplex multipliers (dual solution) that will adjust the cost function of the subproblems. The solution of the subproblems with the adjusted objective function will provide the master problem with new columns to enter the master basis matrix.

**TABLE 13.1    Possibility Distributions for Loads (p.u.)**

| Load Bus | $P_D^{(1)}$ | $P_D^{(2)}$ | $P_D^{(3)}$ | $P_D^{(4)}$ |
|----------|-------|-------|-------|-------|
| 3  | 0.000 | 0.020 | 0.030 | 0.050 |
| 4  | 0.020 | 0.040 | 0.070 | 0.100 |
| 7  | 0.100 | 0.150 | 0.220 | 0.270 |
| 10 | 0.020 | 0.030 | 0.060 | 0.080 |
| 12 | 0.050 | 0.080 | 0.110 | 0.150 |
| 14 | 0.030 | 0.050 | 0.080 | 0.100 |
| 15 | 0.040 | 0.070 | 0.100 | 0.130 |
| 16 | 0.010 | 0.030 | 0.050 | 0.060 |
| 17 | 0.030 | 0.070 | 0.100 | 0.140 |
| 18 | 0.000 | 0.020 | 0.040 | 0.070 |
| 19 | 0.040 | 0.060 | 0.090 | 0.130 |
| 20 | 0.000 | 0.010 | 0.020 | 0.040 |
| 21 | 0.100 | 0.150 | 0.200 | 0.230 |
| 23 | 0.000 | 0.020 | 0.030 | 0.050 |
| 24 | 0.050 | 0.070 | 0.100 | 0.120 |
| 26 | 0.010 | 0.030 | 0.050 | 0.060 |
| 29 | 0.000 | 0.010 | 0.020 | 0.030 |
| 30 | 0.060 | 0.090 | 0.110 | 0.140 |

***Example 13.2:***   The simulation example used here is from reference [20]. Fuzzy economic dispatch method is tested on the modified IEEE 30-bus system. The system has six generators, 41 lines and three phase shifters. All phase shifters have turns ratios equal to 1. Trapezoidal possibility distributions are used to represent the system fuzzy real power loads. The break points of the load possibility distribution are given in Table 13.1. The generators' data are given in Table 13.2 in which each generator cost function is approximated by piecewise linear approximation.

## 13.5.4   Test Case 1

In this case, no line flow constraints are introduced in the problem and the optimal power generation that correspond to the system fuzzy load is found. The break point of the generation possibility distributions are given in Table 13.3. For the sake of comparison, in Table 13.3 we have included the power generation corresponding to the fixed range of load values $P_D^{(1)}$ and $P_D^{(4)}$. This extreme range of loads provides a wider range of line flows than that of the proposed fuzzy model, indicating that the fixed load interval leads to an overestimate of the system behavior in an uncertain environment.

**TABLE 13.2   Generators Data (p.u.)**

| Gen. Bus | Piecewise Section | $P_{G\text{min}}$ | $P_{G\text{max}}$ | Cost Coefficient ($/MWh) |
|---|---|---|---|---|
| G1 | 1 | 0.30 | 0.90 | 25.0 |
| | 2 | 0.00 | 0.35 | 37.5 |
| | 3 | 0.00 | 0.75 | 42.0 |
| G2 | 1 | 0.20 | 0.50 | 28.0 |
| | 2 | 0.00 | 0.30 | 37.0 |
| G5 | 1 | 0.15 | 0.25 | 30.0 |
| | 2 | 0.00 | 0.25 | 36.5 |
| G8 | 1 | 0.10 | 0.15 | 27.0 |
| | 2 | 0.00 | 0.20 | 38.0 |
| G11 | 1 | 0.10 | 0.20 | 27.5 |
| | 2 | 0.00 | 0.10 | 37.0 |
| G13 | 1 | 0.12 | 0.20 | 36.0 |
| | 2 | 0.00 | 0.20 | 39.0 |

**TABLE 13.3   The Results of Fuzzy Economic Dispatch**

| Gen. Bus | $P_G^{(1)}$ | $P_G^{(2)}$ | $P_G^{(3)}$ | $P_G^{(4)}$ | Power Gen. Range for | |
|---|---|---|---|---|---|---|
| | | | | | Min Load | Max Load |
| G1 | 0.900 | 0.900 | 0.968 | 1.217 | 0.900 | 1.250 |
| G2 | 0.478 | 0.500 | 0.800 | 0.800 | 0.466 | 0.800 |
| G5 | 0.150 | 0.488 | 0.500 | 0.500 | 0.150 | 0.500 |
| G8 | 0.150 | 0.150 | 0.150 | 0.150 | 0.150 | 0.272 |
| G11 | 0.200 | 0.200 | 0.300 | 0.300 | 0.200 | 0.300 |
| G13 | 0.120 | 0.200 | 0.200 | 0.200 | 0.120 | 0.200 |

## 13.5.5   Test Case 2

The fuzzy power generations, given in Table 13.3, are used to compute the corresponding line flow possibility distributions. The break points of line 2–6 are 0.2252, 0.2808, 0.4333, and 0.5238 p.u., compared to 0.2248 and 0.5430 p.u. for the fixed load interval, which indicates once again the overestimated results by the fixed interval. Line 2–6 has an overflow as its flow limit is 0.5 p.u. Therefore, the optimal power generation is computed again by considering line 2–6 flow limit. In this case, the phase shifter on line 4–6 alleviates the overflow without any adjustment to the optimal power generation given in Table 13.2. The corresponding break points for the phase shifter on line 4–6 are 0.0, 0.0, 0.0, 0.56°, whereas the phase shifter range for the fixed load interval is between 0.00 and 1.02°. Thus, a smaller range for the phase shifter angle is obtained by utilizing a possibility distribution for the loads.

## 13.6  HYDROTHERMAL SYSTEM OPERATION WITH UNCERTAINTY

There are several complex and interrelated problems associated with the optimization of hydrothermal systems.

- long-term regulation problem (1 to 2 year optimization period);
- intermediate term hydrothermal control (1 month to 6 months planning period);
- short-term hydrothermal dispatch (optimization period is from 1 day to 1 week)

For the short-term optimization problem, the applications of deterministic methods to hydrothermal system operation have been established, in which the water inflows and loads were considered to be deterministic. For the long-term regulation problem, it is necessary to use a stochastic representation for the load and river inflow [22,23]. Since there are the uncertainty factors in the short-term hydrothermal dispatch, the existing methods do not provide the system operators with a convincing answer on how to use the water in each separate reservoir. The following uncertainties should be taken into account in a large hydrothermal system operation.

- Uncertainty of the load
- Uncertainty of the unit availability
- Uncertainty of the river inflow.

The uncertainty of the river inflows, loads, and unit availability can be dealt with in a stochastic representation. The methods to solve ED with uncertainty in the previous section can also be used to solve the uncertainty problem for the hydrothermal system operation.

## 13.7  UNIT COMMITMENT WITH UNCERTAINTIES

### 13.7.1  Introduction

The economy of unit commitment (UC) of power systems is influenced by approximations in the operation planning methods and by the inaccuracies and uncertainties of input data. However, most of the early works on the unit commitment problem (UCP), which are discussed in Chapter 7, use a deterministic formulation neglecting the uncertainties.

As we analyzed before, the uncertain load can be expressed as a normal distribution with a specific correlation structure. Thus, we use a chance-constrained optimization (CCO) formulation for the UCP assuming that the hourly loads follow a multivariate normal distribution [24]. The CCO formulation falls into a class of optimization procedures known as stochastic programming in which the solution methods take into consideration the randomness in input parameters. The advantages of using stochastic programming over the corresponding expected value solution have been demonstrated over a wide spectrum of applications. In the chance-constrained programming, the constraints can be violated with a preassigned (usually very small) level of probability. These probabilistic constraints can often be converted to certain deterministic equivalents and the resulting program can be solved using the general deterministic techniques.

In the stochastic model of UC, the equal constraint of real power balance is expressed by a "chance constraint," which requires that this condition be satisfied at a predetermined level of probability. The reserve constraint is considered in the UC because utilities are required to carry a reserve for many different contingencies such as load peaks, generator failures, scheduled outages, regulation, and local area protection. The reserve is usually referred to as operating reserve, which consists of two parts: spinning reserve (SR) and non-spinning reserve. The additional electricity available (synchronized) to serve load immediately is defined as the SR. In other words, the difference between the total amount of electricity ready to serve the customers and the current demand for electricity is the SR. Generally, the magnitude of the required amount of SR is predetermined and used as an operating constraint in the UC calculation. For example, it is taken to be $1.5-2$ times the capacity of the largest generator or a percentage of the peak load. Instead of using SR as a predetermined constraint, the stochastic method yields as an output the sets of generating units that need to be turned on such that the load is met with a high probability over the entire time horizon. The level of SR can be determined by fuzzy methods, which are similar to SR handling in CCO.

## 13.7.2 Chance-Constrained Optimization Model

***Deterministic UC Model*** The mathematical model for the unit commitment is a mixed integer nonlinear program. The basic deterministic formulation can be written as follows.

$$\min F = \sum_{i=1}^{N} \sum_{t=1}^{T} [F_{i,t}(P_{i,t}, x_{i,t}) + S_{i,t}(P_{i,t}, x_{i,t})] \tag{13.98}$$

s.t.

$$\sum_{i=1}^{N} x_{i,t} P_{i,t} = P_{Dt} \quad t = 1, 2, \ldots, T \tag{13.99}$$

$$P_{i\min} \leq x_{i,t} P_{i,t} \leq P_{i\max} \tag{13.100}$$

$$\sum_{i=1}^{N} x_{i,t} P_{i\max} \geq (1 + \alpha) P_{Dt} \quad t = 1, 2, \ldots, T \tag{13.101}$$

$$x_{i,t} - x_{i,t-1} \leq x_{i,\gamma} \quad \gamma = t + 1, \ldots, \min\{t + t_{up} - 1, T\},$$
$$i = 1, 2, \ldots, N, \quad t = 1, 2, \ldots, T \tag{13.102}$$

$$x_{i,t-1} - x_{i,t} \le 1 - x_{i,\beta} \quad \beta = t + 1, \ldots, \min\{t + t_{\text{down}} - 1, T\},$$

$$i = 1, 2, \ldots, N, \quad t = 1, 2, \ldots, T \tag{13.103}$$

$$x_{i,t} \in \{0, 1\} \quad t = 1, 2, \ldots, T, \quad i = 1, 2, \ldots, N \tag{13.104}$$

where

$F_{i,t}$: the fuel cost of the generator unit $i$ at time $t$.

$S_{i,t}$: the cost of starting up unit $i$ at time $t$.

$P_{Dt}$: the load demand at time $t$.

$P_{i,t}$: the power output of unit $i$ at time $t$.

$T$: the time period.

$x_{i,t}$: the 0-1 variable. 1 if the unit $i$ on at time $t$, 0 otherwise.

$1 - \alpha$: the prescribed probability level for meeting load over the entire time horizon.

$t_{\text{up}}$: minimum number of hours required for a generator to stay up once it is on.

$t_{\text{down}}$: minimum number of hours required for a generator to stay down once it is off.

The objective function consists of the total fuel cost and the starting up cost of the generators. Constraints in equations (13.102) and (13.103) are the uptime/downtime constraints that force the generators to stay up for at least a specified amount of time, $t_{\text{up}}$, once they are turned on and stay down for at least a specified time period, $t_{\text{down}}$, once they are shut down. Constraint (13.100) ensures that the power generated matches the minimum and maximum capacity requirements of the corresponding generators for all time periods. The SR constraint (13.101) attempts to ensure that there is enough power available to meet the demand in the event of an unusual contingency. The power balance constraints (13.99) are the linking constraints that link the decision variables of different generators and time periods. These constraints ensure that the estimated load is satisfied in all time periods. They cause difficulties in solving the problem because adding them to the constraint set makes the problem inseparable, thus requiring sophisticated techniques for finding a solution.

**Stochastic Model**  Let $P_D$, a random variable, denote the load at hour $t$. It can be expressed as a multivariate normal distribution with a specific correlation structure: $P_D \sim N(\mu, \Sigma)$ with mean vector $\mu$ and covariance matrix $\Sigma$ where $\mu_t$ and $\sigma_t$ are the corresponding mean and standard deviation for time period $t$. Changing the equal constraint of the real power balance equation into an inequality constraint and replacing it by the following probabilistic constraint for each hour gives a probability level for satisfying the linking constraint over all time periods.

$$P\left[\sum_{i=1}^{N} x_{i,t} P_{i,t} \ge P_D \quad t = 1, 2, \ldots, T\right] \ge 1 - \alpha \tag{13.105}$$

We replace the probability constraint (13.105) by a set of $T$ separate probability constraints each of which could be inverted to obtain a set of $T$ equivalent deterministic linear inequalities. Initially we choose the $T$ constraints in a manner such that together they are more stringent than constraint (13.105). The initial set of $T$ individual linear constraints (13.110) replacing equation (13.105) are obtained using the following argument.

First we denote the event $\sum_{i=1}^{N} x_{i,t} P_{i,t} \geq P_D$ by $A_t$, and its complementary event $\sum_{i=1}^{N} x_{i,t} P_{i,t} < P_D$ by $A_t^c$. From Boole's inequality of probability theory, it is well known that

$$P\left[\bigcup_{t=1}^{T} A_t\right] \leq \sum_{t=1}^{T} P\left[A_t\right] \tag{13.106}$$

If

$$P[A_t^c] \leq \frac{\alpha}{T}, \quad t = 1, 2, \ldots, T \tag{13.107}$$

then

$$P\left[\bigcap_{t=1}^{T} A_t\right] = 1 - P\left[\bigcup_{t=1}^{T} A_t^c\right] \geq 1 - \sum_{t=1}^{T} P\left[A_t^c\right] \geq 1 - \alpha \tag{13.108}$$

Because $P_D$ is normally distributed with mean $\mu_t$ and standard deviation $\sigma_t$, $P[A_t^c] \leq \frac{\alpha}{T}$ is equivalent to

$$P\left[\sum_{i=1}^{N} x_{i,t} P_{i,t} < P_D\right] \leq \frac{\alpha}{T} \tag{13.109}$$

which is equivalent to

$$\sum_{i=1}^{N} x_{i,t} P_{i,t} \geq \mu_t + (z_\alpha/T)\sigma_t \quad t = 1, 2, \ldots, T \tag{13.110}$$

where $(z_\alpha/T)$ is the $100(1 - \alpha/T)$th percentile of the standard normal distribution.

Setting the initial value of $z$ to be $z = z_\alpha/T$, we get

$$\sum_{i=1}^{N} x_{i,t} P_{i,t} \geq \mu_t + z\sigma_t \quad t = 1, 2, \ldots, T \tag{13.111}$$

### 13.7.3 Chance-Constrained Optimization Algorithm

The deterministic form of the stochastic constraint is used in solving the UCP iteratively by using a different value at each iteration. The steps of the CCO algorithm are as follows:

Step (1)   Choose an initial value in equation (13.111).

Step (2)   Choose a starting set of $\lambda$ multipliers.

Step (3)   For each unit $i$, solve a dynamic program with $4T$ states and $T$ stages; obtain $q^*(\lambda^k)$, which is the objective function value of the optimal solution to the Lagrange dual problem.

Step (4)   Solve the economic dispatch problem for each hour using the scheduled units and obtain $J^*$, which is the objective function value of the optimal solution to the primal problem.

Step (5)   Check the relative duality gap.

Step (6)   Update $\lambda$, using

$$\lambda^{k+1} = \lambda^k + s^k g^k \tag{13.112}$$

where

$$s^k = \frac{\eta^k(J^* - q^*(\lambda^k))}{\|g^k\|^2} \tag{13.113}$$

$$\eta^k = \frac{1 + m}{k + m} \tag{13.114}$$

$g^k$ is the subgradient and $m$ is a constant. If the gap is not small enough, then go back to step 3. Otherwise continue.

Step (7)   If the final solution is feasible, go to step 8. Otherwise, use the heuristic algorithm to derive a feasible solution.

Step (8)   Evaluate the multivariate normal probability using model (13.105); if it differs from the prescribed probability level by more than a preassigned small quantity (see Table 13.4), then update $z$ and go back to step 2, otherwise STOP.

The algorithm starts by choosing a high value for the initial $z$ value as in equation (13.110), which makes the corresponding solution satisfy the load with a probability level higher than $p_{target} = 1 - \alpha$. In step 2, all $\lambda$ multipliers are set to 0.0. Then in step 3 the dual problem is solved using dynamic programming and $q^*(\lambda^k)$, the objective function value for the solution to the Lagrange dual problem, is obtained. In this step, the scheduling problem for each generator is solved separately to decide which generators should be turned on at each time period.

In step 4, an economic dispatch problem is solved for each time period separately. In solving the economic dispatch problem, the algorithm obtains the operating

**TABLE 13.4   Values Used in Checking Convergence of z-Update Algorithm**

| $p_{target}$ | 0.8 | 0.9 | 0.95 | 0.99 | 0.999 | 0.9999 |
|---|---|---|---|---|---|---|
| $\varepsilon$ | 0.005 | 0.005 | 0.005 | 0.005 | 0.0005 | 0.00005 |

levels for all the scheduled generators determined in step 3. $J^*$, the objective function value of the solution to the primal problem, is calculated using the operating levels for the scheduled units in this step. In step 5, the duality gap is checked and if it is less than $\delta$ then the algorithm proceeds to step 7, otherwise the $\lambda$ multipliers are updated using a subgradient method which determines the improving direction in step 6. $\delta$ may be selected as 0.05%. Before proceeding to evaluate the multivariate normal probability, one needs to check whether the final UC schedule is feasible because Lagrange relaxation techniques frequently provide infeasible solutions. If the result is feasible, the algorithm continues to step 8; otherwise, a heuristic is used to derive a feasible solution and the algorithm proceeds to step 8 after this. The heuristic applied here is simply to turn on the cheapest generator available for the time periods that have a shortage of power. After modifying the schedule, the heuristic checks whether the duality gap is still less than $\delta$.

In step 8 of the CCO algorithm one, needs to calculate the multivariate normal probability. This is needed to ensure that the probabilistic constraint, equation (13.105), is satisfied with the prescribed joint probability over the entire time horizon. This calculation can become time consuming especially when the dimension of the time horizon is large. A subregion-adaptive algorithm for carrying out multivariate integration makes this calculation feasible. If the calculated probability level is in the $\varepsilon$ neighborhood of $p_{\text{target}}$ the algorithm terminates as the goal of finding a schedule that satisfies the load with a probability of $p_{\text{target}}$ is accomplished, otherwise the $z$-value is updated and the previous steps are repeated to obtain another schedule.

To update the $z$-value, the following algorithm is used. The goal is to find a $z$-value in equation (13.111) that provides a schedule such that the load can be satisfied with a probability of $p_{\text{target}} = 1 - \alpha$ over the entire time horizon. This $z$-value needs to be obtained iteratively. The following iterative scheme may be used. First, start with two values that are known to be the upper and lower bounds to the needed $z$-value. Then, run steps 2–7 of the CCO algorithm and then find the actual probabilities of meeting the load for these assumed $z$-values. They also indicate the direction and the magnitude by which we should change these $z$-values so that the probability target can be reached through successive iterations using interpolation. The correct $z$-value could be obtained in a few iterations.

The algorithm proceeds as follows. First we choose $z = z_\alpha$ in equation (13.111). Obviously, it yields a lower bound for the correct $z$-value. We call it $z_{\text{lower}}$. We now run steps 2–7 of the CCO algorithm for this lower bound and obtain an estimate of the probability with which the load is being met. We call this probability $p_{\text{lower}}$. Next we choose an arbitrarily large value for $z$. We denote it by $z_{\text{upper}}$. In the next step, we obtain the upper percentiles of the standard normal distribution for these probabilities $p_{\text{upper}}$ and $p_{\text{lower}}$ and denote them by $z_1$ and $z_2$, respectively. We also denote the corresponding percentile for the $p_{\text{target}}$ value by $z_{\text{target}}$. On the basis of these values the updated $z$-value is obtained using the following linear interpolation formula.

$$z_{\text{new}} = z_{\text{lower}} + \frac{z_{\text{target}} - z_1}{z_2 - z_1}(z_{\text{upper}} - z_{\text{lower}}) \qquad (13.115)$$

If the $z_{new}$ value is lower than $z_2$ and higher than $z_{target}$, then replace $z_2$ by $z_{new}$. If it is lower than $z_{target}$ and higher than $z_1$, replace $z_1$ by $z_{new}$. Repeat this process using equation (13.115) until $p_{target}$ is reached.

## 13.8    VAR OPTIMIZATION WITH UNCERTAIN REACTIVE LOAD

### 13.8.1    Linearized VAR Optimization Model

The VAR optimization problem is concerned with minimizing real power transmission losses and improving the system voltage profile by dispatching available reactive power sources in the system. For the purpose of the simplification, the hypersurface of the nonlinear power loss function is approximated by its tangent hyperplane at the current operating point, and the linear programming (LP) is adopted for the VAR control problem. This linear approximation is found to be valid over a small region which is formulated by imposing limits on the deviations of the control variables from their current values. Assume that for each optimization iteration, the voltage phase angles are fixed to disregard the coupling between phase angles and reactive variables. Real power injections at various buses are fixed except at the slack bus, which compensates for power losses. The deterministic operating points are found by executing an AC power flow after each LP iteration, which results in revised system voltage magnitudes and angles. The objective function and constraints are linearized around this new operating point assuming fixed, active power-related variables.

The linearized objective function of VAR optimization can be written as [25,26]

$$\min \Delta P_L = \left[ \frac{\partial P_L}{\partial V_1}, \frac{\partial P_L}{\partial V_2}, \ldots, \frac{\partial P_L}{\partial V_n}, \right] \begin{bmatrix} \Delta V_1 \\ \Delta V_2 \\ \vdots \\ \Delta V_n \end{bmatrix} \quad (13.116)$$

or

$$\min \Delta P_L = M \Delta V \quad (13.117)$$

where, $M$ is the row vector relating to the real power loss increments in the bus voltage increments.

There are $m + l + n$ constraints. The first $m$ constraints are for reactive power sources and tap-changing transformer terminals. We refer to the matrix of reactive power injections at these buses as $Q_l$. The $l$ equality constraints are for loads and junction buses that are not connected to transformer terminals, and we refer to the matrix of reactive power injections at these buses as $Q_2$. The last $n$ constraints are the limits on bus voltages. Therefore, the linearized form of the constraints is given as

$$\Delta Q_{1\min} \le \Delta Q_1 = J_1^* \Delta V \le \Delta Q_{1\max} \quad (13.118)$$

$$\Delta Q_2 = J_2^* \Delta V = 0 \quad (13.119)$$

$$\Delta V_{min} \leq \Delta V \leq \Delta V_{max} \tag{13.120}$$

where, $J_1^*$ and $J_2^*$ are submatrices of $J^*$, which is the modified Jacobian matrix.

Similarly to Section 13.5, the trapezoidal distribution is used to model the uncertainty of reactive power load. The possibility distribution will have a value of 1 for load values that are highly possible, and will drop for low possible loads. A zero possibility is assigned to load values that are rather impossible to occur.

As load changes, the magnitude of voltages at different buses will change accordingly. If the injected power at load bus $i$ is changed by $\Delta Q_{ci}$ as a result of capacitor switching or load change, the corresponding change in load bus voltages is given as

$$\Delta V_{Di} = D \Delta Q_{ci} \tag{13.121}$$

where $D$ is a nonnegative matrix, suggesting that if each $\Delta Q_{ci}$ is positive because of a load reduction, then $\Delta V_{Li}$ will be positive. On the other hand, if the injected power is decreased because of a load increase, then the load bus voltages will decrease.

For generator buses, it is obvious that an increase in the injected load power will cause the generator voltages to decrease and vice versa.

### 13.8.2 Formulation of Fuzzy VAR Optimization Problem

The minimization in the VAR optimization problem is subject to inequality and equality constraints, which are referred to as the operating constraints. The operating constraints will be a set of linking constraints imposed on bus voltages, and four independent sets of constraints corresponding to the breakpoints of the trapezoidal possibility distribution. Using the same approach described in Section 13.5, the formulation of fuzzy VAR optimization problem for determining the possibility distribution of transmission losses for a given possibility distribution of loads can be expressed as follows.

$$\min \Delta P_L = \sum_{i=1}^{n} \sum_{k=1}^{4} \frac{M_i^{(k)} \Delta V_i^{(k)}}{4} \tag{13.122}$$

such that

$$\Delta Q_{1min} \leq \Delta Q_1^{(k)} = J_1^{*(k)} \Delta V^{(k)} \leq \Delta Q_{1max} \tag{13.123}$$

$$\Delta Q_2^{(k)} = J_2^{*(k)} \Delta V^{(k)} = 0 \tag{13.124}$$

$$V_{min} \leq V^{(1)} + \Delta V^{(1)} \leq V^{(2)} + \Delta V^{(2)} \leq V^{(3)} + \Delta V^{(3)} \leq V^{(4)} + \Delta V^{(4)} \leq V_{max} \tag{13.125}$$

where, $k = 1, 2, 3, 4$ and equation (13.122) represents the minimization of fuzzy variables $\Delta P_L$. The $J_1^{*(k)}$ and $J_2^{*(k)}$ are submatrices of matrices of $J^{*(k)}$ which is the modified Jacobian matrix of the $k$ th breakpoint of the possibility distribution. The dimension of the problem is very large and it is reduced through the application of the DWD [27].

# 13.9   PROBABILISTIC OPTIMAL POWER FLOW

## 13.9.1   Introduction

We discussed the deterministic optimal power flow (OPF) problem in Chapter 8. If the uncertain factors such as loads are considered as in the previous sections, we can transform the OPF problem into the *probabilistic optimal power flow (P-OPF)* problem [28,29]. Probabilistic programming, or probabilistic optimization, is concerned with the introduction of probabilistic randomness or uncertainty into conventional linear and nonlinear programs. However, the randomness introduced tends to have some structure to it, and this structure is generally represented by a *probability density function (PDF)*. The goal of the P-OPF problem is to determine the PDFs for all variables in the problem. These PDFs are the distributions of the optimal solutions. This section introduces several P-OPF methods.

## 13.9.2   Two-Point Estimate Method for OPF

Generally, the OPF can be seen as a multivariate nonlinear function

$$Y = h(X) \tag{13.126}$$

where $X$ is the input vector and $Y$ is the output vector.

It must be noted that an uncertain input vector renders all output variables uncertain as well. To account for uncertainties in the P-OPF, a two-point estimate method (TPEM) [30], which is basically a variation of the original point estimate method (PEM), is used to decompose the problem (13.126) into several subproblems by taking only two deterministic values of each uncertain variable placed on both sides of the corresponding mean. The deterministic OPF is then run twice for each uncertain variable, once for the value below the mean and once for the value above the mean, with other variables kept at their means. This method is described in detail in the following.

Suppose that $Y = h(X)$ is a general nonlinear multivariate function. The goal is to find the PDF $f_Y(y)$ of $Y$ when the PDF $f_X(x)$ is known, where $x \in X$ and $y \in Y$. There are several approximate methods to address this problem. The PEM is a simple-to-use numerical method for calculating the moments of the underlying nonlinear function. The method was developed by Rosenblueth in the 1970s [31] and is used to calculate the moments of a random quantity that is a function of one or several random variables. Although the moments of the output variables are calculated, one has no information on the associated probability distribution (PD). Generally speaking, this PD can be any PD with the same first three moments; however, when the PD of the input variables is known, the output variables tend to have the same PD, as showed in the OPF problem, where both input and output variables are normally distributed. However, in some cases, the discrete behavior of the OPF results in PD of the output variables that is no longer normal.

Let $X$ denote a random variable with PDF $f_X(x)$; for $Y = h(X)$, the PEM uses two probability concentrations to replace $h(X)$ by matching the first three moments of $h(X)$. When $Y$ is a function of $n$ random variables, the PEM uses $2^n$ probability concentrations located at $2^n$ points to replace the original joint PDF of the random variables by matching up to the second- and third-order noncrossed moments. The moment of $Y$, that is, $E(y^k)$, $k = 1, 2$, where $E$ is the expectation, is then calculated by weighting the values of $Y$ to the power of $k$ evaluated at each of the $2^n$ points. When $n$ becomes large, the use of $2^n$ probability concentrations is not economical. Hence, a simplified method that makes use of only $2n$ estimates, which is referred to as a TPEM, was used in OPF problem with uncertainty.

***Function of One Variable*** First, a fictitious distribution of $X$ is chosen in such a way that the first three moments exactly match the first three moments of the given PDF of $X$. In order to estimate the first three moments of $Y$, one can choose a distribution of $X$ having only two concentrations placed unsymmetrically around the $X$'s expectation. If that is the case, one has enough parameters to take into account the first three moments of and to obtain a third-order approximation to the first three moments of $Y$. A particularly simple function satisfying these requirements consists in two concentrations, $P_1$ and $P_2$, of the probability density function $f_X(x)$, respectively, at $X = x_1$ and $x_2$

$$f_X(x) = P_1 \delta(x - x_1) + P_2 \delta(x - x_2) \tag{13.127}$$

where the lowercase letters denote specific values of a random variable, and $\delta(\bullet)$ is Dirac's delta function.

Choosing

$$\eta_i = \frac{|x_i - \mu_X|}{\sigma_X}, \quad i = 1, 2 \tag{13.128}$$

where, $\mu_X$ and $\sigma_X$ are the mean and the standard deviation of $X$, respectively, one can calculate the first three moments of $f_X(x)$. Thus, the $j$th moment is defined as

$$M_j(X) = \int_{-\infty}^{\infty} x^j f_X(x) dx \quad j = 1, 2, \ldots \tag{13.129}$$

The central moments are

$$M_j'(X) = \int_{-\infty}^{\infty} (x - \mu_X)^j f_X(x) dx \quad j = 1, 2, \ldots \tag{13.130}$$

The zeroth and the first moment always equal 1 and 0, respectively. The zeroth and the first three central moments of equation (13.127) are then

$$M_0' = 1 = P_1 + P_2 \tag{13.131}$$

$$M_1' = 0 = \eta_1 P_1 - \eta_2 P_2 \tag{13.132}$$

$$M_2' = \sigma_X^2 = \sigma_X^2(\eta_1^2 P_1 + \eta_2^2 P_2) \tag{13.133}$$

$$M'_3 = v_X \sigma_X^3 = \sigma_X^3 (\eta_1^3 P_1 - \eta_2^3 P_2) \tag{13.134}$$

where $v_X$ is the skewness of $X$.

Using the Taylor series expansion of $h(X)$ about $\mu_X$ yields

$$h(X) = h(\mu_X) + \sum_{j=1}^{\infty} \frac{1}{j!} g^{(j)}(\mu_X)(x - \mu_X)^j \tag{13.135}$$

where $g^{(j)}, j = 1, 2, \ldots$, stands for the $j$th derivative of $h$ with respect to $x$. The mean value of $Y$ can be calculated by taking the expectation of the above equation, resulting in

$$\mu_Y = E(h(X)) = \int_{-\infty}^{\infty} h(x) f_X(x) dx = h(\mu_X) + \sum_{j=1}^{\infty} \frac{1}{j!} g^{(j)}(\mu_X) M'_j(X) \tag{13.136}$$

Let

$$x_i = \mu_X + \eta_i \sigma_X, \quad i = 1, 2 \tag{13.137}$$

and $P_i$ be the probability concentrations at location $x_i$, $i = 1, 2$. Multiplying equation (13.135) by $P_i$, and summing them up, we get

$$P_1 h(x_1) + P_2 h(x_2) = h(\mu_X)(P_1 + P_2) + \sum_{j=1}^{\infty} \frac{1}{j!} g^{(j)}(\mu_X)(P_1 \eta_1^j + P_2 \eta_2^j) \sigma_X^j \tag{13.138}$$

From the first four terms of equations (13.136) and (13.138), we get

$$P_1 + P_2 = M'_0(X) = 1 \tag{13.139}$$

$$\eta_1 P_1 + \eta_2 P_2 = M'_1(X)/\sigma_X = \lambda_{X,1} \tag{13.140}$$

$$\eta_1^2 P_1 + \eta_2^2 P_2 = M'_2(X)/\sigma_X^2 = \lambda_{X,2} \tag{13.141}$$

$$\eta_1^3 P_1 + \eta_2^3 P_2 = M'_3(X)/\sigma_X^3 = \lambda_{X,3} \tag{13.142}$$

The above four equations have four unknowns, that is, $P_1$, $P_2$, $\eta_1$ and $\eta_2$. Their solutions are

$$\eta_1 = \lambda_{X,3}/2 + \sqrt{1 + (\lambda_{X,3}/2)^2} \tag{13.143}$$

$$\eta_2 = \lambda_{X,3}/2 - \sqrt{1 + (\lambda_{X,3}/2)^2} \tag{13.144}$$

$$P_1 = -\eta_2/\varepsilon \tag{13.145}$$

$$P_2 = \eta_1/\varepsilon \tag{13.146}$$

where

$$\varepsilon = \eta_1 - \eta_2 = 2\sqrt{1 + (\lambda_{X,3}/2)^2} \tag{13.147}$$

For a normal distribution, $\lambda_{X,3} = 0$, then equations (13.143)–(13.146) can be simplified as

$$\eta_1 = 1 \tag{13.148}$$

$$\eta_2 = -1 \tag{13.149}$$

$$P_1 = P_2 = \frac{1}{2} \tag{13.150}$$

From equations (13.138)–(13.142), and equations (13.148)–(13.150), we get

$$h(\mu_X) + \sum_{j=1}^{3} \frac{1}{j!} g^{(j)}(\mu_X)\lambda_{X,j}\eta_X^j = P_1 h(x_1) + P_2 h(x_2)$$

$$-\sum_{j=4}^{\infty} \frac{1}{j!} g^{(j)}(\mu_X)(P_1\eta_1^j + P_2\eta_2^j)\sigma_X^j \tag{13.151}$$

Substituting equation (13.151) in equation (13.136),

$$\mu_Y = P_1 h(x_1) + P_2 h(x_2) + \sum_{j=4}^{\infty} \frac{1}{j!} g^{(j)}(\mu_X)(\lambda_{X,j} - P_1\eta_1^j - P_2\eta_2^j)\sigma_X^j \tag{13.152}$$

and neglecting the third term in equation (13.152), we get

$$\mu_Y \approx P_1 h(x_1) + P_2 h(x_2) \tag{13.153}$$

This is a third-order approximation. If $h(X)$ is a third-order polynomial, that is, the derivatives of order higher than three are zero, TPEM gives the exact solution to $\mu_Y$.

Similarly, the second- and the third-order moment of $Y$ can be approximated by

$$E(Y^2) \approx P_1 h(x_1)^2 + P_2 h(x_2)^2 \tag{13.154}$$

$$E(Y^3) \approx P_1 h(x_1)^3 + P_2 h(x_2)^3 \tag{13.155}$$

***Function of Several Variables***  Let $Y$ be a random quantity that is a function of $n$ random variables, that is,

$$Y = h(X) = h(x_1, x_2, \dots, x_n) \tag{13.156}$$

Let $\mu_{X,k}, \sigma_{X,k}, \nu_{X,k}$ stand for the mean, standard deviation, and skewness of $X_k$, respectively. Let $P_{k,i}$ stand for the concentrations (or weights) located at

$$X = [\mu_{X,1}, \mu_{X,2}, \dots, \mu_{X,n}] \tag{13.157}$$

and

$$x_{k,i} = \mu_{X,k} + \eta_{k,i}\sigma_{X,k}, \quad i = 1, 2, \ldots, n \tag{13.158}$$

Expand equation (13.156) in a multivariable Taylor series about the mean value of $X$. Similar to the case of a function of one variable, the following three equations can be obtained by matching the first three moments of the PDF of $X_k$.

$$\sum_{k=1}^{n}(P_{k,1} + P_{k,2}) = 1 \tag{13.159}$$

$$\eta_{k,1}P_{k,1} + \eta_{k,2}P_{k,2} = M_1'(X_k)/\sigma_{X,k} = \lambda_{X,k,1} \tag{13.160}$$

$$\eta_{k,1}^2 P_{k,1} + \eta_{k,2}^2 P_{k,2} = M_2'(X_k)/\sigma_{X,k}^2 = \lambda_{X,k,2} \tag{13.161}$$

$$\eta_{k,1}^3 P_{k,1} + \eta_{k,2}^3 P_{k,2} = M_3'(X_k)/\sigma_{X,k}^3 = \lambda_{X,k,3} \tag{13.162}$$

Equation (13.159) can also be expressed as

$$P_{k,1} + P_{k,2} = 1/n \tag{13.163}$$

We also can get the solution for the random variable $X_k$.

$$\eta_{k,1} = \lambda_{k,3}/2 + \sqrt{n + (\lambda_{k,3}/2)^2} \tag{13.164}$$

$$\eta_{k,2} = \lambda_{k,3}/2 - \sqrt{n + (\lambda_{k,3}/2)^2} \tag{13.165}$$

$$P_{k,1} = -\eta_{k,2}/(n\varepsilon_k) \tag{13.166}$$

$$P_{k,2} = \eta_{k,1}/(n\varepsilon_k) \tag{13.167}$$

where

$$\varepsilon_k = \eta_{k,1} - \eta_{k,2} = 2\sqrt{n + (\lambda_{k,3}/2)^2}, \quad k = 1, 2, \ldots, n \tag{13.168}$$

For symmetric probability distributions, $\lambda_{k,3} = 0$, equations (13.164)–(13.167) can then be simplified as below.

$$\eta_{k,1} = \sqrt{n} \tag{13.169}$$

$$\eta_{k,2} = -\sqrt{n} \tag{13.170}$$

$$P_{k,1} = P_{k,2} = 1/(2n) \tag{13.171}$$

Thus, the first three moments can then be approximated by

$$E(Y) \approx \sum_{k=1}^{n} \sum_{i=1}^{2} \left( P_{k,i} h\left( \left[ \mu_{X,1}, \ldots, \mu_{k,i}, \ldots, \mu_{X,n} \right] \right) \right) \tag{13.172}$$

$$E(Y^2) \approx \sum_{k=1}^{n} \sum_{i=1}^{2} \left( P_{k,i} h\left( \left[ \mu_{X,1}, \ldots, \mu_{k,i}, \ldots, \mu_{X,n} \right] \right)^2 \right) \tag{13.173}$$

$$E(Y^3) \approx \sum_{k=1}^{n} \sum_{i=1}^{2} \left( P_{k,i} h\left( \left[ \mu_{X,1}, \ldots, \mu_{k,i}, \ldots, \mu_{X,n} \right] \right)^3 \right) \tag{13.174}$$

**Computational Procedure**  The procedure for computing the moments of the output variables for the OPF problem can be summarized in the following steps [29].

**(1)** Determine the number of uncertain variables.

**(2)** Set $E(Y) = 0$ and $E(Y^2) = 0$.

**(3)** Set $k = 1$.

**(4)** Determine the locations of concentrations $\eta_{k,1}, \eta_{k,2}$ and the probabilities of concentrations $P_{k,1}, P_{k,2}$ from equations (13.169)–(13.171).

**(5)** Determine the two concentrations $x_{k,1}, x_{k,2}$

$$x_{k,1} = \mu_{X,k} + \eta_{k,1} \sigma_{X,k} \tag{13.175}$$

$$x_{k,2} = \mu_{X,k} + \eta_{k,2} \sigma_{X,k} \tag{13.176}$$

where $\mu_{X,k}, \sigma_{X,k}$ are the mean and standard derivation of $X_k$, respectively.

**(6)** Run the deterministic OPF for both concentrations $x_{k,i}$ using $X = [\mu_{X,1}, \mu_{X,2}, \ldots, \mu_{X,n}]$.

**(7)** Update $E(Y)$ and $E(Y^2)$ using equations (13.172)–(13.173).

**(8)** Calculate the mean and standard deviation

$$\mu_Y = E(Y) \tag{13.177}$$

$$\sigma_Y = \sqrt{E(Y)^2 - \mu_Y^2} \tag{13.178}$$

**(9)** Repeat steps (4)–(8) for $k = k + 1$ until the list of uncertain variables is exhausted.

**Comparison TPEM with MCS**  Since OPF is a deterministic tool, it would have to be run many times to encompass all, or at least the majority of, possible operating conditions. More accurate Monte Carlo simulations (MCSs), which are able to handle "complex" random variables, are an option but are computationally more demanding and, as such, of limited use for online types of applications. Herein, the mean and standard deviation of the TPEM are compared with the corresponding values obtained with the MCS, which are calculated as

$$\mu_{\text{MCS}} = \frac{1}{N} \sum_{i=1}^{N} x_i \qquad (13.179)$$

$$\sigma_{\text{MCS}} = \sqrt{\frac{1}{N} \sum_{i=1}^{N} (x_i - \mu_{\text{MCS}})^2} \qquad (13.180)$$

where $N$ is the number of Monte Carlo samples, and $x$ is the variable for which the mean $\mu_{\text{MCS}}$ and standard deviation $\sigma_{\text{MCS}}$ are calculated. The errors for the mean and standard deviation, respectively, are therefore defined as

$$\varepsilon_\mu = \frac{\mu_{\text{MCS}} - \mu_{\text{TPEM}}}{\mu_{\text{MCS}}} \times 100\% \qquad (13.181)$$

$$\varepsilon_\sigma = \frac{\sigma_{\text{MCS}} - \sigma_{\text{TPEM}}}{\sigma_{\text{MCS}}} \times 100\% \qquad (13.182)$$

The investigation and tests show that the output variables tend to have the same PD as the input variables, which is a normal distribution. Thus, the corresponding mean and standard deviation of the TPEM and MCS works reasonably well in most cases, given the fact that output variables tend to be normally distributed.

It is noted that the TPEM approach is accurate provided that the OPF is "well behaved" and that the number of uncertain parameters is not "too large." In larger systems, the TPEM does not perform well if the number of uncertain variables is too large. With lower numbers of uncertain variables, the performance is adequate. The TPEM method is computationally significantly faster than using an MCS approach. This is especially true when the number of uncertain parameters is low, as the computational time is directly proportional to the number of uncertain variables. When the number of random variables is large, MCS is a better alternative, given its accuracy.

### 13.9.3 Cumulant-Based Probabilistic Optimal Power Flow [32]

***Gram–Charlier A Series*** The Gram–Charlier A Series allows many PDFs, including Gaussian and gamma distributions, to be expressed as a series composed of a standard normal distribution and its derivatives. As a part of the proposed P-OPF method, distributions are reconstructed with the use of the Gram–Charlier A Series. The series can be stated as follows:

$$f(x) = \sum_{j=0}^{\infty} c_j He_j(x)\alpha(x) \qquad (13.183)$$

where $f(x)$ is the PDF for the random variable $X$. $c_j$ is the $j$th series coefficient. $He_j(x)$ is the $j$th Tchebycheff–Hermite, or Hermite, polynomial, and $\alpha(x)$ is the standard normal distribution function.

The Gram–Charlier form uses moments to compute series coefficients, while the Edgeworth form uses cumulants, which is discussed here.

Since the PDF for a normal distribution is an exponential term, taking derivatives successively returns the original function with a polynomial coefficient multiplier. These coefficients are referred to as Tchebycheff–Hermite, or Hermite, polynomials.

To illustrate how the Hermite polynomials are generated, the first four derivatives of the standard unit normal distribution are taken as follows.

$$D^0\alpha(x) = D^0 e^{-\frac{1}{2}x^2} = e^{-\frac{1}{2}x^2} \tag{13.184}$$

$$D^1\alpha(x) = D^1 e^{-\frac{1}{2}x^2} = -x e^{-\frac{1}{2}x^2} \tag{13.185}$$

$$D^2\alpha(x) = D^2 e^{-\frac{1}{2}x^2} = (x^2 - 1) e^{-\frac{1}{2}x^2} \tag{13.186}$$

$$D^3\alpha(x) = D^3 e^{-\frac{1}{2}x^2} = (3x - x^3) e^{-\frac{1}{2}x^2} \tag{13.187}$$

$$D^4\alpha(x) = D^4 e^{-\frac{1}{2}x^2} = (x^4 - 6x^2 + 3) e^{-\frac{1}{2}x^2} \tag{13.188}$$

where $D^n$ is the $n$th derivative.

The Tchebycheff–Hermite polynomials are the polynomial coefficients in the derivatives. Using the results of the first four derivatives in equations (13.184)–(13.188), the first five Tchebycheff–Hermite polynomials are written as follows:

$$He_0(x) = 1 \tag{13.189}$$

$$He_1(x) = x \tag{13.190}$$

$$He_2(x) = x^2 - 1 \tag{13.191}$$

$$He_3(x) = x^3 - 3x \tag{13.192}$$

$$He_4(x) = x^4 - 6x^2 + 3 \tag{13.193}$$

Because of the structure of equations (13.184)–(13.188), the highest power coefficient of the odd derivatives, that is, the third, fifth, seventh, etc., are negative. Equations (13.189)–(13.193) have been formed following the convention that the equations relating to the odd derivatives are multiplied by negative one, such that the coefficient of the highest power is positive [33].

Therefore, the $n$th Tchebycheff–Hermite polynomial can be symbolically written as

$$He_n(x)\alpha(x) = (-D)^n\alpha(x) \tag{13.194}$$

In addition, a recursive relationship is available to determine third-order and higher polynomials

$$He_n(x) = xHe_{n-1}(x) - (n-1)He_{n-2} \tag{13.195}$$

**Edgeworth A-Series Coefficients**   Given the cumulants for a distribution in standard form, that is, zero mean and unit variance, the coefficients for the Edgeworth form of the A series can be computed. In order to find the equations for the A series coefficients, an exponential representation of the PDF is broken into its series representation and equated with the Gram–Charlier A series in equation (13.183).

The PDF, as an exponential, is written in the following form using cumulants [9]:

$$f(x) = e^{\left(-\frac{K_3}{3!}D^3 + \frac{K_4}{4!}D^4 - \frac{K_5}{5!}D^5 + \cdots\right)}\alpha(x) \tag{13.196}$$

where $D^n$ is the $n$th derivative of the unit normal distribution, $K_n$ is the $n$th cumulant, and $\alpha(x)$ is the standard unit normal PDF.

Expanding equation (13.196) as an exponential series yields

$$
f(x) = \left[ 1 + \frac{\left(-\frac{K_3}{3!}D^3 + \frac{K_4}{4!}D^4 - \frac{K_5}{5!}D^5 + \cdots\right)}{1!} \right.
$$
$$
+ \frac{\left(-\frac{K_3}{3!}D^3 + \frac{K_4}{4!}D^4 - \frac{K_5}{5!}D^5 + \cdots\right)^2}{2!} \tag{13.196}
$$
$$
\left. + \frac{\left(-\frac{K_3}{3!}D^3 + \frac{K_4}{4!}D^4 - \frac{K_5}{5!}D^5 + \cdots\right)^3}{3!} + \cdots \right]\alpha(x)
$$

If each of the terms is expanded individually and grouped on the basis of powers of $D$, the following result is obtained:

$$
f(x) = \left[ 1 - \frac{K_3}{3!}D^3 + \frac{K_4}{4!}D^4 - \frac{K_5}{5!}D^5 + \left(\frac{K_6}{6!} + \frac{K_3^2}{2!3!^2}\right)D^6 \right.
$$
$$
\left. + \left(\frac{K_7}{7!} + \frac{2K_3K_4}{2!3!4!}\right)D^7 + \cdots \right]\alpha(x) \tag{13.197}
$$

Returning to the definition for the Gram–Charlier A series in equation (13.183) and expanding the summation yields

$$f(x) = c_0He_0(x)\alpha(x) + c_1He_1(x)\alpha(x) + c_2He_2(x)\alpha(x) + \cdots \tag{13.198}$$

**TABLE 13.5 A Series Coefficient Equation**

| Coefficient | Equation |
|---|---|
| 0 | 1 |
| 1 | 0 |
| 2 | 0 |
| 3 | $\dfrac{K_3}{6}$ |
| 4 | $\dfrac{K_4}{24}$ |
| 5 | $\dfrac{K_5}{120}$ |
| 6 | $\dfrac{1}{720}\left(K_6 + 10K_3^2\right)$ |
| 7 | $\dfrac{1}{5040}\left(K_7 + 35K_3K_4\right)$ |

Comparing equations (13.197) and (13.198), the values for the coefficients can be determined. On the basis of the equations presented, the first seven terms of the Edgeworth form of the A series are presented in Table 13.5

***Adaptation of the Cumulant Method to P-OPF Problem*** The cumulant method relies on the behavior of random variables and their associated cumulants when they are combined in a linear manner. This section discusses the formation of random variables from a linear combination of others and the role cumulants play in this combination.

Given a new random variable $z$, which is the linear combination of independent random variables, $c_1, c_1, \ldots, c_n$

$$z = a_1c_1 + a_2c_2 + \cdots + a_nc_n \tag{13.199}$$

the moment generating function $\Phi_z(s)$ for the random variable $z$ can be written as

$$\Phi_z(s) = E[e^{sz}] = E[e^{s(a_1c_1 + a_2c_2 + \cdots + a_nc_n)}]$$

$$= E[e^{s(a_1c_1)}e^{s(a_2c_2)}\ldots\ldots e^{s(a_nc_n)}] \tag{13.200a}$$

Since $c_1, c_1, \ldots, c_n$ are independent, the above equation can be written as

$$\Phi_z(s) = E[e^{s(a_1c_1)}]E[e^{s(a_2c_2)}] \cdot \cdots \cdot E[e^{s(a_nc_n)}]$$

$$= \Phi_{c_1}(a_1s)\Phi_{c_2}(a_2s) \cdot \cdots \cdot \Phi_{c_n}(a_ns) \tag{13.200b}$$

The cumulants for the variable $z$ can be computed using the cumulant-generating function, in terms of the component variables as follows:

$$\Psi_z(s) = \ln(\Phi_z(s)) = \ln(\Phi_{c_1}(a_1 s)\Phi_{c_2}(a_2 s)\cdots\cdots\Phi_{c_n}(a_n s))$$

$$= \ln(\Phi_{c_1}(a_1 s)) + \ln(\Phi_{c_2}(a_2 s)) + \cdots\cdots + \ln(\Phi_{c_n}(a_n s)) \qquad (13.201)$$

$$= \Psi_{c_1}(a_1 s) + \Psi_{c_2}(a_2 s) + \cdots\cdots + \Psi_{c_n}(a_n s)$$

To compute the second-order cumulant, the first-, and second-order derivatives of the cumulant generating function for the random variable $z$ are computed as

$$\Psi_z'(s) = a_1 \Psi_{c_1}'(a_1 s) + a_2 \Psi_{c_2}'(a_2 s) + \cdots\cdots + a_n \Psi_{c_n}'(a_n s) \qquad (13.202)$$

$$\Psi_z''(s) = a_1^2 \Psi_{c_1}''(a_1 s) + a_2^2 \Psi_{c_2}''(a_2 s) + \ldots\ldots + a_n^2 \Psi_{c_n}''(a_n s) \qquad (13.203)$$

Evaluating equation (13.203) at $s = 0$ gives

$$\Psi_z''(0) = a_1^2 \Psi_{c_1}''(0) + a_2^2 \Psi_{c_2}''(0) + \cdots\cdots + a_n^2 \Psi_{c_n}''(0) \qquad (13.204)$$

Similarly, the $n$th-order cumulant for $z$, a linear combination of independent random variables, can be determined with the following equation.

$$\lambda_n = \Psi_z^{(n)}(0) = a_1^n \Psi_{c_1}^{(n)}(0) + a_2^n \Psi_{c_2}^{(n)}(0) + \cdots\cdots + a_n^n \Psi_{c_n}^{(n)}(0) \qquad (13.205)$$

where the exponent $(n)$ denotes the $n$th derivative with respect to $s$.

The cumulant method is adapted from the basic derivation above to accommodate the P-OPF problem when a logarithmic barrier interior point method (LBIPM)-type solution is used. The Hessian of the Lagrange function is necessary for the computation of the Newton step in the LBIPM. The inverse of the Hessian, however, can be used as the coefficients for the linear combination of random bus loading variables. The pure Newton step is computed in iteration $k$ of the LBIPM using the following equation:

$$y_{k+1} = y_k - H^{-1}(y_k)G(y_k) \qquad (13.206)$$

where, $y$ is the vector of variables; $G(y_k)$ is the gradient of the Lagrange function; and $H^{-1}(y_k)$ is the inverse Hessian matrix, which contains the multipliers for a linear combination of PDFs for random bus loads.

It is necessary to introduce the cumulants related to the random loads into the system in such a way that the cumulants for all other system variables can be computed. Some characteristics of the gradient of the Lagrangian are used to accomplish this. When the gradient of the Lagrangian is taken, the power flow equations appear unmodified in this vector. Therefore, cumulant models in the bus loads map directly into the gradient of the Lagrangian. For the purposes of mapping, the mismatch vector, in equation (13.206), is replaced by a new vector containing the cumulants of the random loads in the rows corresponding to their associated power flow equations.

The linear mapping information contained in the inverse Hessian can be used to determine cumulants for other variables when bus loading is treated as a random variable. If $-H^{-1}(y_k)$ is written in the following form

$$-H^{-1} = \begin{bmatrix} a_{1,1} & a_{1,2} & \cdots & a_{1,n} \\ a_{2,1} & a_{2,1} & \cdots & a_{2,n} \\ \cdots & \cdots & \cdots & \cdots \\ a_{n,1} & a_{n,,2} & \cdots & a_{n,n} \end{bmatrix} \qquad (13.207)$$

then the $n$ th cumulant for the $i$ th variables in $y$ is computed using the following equation:

$$\lambda_{yi,n} = a_{i,1}^n \lambda_{x1,n} + a_{i,2}^n \lambda_{x2,n} + \cdots + a_{i,n}^n \lambda_{xn,n} \qquad (13.208)$$

where $y_i$ is the $i$ th element in $y$ and $\lambda_{xj,n}$ is the $n$ th cumulant for th $j$ th component variable.

For the cumulant method used for P-OPF, the cumulants for unknown random variables are computed from known random variables, and PDFs are reconstructed using the Gram–Charlier/Edgeworth expansion theory.

## 13.10 COMPARISON OF DETERMINISTIC AND PROBABILISTIC METHODS

As we analyzed in this chapter, it is impossible to obtain all available data in the real time operation because of the aforementioned uncertainties of power systems and competitive environment. Nevertheless, it is important to select an appropriate technique to handle these uncertainties. The existing deterministic methods and tools are not adequate to handle them. The probabilistic methods, Gray Mathematics, fuzzy theory, and analytic hierarchy process (AHP) [34–37] are very useful to compute the unavailable or uncertain data so that power system operation problems such as the economic dispatch, optimal power flow, and state estimation can be solvable even when some data are not available.

The deterministic and probabilistic methods are compared Table 13.6.

Through comparing the various approaches, the following methods to handle uncertainties are recommended:

- Characterization and probabilistic methods
- Probabilistic methods/tools for evaluating the contingencies
- Fuzzy/ANN/AHP methods to handle uncertainties (e.g., contingency ranking)
- Risk management tools to optimize energy utilization while maintaining the required levels of reliability
- Cost–benefit-analysis (CBA) for quantifying the impact of uncertainty.

**TABLE 13.6  Deterministic Versus Probabilistic Methods**

| Methods Comparison | Deterministic Method | Probabilistic Method |
|---|---|---|
| Contingency selection | Typically a few probable and extreme contingencies | More exhaustive list of contingencies; ranking based on fuzzy/AHP methods |
| Contingency probabilistic | Based on judgment | Based on inadequate or uncertain data (ANN, fuzzy, and AHP methods) |
| Load levels (forecast) | Typically seasonal peaks and selected off-peak loads | Multiple levels with uncertain factors (fuzzy, ANN) |
| Unit commitment | Traditional optimization technology | Optimization technology and AHP/fuzzy/ANN |
| Security regions | Deterministic security region | Variable security regions |
| Criteria for decision | Well established | Need a suitable method/criteria to make decision (ANN, fuzzy, and AHP methods) |

# PROBLEMS AND EXERCISES

1. List some uncertainties occurred in power systems operation and planning.

2. List several major methods to handle uncertainties.

3. List several probabilistic OPF methods.

4. What is the chance-constrained optimization method?

5. What uncertainties should be taken into account in a large hydrothermal system operation?

6. How is the probabilistic method used in power system operation and planning?

7. If the uncertain load $P_D$ is expressed as a normal distribution, write the probability density function of this load.

8. There are two uncertainty loads that are expressed by a triangular fuzzy number, $P_{D1} = (25, 4, 6)_{LR}$ and $P_{D2} = (28, 9, 5)_{LR}$.

    (1) Use a diagram represent these two loads.

    (2) Compute the total of the two loads.

9. There are three uncertainty loads that are expressed by a triangular fuzzy number, $P_{D1} = (20, 3, 6)_{LR}$, $P_{D2} = (18, 4, 3)_{LR}$, and $P_{D3} = (23, 7, 5)_{LR}$

    (1) Use a diagram represent these three loads.

    (2) Compute the total of the three loads.

**10.** Use the same data as in exercise 4. If we represent two uncertainty loads by using an interval format of a fuzzy number and 0.8-cut of the fuzzy number, what is the sum of the two uncertainty loads $P_{D1}$ and $P_{D2}$?

**11.** Use the same data as in exercise 5. If we represent three uncertainty loads by using an interval format of a fuzzy number, and 0.7-cut of the fuzzy number, what is the sum of three uncertainty loads $P_{D1}$, $P_{D2}$ and $P_{D3}$?

# REFERENCES

1. Merlin AS. Latest developments and future prospects of power system operation and control. Int. J. of Elec. Power 1994;16(3):137–139.
2. Rau N, Fong CC, Grigg CH, Silverstein B. Living with uncertainty. IEEE Power Eng. Rev. 1994;14(11):24–26.
3. Abdul-Rahman KH, Shahidehpour SM, Deeb NI. Effect of EMF on minimum cost power transmission. IEEE Trans. Power Syst. 1995;10(1):347–353.
4. Ivey M. Accommodating uncertainty in planning and operation. Workshop on Electric Transmission Reliability, Washington, DC September 17, 1999.
5. Leite da Silva AM, Arienti VL, Allan RN. Probabilistic load flow considering dependence between input nodal powers. IEEE Trans. PAS 1984;103(6):1524–1530.
6. El-Hawary ME, Mbamalu GAN. A comparison: probabilistic perturbation and deterministic based optimal power flow solutions. IEEE Trans. Power Syst. 1991;6(3):1099–1105.
7. Sauer PW, Hoveida B. Constrained stochastic power flow analysis. Electr. Pow. Syst. Res. 1982;5:87–95.
8. Karakatsanis TS, Hatziargyriou ND. Probabilistic constrained load flow based on sensitivity analysis. IEEE Trans. Power Syst. 1994;9(4):1853–1860.
9. Abdul-Rahman KH, Shahidehpour SM, Deeb NI. AI approach to optimal VAR control with fuzzy reactive loads. IEEE Trans. Power Syst. 1995;10(1):88–97.
10. Miranda V, Saraiva JT. Fuzzy modeling of power system optimal load flow. IEEE Trans. Power Syst. 1992;7(2):843–849.
11. Wang Z, Alvarado FL. Interval arithmetic in power flow analysis. IEEE Trans. Power Syst. 1992;7(3):1341–1349.
12. *NIST/SEMATECH e-Handbook of Statistical Methods*, http://www.itl.nist.gov/div898/handbook.
13. More RE. *Interval Analysis*. Englewood Gliffs, NJ: Prentice-Hall; 1996.
14. F.N. Ris, Interval analysis and applications to linear algebra, D. Phil, Thesis, Oxford, 1972.
15. Hansen E, Smith R. Interval arithmetic in matrix computations, Part II. SIAM J. Numer. Anal. 1967;4:1–9.
16. Hansen E, Sengupta S. Bounding solutions of systems of equations using interval analysis. BIT 1981;21:203–211.
17. Valdma M, Keel M, Liik O, Tammoja H. Method for minimax optimization of power system operation. Proceedings of IEEE Bologna PowerTech 2003, 23–26 June 2003, Bologna, Italy. Paper 252. p. 1–6.
18. Valdma M, Keel M, Liik O. Optimization of active power generation in electric power system under incomplete information, Proceedings of Tenth Power Systems Computation Conference, 1990, Graz, Austria, p. 1171–1176.
19. Selvi K, Ramaraj N, Umayal SP. Genetic algorithm applications to stochastic thermal power dispatch. IE(I) Journal EL June 2004;85:43–48.
20. Abdul-Rahman KH, Shahidehpour SM. Static security in power system operation with fuzzy real load constraints. IEEE Trans. Power Syst. 1995;10(1):77–87.
21. Kaufmann A, Gupta MM. *Fuzzy Mathematical Models in Engineering and Management Science*. Amsterdam: North-Holl and Publishing Company; 1988.
22. Dillion TS, Tun T. Integration of the sub-problems involved in the optimal economic operation of hydro-thermal system. Proceedins of IFAC Symposium Control and Management of Integrated Industry. France, September 1977, p. 171–180.

23. Tun T, Dillion TS. Sensitivity analysis of the problem of economic dispatch of hydro-thermal system. Proceedings of IFAC Symposium Auto Control and Protection of Electric Power System: Melbourne; 1977.

24. Ozturk UA, Mazumdar M, Norman BA. A solution to the stochastic unit commitment problem using chance constrained programming. IEEE Trans. Power Syst. 2004;19(3):1589–1598.

25. Qiu J, Shahidehpour SM. A new approach for minimizing power losses and improving voltage profile. IEEE Trans. Power Syst. 1987;2(2):287–295.

26. Abdul-Rahman KH, Shahidehpour SM. Application of fuzzy sets to optimal reactive power planning with security constraints. IEEE Trans. Power Syst. 1994;9(2):589–597.

27. Dantzig GB, Wolfe P. The decomposition algorithm for linear programs. Econometrica 1961;29(4):767–778.

28. Madrigal M, Ponnambalam K, Quintana VH. Probabilistic optimal power flow. Proceedingsof IEEE Canada Conference on Electrical Computer Engineering, Waterloo, ON, Canada, May 1998, p. 385–388.

29. Verbič G, Cañizares CA. Probabilistic optimal power flow in electricity markets based on a two-point estimate method. IEEE Trans. Power Syst. 2006;21(4):1883–1993.

30. Hong HP. An efficient point estimate method for probabilistic analysis. Reliab. Eng. Syst. Saf. 1998;59:261–267.

31. Rosenblueth E. Point estimation for probability moments. Proc. Nat. Acad. Sci. USA 1975;72(10):3812–3814.

32. Schellenberg A, Rosehart W, Aguado J. Cumulant based probabilistic optimal power flow (P-OPF). IEEE Trans. Power Syst. 2005;20(2):773–781.

33. Kendall MG, Stuart A. The Advanced Thoery of Statistics. 4th ed. New York: Macmillan; 1977.

34. Satty TL. *The Analytic Hierarchy Process*. New York: McGraw Hill, Inc.; 1980.

35. Zhu JZ, Irving MR. Combined active and reactive dispatch with multiple objectives using an analytic hierarchical process. IEE Proc., Part C 1996;143:344–352.

36. Zhu JZ, Irving MR, Xu GY. Automatic contingency selection and ranking using an analytic hierarchical process. Electr. Mach. Pow Syst J. 1998;(4).

37. Zhu JZ, Momoh JA. Optimal VAR pricing and VAR placement using analytic hierarchy process. Electr. Pow. Syst. Res. 1998;48(1):11–17.

# CHAPTER 14

# OPERATION OF SMART GRID

## 14.1 INTRODUCTION

Traditionally, the term *grid* is used for an electricity system that may support all or some of the following four operations: electricity generation, electricity transmission, electricity distribution, and electricity control.

A smart grid is a set of disparate goals, including facilitating better competition among suppliers; enabling better use of different energy sources; and setting up the automation and monitoring abilities needed for the grid at cross continent.

In 2009, the US President, Barack Obama, asked the United States Congress "to act without delay" to pass legislation that included doubling alternative energy production in the next 3 years and building a new electricity "**smart grid**." Europe and Australia are also following similar visions. In those countries the integration of communications and power control, both of which have generally fallen under greater government supervision, is more advanced, with utilities often required or asked to provide competitive access to communications transit exchanges and distributed power co-generation connection points. The smart grid in China focuses more on the transmission side than the distribution side at present. China is constructing ultra-high and extra-high voltage direct current $(+/-800$ kV, $+/-500$ kV$)$ and alternating current transmission systems (1000 kV, 500 kV, 220 kV), and coordinating the development of a smart grid based on information technology and automation technology.

To reduce power demand during peak usage periods, communications and metering technologies inform smart devices in the house, factory, or business building when energy demand is high and track how much electricity is used and when it is used. Electricity prices increase during peak usage periods and decrease during low-demand periods. The end user will tend to consume less during peak usage periods if it is made possible for users and user devices to be familiar with the high price premium for using electricity at peak periods. When end users see a direct economic profit to become more energy efficient, it is more likely that they will make wise decisions on consumption.

This chapter will introduce some basic concepts and technologies of the smart grid, as well as applications of smart grid operation [1–32].

*Optimization of Power System Operation*, Second Edition. Jizhong Zhu.
© 2015 The Institute of Electrical and Electronics Engineers, Inc. Published 2015 by John Wiley & Sons, Inc.

## 14.2 DEFINITION OF SMART GRID

As we mentioned in the previous section, there are different definitions of smart grid from different viewpoints. Some people call the intelligent transmission and distribution automation network a smart grid. Some people think the smart grid refers to distributed generation and storage, which includes solar energy, wind power, micro turbines, compressed air, energy storage, and so on. Looking the end-user side, there is another aspect of the smart grid. We call it demand response and load control. Demand response relates to how the end user reacts to different price signals, different availability signals, and so on. In addition, the advanced metering infrastructure (AMI) is also important. It is the interface between the home or end user and the smart grid. AMI technology uses remote two-way wireless communication to retrieve customer energy usage information at frequent intervals from customers' electric smart meters and/or natural gas meters via a radio frequency (RF) fixed network. A meter data management system receives and houses the data for analysis and use by other systems such as customer information and billing, power outage management, load research, and delivery system planning. All these are related to smart grid. Then, what is the official definition of a smart grid? According to the US Department of Energy's Modern Grid Initiative, "an intelligent or a smart grid integrates advanced sensing technologies, control methods and integrated communications into the current electricity grid." It has seven characteristics:

1. *Consumer participation*: Enables and motivates active participation by consumers
2. *Accommodate generation options*: Accommodates all generation and energy storage options
3. *Enable electricity market*: Enables new products, services, and markets
4. *High-quality power*: Provides the quality of power required for the digital, computer, and communication-based economy
5. *Optimize assets*: Operates efficiently and optimizes the utilization of existing and new assets
6. *Self-healing*: Anticipates and reponds to system disturbances in a self-healing manner
7. *Resist attack*: Operates resiliently against attack and natural disaster.

## 14.3 SMART GRID TECHNOLOGIES

Making the smart grid work will require a series of reliable technologies, which include integrated communications systems, sensors, advanced meters, and storage devices. Many of these already exist; others are being adapted to synchronize with a modern power grid. The US Department of Energy has identified five key technology areas for the smart grid as follows:

1. Integrated communications to allow every part of the grid to both "talk" and "listen," that is, two-way communication technology.

   Some communications are up to date, but are not uniform and not fully integrated into the grid. Areas for improvement include substation automation, demand response, distribution automation, supervisory control and data acquisition (SCADA), energy management systems, wireless mesh networks and other technologies, power-line carrier communications, and fiber optics. Integrated communications will allow for real-time control, information and data exchange to optimize system reliability, asset utilization, and security.

2. Sensing and measurement technologies to support faster and more accurate response.

   The technologies of sensing and measurement include smart meters, meter reading equipment, phasor measurement units (PMUs), dynamic line rating, advanced switches and cables, and digital protective relays. Especially, the smart meters, which replace the analog mechanical meters, record usage in real time. A wide-area measurement system (WAMS) is a network of PMUS that can provide real-time monitoring on a regional and national scale.

3. Advanced components to apply the latest research in superconductivity, power electronics, storage, and diagnostics.

   Innovations in superconductivity, fault tolerance, storage, power electronics, and diagnostics components are changing fundamental abilities and characteristics of grids. The related technologies include flexible alternating current transmission system devices, high-voltage direct current, first- and second-generation superconducting wire, high-temperature superconducting cable, distributed energy generation and storage devices, composite conductors, and "intelligent" appliances.

4. Advanced control methods for monitoring, diagnosing, and addressing any event.

   The technology categories for advanced control methods are distributed intelligent agents (control systems), analytical tools (software algorithms and high-speed computers), and operational applications (SCADA, substation automation, demand response, etc.). The advanced algorithms have been discussed in the earlier chapters in the book.

5. Improved interfaces and decision support to enhance human decision making. Technologies include visualization techniques that reduce large quantities of data into easily understood visual formats, software systems that provide multiple options when systems operator actions are required, and simulators for operational training and "what-if" analysis.

## 14.4    SMART GRID OPERATION

A smart grid is typically reliable, secure, efficient, economic, environment friendly, and safe to the extreme extent as feasible. It is the application of technologies to all aspects of the energy transmission and delivery system that provide better monitoring and control, and efficient use of the system. The objectives of smart grid operation and control are

- to address the challenges that secure and reliable operation of the power grids will face in the future;
- to develop a solid interdisciplinary theoretical foundation supporting development of better tools for planning, operation, and control of power grids interconnected at various voltage levels;
- to innovate in power distribution monitoring and control;
- to enable consumers to react to grid conditions making them active participants in their energy use;
- to leverage conventional generation and emerging technologies when possible including distributed energy resources, demand response, and energy storage, to address the challenges introduced by variable renewable resources.

For achieving the aforementioned objectives, it requires on the one hand, a smart grid to adjust with generation and its possible storage with availability whenever and wherever called for, self-healing mechanism in the face of disturbance, optimum utilization of assets achieving high level of efficiency in operation, while on the other hand, consumer should get quality electricity as per quantitative requirement, through successfully enabled provision of services, products marketed, etc. Extensive usage of digital technology in terms of communication and information technology on real-time basis is an essential feature for achieving success in the matter considering the demand–supply scenario accurately at every instant.

A key factor of smart grid operation will be *distributed generation* (DG). DG takes advantage of *distributed energy resource* (DER) systems (e.g., solar panels and small wind turbines), which are often small-scale power generators (typically in the range of 3 to 10,000 kW), in order to improve the power quality and reliability. However, implementing DG(s) in practice is not an easy thing. First, DG involves large-scale deployments for generation from renewable resources, such as solar and wind. As we mentioned in Chapter 10, the operation of renewable energy is subject to wide fluctuations. Precise wind or solar forecast is required. Therefore, effective utilization of the DG in a way that is cognizant of the variability of the yield from renewable sources is important. Second, with the current technologies, the usual operation costs of distributed generators for generating one unit of electricity are high compared with that of traditional large-scale central power plants. The development and deployment of DG further lead to a concept, namely the *virtual power plant* (VPP), which manages a large group of distributed generators with a total capacity comparable to that of a conventional power plant. This cluster of distributed generators is collectively run by a central controller. The concerted operational mode delivers extra benefits such as the ability to deliver peak load electricity or load-aware power generation at short notice. Such a VPP can replace a conventional power plant while providing higher efficiency and more flexibility. DG and VPP will be further discussed in the following sections.

## 14.4.1    Demand Response

Demand response (DR) encompasses many customer-level actions that can help to smooth the electric power load shape and reduce energy consumption. There are two issues: one is reducing the peak load to keep the utility system run more efficiently, and at the same time using the energy savings by practicing or deploying demand response programs so that the overall demand for electricity is reduced. Therefore, demand response has two components. One is the load component or the kilowatt component, which is applying demand response to reduce the peak load. The other is the energy component or the kilowatt hour component, which is applying demand response to save energy by using less or using more efficient devices, appliances, and so on.

The formal definition of demand response is given by the US Federal Energy Regulatory Commission (FERC). According to FERC, demand response is "a reduction in the consumption of electric energy by customers from their expected consumption in response to an increase in the price of electric energy or to incentive payments designed to induce lower consumption of electric energy."

The definition of demand response is a little different from that of demand-side management (DSM). In the DSM scenario, the load is controlled by the electric utility and once the customer gives their consent to the electric utility to control their load, it could be air conditioner, water heater and the like, the customer has no control, that is, the customer cannot choose what to control for how long. The power company will choose for them. For example, they will turn the water for 30 minutes, change the air conditioner thermostat temperature setting or turn the AC off for half an hour or 10 minutes, whatever the case may be. So the customer has no control once they have given consent to the power company to control the load. On the other hand, in the demand response concept, the customer has full control. They will decide what load to control for how long depending on the incentive they get and what their situation is at home or at business to effect the control.

Demand response needs are driving infrastructure needs, which include smart devices and control systems that can collect data, present it to the power user, and then relay their decisions back to the utilities or third party aggregators (also called curtailment service providers). The enabling technologies include but are not limited to:

- Building automation systems—the software and hardware needed to monitor and control the mechanical, heating and cooling, and lighting systems in buildings that can also interface with smart grid technologies.

- Home Area Networks—similar to smart building technologies, except for the home where devices communicate with the smart grid to receive and present energy use and costs, as well as enable energy users to reduce or shift their use and communicate those decisions to the load-serving entities.

## 14.4.2 Devices Used in Smart Grid

The Smart Grid promises to improve the quality, resiliency, and integrity of the grid through the optimization of the existing energy delivery infrastructure and integration of new renewable generation sources. Thus, the smart grid requires seamlessly integrated products and services to deliver the highest performance possible. With all kinds of smart devices, utilities can select and confidently use the products they want knowing everything will operate together as it should.

The main devices that are used in the smart grid include:

- *Advanced metering (or smart metering) devices*—Advanced metering is often the starting point for a smart grid deployment. Advanced metering devices support acquiring data to evaluate the health and integrity of the grid and support automatic meter reading, elimination of billing estimates, and prevent energy theft.

- *Integrated communications devices*—These include data acquisition, protection, and control, and enable users to interact with intelligent electronic devices in an integrated system.

- *Home area network (HAN) devices*—A primary element of the smart grid is the enhanced communications capabilities that enable consumers to better manage their electricity consumption and costs via new smart appliances and devices located at the customer's premises. These are commonly referred to as home area network (HAN) devices.

- *In-home devices*—These include communicating thermostats, load control switches, and electric vehicle (EV) charging stations, which help consumers to manage their energy use.

- *Network infrastructure*—This comprises grid routers and signal repeaters, for example, which enable utilities to cost-effectively network their grid devices. In addition, as the smart grid extends out to homes and businesses, wireless sensors and mobile control devices become important elements in monitoring and managing energy use.

- *Geographic information system (GIS)*—With the smart grid's promises of a more reliable, robust electric delivery system come the virtual representation of that system used to make operational decisions. The source of the base data for this virtual representation is the GIS. The GIS must be able to efficiently and effectively export the required data to the systems that need it, preferably in a format that is easily imported by those receiving systems.

- *Energy storage devices*—Energy storage is accomplished by devices or physical media that store energy to perform useful operation at a later time. A device that stores energy is sometimes called an accumulator.

## 14.4.3 Distributed Generation

Distributed generation (DG) means that power sources are widely distributed, so that power is generated close to the place where it is being used. This includes all generation installed on sites owned and operated by utility customers. Most of renewable power supplies such as solar energy and wind power are distributed generation. Renewable energy and distributed generation technologies (REDG) are very important to the smart grid operation. Energy access, energy security, poverty alleviation, and environmental considerations, combined with increasing fossil fuel prices, are key drivers for accelerating the adoption of affordable and reliable renewable energy and distributed generation.

Generally, distributed generation is connected to the grid through the distribution system. This is called a grid-connected distributed generation system, which can make the whole grid more secure because there is less reliance on any particular source of power in the system. With several smaller distributed generation sources, if something goes wrong, it is easier for another source of power to step in and fill the gap. This is essential for many renewable technologies such as solar and wind, which produce intermittent power and for other technologies that may need to be shut down for periodic maintenance.

Since distributed generation is typically sited close to customer loads, it can help reduce the number of transmission and distribution lines that need to be upgraded or built. Obviously, it reduces transmission and distribution losses.

Distributed generation has the potential to mitigate congestion in transmission lines, reduce the impact of electricity price fluctuations, strengthen energy security, and provide greater stability to the smart grid.

Distributed generation encompasses a wide range of technologies including solar power, wind turbines, fuel cells, microturbines, reciprocating engines, load reduction technologies, and battery storage systems. The effective use of grid-connected distributed energy resources can also require power electronic interfaces and communications and control devices for efficient dispatch and operation of generating units. The main distribution generation technologies are summarized as follows [9].

(1) Reciprocating engine
   A reciprocating engine, also often known as a piston engine, is a heat engine that uses one or more reciprocating pistons to convert pressure into a rotating motion. Diesel- or petrol-fueled reciprocating engine is one of the most common distributed energy technologies in use today, especially for standby power applications. However, it creates significant pollution (in terms of both emissions and noise) relative to natural-gas- and renewable-fueled generators. As a result, they are subject to severe operational limitations not faced by other distributed generating technologies.

(2) Solar photovoltaic cells
   Solar photovoltaic (PV) cells is discussed in Chapter 10.

(3) Wind turbine
   Wind turbine is also discussed in Chapter 10.

(4) Fuel cells
   A fuel cell is an electrochemical cell that converts a source fuel into an electrical current through a chemical reaction in a fuel. It generates electricity inside a cell

through reactions between a fuel and an oxidant, triggered in the presence of an electrolyte. The reactants flow into the cell, and the reaction products flow out of it, while the electrolyte remains within it. Fuel cells can operate continuously as long as the necessary reactant and oxidant flows are maintained. For utilities, most fuel cells currently use natural gas, which is not renewable.

(5) Microturbine

A microturbine is a small turbine (about the size of refrigerator, generally less than 300 kW) that makes both electricity and heat in small amounts. Microturbines are becoming widespread for distributed power and combined heat and power applications. They are one of the most promising technologies for powering hybrid electric vehicles. They range from handheld units producing less than a kilowatt, to commercial-sized systems that produce tens or hundreds of kilowatts. They can run on nonrenewable fuels such as natural gas, but can also use waste fuels.

(6) Internal combustion engine

The internal combustion engine is an engine in which the combustion of a fuel (normally a fossil fuel) occurs with an oxidizer (usually air) in a combustion chamber. In an internal combustion engine the expansion of the high-temperature and high-pressure gases produced by combustion apply direct force to some component of the engine, such as pistons, turbine blades, or a nozzle.

(7) CHP technology

CHP is a combined heat and power technology. Conventional electricity generation is inherently inefficient, using only about a third of the fuel's potential energy. In applications where heating or cooling is needed as well, the total efficiency of separate thermal and power systems is still only about 45%, despite the higher efficiencies of thermal conversion equipment.

Combined cooling, heating, and power systems are significantly more efficient. CHP technologies produce both electricity and thermal energy from a single energy source. These systems recover heat that normally would be wasted in an electricity generator, then use it to produce one or more of the following: steam, hot water, space heating, humidity control, or cooling. By using a CHP system, the fuel that would otherwise be used to produce heat or steam in a separate unit is saved.

Recent technological advances have resulted in the development of a range of efficient and versatile systems for industrial and other applications. Especially, with the wide use of the renewable resources today, CHP technologies are becoming more important.

(8) Distributed energy management

Distributed energy management technologies include energy storage devices and various methods for reducing overall electrical load.

Energy storage technologies are essential for meeting the levels of power quality and reliability required by high-tech industries. Energy storage is important for other distributed energy devices by giving them more load-following capability, and also supporting renewable technologies such as wind and solar electricity by making them dispatchable.

In the smart grid, reducing electrical load can be accomplished by improving the efficiency of end-use equipment and devices, or by switching an electrical load to an alternative energy source—heating water or building interiors with heat from the earth or sun.

## 14.4.4 Simple Smart Grid Economic Dispatch with Single Generator

The economic dispatch (ED) problem is one of the fundamental problems in the power system. The objective of ED is to reduce the total power generation cost, subject to system security constraints. Previous chapters discussed various numerical methods and optimization techniques to solve the ED problem. Owing to the addition of uncertain wind power and chargeable and dischargeable storage in the smart grid, economic dispatch problem in the smart grid environment is more complicated. This section describes a simple smart grid economic dispatch (SGED) approach without considering the network security constraint.

The simplest SGED problem is a single generator single load with one battery storage device [10]. As we mentioned before, generator cost function is quadratic and can be simply expressed as follows.

$$f(P_g) = \frac{1}{2}\alpha P_g^2 + \beta P_g + \gamma \tag{14.1}$$

The cost function of the battery can be expressed as follows.

$$h(P_b) = \eta(P_{b\max} - P_b) \tag{14.2}$$

For simplifying the analysis, assume the load is constant for every time period, that is,

$$P_d(t) = D \quad t = 1, 2, \dots, T \tag{14.3}$$

Thus, the simplest SGED problem can be expressed as follows.

$$\min J = \sum_{t=1}^{T}[f(P_g(t)) + h(P_b(t))] \tag{14.4}$$

such that

$$P_b(t) = P_b(t-1) + P_g(t) - D \tag{14.5}$$

$$0 \le P_b(t) \le P_{b\max} \tag{14.6}$$

$$0 \le P_g(t) \le P_{g\max} \tag{14.7}$$

where

$P_g$: the generator power output

$P_{gmax}$: the maximal power output of the generator

$P_b$: the power value of the battery (charge or discharge)

$P_{bmax}$: the maximal capacity of the battery

$D$: the constant load value

$T$: the time period of the smart grid operation

$\alpha, \beta, \gamma$: the coefficients of the generation cost function

$\eta$: the coefficient of the battery cost function.

**Neglecting Constraint**  If the battery constraint and generator constraint are inactive, that is, the inequality constraint is ignored, from the objective function and power balance equation, we can get the following optimality condition:

$$\alpha P'_g(t) + \beta = \eta[T - (t - 1)] \tag{14.8}$$

or

$$\alpha P'_g(t) + \beta = \eta(T + 1 - t) \tag{14.9}$$

From above equation, we get

$$P'_g(t) = \frac{\eta}{\alpha}(T + 1 - t - \beta) \tag{14.10}$$

If the generator cost function is simplified as follows,

$$f(P_g) = \frac{1}{2}\alpha P_g^2 \tag{14.11}$$

then the optimal generation becomes

$$P'_g(t) = \frac{\eta}{\alpha}(T + 1 - t) \tag{14.12}$$

The power change of the battery can be obtained as

$$P'_b(t) = P_b(t - 1) + \frac{\eta}{\alpha}(T + 1 - t) - D \tag{14.13}$$

It can be observed from equation (14.12) that the optimal generation will decrease linearly over time. From equation (14.13), the battery charges initially, and then discharges. The battery changes from charging to discharging when

$$\frac{\eta}{\alpha}(T + 1 - t) - D = 0 \tag{14.14}$$

that is,

$$P'_b(t) = P_b(t - 1) \tag{14.15}$$

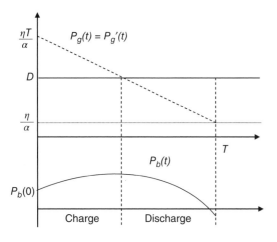

Figure 14.1  Simple SGED without constraint.

when

$$t_D = T + 1 - \frac{\alpha}{\eta}D \tag{14.16}$$

where $t_D$ is the time stage that the battery starts to discharge.

The simple SGED can be illustrated as in Figure 14.1.

***Considering Constraint***  It can be observed from equation (14.12) that the maximal generation output appears at initial time, and minimal generation output is appears at the end of operation period $T$, that is,

$$\frac{\eta}{\alpha} \le P_g^*(t) \le \frac{\eta}{\alpha}T \tag{14.17}$$

It means that the following equation should be satisfied in order to meet the generation constraint (14.7):

$$\frac{\eta T}{\alpha} \le P_{gmax} \tag{14.18}$$

Obviously, there is no generation constraint problem if the above equation is met. Since the generator supplies both load and battery, the generator's capacity must be greater than the load. Thus, the simple SGED will become a constrained problem when the generator's capacity is given as

$$D \le P_{gmax} \le \frac{\eta T}{\alpha} \tag{14.19}$$

If the optimal generation exceeds the generator's capacity at the initial time $t_g$, the generation will be set to the limit value of the generator, that is,

$$P_g'(t) = \frac{\eta}{\alpha}(T + 1 - t) = P_{gmax} \tag{14.20}$$

$$t_g = T + 1 - \frac{\alpha}{\eta}P_{gmax} \tag{14.21}$$

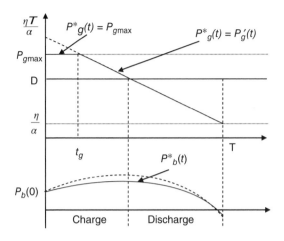

Figure 14.2   Simple SGED with constraint.

The optimal generation over time will be

$$P_g^*(t) = \begin{cases} P_{gmax}, & \text{if } t \leq t_g \\ \dfrac{\eta}{\alpha}(T+1-t), & \text{if } t > t_g \end{cases} \tag{14.22}$$

Similarly, the optimal battery value will be

$$P_b^*(t) = \begin{cases} P_b(t-1) + P_{gmax} - D, & \text{if } t \leq t_g \\ P_b(t-1) + \dfrac{\eta}{\alpha}(T+1-t) - D, & \text{if } t > t_g \end{cases} \tag{14.23}$$

Figure 14.2 demonstrated the constrained simple SGED case.

If battery power constraint (14.7) is considered and the computed optimal charging value exceeds the battery capacity at time $t_B$, the actual charging power must be set to the maximal limit, that is,

$$P_b(t_B) = P_{bmax}, \quad \text{if } P_b(t_B) > P_{bmax} \tag{14.24}$$

Owing to the reduction of the battery charging, the optimal generator output will be reduced and can be computed as follows:

$$P_b(t_B) = P_{bmax} = P_b(t_B - 1) + P_g(t_B) - D \tag{14.25}$$

$$P_g(t_B) = P_{bmax} - P_b(t_B - 1) + D \tag{14.26}$$

Thus, the optimal generation with battery capacity constraint over time will be

$$P_g^*(t) = \begin{cases} P_{bmax} - P_b(t_B - 1) + D, & \text{if } t = t_B \\ \dfrac{\eta}{\alpha}(T+1-t), & \text{if } t \neq t_B \end{cases} \tag{14.27}$$

Similarly, the optimal battery value will be

$$P_b^*(t) = \begin{cases} P_{bmax}, & \text{if } t = t_B \\ P_b(t-1) + \frac{\eta}{\alpha}(T+1-t) - D, & \text{if } t \neq t_B \end{cases} \quad (14.28)$$

On the other hand, if battery power constraint (14.7) is considered and the computed power value of the battery is negative at the discharging time $t_b$, the actual power must be set to zero, that is,

$$P_b(t_b) = 0, \quad \text{if } P_b(t) < 0 \quad (14.29)$$

Since the battery cannot discharge enough power, the optimal generator output will increase to meet the power balance of smart grid. This can be computed as follows:

$$P_b(t_b) = 0 = P_b(t_b - 1) + P_g(t_b) - D \quad (14.30)$$

$$P_g(t_b) = D - P_b(t_b - 1) \quad (14.31)$$

In this case, the optimal generation over time will be

$$P_g^*(t) = \begin{cases} P_g(t_b) = D - P_b(t_b - 1), & \text{if } t = t_b \\ \frac{\eta}{\alpha}(T+1-t), & \text{if } t \neq t_b \end{cases} \quad (14.32)$$

Similarly, the optimal battery value will be

$$P_b^*(t) = \begin{cases} 0, & \text{if } t = t_b \\ P_b(t-1) + \frac{\eta}{\alpha}(T+1-t) - D, & \text{if } t \neq t_b \end{cases} \quad (14.33)$$

In summary, the optimal SGED with battery capacity constraint can be expressed as follows.

$$P_b^*(t) = \begin{cases} P_{bmax}, & \text{if } P_b(t) > P_{bmax} \\ 0 & \text{if } P_b(t) < 0 \\ P_b(t-1) + \frac{\eta}{\alpha}(T+1-t) - D, & \text{if } 0 \leq P_b(t) \leq P_{bmax} \end{cases} \quad (14.34)$$

$$P_g^*(t) = \begin{cases} P_{bmax} + D - P_b(t-1), & \text{if } P_b(t) > P_{bmax} \\ D - P_b(t-1), & \text{if } P_b(t) < 0 \\ \frac{\eta}{\alpha}(T+1-t), & \text{if } 0 \leq P_b(t) \leq P_{bmax} \end{cases} \quad (14.35)$$

***Example 14.1:*** A simple smart grid has one generator and one storage battery. The load of 8.0 MW is assumed to be constant over time. The generator cost function is quadratic, given by

$$f(P_g) = \frac{1}{2}\alpha P_g^2 = \frac{1}{2}(0.04 P_g^2)$$

**TABLE 14.1 Results of the Simple SGED**

| Time $t$ | 1 | 2 | 3 | 4 | 5 | 6 | 7 |
|---|---|---|---|---|---|---|---|
| Generation power | 14 | 12 | 10 | 8 | 6 | 4 | 2 |
| Battery power | 8 | 12 | 14 | 14 | 12 | 8 | 2 |

The battery has initial power 2MW and the unit coefficient of battery storage is $\eta = 0.08$. The capacity of the generator is 25MW. Let us compute the optimal generation and batter power over the time period 7 hours.

According to the given parameters, we get $\alpha = 0.04$, $\eta = 0.08$, $T = 7$, and $D = 8$. We can first compute the key time point that the battery changes from charging to discharging, that is,

$$t_D = T + 1 - \frac{\alpha}{\eta}D = 7 + 1 - \frac{0.04}{0.08} \times 8 = 4$$

It means the battery will be charged until hour 4, and after that it will discharge. The optimal generation can be computed through equation (14.22), that is,

$$P'_g(t) = \frac{\eta}{\alpha}(T + 1 - t) = \frac{0.08}{0.04}(7 + 1 - t) = 16 - 2t$$

The power change of the battery can be obtained on the basis of equation (14.23), that is,

$$P'_b(t) = P_b(t - 1) + \frac{\eta}{\alpha}(T + 1 - t) - D = P_b(t - 1) + 8 - 2t$$

The calculation results are shown in Table 14.1. It can be observed from Table 14.1 that the generation decreases linearly and battery power variation is quadratic.

***Example 14.2:*** For Example 14.1, if the initial power of the battery is changed to 0.4 MW, and the capacity of the generator is 11.0 MW, then, the SGED becomes a constrained problem. The time that the generator output will be limited can be computed using equation (14.31).

$$t_g = T + 1 - \frac{\alpha}{\eta}P_{gmax} = 7 + 1 - \frac{0.04}{0.08} \times 11 = 2.5$$

This means the generator's power will be set to 11.0 MW before hour 2.5. The optimal generation over time will be computed through equation (14.31)

$$P_g^*(t) = \begin{cases} 11, & \text{if } t \le 2.5 \\ 16 - 2t, & \text{if } t > 2.5 \end{cases}$$

**TABLE 14.2    Results of the Simple SGED with Generation Constraint**

| Time $t$ | 1 | 2 | 3 | 4 | 5 | 6 | 7 |
|---|---|---|---|---|---|---|---|
| Generation power | **11** | **11** | 10 | 8 | 6 | 4 | 2 |
| Battery power | 7 | **10** | **12** | 12 | 10 | 6 | 0 |

The optimal battery value will be computed through equation (14.32)

$$P_b^*(t) = \begin{cases} P_b(t-1) + 3, & \text{if } t \le 2.5 \\ P_b(t-1) + 8 - 2t, & \text{if } t > 2.5 \end{cases}$$

The calculation results are shown in Table 14.2. It can be observed from Table 14.1 that the battery power still varies quadratically, and the generation is constant in initial hours and then decreases linearly.

The numbers in bold in Table 14.2 show that the values change because of the introduction of generation constraint compared with unconstrained results of Table 14.1 in Example 14.1.

**Example 14.3:**   For Example 14.1, the capacity of the battery is 12.0 MWh, which will be added in the problem. The battery power over time can be computed using equation (14.28).

$$P_b^*(t) = \begin{cases} 12, & \text{if } P_b(t) > 12 \\ P_b(t-1) + 8 - 2t, & \text{if } P_b(t) \le 12 \end{cases}$$

Thus, the optimal generation with battery capacity constraint over time will be

$$P_g^*(t) = \begin{cases} 20 - P_b(t_B - 1), & \text{if } P_b(t) > 12 \\ 16 - 2t, & \text{if } P_b(t) \le 12 \end{cases}$$

The calculation results are shown in Table 14.3.

The numbers in bold in Table 14.3 show that the values change because of the introduction of battery constraint compared with unconstrained results in Table 14.1 of Example 14.1.

**TABLE 14.3    Results of the Simple SGED with Battery Constraint**

| Time $t$ | 1 | 2 | 3 | 4 | 5 | 6 | 7 |
|---|---|---|---|---|---|---|---|
| Generation power | 14 | 12 | **8** | 8 | 6 | 4 | 2 |
| Battery power | 8 | 12 | **12** | 12 | 10 | 6 | 0 |

**TABLE 14.4 Results of the Simple SGED with Battery Upper and Lower Limits**

| Time $t$ | 1 | 2 | 3 | 4 | 5 | 6 | 7 |
|---|---|---|---|---|---|---|---|
| Generation power | 14 | **11** | **8** | 8 | 6 | 4 | **6** |
| Battery power | 8 | **11** | **11** | **11** | 6 | 2 | **0** |

**Example 14.4:** For Example 14.3, the capacity of the battery is changed to 11.0 MWh. In this case, the computed discharging value at the end of time period will become negative, which should be set to zero.

The optimal generator power output and the battery power over time can be computed as follows.

$$P_b^*(t) = \begin{cases} 11, & \text{if } P_b(t) > 11 \\ 0 & \text{if } P_b(t) < 0 \\ P_b(t-1) + 8 - t, & \text{if } 0 \le P_b(t) \le 11 \end{cases}$$

$$P_g^*(t) = \begin{cases} 19 - P_b(t-1), & \text{if } P_b(t) > 11 \\ 8 - P_b(t-1), & \text{if } P_b(t) < 0 \\ 16 - 2t, & \text{if } 0 \le P_b(t) \le 11 \end{cases}$$

The calculation results are shown in Table 14.4.

The numbers in bold in Table 14.4 show that the values are change because of the introduction of generation constraint compared with unconstrained results Table 14.1 in Example 14.1.

### 14.4.5 Simple Smart Grid Economic Dispatch with Multiple Generators

If the smart grid has multiple generators, the SGED problem can be expressed as follows.

$$\min J = \sum_{t=1}^{T} \left[ \sum_{i=1}^{NG} f_i \left( P_{gi}(t) \right) + h(P_b(t)) \right] \tag{14.36}$$

such that

$$P_b(t) = P_b(t-1) + \sum_{i=1}^{NG} P_{gi}(t) - D \tag{14.37}$$

$$0 \le P_b(t) \le P_{b\max} \tag{14.38}$$

$$0 \le P_{gi}(t) \le P_{gi\max} \tag{14.39}$$

where

$P_{gi}$: the power output of generator $i$
$P_{gimax}$: the maximal power output of generator $i$
$NG$: the number of generators in the grid.

Similar to the single SGED case, the inequality constraints are ignored first. A Lagrange function can be formed from the objective function and power balance equation.

The necessary conditions for an extreme value of the Lagrange function are to set the first derivative of the Lagrange function with respect to each of the independent variables equal to zero. Let the cost functions of the generators be

$$f_i(P_{gi}) = \frac{1}{2}\alpha_i P_{gi}^2 \quad i = 1, 2, \dots, NG \tag{14.40}$$

The optimality conditions for the SGED with multiple generators can be obtained:

$$\alpha_i P'_{gi}(t) = \eta(T + 1 - t) \quad i = 1, 2, \dots, NG \tag{14.41}$$

$$P'_b(t) = P_b(t - 1) + \sum_{i=1}^{NG} P'_{gi}(t) - D \tag{14.42}$$

From the above equations, we get

$$P'_{gi}(t) = \frac{\eta}{\alpha_i}(T + 1 - t) \quad i = 1, 2, \dots, NG \tag{14.43}$$

$$P_b'(t) = P_b(t - 1) + \sum_{i=1}^{NG} \left[ \frac{\eta}{\alpha_i}(T + 1 - t) \right] - D \tag{14.44}$$

It can be observed from equation (14.43) that the optimal generation of each unit will decrease linearly over time. From equation (14.44), the battery charges initially, and then discharges. The battery changes from charging to discharging when

$$\sum_{i=1}^{NG} \frac{\eta}{\alpha}(T + 1 - t) - D = 0 \tag{14.45}$$

that is

$$P'_b(t) = P_b(t - 1) \tag{14.46}$$

when

$$t_D = T + 1 - \frac{D}{\displaystyle\sum_{i=1}^{NG} \frac{\eta}{\alpha_i}} \tag{14.47}$$

From equation (14.41), we get

$$\alpha_1 P_{g1}(t) = \frac{\partial f_1}{\partial P_{g1}} = \alpha_2 P_{g2}(t) = \frac{\partial f_2}{\partial P_{g2}} = \cdots = \alpha_{NG} P_{gNG}(t) = \frac{\partial f_{NG}}{\partial P_{gNG}} \qquad (14.48)$$

This corresponds to the principle of equal incremental rate of economic dispatch for multiple generators mentioned in Chapter 4.

If the battery capacity constraint is considered, the optimal SGED with multiple generators can be expressed as follows.

$$P_b^*(t) = \begin{cases} P_{bmax}, & \text{if } P_b(t) > P_{bmax} \\ 0 & \text{if } P_b(t) < 0 \\ P_b(t-1) + \sum\limits_{i=1}^{NG} \dfrac{\eta}{\alpha_i}(T+1-t) - D, & \text{if } 0 \le P_b(t) \le P_{bmax} \end{cases} \qquad (14.49)$$

$$P_{gk}^*(t) = \begin{cases} P_{bmax} + D - P_b(t-1) - \sum\limits_{i=1,i\neq k}^{NG} P_{gi}(t), & \text{if } P_b(t) > P_{bmax} \\ D - P_b(t-1) - \sum\limits_{i=1,i\neq k}^{NG} P_{gi}(t), & \text{if } P_b(t) < 0 \\ \dfrac{\eta}{\alpha_k}(T+1-t), & \text{if } 0 \le P_b(t) \le P_{bmax} \end{cases}$$

$$\qquad (14.50)$$

It may be noted that the load is assumed to be constant over time in the above analysis. If the load is varies as time, that is, $D(t)$, the above mentioned method can still be adopted. In this case, the optimal SGED with multiple generators can be expressed as follows.

$$P_b^*(t) = \begin{cases} P_{bmax}, & \text{if } P_b(t) > P_{bmax} \\ 0 & \text{if } P_b(t) < 0 \\ P_b(t-1) + \sum\limits_{i=1}^{NG} \dfrac{\eta}{\alpha_i}(T+1-t) - D(t), & \text{if } 0 \le P_b(t) \le P_{bmax} \end{cases} \qquad (14.51)$$

$$P_{gk}^*(t) = \begin{cases} P_{bmax} + D(t) - P_b(t-1) - \sum\limits_{i=1,i\neq k}^{NG} P_{gi}(t), & \text{if } P_b(t) > P_{bmax} \\ D(t) - P_b(t-1) - \sum\limits_{i=1,i\neq k}^{NG} P_{gi}(t), & \text{if } P_b(t) < 0 \\ \dfrac{\eta}{\alpha_k}(T+1-t), & \text{if } 0 \le P_b(t) \le P_{bmax} \end{cases}$$

$$\qquad (14.52)$$

*Example 14.5:*    A simple smart grid has two generators and one storage battery. The load of 8.0 MW is assumed to be constant over time. The cost functions of two generators are given as follows.

$$f_1(P_{g1}) = \frac{1}{2}\alpha_1 P_{g1}^2 = \frac{1}{2}(0.04P_{g1}^2)$$

$$f_2(P_{g2}) = \frac{1}{2}\alpha_2 P_{g2}^2 = \frac{1}{2}(0.02P_{g2}^2)$$

The battery has no initial power and the unit coefficient of battery storage is $\eta = 0.08$. The capacities of the two generators are 25 MW and 35 MW, respectively. Let us compute the optimal generation and battery power over a time period of 7 hours.

According to the given parameters, we get $\alpha_1 = 0.04$, $\alpha_2 = 0.02$, $\eta = 0.08$, $T = 7$, and $D = 30$. We can first compute the key time point that the battery changes from charging to discharging, that is,

$$t_D = T + 1 - \frac{D}{\sum\limits_{i=}^{NG} \frac{\eta}{\alpha_i}} = 7 + 1 - \frac{30}{\left(\frac{0.08}{0.04} + \frac{0.08}{0.02}\right)} = 3$$

This means the battery will be charged until hour 3, and after that it will discharge. The optimal generation can be computed through the following equations.

$$P'_{g1}(t) = \frac{\eta}{\alpha_1}(T + 1 - t) = \frac{0.08}{0.04}(7 + 1 - t) = 16 - 2t$$

$$P'_{g2}(t) = \frac{\eta}{\alpha_2}(T + 1 - t) = \frac{0.08}{0.02}(7 + 1 - t) = 32 - 4t$$

The power change of the battery can be obtained as follows.

$$P'_b(t) = P_b(t - 1) + \sum_{i=1}^{2} \frac{\eta}{\alpha_i}(T + 1 - t) - D$$

$$= P_b(t - 1) + 16 - 2t + 32 - 4t - 30 = P_b(t - 1) + 18 - 6t$$

The calculation results are shown in Table 14.5. It can be observed from Table 14.5 that the generations of the two units decrease linearly and then increase linearly from hours 6 and 7 as the battery was fully discharged after hour 5, and the units must make up the power mismatch to meet the power balance of the grid.

# 14.5    TWO-STAGE APPROACH FOR SMART GRID DISPATCH

Chapter 5 introduces the two-stage economic dispatch approach. A similar idea will be used for smart grid dispatch [12].

**TABLE 14.5 Results of the Simple SGED with Multiple Generators**

| Time $t$ | 1 | 2 | 3 | 4 | 5 | 6 | 7 |
|---|---|---|---|---|---|---|---|
| Power of unit 1 | 14 | 12 | 10 | 8 | 6 | 10 | 10 |
| Power of unit 2 | 28 | 24 | 20 | 16 | 12 | 20 | 20 |
| Battery power | 12 | 18 | 18 | 12 | 0 | 0 | 0 |

Given that a smart grid consists of integrated electricity and natural gas system, and also that renewable energy sources such as photovoltaic (PV) and wind installations are available. The gas pipeline network is modeled on a single pressure level, which supplies all gas demands. The system takes into account the increased penetration of distributed generation and of renewable resources (wind and PV). Each household or user decides autonomously when to produce electricity locally, when to store energy, and to feed it back to the grid at a later instant, or when to consume energy from higher network levels.

In the first stage of smart grid dispatch, all the loads and renewable energy resources are fixed. This means the wind and PV power are not controllable or adjustable during the first stage. In the second stage of smart grid dispatch, the renewable energy resources are variable at some range. Three objectives can be used for the second stage. They are (1) to minimize the fuel consumption, (2) to minimize system loss, and (3) to minimize the movement of generator output from the initial generation plans.

## 14.5.1 Smart Grid Dispatch — Stage One

Given the input–output characteristics of $NG$ generating units are $F_{G1}(P_{G1})$, $F_{G2}(P_{G2})$, ... , $F_{Gn}(P_{Gn})$. The input–output characteristics of $Ng$ natural gas units are $F_{g1}(P_{g1})$, $F_{g2}(P_{g2})$, ... , $F_{gn}(P_{gn})$. There are NR renewable resources $(P_{R1}, P_{R2}, ... , P_{Rn})$, $NE$ storage devices $(P_{E1}, P_{E2}, ... , P_{En})$, and $ND$ loads $(P_{D1}, P_{D2}, ... , P_{Dn})$. The values $P_{Ri}$ and $P_{Di}$ are fixed during first stage. The problem is to minimize the total operation cost subject to the components and network security constraints for a time period (for example, 24 h), that is,

$$\min J = \sum_{t=1}^{24} \sum_{Gi=1}^{NG} F_{Gi}(P_{Gi}(t)) + \sum_{t=1}^{24} \sum_{gi=1}^{Ng} F_{gi}(P_{gi}(t)) + \sum_{t=1}^{24} h_{Ei} \sum_{Ei=1}^{NE} (|P_{Ei}(t) - P_{Ei\,\text{Cap}}(t)|)^2$$

(14.53)

such that

$$\sum_{Gi=1}^{NG} P_{Gi}(t) + \sum_{gi=1}^{Ng} P_{gi}(t) + \sum_{Ri=1}^{NR} P_{Ri}(t) = \sum_{Di=1}^{ND} P_{Di}(t) + \sum_{Ei=1}^{NE} P_{Ei}(t) + P_L(t)$$

$$t = 1, 2, ... , 24$$

(14.54)

$$P_{Gi\min} \leq P_{Gi}(t) \leq P_{Gi\max} \quad i \in NG, \quad t = 1, 2, ... , 24$$

(14.55)

$$P_{gimin} \leq P_{gi}(t) \leq P_{gimax} \quad i \in Ng, \quad t = 1, 2, \dots, 24 \tag{14.56}$$

$$|P_{Ei}(t)| \leq P_{Ei\,Cap}(t) \quad i \in NE, \quad t = 1, 2, \dots, 24 \tag{14.57}$$

$$|P_{ij}(t)| \leq P_{ijmax} \quad ij \in NT, \quad t = 1, 2, \dots, 24 \tag{14.58}$$

where

$P_{Di}$: the real power load at bus $i$

$P_{Ri}$: the power output at renewable resource (wind or PV) bus $i$

$P_{Ei}$: the stored or released power at storage device bus $i$. It means that the storage device is in store mode if the value is positive, otherwise, in release mode

$P_{Ei\,Cap}$: the operation limitation of the storage device

$P_{Gi}$: the real power output at generator bus $i$

$P_{Gimin}$: the minimal real power output at generator $i$

$P_{Gimax}$: the maximal real power output at generator $i$

$P_{gi}$: the real power output at gas bus $i$

$P_{gimin}$: the minimal real power output at gas bus $i$

$P_{gimax}$: the maximal real power output at gas bus $i$

$P_{ij}$: the power flow of transmission line $ij$

$P_{ijmax}$: the power limits of transmission line $ij$

$P_L$: the network losses

$h_{Ei}$: the penalty factor of storage device

$NT$: the number of transmission lines

$NG$: the number of generators

$Ng$: the number of gas units

$NR$: the number of renewable energy resources

$NE$: the number of storage devices

$t$: the hourly time period.

The first term in equation (14.53) corresponds to the overall cost of electricity, and the generation cost function is generally quadratic. The second term in equation (14.53) corresponds to natural gas consumption, and the gas cost function can be linear. The last term in equation (14.53) represents penalties for all storage devices when they are passing their optimal operation limits.

## 14.5.2 Smart Grid Dispatch — Stage Two

The energy forecasts of the renewable resources at stage two will be more precise than those at stage one. Furthermore, these DG units may be adjustable during stage two according to the practical needs of the end users. To implement the optimal smart grid dispatch for stage two, several objectives may be selected. On one hand, the system loss minimization, or the system operation cost minimization can be selected as the objective function. On the other hand, the operators expect the optimal dispatch points close to the economic operation points $P^0_{Gi}$ of the first stage. Thus, the following three objectives may be adopted in the second stage.

**(1)** Minimize the fuel consumption

$$\min J_1 = \sum_{t=1}^{24}\sum_{Gi=1}^{NG} F_{Gi}(P_{Gi}(t)) + \sum_{t=1}^{24}\sum_{gi=1}^{Ng} F_{gi}(P_{gi}(t)) + \sum_{t=1}^{24}\sum_{Ri=1}^{NR} F_{Ri}(P_{Ri}(t))$$

$$+ \sum_{t=1}^{24} h_{Ei} \sum_{Ei=1}^{NE} (|P_{Ei}(t) - P_{Ei\,Cap}(t)|)^2 \qquad (14.59)$$

**(2)** Minimize the system loss

$$\min J_2 = \sum_{t=1}^{24} P_L(t) \qquad (14.60)$$

**(3)** Minimize the adjustment of generation output

$$\min J_3 = \sum_{t=1}^{24}\left[\sum_{Gi=1}^{NG} \left(P_{Gi}(t) - P^0_{Gi}(t)\right)^2 + \sum_{gi=1}^{Ng} (P_{gi}(t) - P^0_{gi}(t))^2\right.$$

$$\left. + \sum_{Ei=1}^{NE} \left(P_{Ei}(t) - P^0_{Ei}(t)\right)^2\right] \qquad (14.61)$$

In order to actualize the transition from the time point $t$ to $t+1$ schedule successfully, the real power generation regulations constraint, $\Delta P_{GRC\,imax}$ must be considered. These are determined from the product of the relevant regulating speed and regulating time specified.

$$|P_{Gi}(t) - P_{Gi}(t-1)| \le \Delta P_{GRC\,imax} \quad i \in NG, \ t = 1, 2, \dots, 24 \qquad (14.62)$$

or

$$-\Delta P_{GRC\,imax} + P_{Gi}(t-1) \le P_{Gi}(t) \le \Delta P_{GRC\,imax} + P_{Gi}(t-1)$$

$$i \in NG, \quad t = 1, 2, \dots, 24 \qquad (14.63)$$

Thus, the regulating value of the generation is restricted by the two inequality equations (14.55) and (14.63), which can be combined into one expression:

$$\max\{-\Delta P_{GRC\,imax} + P_{Gi}(t-1), \ P_{Gimin}\} \le P_{Gi}(t) \le \min\{\Delta P_{GRC\,imax}$$

$$+ P_{Gi}(t-1), \ P_{Gimax}\} \qquad i \in NG, \quad t = 1, 2, \dots 24 \qquad (14.64)$$

The optimal smart grid dispatch model for the second stage can be written as

$$\min J = h_1 J_1 + h_2 J_2 + h_3 J_3 \tag{14.65}$$

such that.

$$\sum_{Gi=1}^{NG} P_{Gi}(t) + \sum_{gi=1}^{Ng} P_{gi}(t) + \sum_{Ri=1}^{NR} P_{Ri}(t) = \sum_{Di=1}^{ND} P_{Di}(t) + \sum_{Ei=1}^{NE} P_{Ei}(t) + P_L(t)$$

$$t = 1, 2, \ldots, 24 \tag{14.66}$$

$$\max\{-\Delta P_{GRCimax} + P_{Gi}(t-1), \ P_{Gimin}\} \le P_{Gi}(t) \le \min\{\Delta P_{GRCimax}$$

$$+ P_{Gi}(t-1), \ P_{Gimax}\} \qquad i \in NG, \quad t = 1, 2, \ldots 24 \tag{14.67}$$

$$P_{Gimin} \le P_{Gi}(t) \le P_{Gimax} \quad i \in NG, \ t = 1, 2, \ldots, 24 \tag{14.68}$$

$$P_{gimin} \le P_{gi}(t) \le P_{gimax} \quad i \in Ng, \ t = 1, 2, \ldots, 24 \tag{14.69}$$

$$P_{Rimin} \le P_{Ri}(t) \le P_{Rimax} \quad i \in NR, \ t = 1, 2, \ldots, 24 \tag{14.70}$$

$$|P_{Ei}(t)| \le P_{EiCap}(t) \quad i \in NE, \ t = 1, 2, \ldots, 24 \tag{14.71}$$

$$|P_{ij}(t)| \le P_{ijmax} \quad ij \in NT, \ t = 1, 2, \ldots, 24 \tag{14.72}$$

where

$\Delta P_{GRCimax}$: the real power generation regulation rate. It is also called unit ramp up or down rate

$P_{Eimin}$: the minimal real power output at renewable resource $i$

$P_{Eimax}$: the maximal real power output at renewable resource $i$

$h_1$: the weighting factor of the operation cost objective function

$h_2$: the weighting factor of the loss minimization objective function

$h_3$: the weighting factor of the generation output adjustment objective function.

The weighting factors $(h_1 + h_2 + h_3 = 1)$, which have been discussed in Chapter 5, can be determined according to the practical situation of the specific system.

The economic dispatch model for the second stage can be solved by any optimization algorithm mentioned in the preceding chapters.

***Example 14.6:*** A smart grid example based on the IEEE 30-bus system is formed with some data change. The modified 30 bus system consists of five traditional generation units, a wind farm, a storage device, 21 loads, and 41 transmission lines and transformers. A wind farm with 13 MW capacity is connected to bus 9, and the cost of wind power is $40 MWh. The storage device is connected to bus 4 (the capacity is 20 MW an hour). For simplification, only one hour smart grid economic dispatch

**TABLE 14.6 The Cost Functions of Generators for Modified IEEE 30 Bus-System (p.u.)**

| Gen. No. | $a$ | $b$ | $c$ |
|---|---|---|---|
| 1 | 0.00984 | 0.33500 | 0.00000 |
| 2 | 0.00834 | 0.22500 | 0.00000 |
| 5 | 0.00850 | 0.18500 | 0.00000 |
| 11 | 0.00884 | 0.13500 | 0.00000 |
| 13 | 0.00834 | 0.22500 | 0.00000 |
| Wind power | 0.00000 | 0.40000 | 0.00000 |

where: $F_1 = a_i P_{Gi}^2 + b_i P_{Gi} + c_i$

**TABLE 14.7 The Results of Generation Scheduling for IEEE 30-Bus System (p.u.)**

| Gen. no. | Stage one for SGED | Stage two for SGED |
|---|---|---|
| 1 | 0.60306 | 0.76099 |
| 2 | 0.59634 | 0.37911 |
| 5 | 0.60384 | 0.66204 |
| 11 | 0.57580 | 0.56390 |
| 13 | 0.59523 | 0.59998 |
| Wind farm | 0.10000 | 0.10816 |

**TABLE 14.8 The Results of System Cost for IEEE 30-Bus System (p.u.)**

| Stage | Stage one for SGED | Stage two for SGED |
|---|---|---|
| Total system loss | 0.04038 | 0.04018 |
| Generation cost | 0.7342592 | 0.7291313 |

(SGED) calculation using a two-stage approach is demonstrated, and the batteries store the power during this hour. The wind power is estimated as 10 MW at the first stage, and is adjustable at the second stage with the range of 9–13 MW.

The cost functions of the generators are quadratic curves and are shown in Table 14.6. The cost function of wind power is linear and is also listed in Table 14.6. The two stage SGED results are shown in Tables 14.7 and 14.8.

Table 14.7 shows the generation plans for two stages. Table 14.8 shows system total losses and the generation costs for the two stages.

It can be observed from Table 14.8 that the system losses and the generation cost of the second stage are lower than those from the first stage, where loss is about 0.495% reduction, and the generation cost is about 0.698% reduction.

# 14.6    OPERATION OF VIRTUAL POWER PLANTS

A virtual power plant (VPP) can be understood as a coalition composed of multiple energy producers (such as renewable sources) and, possibly, energy storage providers (such as electric vehicles (EVs) or Vehicle-to-Grid (V2G)) that come together to sell electricity as an aggregate [14–19]. For the sake of simplicity, we assume that there is a unique VPP in the system, which is the VPP leader. It can contain multiple generation sites such as PV plants and wind farms. The local loads supplied by the VPP are constant.

Let the estimated electricity generated by the VPP on the next day at the time stage $t$ be $P(t)$. This estimated quantity $P(t)$ is produced by all PV plants and wind farms in the VPP at the time stage $t$ and can be supplied directly to the grid, stored in the batteries of the electric vehicles, or both. Furthermore, for the same time stage, the VPP leader may want to transfer to the grid an additional quantity of electricity that was stored in the batteries of the vehicles in previous time steps. These decisions depend on the market prices and also on the cost of using storage. Before we discuss the optimal model of VPP, we assume the method of storage payment as follows.

The payment for storage is provided to the EVs in the form of charging entitlements rather than money, that is, the storage payment scheme is in the form of energy given away to the EVs by the generators. The amount of energy given away is measured as a proportion of the amount of storage used, which thereby acts as the representative of the storage cost. In this way, the agent leading the VPP computes the optimal schedule that maximizes its profit on the basis of predictions of energy production and storage capacity for the next day, and uses this schedule to place bids in the day-ahead market. Then, on the actual day of delivery, the leader continuously re-optimizes the schedule for the remainder of the day to take into account the contracted energy supply and the latest predictions of the energy production and available storage [19].

To place the bid in day-ahead optimization, the VPP leader has to compute the following five parameters that determine the supply schedule: (i) the amount supplied directly into the grid, (ii) the amount of energy transferred to the storage devices, (iii) the energy transferred from the storage devices such as batteries to the grid, (iv) the amount of the storage capacity needed. For example, if the storage provider is EV, the amount of the storage capacity is used to determine the numbers of EVs needed in the VPP, and (v) the amount of energy transferred to the EVs as payment.

Let us suppose that the electricity supplied to the grid (either directly or drained from the storage devices) is paid for at price $c(t)$. Also, let the ratio between the amount of energy given to the EVs as payment and the amount of storage used be denoted by $\alpha$ and, let $\beta$ be the storage device's overall conversion loss, which takes into account the percentage of electricity that is lost when electricity flows from the grid to the storage device and vice versa. Therefore, it is necessary to store $1 + \beta$ units of energy to have 1 unit actually delivered from the storage device. Suppose that the transmission losses are ignored, and the local load supplied directly by the VPP is $P_l(t)$. Then, the optimal VPP model can be expressed as follows [19,20].

$$\max J(P_g, P_d) = \sum_{t=1}^{24} c(t)[P_g(t) + P_d(t)] \tag{14.73}$$

subject to

$$P_g(t) + (1 + \beta)P_b(t) + P_e(t) + P_l(t) = P(t) \quad \beta \in [0, 1] \tag{14.74}$$

$$P(t) = \sum_{i \in NW} P_{wi}(t) + \sum_{k \in NV} P_{vk}(t) \tag{14.75}$$

$$\Delta E(t) + P_b(t) \leq P_{bmax}(t) \tag{14.76}$$

$$\Delta E(t) - P_d(t) \geq 0 \tag{14.77}$$

$$\Delta E(t) = \begin{cases} 0 & \text{if } t = 0 \\ \sum_{i=0}^{t-1} \left( P_b(i) - P_d(i) \right) & \text{otherwise} \end{cases} \tag{14.78}$$

$$P_e(t) \geq \alpha P_{bmax}(t) \quad \alpha \in [0, 1] \tag{14.79}$$

$$0 \leq P_e(t) + P_{bmax}(t) \leq S_{max}(t) \tag{14.80}$$

$$P_g(t) \geq 0, \quad P_b(t) \geq 0, \quad P_d(t) \geq 0, \quad P_{bmax}(t) \geq 0 \tag{14.81}$$

where

$t$: the hourly time stages

$P(t)$: the estimated electricity generated by the VPP on the next-day at the time stage $t$

$P_{wi}(t)$: the power output of wind farm $i$ in the VPP at the time stage $t$.

$P_{vi}(t)$: the power output of PV plant $k$ in the VPP at the time stage $t$

$P_l(t)$: the local load supplied directly by the VPP at the time stage $t$

$P_g(t)$: the power or energy supplied directly into the grid at the time stage $t$

$P_b(t)$: the amount of energy transferred to the storage devices (batteries) at the time stage $t$.

$P_d(t)$: the energy transferred from the batteries to the grid at the time stage $t$

$P_{bmax}(t)$: the amount of the storage capacity needed at the time stage $t$

$P_e(t)$: the amount of energy transferred to the EVs at the time stage $t$.

$S_{max}(t)$: the maximum total storage (upper bound for the storage capacity) available to the VPP at the time stage $t$.

$c(t)$: the price for the electricity supplied to the grid (either directly or drained from the storage devices) at the time stage $t$

$J(P_g, P_d)$: the revenues raised by the VPP from the electricity sold in the market, based on the estimated generations for the next day

$\Delta E(t)$: the net energy stored in the EVs' batteries at the beginning of time slot $t$

$NW$: the number of wind farms in the VPP

$NV$: the number of PV plants in the VPP.

Equations (14.74) and (14.75) represent the power balance at the time stage $t$. Equation (14.76) guarantees that the electricity that is stored in the batteries fits the available storage. The constraint Equation (14.77) guarantees that the electricity that is drained from the batteries does not exceed the energy that is actually stored in the batteries. Equation (14.79) is the storage payment constraint. By solving this optimization problem, the day-ahead bid $X$ is given by $X = P_g + P_d$.

## 14.7    SMART DISTRIBUTION GRID

### 14.7.1    Definition of Smart Distribution Grid

Distribution systems are responsible for transferring electricity from the high-voltage power grid to commercial, industrial, and residential customers. Distribution lines consist of medium- and low-voltage circuits ranging from 35 down to 110V. Since almost 90% of all power outages and disturbances have their roots in the distribution network, the smart devices and technologies must be applied to the distribution system.

At present, the smart grid incorporates distributed intelligence at all levels of the electric grid to improve reliability, security, and efficiency. To fully realize the potential of the smart grid, it is necessary to examine the distribution system at length. The traditional distribution system is largely passive and radial, whereas the "smart" distribution system is expected to be active and networked [21]. Since this smart grid mainly involves the distribution level, it is generally known as a smart distribution grid."

The definition of the smart distribution grid is evolving depending on the level of deployment of automation technology. The goals of the smart distribution grid are incremental efficiency and reliability improvements over the present level of automation technology deployment [21–28]. Better communications, computing and control schemes, distributed energy sources including microgrids and power electronic equipment are being introduced in the smart distribution grid at an unprecedented pace. New topologies such as looped and network structures are being adopted to provide increased reliability and efficiency to customers.

The emerging smart grid promises incremental efficiency and reliability improvement for the electric distribution system. The next-generation distribution management system (DMS) will be based on a connected model imported from a geographical information system (GIS). DMS is a collection of applications designed to monitor and control the entire distribution network efficiently and reliably. It acts as a decision support system to assist the control room and field-operating personnel with the monitoring and control of the electric distribution system. Improving the reliability and quality of service in terms of reducing outages, minimizing outage time, maintaining acceptable frequency and voltage levels are the key deliverables of a DMS.

DMS is the key to integrating emerging and mature smart grid technologies and applications focused on automation, consumer enablement, distributed energy resources, and controllable demand, while effectively balancing optimal network operations with environmental and open-market objectives. The DMS will maintain the connectivity and interconnected relationship of the distribution SCADA substation and its associated distribution automation sites at the discrete locations along the distribution circuit. The operator will be presented with an integrated view of the electric distribution system including outage management system (OMS) information and customer information system (CIS) data. Navigation, data presentation, and analysis techniques are being developed to facilitate the operator's response to the dynamics of the distribution system and to system disturbances.

## 14.7.2 Requirements of Smart Distribution Grid

Information gathering to support decision and control actions in the smart distribution grid will be distributed, requiring new two-way communications infrastructure and associated data management framework. In order to make the distribution grid "smart," it requires the following:

- Smart infrastructure, low cost sensors, and smart meters
- Smart planning and design, smart operations, smart customers, and smart customer appliances
- Distributed energy resources, distributed information, and intelligence
- High-efficiency transformers, new storage devices, and improved fault limiting and protective devices
- New materials such as high-temperature superconducting materials.

## 14.7.3 Smart Distribution Operations

Since there has been no comprehensive approach to automation of distribution systems, distribution management system, which in general can be defined as a computer- and communication-based system to operate and manage the distribution systems, has had a different meaning to different utilities. It could be a system for distribution automation (DA), outage management, or facilities and work order management utilizing the GIS. In some cases, it is SCADA with enhanced DA functionality. In many instances, we find different systems within the same utility addressing different system management issues. These systems employ application interfaces between dissimilar applications and frequently these applications run on separate noncompatible databases. Synchronization of databases is a constant concern and maintenance issue for the existing DMS. The synchronization issue has been overcome, but it requires constant attention. Although different utilities implemented different approaches for automation over the years starting from 1970s, the boundaries between these systems have become blurred now [23].

A DMS provides the foundation and technology for the emerging automation technology that is being deployed along the distribution circuits. It also provides an efficient visual interactive work environment that integrates all information sources within a common real-time workspace. It reduces the number of systems on the operator's desk, predicts operating issues, provides greater clarity in emergency situations and improves operator response time. DMS also supports management of the system with less experienced users and promotes improved staff retention.

With SCADA remote terminal units (RTUs) at the substation, the SCADA system is immediately aware of faults that cause both temporary and permanent breaker trips. Utilizing the capabilities of the advanced RTU and line-post sensors, DMS supports fault detection techniques and reports power quality measurements at discrete locations along the distribution circuit. The advanced RTU is integrated into the automation system for motor-operated gang switches, pole-mounted reclosers, pole-mounted regulators, and switched capacitor banks.

Advanced automation needed for smart distribution systems requires faster decisions and thus real-time analysis of distribution systems. The robust distribution state estimator is an example of the analysis tools needed for advanced automation. The input data for analysis includes system topology, parameters of different components in the system, status of switches and breakers, and measured data from various points in the system. Since more data can be measured, the analysis becomes more complex. The tools should be able to use these data effectively. Real-time analysis will allow faster control of distribution systems. Real-time monitoring and analysis not only provide the status on loading of equipment but also allows determination of the next step, such as location and time of the next switch to be closed to restore a group of customers. With judicious selection, restoration can be accomplished in minimum time, thus improving reliability of electricity supply to the customers.

Application integration envisioned for the next generation integrated DMS includes [23]:

- Optimal volt/var optimization
- Online power flow and short circuit analysis
- Advanced and adaptive protection
- $(N-2)$ Contingency analysis
- Advanced fault detection and location
- Advanced fault isolation and service restoration
- Automated vehicle management system

- Dynamic derating of power equipment due to harmonic content in the load
- Distribution operator training simulator
- System operation with large penetration of customer-owned renewable generation
- Distribution system operation as a microgrid
- Real-time pricing and demand response applications.

## 14.8 MICROGRID OPERATION

### 14.8.1 Application of Microgrid

A microgrid is defined as a low-to-medium voltage network of small load clusters with distributed generation sources and storage. It is characterized by the following:

- It is locally controlled.
- It is a section of distribution system, usually connected to the primary or secondary distribution system depending on the capacity.
- It contains multiple distributed energy resources (DERs), which include photovoltaic (PV), small wind turbines (WT), heat or electricity storage, combined heat and power (CHP), and controllable loads.
- It is seen as an aggregate source or load by the system, which can be dispatched if seen as a source.
- It has a capacity of less than 10 MVA.

DER applications play important role in the operation of a microgrid, which would increase the efficiency of energy supply and reduce the electricity delivery cost and carbon footprint in the microgrid. In addition, DER applications would also make it possible to impose intentional islanding in microgrids. Among the DERs, electricity storage is paid increasing attention in the smart grid. Storage devices including batteries, supercapacitors, and flywheels could be used to match generation with demand in microgrids. Storage can supply generation deficiencies, reduce load surges by providing ride-through capability for short periods, reduce network losses, and improve the protection system by contributing to fault currents. Vehicle-to-grid (V2G) communicates with the power grid to sell demand response services by either delivering electricity into the grid or by throttling their charging rate. V2G and electric vehicle (EV) technologies can reduce the microgrid reliance on the grid supply.

Owing to the DER applications and the flexibilities of microgrid operation, the microgrid has the following advantages:

(1) For utilities
   ○ The microgrid has hierarchical control of DERs;
   ○ It ensures decreased transmission losses and increased efficiency.
   ○ It behaves as either an interruptible load or a dispatchable source.

- With fewer load sources, the demand on the microgrid infrastructure is less than in a typical grid.
- By being smaller and closer to source and demand and being able to use power generation more specific to the location, the system has higher reliability and is able to respond to demand more quickly.
- Microgrids are laid out in a modular manner making expansion and updating more efficient.
- With local control, both design and future planning are specific to the needs of the entities participating in the microgrid.

**(2)** For customers

- There is a more diverse generation mix.
- There is increased reliability through islanding.
- Power quality and reliability are increased.

**(3)** For society and environment

- There is an increased ability for renewable energy integration.
- Emissions are reduced.
- There is potential for increased fuel efficiency (CHP).

## 14.8.2    Microgrid Operation with Wind and PV Resources

A "microgrid" is a cluster of distributed energy resource units, both distributed generation and distributed storage units, serviced by a distribution system, and can operate:

- in the grid-connected mode;
- in the islanded (autonomous) mode;
- dynamically between the two modes.

In the normal operation, the microgrid is connected to a traditional power grid (main grid or macrogrid). The users in a microgrid can generate low-voltage electricity using distributed generation, such as solar panels, wind turbines, and fuel cells. The single point of common coupling with the main grid can be disconnected, with the microgrid functioning autonomously. This operation will result in an *islanded* microgrid, in which distributed generators continue to power the users in this microgrid without obtaining power from the electric utility located in the main grid. Thus, the multiple distributed generators and the ability to isolate the microgrid from a larger network in disturbance will provide highly reliable electricity supply. This islanding operation of microgrid is good for the users under the emergency condition. The users will reconnect to the main grid and obtain power from the electric utility once the whole system is back to normal status.

Major modeling components in microgrid operation with wind and PV resources are discussed in the following sections.

***Wind Speed Model***    Wind speed is a variable and uncertain factor. Fuzzy numbers are one of the methods used to represent the uncertainty of wind speed. The wind speed is said to be an *LR*-type fuzzy number if

$$
\mu_{P_D(x)} = \begin{cases} L\left(\dfrac{m-x}{a}\right), & x \le m, \ a > 0 \\ R\left(\dfrac{x-m}{b}\right), & x \ge m, \ b > 0 \end{cases} \tag{14.82}
$$

The LR-type fuzzy number of the uncertainty wind speed can be written as

$$
v = (m, a, b)_{LR} \tag{14.83}
$$

where

$v$: the wind speed

$m$: the mean value of uncertainty wind speed

$a$: the inferior dispersion of uncertainty wind speed

$b$: the superior dispersion of uncertainty wind speed.

**Wind Power Model**  Wind model input assumptions vary from constant torque to constant power. The frequently made assumption of constant torque means any changes in shaft speed will result in a change in captured mechanical power, and consequently, a change in power output of the wind plant. A simple relationship exists between the power generated by a wind turbine and the wind parameters.

$$
P_w = \frac{1}{2}\rho A C_p \eta_g \eta_b v^3 \tag{14.84}
$$

where

$P_w$: the power generated by a wind turbine

$\rho$: the air density (about 1.225 kg/m$^3$ at sea level, less at higher elevations)

$A$: the rotor-swept area, exposed to the wind (m$^2$)

$C_p$: the coefficient of performance (0.59 to 0.35 depending on turbine)

$\eta_g$: the generator efficiency

$\eta_b$: the gearbox/bearings efficiency

$v$: the wind speed in m/s.

**PV Array Model**  Solar cells, also called photovoltaic (PV) cells by scientists, convert sunlight directly into electricity. PV panels used to power homes and businesses are typically made from solar cells combined into modules that hold about 40 cells. Many PV panels combined together to create one system called a PV array. For large electric utility or industrial applications, hundreds of PV arrays are interconnected to form a large utility-scale PV system.

In the actual utility, the controlled current source is generally used for modeling the PV array. For a PV array with $N_S$ PV cells in series and $N_P$ PV cells in parallel, the terminal current $I_A$ is

$$
I_A = N_P \cdot I_L - N_P \cdot I_0 \cdot \left[ \exp\left( \frac{q \cdot (V_A + I_A \cdot R_{sa})}{N_S \cdot n \cdot m \cdot k \cdot T} \right) - 1 \right] \tag{14.85}
$$

where

$N_S$: the number of PV modules in series

$N_P$: the number of PV modules in parallel

$I_0$: the diode saturation current

$I_L$: the short-circuit current of the PV cell

$n$: the ideal constant of diode

$V_A$: the terminal voltage of the PV array

$I_A$: the terminal current of the PV array

$R_{sa}$: the equivalent series resistance of the PV array.

***Energy Storage System Model***   Energy storage system (ESS) applications are classified according to power, energy capacity, usage time, etc. Applications include megawatt-scale power storage for frequency regulations, large-capacity energy storage (MWh scale) for peak-time demand response, and residential energy storage with medium capacity (kWh scale).

If a storage device is modeled as ideal storage in combination with a storage interface, the power exchange $P_\alpha(k)$ is defined as the difference between the amounts of stored energy $E_\alpha(k)$ at two consecutive time steps, plus some standby energy losses $E_\alpha^{stb}$, which must be covered at each time period ($E_\alpha^{stb} \geq 0$) [13]:

$$P_\alpha(k) = \frac{\dot{E}_\alpha}{e_\alpha} = \frac{1}{e_\alpha}\frac{dE_\alpha}{dt} \approx \frac{1}{e_\alpha}\frac{\Delta E_\alpha}{\Delta t} = \frac{1}{e_\alpha}\left(\frac{E_\alpha(k) - E_\alpha(k-1)}{\Delta t} + E_\alpha^{stb}\right) \quad (14.86)$$

The parameter $e_\alpha$ stands for the charging ($e_\alpha^+$) or discharging ($e_\alpha^-$) efficiency of the storage device. The subscript $\alpha$ stands for the storage mediums such as heat or electricity.

The stored energy $E_\alpha(k)$ and the power exchange $P_\alpha(k)$ have to remain within limits, resulting in the following constraints for the storage device:

$$P_{\alpha,\min}(k) < P_\alpha(k) < P_{\alpha,\max}(k) \quad (14.87)$$

$$E_{\alpha,\min}(k) - \varepsilon(k) < E_\alpha(k) < E_{\alpha,\max}(k) + \varepsilon(k) \quad (14.88)$$

$$\varepsilon(k) \geq 0 \quad (14.89)$$

### 14.8.3   Optimal Power Flow for Smart Microgrid

***Distributed Optimal Power Flow Model [29–31]***   As we mentioned above, a microgrid is a portion of an electric distribution network located downstream of the distribution substation that supplies a number of industrial and residential loads through distributed energy resources such as distributed generation (DG) and distributed storage (DS) units. To achieve the goal of economic operation of the entire system, it is necessary to optimally operate all energy resources (traditional generation units in the main grid and DG units in the microgrid) for the whole system. However, for real-time network management, it is generally required to find a new network operational setup rapidly (e.g., in a few seconds or minutes) in order to promptly respond to abrupt local load variations and to cope with the intermittent power generation that is typical of renewable-based DG units. It is then of paramount importance to solve the distributed OPF problem in a distributed manner, by decomposing the main problem into multiple sub-instances that can be solved efficiently and in parallel. The microgrid optimization operation may have some differences compared with the general optimization problems that are discussed in Chapters 8 and 12, where the distribution system has been typically assumed to be a balanced three-phase system, and hence single-phase equivalents are used to reduce the computational burden. However, such an assumption for distribution feeders is not very realistic because of untransposed three-phase feeders, existence of single-phase laterals, and unbalanced loads. In addition, single-phase DG units may worsen the network imbalance. Thus, there is a need to consider three-phase models of distribution systems for more precise operational decisions in optimal microgrid operation [29–31].

There are two types of components in a microgrid, namely, the series and shunt components. Conductors/cables, transformers, Load tap changers (LTCs), and switches are series components. Conductors and cables can be modeled as $\pi$-equivalent circuits. Switches are modeled as zero-impedance series components. Three-phase transformer models depend on the connection type (wye or delta), with the most common types of distribution system transformers being considered, namely, single-phase and three-phase wye grounded–wye grounded, delta–wye grounded, and open wye–open delta connections. Voltage-regulating transformers in distribution systems are equipped with LTCs.

DG units, loads, and capacitors are shunt components, which are modeled for individual phases separately to represent unbalanced three-phase loads, as single-phase loads and single-phase capacitors are common in distribution feeders. A polynomial load model is adopted, where each load is modeled as a mix of constant-impedance, constant-current, and constant-power components. Capacitors are modeled as constant-impedance loads. Capacitor banks are modeled as multiple capacitor units with switching options. Wye-connected and delta-connected loads and capacitors are often adopted.

For each series element, a set of equations based on the ABCD parameters are used, which relate the three-phase voltages and currents of the sending-end and receiving-end as follows:

$$\begin{bmatrix} \overline{V}_{s,f} \\ \overline{I}_{s,f} \end{bmatrix} = \begin{bmatrix} A & B \\ C & D \end{bmatrix} \begin{bmatrix} \overline{V}_{s,r} \\ \overline{I}_{s,r} \end{bmatrix} \qquad \forall s \qquad (14.90)$$

where

$s$: the series elements, $s = 1, \ 2, \ \ldots Ns$

$A, B, C, D$: the three-phase ABCD parameter matrices, p.u.

$\overline{V}$: the vector of three-phase line voltage phasors, p.u.

$\bar{I}$: the vector of three-phase line current phasors, p.u.

$r$: the receiving-end of the component

$f$: from-end or sending-end of the component

The ABCD parameters of all series elements are constant except for LTCs, which depend on the setting of tap positions during operation. The following additional set of equations is needed to represent the $A$ and $D$ matrices in (14.88) for each LTC:

$$A_t = W \begin{bmatrix} 1 + \Delta S_t \text{tap}_{a,t} \\ 1 + \Delta S_t \text{tap}_{b,t} \\ 1 + \Delta S_t \text{tap}_{c,t} \end{bmatrix} \quad \forall t \tag{14.91}$$

$$D_t = A_t^{-1} \quad \forall t \tag{14.92}$$

where

tap: the tap position

$t$: the controllable tap changers, $t = 1, \ 2, \ \dots Nt$

$\Delta S$: the percentage voltage change for each LTC tap

$a, b, c$: the phases

$W$: the $3 \times 3$ identity matrix.

Equations (14.91) and (14.92) are for a tap changer with per-phase tap controls. For a three-phase tap changer, the following additional equation is used to make sure that all tap operations are the same:

$$\text{tap}_{a,t} = \text{tap}_{b,t} = \text{tap}_{c,t} \tag{14.93}$$

If the load is wye-connected on a per-phase basis, the load can be represented as follows.

For constant power loads:

$$V_{p,d} I_{p,d}^* = P_{p,d} + jQ_{p,d} \quad \forall p, \forall d \tag{14.94}$$

For constant impedance loads:

$$V_{p,d} = Z_{p,d} I_{p,d} \quad \forall p, \forall d \tag{14.95}$$

For constant current loads:

$$|I_{p,d}|(\angle V_{p,d} - \angle I_{p,d}) = |I_{p.d}^0|\angle \theta_{p,d} \quad \forall p, \forall d \tag{14.96}$$

where

$p$: the phases, $p = a,\ b,\ c$
$d$: the demand loads, $d = 1,\ 2,\ \dots Nd$
$\theta$: the load power factor angle, rad
$I^0$: the load phase current at specified power and nominal voltage, p.u.
$P$: the active power, p.u.
$Q$: the reactive power, p.u.
$Z$: the load impedance at specified power and nominal voltage, p.u.

For each wye-connected capacitor bank with multiple capacitor blocks, the corresponding model are represented by following mathematical models:

$$V_{p,c} = X_{p,c} I_{p,c} \quad \forall p, \forall c \tag{14.97}$$

$$X_{p,c} = \frac{-j(I^0_{p,c})^2}{C_{p,c} \Delta Q_{p,c}} \quad \forall p, \forall c \tag{14.98}$$

$$Q_{p,c} = N\mathrm{max}_{p,c} \Delta Q_{p,c} \quad \forall p, \forall c \tag{14.99}$$

where

$c$: the controllable capacitor banks, $c = 1,\ 2,\ \dots Nc$
$V^0$: the nominal phase voltage, p.u.
$X$: the reactance of capacitor, p.u.
$N_{\mathrm{max}}$: the number of capacitor blocks available in capacitor banks
$C$: the number of capacitor blocks switched in capacitor banks
$\Delta Q$: the size of each capacitor block in capacitor banks, p.u.

If the loads and capacitors banks are delta-connected, line-to-line voltages and currents need to be used. In that case, equations (14.95)–(14.99) can be used by replacing the line variables with line-to-line variables. The relationships of line-to-line variables to line variables are as follows:

$$\begin{bmatrix} V_{a,b} \\ V_{b,c} \\ V_{c,a} \end{bmatrix} = \begin{bmatrix} 1 & -1 & 0 \\ 0 & 1 & -1 \\ -1 & 0 & 1 \end{bmatrix} \begin{bmatrix} V_a \\ V_b \\ V_c \end{bmatrix} \tag{14.100}$$

$$\begin{bmatrix} I_a \\ I_b \\ I_c \end{bmatrix} = \begin{bmatrix} 1 & -1 & 0 \\ 0 & 1 & -1 \\ -1 & 0 & 1 \end{bmatrix} \begin{bmatrix} I_{a,c} \\ I_{b,a} \\ I_{c,b} \end{bmatrix} \tag{14.101}$$

Equations (14.90)–(14.101) correspond to the component models in the microgrid. If the objective of the optimal microgrid operation is to minimize the power losses of the microgrid, the distributed optimal power flow model for the smart microgrid can be expressed as follows.

$$\min f = \sum_{s=1}^{NS} R_{p,s} I_{p,s}^2 \quad s \in NS \tag{14.102}$$

such that

Equations (14.90)–(14.101), and

$$\sum_{DG \to i} I_{p,DG} + \sum_{c \to i} I_{p,c} + \sum_{r \to i} I_{p,s,r} = \sum_{f \to i} I_{p,s,f} + \sum_{d \to i} I_{p,d} \quad \forall p, \forall d, \forall DG, \forall c \tag{14.103}$$

$$V_{p,DGi} = V_{p,ci} = V_{p,s,ri} = V_{p,s,fi} = V_{p,di} = V_{p,i} \quad \forall p, \forall d, \forall DG, \forall c \tag{14.104}$$

$$|I_{p,s}| \le I_{p,s\ max} \quad s \in NS \tag{14.105}$$

$$V_{p,imin} \le V_{p,i} \le V_{p,imax} \quad i \in N \tag{14.106}$$

where

$I_{p,s}$: the plural current in the series component $s$
$R_{p,s}$: the resistance of the series component $s$
$V_{p,i}$: the node voltage at node $i$
$DG$: the $DG$ units in microgrid
$N$: the set of nodes in microgrid
$NS$: the set of the series components.

In the above model, subscripts "*min*" and "*max*" represent the lower and upper bounds of the constraint. The symbol $x \to i$ means that $x$ is connected to node $i$. Equation (14.103) represents Kirchhoff's first law (KCL) for each node and phase. Obviously, at each node and phase, the voltages of the elements connected to that node are equal to the corresponding nodal voltage, which is shown in equation (14.104).

**Solution Method** In the above three-phase DOPF model, LTC and capacitor-switching actions are discrete operations. Thus, this is a mixed integer nonlinear programming (MINLP) problem. In order to simplify the solution of such an MINLP problem, we may relax the integer variables and convert the problem into a nonlinear programming (NLP) one.

To alleviate the use of integer variables, a quadratic penalty term is augmented to the objective function, resulting in the following modified objective function:

$$\min f' = f + \sum_{ki} h_{ki}(x_{ki} - \text{round}(x_{ki}))^2 \tag{14.107}$$

where

  $ki$: the integer variables, $ki = 1, 2, \ldots Ni$

  $x_{ki}$: the tap and cap variables

  $h_{ki}$: the penalty value, which is a big number.

The quadratic term adds a high penalty value to the objective function at noninteger solutions, and thus drives $x_{ki}$ close to its corresponding integer value round $(x_{ki})$. By employing the above method, the MINLP problem of DOPF in Section 14.6.2 is converted into an NLP problem. The optimization algorithms presented in Chapter 8 can be used to solve the DOPF.

## 14.9   A NEW PHASE ANGLE MEASUREMENT ALGORITHM

With the adoption of PMU and other intelligent devices in the smart grid, all kinds of power system parameters such as phase angle are easily measured. To ensure secure, reliable, and stable operation the smart grid, as well as to reduce the impact of the measurement error, proper methods to analyze and handle the measurement data are needed. This section introduces a new phase angle measurement algorithm [32].

### 14.9.1   Error Analysis of Phase Angle Measurement Algorithm

Owing to its superiority in the aspect of harmonic restraining, the discrete Fourier transform (DFT) is generally usedto study phase angle measurement. This has been proven to be sufficiently accurate in a variety of power system applications when the input signal is three-phase voltage and the voltage frequency is near the nominal value.

When the frequency of the input signal deviates from the nominal value, the traditional DFT may cause a fence effect and spectrum leakage due to the asynchronous sampling, which brings large error in the phase angle measurement. Considering a sinusoidal input signal $x(t)$, which is sampled $N$ times per cycle of the $f_0$ waveform, then the error of measured phase angle by DFT method should be as follows:

$$\Delta\varphi \approx \frac{(N-1)\pi\Delta f}{Nf_0} - \frac{\Delta f}{2f_0 + \Delta f} \sin\left(2\varphi + \frac{2\pi(N-1)(f_0 + \Delta f)}{Nf_0}\right) = \Delta\varphi_0 + \Delta\varphi_s$$

$$(14.108)$$

According to the above equation, when the frequency deviation of the input stays at a constant value, the phase angle measurement error consists of the invariant part $\Delta\varphi_0$ and the sine variation part $\Delta\varphi_s$. The frequency deviation is practically deemed as unchanged during only one or two cycles in power systems. A sample $x(t)$ of frequency deviation $\Delta f$ with sampling frequency $Nf_0$ can be expressed by Euler's formula as follows:

$$x(n) = \sqrt{2}X \frac{\left(e^{j\left(\frac{2n\pi(f_0 + \Delta f)}{Nf_0} + \varphi\right)} + e^{-j\left(\frac{2n\pi(f_0 + \Delta f)}{Nf_0} + \varphi\right)}\right)}{2}$$

$$(14.109)$$

Considering a sinusoidal input signal of angular frequency $\omega$ given by

$$x(t) = \sqrt{2}X\cos(\omega t + \varphi) \tag{14.110}$$

where, $\omega = 2\pi(\Delta f + f_0), f_0$ is the nominal frequency, and $\Delta f$ is the frequency deviation from the nominal value.

According to electrical engineering convention, this signal is usually represented by a complex number:

$$\dot{x} = Xe^{j\varphi} = X\cos\varphi + jX\sin\varphi \tag{14.111}$$

which is called the phasor of the input signal.

If the input signal is sampled $N$ times per cycle of the $f_0$ waveform, we can get

$$x(n) = x(t)|_{t=nT_s} = \sqrt{2}X\cos\left(\frac{2n\pi\left(f_0 + \Delta f\right)}{Nf_0} + \varphi\right) \tag{14.112}$$

Equation (14.112) can be expressed with the help of Euler's formula as follows:

$$x(n) = \sqrt{2}X\frac{\left(e^{j\left(\frac{2n\pi(f_0+\Delta f)}{Nf_0}+\varphi\right)} + e^{-j\left(\frac{2n\pi(f_0+\Delta f)}{Nf_0}+\varphi\right)}\right)}{2} \tag{14.113}$$

Windowing $x(n)$ with a rectangular window $d(n)$, we can obtain

$$x_d(n) = x(n)d(n) \tag{14.114}$$

where

$$d(n) = \begin{cases} 1, 0 \le n \le N-1 \\ 0, n < 0, n \ge N \end{cases} \tag{14.115}$$

The DFT of $d(n)$ is

$$D(e^{j\omega}) = \sum_{n=0}^{N-1} e^{-jn\omega} = e^{-j\omega\frac{N-1}{2}}\frac{\sin(\omega N/2)}{\sin(\omega/2)} \tag{14.116}$$

So we can deduce the DFT of $x_d(n)$ as

$$\dot{X} = \sum_{n=0}^{N-1} x_d(n)e^{-j\frac{2\pi n}{N}}$$

$$= \sqrt{2}Xe^{j\varphi}e^{j\left(\frac{(N-1)\pi(\Delta f)}{Nf_0}\right)}\frac{\sin\left(\frac{\pi\Delta f}{f_0}\right)}{N\sin\left(\frac{\pi\Delta f}{Nf_0}\right)}$$

$$+ \sqrt{2} X e^{-j\varphi} e^{-j\left(\frac{2n(N-1)}{N} + \frac{(N-1)\pi(\Delta f)}{Nf_0}\right)} \frac{\sin\left(\frac{\pi\Delta f}{f_0}\right)}{N \sin\left(\frac{\pi\Delta f}{Nf_0} + \frac{2\pi}{N}\right)}$$

$$= A e^{j\varphi} + B e^{-j\varphi} \tag{14.117}$$

where

$$A = \sqrt{2} X \frac{\sin\left(\frac{\pi\Delta f}{f_0}\right)}{N \sin\left(\frac{\pi\Delta f}{Nf_0}\right)} e^{j\left(\frac{(N-1)\pi\Delta f}{Nf_0}\right)} \tag{14.118}$$

$$B = \sqrt{2} X \frac{\sin\left(\frac{\pi\Delta f}{f_0}\right)}{N \sin\left(\frac{\pi(2f_0+\Delta f)}{Nf_0}\right)} e^{-j\left(\frac{(N-1)\pi(2f_0+\Delta f)}{Nf_0}\right)} \tag{14.119}$$

For the convenience of analyzing the phase angle, equation (14.117) can be written as

$$\dot{x} = A e^{j\varphi} \left( 1 + e^{-j\left(2\varphi + \frac{2\pi(N-1)(f_0+\Delta f)}{Nf_0}\right)} \frac{\sin\left(\frac{\pi\Delta f}{Nf_0}\right)}{\sin\left(\frac{\pi\Delta f}{Nf_0} + \frac{2\pi}{N}\right)} \right) \tag{14.120}$$

In equation (14.120), the phase angle of the term $A$ is just the invariant error part $\Delta\varphi_0$ in equation (14.108) and the angle of phasor in the bracket is the sine variation error part $\Delta\varphi_s$ in (14.108) according to the definition of phasor addition. The above derivation shows that the existence of $B^*(\exp(-j(\varphi)))$ in (14.117) is the essential reason that induces the $\Delta\varphi_s$, so the $B^*(\exp(-j(\varphi)))$ could be the key point in maximally eliminating the sinusoidal function error when improving the algorithm.

Considering that the coordinate axis has the orthogonality property in $\alpha\beta$ stationary coordinates system, and two expression forms of $\beta$-axis that phase reversed with each other can be obtained according to leading or lagging $\alpha$-axis 90°, these characteristics could be applied to the calculation of eliminating the $B^*(\exp(-j(\varphi)))$. Let us imitate an $\alpha\beta$ stationary coordinate including two reversed-phase $\beta$-axis, and set input signal $\dot{x}$ as phasor $\alpha$, $\pi/2$ after the phasor $\alpha$ as an imitation of phasor $\beta1$, and $\pi/2$ before the phase $\alpha$ as an imitation of phasor $\beta2$. Since the input frequency is not exactly the nominal one, the expressions of phasor $\alpha$, phasor $\beta1$, and phasor $\beta2$ should be written as follows:

$$\dot{x}_\alpha = A e^{j\varphi} + B e^{-j\varphi} \tag{14.121}$$

$$\dot{x}_{\beta1} = A e^{j\left(\varphi + \frac{\pi(f_0+\Delta f)}{2f_0}\right)} + B e^{-j\left(\varphi + \frac{\pi(f_0+\Delta f)}{2f_0}\right)} \tag{14.122}$$

$$\dot{x}_{\beta2} = A e^{j\left(\varphi - \frac{\pi(f_0+\Delta f)}{2f_0}\right)} + B e^{-j\left(\varphi - \frac{\pi(f_0+\Delta f)}{2f_0}\right)} \tag{14.123}$$

By convention, the positive sequence component of phase $\alpha$ is defined as the positive sequence phasor of the $\alpha\beta$ stationary coordinate, which can be derived by phasor $\alpha$ and phasor $\beta 1$ as well as phasor $\alpha$ and phasor $\beta 2$.

$$\dot{x}_{1+} = \dot{x}_{\alpha 1+}$$

$$= \frac{1}{2}\left(\dot{x}_\alpha + e^{-j\frac{\pi}{2}}\dot{x}_{\beta 1}\right)$$

$$= \frac{1}{2}\left(Ae^{j\varphi}\left(1 + e^{j\frac{\pi\Delta f}{2f_0}}\right) + Be^{-j\varphi}\left(1 + e^{-j\left(\pi + \frac{\pi\Delta f}{2f_0}\right)}\right)\right) \tag{14.124}$$

$$\dot{x}_{2+} = \dot{x}_{\alpha 2+}$$

$$= \frac{1}{2}\left(\dot{x}_\alpha + e^{j\frac{\pi}{2}}\dot{x}_{\beta 2}\right)$$

$$= \frac{1}{2}\left(Ae^{j\varphi}\left(1 + e^{-j\frac{\pi\Delta f}{2f_0}}\right) + Be^{-j\varphi}\left(1 + e^{j\left(\pi + \frac{\pi\Delta f}{2f_0}\right)}\right)\right) \tag{14.125}$$

What is expressed in equations (14.124) and (14.125) is the same positive sequence phasor $\dot{x}_+$. When both phasor $\beta 1$ and phasor $\beta 2$ are orthogonal with phasor $\alpha$ and we can deduce the following expression.

$$2\dot{x}_+ = \frac{1}{2}\left(Ae^{j\varphi}\left(1 + e^{j\frac{\pi\Delta f}{2f_0}}\right) + Be^{-j\varphi}\left(1 + e^{-j\left(\pi + \frac{\pi\Delta f}{2f_0}\right)}\right)\right)$$

$$+ \frac{1}{2}\left(Ae^{j\varphi}\left(1 + e^{-j\frac{\pi\Delta f}{2f_0}}\right) + Be^{-j\varphi}\left(1 + e^{j\left(\pi + \frac{\pi\Delta f}{2f_0}\right)}\right)\right) \tag{14.126}$$

The variable $\pi\Delta f/2f_0$ is very small in practical applications, so equation (14.86) can be simplified as (14.131) through performing a series of transformation shown below.

$$1 + e^{j\frac{\pi\Delta f}{2f_0}} = 1 + \left(\cos\frac{\pi\Delta f}{2f_0} + j\sin\frac{\pi\Delta f}{2f_0}\right) \tag{14.127}$$

$$1 + e^{-j\left(\pi + \frac{\pi\Delta f}{2f_0}\right)} = 1 + \cos\left(\pi + \frac{\pi\Delta f}{2f_0}\right) - j\sin\left(\pi + \frac{\pi\Delta f}{2f_0}\right)$$

$$= 1 - \cos\frac{\pi\Delta f}{2f_0} + j\sin\frac{\pi\Delta f}{2f_0} \tag{14.128}$$

$$1 + e^{-j\frac{\pi\Delta f}{2f_0}} = 1 + \left(\cos\frac{\pi\Delta f}{2f_0} - j\sin\frac{\pi\Delta f}{2f_0}\right) \tag{14.129}$$

$$1 + e^{j\left(\pi + \frac{\pi\Delta f}{2f_0}\right)} = 1 + \cos\left(\pi + \frac{\pi\Delta f}{2f_0}\right) + j\sin\left(\pi + \frac{\pi\Delta f}{2f_0}\right)$$

$$= 1 - \cos\frac{\pi\Delta f}{2f_0} - j\sin\frac{\pi\Delta f}{2f_0} \tag{14.130}$$

$$\dot{x}_+ = \frac{1}{2}\left(Ae^{j\varphi}\left(1 + \cos\frac{\pi\Delta f}{2f_0}\right) + Be^{-j\varphi}\left(1 - \cos\frac{\pi\Delta f}{2f_0}\right)\right)$$

$$= \frac{1}{2}\left(Ae^{j\varphi}\left(2 - \left(1 - \cos\frac{\pi\Delta f}{2f_0}\right)\right) + Be^{-j\varphi}\left(2\sin^2\frac{\pi\Delta f}{4f_0}\right)\right)$$

$$= \frac{1}{2}\left(Ae^{j\varphi}\left(2 - 2\sin^2\frac{\pi\Delta f}{4f_0}\right) + Be^{-j\varphi}\left(2\sin^2\frac{\pi\Delta f}{4f_0}\right)\right) \tag{14.131}$$

If the frequency deviation is small, the following simplification could be used:

$$2 - 2\sin^2\frac{\pi\Delta f}{4f_0} \approx 2 \tag{14.132}$$

$$2\sin^2\frac{\pi\Delta f}{4f_0} \approx \frac{(\pi\Delta f)^2}{8f_0{}^2} \tag{14.133}$$

From (14.91)–(14.93), we get

$$\dot{x}_+ = Ae^{j\varphi} + \left(\frac{\pi\Delta f}{4f_0}\right)^2 Be^{-j\varphi} \tag{14.134}$$

According to equations (14.108)–(14.120), the phase angle measurement error of the proposed algorithm can be specified as

$$\Delta\varphi = \frac{(N-1)\pi\Delta f}{Nf_0} - \left(\frac{\pi\Delta f}{4f_0}\right)^2\frac{\Delta f}{2f_0 + \Delta f}\sin\left(2\varphi + \frac{(N-1)\,2\pi(f_0 + \Delta f)}{Nf_0}\right) \tag{14.135}$$

Equation (14.135) shows that the sinusoidal function error was multiplied by an attenuation coefficient $(\pi\Delta f/4f_0)^2$, which effectively eliminates the sine variation part $\Delta\varphi_s$.

When the sampling rate $N$ is an integer multiple of 4, and the vector of the samples is $x_k$, the actual phasor $\dot{x}_\alpha$ and the simulative phasor $\dot{x}_{\beta 1}$, and $\dot{x}_{\beta 2}$ can be expressed in DFT form as follows:

$$\dot{x}_\alpha = \frac{2}{N}\sum_{k=0}^{N-1} x_k e^{-jk\frac{2\pi}{N}} \tag{14.136}$$

$$\dot{x}_{\beta 1} = \frac{2}{N}\sum_{k=\frac{N}{4}}^{\frac{5N}{4}-1} x_k e^{-jk\frac{2\pi}{N}} \tag{14.137}$$

$$\overset{\bullet}{x}_{\beta 2} = \frac{2}{N} \sum_{k=-\frac{N}{4}}^{\frac{3N}{4}-1} x_k e^{-jk\frac{2\pi}{N}} \tag{14.138}$$

According to the discussion above, we can get the practical equation to implement the algorithm as equation (14.139).

$$\overset{\bullet}{x}_+ = \frac{1}{N}\left( e^{j\frac{\pi}{2}} \sum_{k=-\frac{N}{4}}^{\frac{3N}{4}-1} x_k e^{-jk\frac{2\pi}{N}} + 2\sum_{k=0}^{N-1} x_k e^{-jk\frac{2\pi}{N}} + e^{-j\frac{\pi}{2}} \sum_{k=\frac{N}{4}}^{\frac{5N}{4}-1} x_k e^{-jk\frac{2\pi}{N}} \right) \tag{14.139}$$

## 14.9.2 Simulation Results

MATLAB-based simulation examples are presented to verify the effectiveness of the algorithm. The frequency of the input signal is set to 48Hz, and the number of samples $N = 36$. The curves in Figure 14.3 show the measurement errors obtained by the proposed method, the Qps–DFT algorithm and the conventional DFT algorithm, respectively, with the phase angle of input signal varying from $-180°$ to $180°$. It can be observed from Figure14.3 that the average deviations of the three methods are equal, whereas the peak–peak value of sine variation error of the proposed method, which is remarkably eliminated compared to that of DFT algorithm, is only 0.00232.

To clarify the measurement accuracy of the proposed method compared with the Qps–DFT algorithm more clearly, Figure 14.4 shows the phase angle errors of the two methods with offset compensation, in which a pure sinusoidal continuous-time

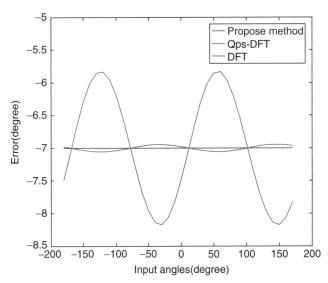

Figure 14.3   The error comparison between proposed method and conventional DFT for 48-Hz input.

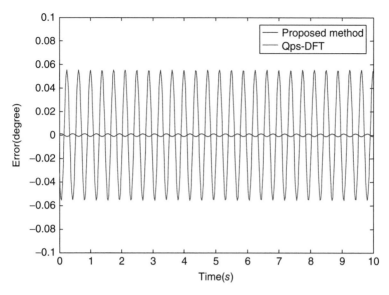

Figure 14.4 The error comparison between proposed method and Qps-DFT for 48-Hz input.

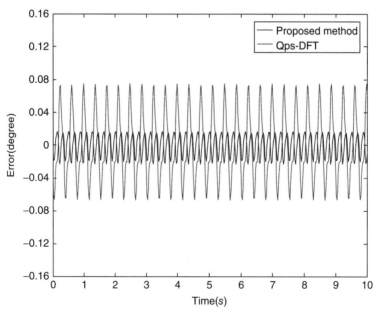

Figure 14.5 The error comparison between proposed method and Qps-DFT for 48-Hz input under the existence of harmonics.

signal is employed as the input signal. From Figure14.4, the peak–peak value of sinusoidal function error of Qps–DFT is 0.11102, which is smaller than that of DFT, but still much bigger than that of the proposed method, 0.00232.

Considering the large amount of harmonics in practical power system, which greatly affect the precision of the phase angle measurement, Figure 14.5 shows the angle errors of the proposed method and the Qps–DFT algorithm with offset compensation under the existence of the third, fifth, and seventh harmonics, and the THD (total harmonic distortion) of the input signal to be 10.11%. As seen from the Figure 14.5, the accuracy of both methods is decreased compared with Figure 14.4. The peak–peak value of sinusoidal variation error presented is 0.03964, which is much smaller than 0.141625 of Qps–DFT, still within the precision allowed in power system.

## PROBLEMS AND EXERCISES

1. What is a "smart grid"?

2. What are the major characteristics of the smart grid?

3. What is a "smart distribution grid"?

4. What is meant by "demand response"?

5. What is DMS?

6. What is VPP?

7. Will smart grid reduce outages?

8. Will smart grid give utility companies control of customers' electric use?

9. How can a customer use or take advantage of the smart grid?

10. What is AMI?

11. State "True" or "False"

    11.1 The smart grid does not contain a generation system.

    11.2 The microgrid is part of a distribution system.

    11.3 Islanding operation for a microgrid is not allowed.

    11.4 A battery can supply power to a grid.

    11.5 There is no load curtailment in the smart grid.

    11.6 The PMU will completely replace the traditional SCADA system in the smart grid.

    11.7 Transmission losses can be reduced in the smart grid.

    11.8 The smart grid can reduce the chance of system fault.

12. Multiple choices

    12.1 Which ones are smart devices?

        (a) PMU           (b) Smart meter

        (c) Transmission line    (d) Digital protective relay

12.2 Which ones are components of a virtual power plant?

(a) Energy storage    (b) Hydro plant
(c) Wind farm    (d) PV plant

12.3 Which ones are energy storage providers?

(a) Electric vehicles (EVs)    (b) Renewable sources
(c) Wind farm    (d) Vehicle-to-Grid (V2G)

12.4 Which ones are storage devices?

(a) Super-capacitors    (b) Batteries
(c) Flywheels    (d) Generator

12.5 Which ones are distributed energy resources?

(a) Photovoltaic    (b) Small wind turbines
(c) Electricity storage    (d) Combined heat and power

13. A simple smart grid has one generator and one storage battery. The load is assumed as constant over time, 12.0MW. The generator cost function is quadratic:

$$f(P_g) = \frac{1}{2}\alpha P_g^2 = \frac{1}{2}(0.02P_g^2)$$

The unit coefficient of battery storage is $\eta = 0.08$. The capacity of generator is 30MW. The time period is 5 h.

(1) if the battery has initial power 2 MW, compute the optimal generation and battery power over time.

(2) if the battery has no initial power, compute the optimal generation and battery power over time.

(3) Does the battery starts to discharge at the same time for the above two cases?

14. A simple smart grid has one generator and one storage battery taking generator constraint into consideration. If all data are the same as those in Exercise 14, but the generator capacity, which is 18MW.

(1) if the battery has initial power 2 MW, compute the optimal generation and battery power over time.

(2) if the battery has no initial power, compute the optimal generation and battery power over time.

15. A simple smart grid has one generator and one storage battery. The load is assumed as constant over time, 8.0MW. The generator cost function is quadratic:

$$f(P_g) = \frac{1}{2}\alpha P_g^2 = \frac{1}{2}(0.03P_g^2)$$

The unit coefficient of battery storage is $\eta = 0.06$. The capacity of generator is 30MW. The time period is 7 h.

**(1)** When does the battery start to discharge?

**(2)** If the battery has initial power 3 MW, compute the optimal generation and battery power over time.

**(3)** If the battery has no initial power, compute the optimal generation and battery power over time.

16. A simple smart grid has one generator and one storage battery taking both generator and battery constraints into consideration. If all data are the same as those in Exercise 16, but the generator capacity and the battery capacity. The generator limit is 12 MW.

   **(1)** Considering only generation constraint. If the battery has initial power 2 MW and the generator limit is 12 MW, compute the optimal generation and battery power over time $(T = 7)$.

   **(2)** Considering only generation constraint. If the battery has no initial power and the generator limit is 12 MW, compute the optimal generation and battery power over time $(T = 7)$.

   **(3)** Considering only battery constraint. If the battery has initial power 2 MW and the battery capacity limit is 12 MW, compute the optimal generation and battery power over time $(T = 7)$.

   **(4)** Considering both generator and battery constraints. If the battery has initial power 4MW, the generator limit is 11MW, and the battery capacity is 10MWh, compute the optimal generation and battery power over time $(T = 7)$.

17. A simple smart grid has two generators and one storage battery. The load is assumed as constant over time, 28.0 MW. The cost functions of two generators are

$$f(P_{g1}) = \frac{1}{2}\alpha_1 P_{g1}^2 = \frac{1}{2}(0.06P_{g1}^2)$$

$$f(P_{g2}) = \frac{1}{2}\alpha_2 P_{g2}^2 = \frac{1}{2}(0.03P_{g2}^2)$$

The unit coefficient of battery storage is $\eta = 0.12$. The time period is 7 h.

   **(1)** When does the battery start to discharge?

   **(2)** If the battery has initial power 4 MW, compute the optimal generation and battery power over time.

   **(3)** If the battery has no initial power, compute the optimal generation and battery power over time.

   **(4)** If the battery has no initial power, and the limits of two generators are 25 MW, compute the optimal generation and battery power over time.

   **(5)** If the battery has no initial power, and the battery capacity is 20 MW, compute the optimal generation and battery power over time.

   **(6)** If the battery has no initial power, the battery capacity is 20 MW, and the limits of two generators are 25 MW, compute the optimal generation and battery power over time.

# REFERENCES

1. Gharavi H, Ghafurian R. Smart grid: the electric energy system of the future. Proc. of IEEE 2011;99(6):917–921.
2. Fang X, Misra S, Xue GL, Yang DJ. Smart grid—the new and improved power grid: a survey. IEEE Commun. Surv. Tutor. 2012;14(4):944–980.
3. Molderink A, Bakker V, Bosman MGC, Hurink JL, Smit GJM. Management and control of domestic smart grid technology. IEEE Trans. Smart Grid 2010;1(2):109–119.
4. U.S. Department of Energy. Smart Grid/Department of Energy, Retrieved June 18, 2012.
5. Rahman S. *Smart Grid From Concept to Reality. IEEE Educational Activities*; New Jersey, 2012.
6. Mukhopadhyay S, Soonee SK, Joshi R. Plant operation and control within smart grid concept: Indian approach, 2011 IEEE PES general meeting, San Diego, CA, July 24–29, 2011.
7. CAISO, SMART GRID Roadmap and Architecture, http://www.caiso.com/green/greensmartgrid.html, December 2010.
8. European Smart Grid Technology Platform. Vision and Strategy for Europe's Electricity Networks of the Future. Brussels, Belgium: *EU Commission, Directorate-General for Research, Information and Communication Unit*; 2006.
9. Zhu JZ. Renewable Energy Applications in Power Systems. New York: Nova Press; 2012.
10. Low S. Smart Grid Intro Economic Dispatch with Battery, www.lccc.lth.se, March 2010.
11. Zhu JZ, Cheung K, Hwang D, Sadjadpour A. Operation Strategy for improving voltage profile and reducing system loss. IEEE Trans. on Power deliver. 2010;25(1):390–397.
12. Zhu JZ. An optimal approach for smart grid economic dispatch, 2014 IEEE PES general meeting, MD, USA, July 27–31, 2014.
13. Arnold M, Anderson G. Model Predictive Control of Energy Storage including Uncertain Forecasts, 2011 PSCC 2011.
14. Pudjianto D, Ramsay C, Strbac G. Virtual power plant and system integration of distributed energy resources. Renew. Pow. Gen. 2007;1(1):10–16.
15. Lassila J, Haakana J, Tikka V, Partanen J. Methodology to analyze the economic effects of electric cars as energy storages. IEEE Trans. on Smart Grid 2012;3(1):506–516.
16. Sortomme E, El-Sharkawi MA. Optimal scheduling of vehicle-to-grid energy and ancillary services. IEEE Trans. on Smart Grid 2012;3(1):351–359.
17. Sekyung H, Soohee H, Sezaki K. Estimation of achievable power capacity from plug-in electric vehicles for v2g frequency regulation: Case studies for market participation. IEEE Trans. on Smart Grid 2011;2(4):632–641.
18. Tuttle DP, Baldick R. The evolution of plug-in electric vehicle grid interactions. IEEE Trans. on Smart Grid 2012;3(1):500–505.
19. Vasirani M, Kota R, Cavalcante RLG, Ossowski S, Jennings NR. An Agent-Based Approach to Virtual Power Plants of Wind Power Generators and Electric Vehicles. IEEE Trans. Smart Grid 2013;4(3):1314–1322.
20. Ramchurn SD, Vytelingum P, Rogers A, Jennings NR. Putting the 'smarts' into the smart grid: A grand challenge for artificial intelligence. Commun. ACM 2012;55(4):86–97.
21. Brown RE. Impact of Smart Grid on distribution system design, 2008 IEEE PES General Meeting; Pittsburgh, USA; July 20–24, 2008.
22. Venkata SS. Basics of power delivery systems, NSF/ECEDHA education workshop. Atlanta, GA; July 9, 2011.
23. Venkata SS, Roy S, Pahwa A, Clark GL, Boardman EC. Realizing the "Smart" in Smart Distribution Grid: A Vision and Roadmap, 2011.
24. Zhu JZ, Cheung K. Voltage impact of photovoltaic plant in distributed network. 2012 IEEE APPEEC Conference, Shanghai, China. March 27–29; 2012.
25. Zhou NC, Zhu JZ. Voltage assessment in distributed network with photovoltaic plan. ISRN Renewable Energy Journal 2011;1(1).
26. Lin L, Guo W, Wang J, Zhu JZ. Real-time voltage control model with power and voltage characteristics in the distribution substation. Int. J. Pow. Ener. Syst. 2013;33(1):8–14.
27. Zhu JZ. A rule based comprehensive approach for reconfiguration of electrical distribution network. Electr. Pow. Syst. Res. 2009;78(2):311–315.

28. Xu BY, Li TY, Xue YD. Smart distribution grid and distribution automation. Autom. Electr. Power Syst. 2009;33(17):38–41.

29. Dall'Anese E, Zhu H, Giannakis GB. Distributed optimal power flow for smart microgrids. IEEE Trans. Smart Grid 2013;4(3):1464–1475.

30. Paudyal S, Caῡnizares CA, Bhattacharya K. Optimal operation of distribution feeders in smart grids. IEEE Trans. Ind. Electronics 2011;58(10):4495–4503.

31. Bae S, Kwasinski A. Dynamic modeling and operation strategy for a microgrid with wind and photovoltaic resources. IEEE Trans. Smart Grid 2013;3(4):1867–1876.

32. Zhou NC, Zhu JZ, Liao Y. A new phase angle measurement algorithm based on virtual $\alpha\beta$ stationary coordinate. IEEE Trans. on Power delivery 2012;27(4):2418–2410.

# INDEX

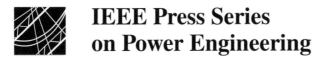

# IEEE Press Series
# on Power Engineering

**Series Editor: M. E. El-Hawary,** Dalhousie University, Halifax, Nova Scotia, Canada

The mission of IEEE Press Series on Power Engineering is to publish leading-edge books that cover the broad spectrum of current and forward-looking technologies in this fast-moving area. The series attracts highly acclaimed authors from industry/academia to provide accessible coverage of current and emerging topics in power engineering and allied fields. Our target audience includes the power engineering professional who is interested in enhancing their knowledge and perspective in their areas of interest.